T0203601

Volume 1

William J. Rea, M.D.
Environmental Health Center
Dallas, Texas

CRC Press
Taylor & Francis Group
Boca Raton London New York

CRC Press is an imprint of the
Taylor & Francis Group, an **informa** business

First published 1992 by Taylor & Francis

Published 2019 by CRC Press
Taylor & Francis Group
6000 Broken Sound Parkway NW, Suite 300
Boca Raton, FL 33487-2742

© 1992 by Taylor & Francis Group, LLC
CRC Press is an imprint of Taylor & Francis Group, an Informa business

First issued in paperback 2019

No claim to original U.S. Government works

ISBN-13: 978-0-367-45022-9 (pbk)
ISBN-13: 978-0-87371-541-6 (hbk)

**Visit the Taylor & Francis Web site at
http://www.taylorandfrancis.com**

**and the CRC Press Web site at
http://www.crcpress.com**

Library of Congress Cataloging-in-Publication Data

Rea, William J.
  Chemical sensitivity / William J. Rea
      p. cm.
  Includes bibliographical references and index.
  ISBN 0-87371-541-1
  1. Environmentally induced diseases. 2. Chemicals—Health aspects
  3. Toxicology. I. Title.
RB152.R38   1992
616.9′8—dc20   92-9436

Library of Congress Card Number 92-9436

DEDICATION

To the fathers of modern clinical ecology, Drs. Theron Randolph, Larry Dickey, Dor Brown, Herbert Rinkel, Carleton Lee, Russ Williams, Jim Willoughby, Joe Miller, and many more whose names space will not permit, but whose precepts inspired our work, which then led to the data presented in this book.

To the chemically sensitive patient, who has often been mistreated and maligned and who has longed not only for help, but for understanding from the medical and scientific community.

To the general public, who will benefit from the compilation of all the clinical and scientific data collected in this study.

To my family, especially my wife, Vera, and my children, Joe, Chris, Tim, and Andrea, and loved ones, who have suffered from this problem, often to the point of severe frustration, and who have provided encouragement and understanding when they were desperately needed.

To the surgical mentors of Dr. Rea, Dr. Tom Shires and Dr. Watts Webb, who taught and allowed him to learn, to innovate, and to provide for patients care that was in their best interest in spite of all adversity and criticism.

FOREWORD

The field of chemical sensitivity has begun to capture the interest and fascination of mainstream physicians, environmental and occupational physicians and scientists, and others interested in the effects of chemicals on health. Sensitivity to chemicals manifests itself in the context of occupational exposure, indoor air, contaminated communities and in persons with exposures to consumer chemicals and pharmaceuticals/anesthesia. A number of recent volumes on this elusive and difficult problem, written by newcomers to the field have suggested that chemical sensitivity ought to receive more thorough attention. The National Research Council recently sponsored a workshop on the subject and published a monograph entitled "Multiple Chemical Sensitivities" suggesting future areas for research. The American Occupational and Environmental Health Clinics similarly held a workshop exploring the role of environmental and occupational physicians in recognizing and understanding the problems of patients with nonclassical sensitivities to chemicals.

This first of four planned volumes on chemical sensitivity places for the first time the long and varied experience of one of the innovative pioneers in the field of clinical ecology, founded by Dr. Theron Randolph. Dr. William Rea provides the much needed clinical perspective derived from observing or treating over 20,000 environmentally sensitive patients. He offers his wide experience concerning the identification, diagnosis, and treatment of chemically sensitive patients. Much of the literature describing the experience of clinicians is dispersed and unfamiliar to most of mainstream medicine. This work provides a unique opportunity to follow the perspective of a major clinician in the field. Some will disagree with his observations, explanations, and conclusions. However, the work represents the coalescence of many patient-years of experience and should be reviewed with an open mind. There is no doubt that our understanding of this difficult area will change over the next few years. This volume provides a valuable time capsule against which to compare our future observations.

Nicholas A. Ashford
Professor of Technology and Policy
Massachusetts Institute of Technology

ACKNOWLEDGMENTS

Thanks to Christine Bishop and Dr. Hsueh-Chia Liang, whose help in preparing the manuscript and illustrations was invaluable; their efforts were herculean and the book could not have been completed without them; to Drs. Alfred Johnson, Gerald Ross, Ralph Smiley, Thomas Buckley, Nancy Didriksen, Joel Butler, Ervin Fenyves, John Laseter, and Jon Pangborn, who supplied cases, data, reports, and criticisms of what should and should not have been done; to Dr. Ya Qin Pan for help in analyzing data; to Drs. Sherry Rogers, Allan Lieberman, Bertie Griffiths, and Kalpana Patel, who proofread and helped compile sections of the book; to the staff at the EHC-Dallas for all their support; to the members of the American Academy of Environmental Medicine for their contribution and support to the EHC-Dallas and to the American Environmental Health Foundation, who lent financial support to this 7-year effort; to Drs. Doris Rapp, Theron Randolph, Lawrence Dickey, John MaClennen, Dor Brown, Carlton Lee, James Willoughby, George Kroker, Jean Monro, Jonathan Maberly, Klaus Runow, Colin Little, Marshall Mandell, Jozef Krop, Hongyu Zhang, Satoshi Ishikawa, Mikio Miyata, Joe Miller, and Ronald Finn for advice and for freely exchanging information.

I am especially indebted to Dr. Jon Pangborn, William B. Jakoby, Andrew L. Reeves, Thad Godish, Steve Levine, Alan Levin, Felix Gad Sulman, and Eduardo Gaitan, whose research, books, and papers provided an invaluable foundation for the preparation of this text.

William J. Rea

PREFACE

This book is the result of our study of more than 20,000 environmentally sensitive patients under various degrees of environmental control at the Environmental Health Center (EHC)-Dallas. It focuses on one aspect of environmental overload, chemical sensitivity. It integrates our experience with the effects of environmental pollutants on known mechanisms of immune and nonimmune detoxication systems, and it emphasizes the importance of maintaining a balance between endocrine, immunological, and neurological systems and their nutrient fuels (Figure 1).

The principles of diagnosis and treatment outlined here have been developed from the combined experience of the EHC-Dallas and information accumulated from treatment and study of an estimated 100,000 patients by other environmentally oriented physicians and scientists around the world.

This work on chemical sensitivity will be published in four volumes. The first volume includes chapters that define the field of chemical sensitivity and identify the basic principles used for its diagnosis and treatment. Chapters on immune and nonimmune mechanisms explain the body's processing of pollutants. The final chapter in this volume is on nutrition, which provides fuel for the endocrine, immunological, and neurological systems and equips the body to respond to pollutant exposure.

This work is intended for medical students, novice physicians, and practicing clinicians. While the myriad of facts supplied here may initially seem cumbersome, they provide evidence for a new, rational, and exciting way to understand chemical sensitivity, practice medicine, and prevent disease. No longer is symptom suppression and/or intervention after the onset of end-organ damage the focus of healing. Instead, our approach emphasizes the importance of isolating and eradicating root causes of illness before fixed-named disease and permanent damage can occur.

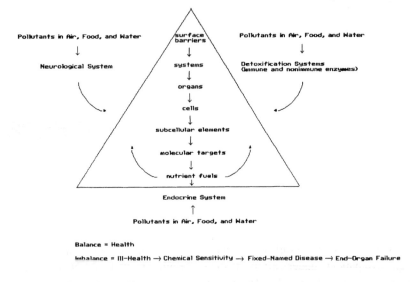

Figure 1 layout text:

Pollutants in Air, Food, and Water

surface barriers

Pollutants in Air, Food, and Water

Neurological System

systems

Detoxification Systems
(Immune and nonimmune enzymes)

organs

cells

subcellular elements

molecular targets

nutrient fuels

Endocrine System

Pollutants in Air, Food, and Water

Balance = Health

Imbalance = Ill-Health → Chemical Sensitivity → Fixed-Named Disease → End-Organ Failure

Figure 1. Effects of environment on bodily functions.

Our study of chemically sensitive individuals has convinced us that after exposure to certain sets of environmental pollution many people experience a lowered resistance to illness that is followed by periods of vulnerability during which their susceptibility to illness increases. We believe that this is the time when intervention is critical. If, under controlled conditions, these pollutants can be identified along with their route of entry and their effect on individual systems, both intervention and prevention programs can be implemented to combat the threat of chronic illness with multiple symptom manifestation that could lead to fixed-named disease and permanent organ damage.

Our work with the chemically sensitive has led to the development of a less polluted environment (ECU), which we use therapeutically to reduce our patients' total pollutant load. We also use this unit to study precisely the effects of environmental pollutants on our individual patients. Use of the less polluted environment helps us to identify specific environmental causes of our patients's symptoms. We can then intervene in the disease process by reducing or eliminating these causative environmental factors (triggers). We can also offset or reverse pollutant damage by bolstering the nutritional state of the patient, thereby fueling and strengthening the systems responsible for protecting the body from pollutant damage.

Attention to the principles (total body load, adaptation, bipolarity, biochemical individuality, spreading, and switch phenomenon) that we have developed along with implementation of our method of exposure and challenge followed by diagnosis and treatment has many advantages. Through the identification of causative factors (triggering agents), their elimination or reduction, and the education of patients as to the resourcefulness of good nutrition and the effects of environmental pollutants upon it, physicians can help their patients gain relief from much unnecessary suffering. Patients can achieve improved health, and they can exert control over their life through their own active management of their illness. They can expect a marked decrease in their health care expenses as their chronic symptoms dissipate and end-organ damage or fixed-named disease is avoided.

William J. Rea, M.D., is a practicing thoracic and cardiovascular surgeon with an added interest in the environmental aspects of health and disease. Founder of the Environmental Health Center–Dallas, Dr. Rea is currently director of this highly specialized medical facility.

In 1988, Dr. Rea was named to the world's first professorial chair of environmental medicine at the Robens Institute of Industrial and Environmental Health and Safety at the University of Surrey in Guildford, England. He was also awarded the Jonathan Forman Gold Medal Award by the American Academy of Environmental Medicine in 1987. Co-author of *Your Home, Your Health and Well-Being,* Dr. Rea has published more than 80 research papers related to the topics of thoracic and cardiovascular surgery as well as environmental medicine.

Dr. Rea currently serves on the Board of Directors of the Pan American Allergy Society and on the American Environmental Health Foundation. Previously, he has held the position of chief of surgery at Brookhaven Medical Center and is past president of the American Academy of Environmental Medicine. He has also served on the Science Advisory Board for the U.S. Environmental Protection Agency, on the Research Committee for the American Academy of Otolaryngic Allergy, on the Committee on Aspects of Cardiovascular Endocrine and Autoimmune Diseases of the American College of Allergists, Committee on Immunotoxicology for the Office of Technology Assessment and on the panel on Chemical Sensitivity of the National Academy of Sciences. Dr. Rea is a fellow of the American College of Surgeons, the American Academy of Environmental Medicine, the American College of Allergists, the American College of Preventive Medicine and the American College of Nutrition.

Born in Jefferson, Ohio, Dr. Rea was graduated from Otterbein College in Westerville, Ohio, and Ohio State University College of Medicine in Columbus, Ohio. He then completed a rotating internship at Parkland Memorial

Hospital in Dallas, Texas. He held a general surgery residency from 1963 to 1967 and a cardiovascular surgery fellowship and residency from 1967 to 1969 with The University of Texas Southwestern Medical School system, which includes Parkland Memorial Hospital, Baylor Medical Center, Veteran's Hospital, and Children's Medical Center.

From 1984 to 1985, Dr. Rea held the position of adjunct professor of environmental sciences and mathematics at the University of Texas, while from 1972 to 1982, he acted as clinical associate professor of thoracic surgery at The University of Texas Southwestern Medical School. He has also served as chief of thoracic surgery at Veteran's Hospital, as adjunct professor of psychology and guest lecturer at North Texas State University. Dr. Rea is currently affiliated with Tri City Health Centre of Dallas and Garland Community Hospital in Garland.

TABLE OF CONTENTS, VOLUMES I–IV

Volume I   Chemical Sensitivity: Principles and Mechanisms

1. Introduction
2. Definition of Chemical Sensitivity
3. Principles of Chemical Sensitivity
4. Nonimmune Mechanisms
5. Pollutant Effects on the Blood and Reticuloendothelial System (Lymphatic and Immune System)
6. Nutritional Status and Pollutant Overload

Volume II   Sources of Total Body Load

7. Water Pollution
8. Food, Food Toxins, and Sensitivity
9. Outdoor Air Pollution
10. Indoor Air Pollution
11. Inorganic Chemical Pollutants
12. Organic Chemical Pollutants
13. Pesticides
14. Formaldehyde
15. Terpenes
16. Drugs
17. Compounding Factors

Volume III   Clinical Manifestations of Pollutant Overload

18. ENT/Upper Respiratory
19. Lower Respiratory System, Chest Wall, and Breast
20. Cardiovascular System
21. Gastrointestinal System
22. Genitourinary System
23. Musculoskeletal System
24. Endocrine System
25. Integument
26. Nervous System
27. Ophthalmological System
28. Children

Volume IV   Tools of Diagnosis and Methods of Treatment

29. History and Physical
30. Laboratory
31. ECU
32. Avoidance — Air
33. Avoidance — Water
34. Avoidance — Food
35. Sauna Depuration
36. Injection Therapy
37. Nutrition Replacement
38. Endocrine Treatment
39. Tolerance Modulators
40. Behavior Therapy
41. Surgery
42. Long-term Results

CONTENTS

1.   **Introduction**   1

2.   **Definition of Chemical Sensitivity**   7

3.   **Principles of Chemical Sensitivity**   17

4.   **Nonimmune Mechanisms**   47

5.   **Pollutant Effects on the Blood and Reticuloendothelial System (Lymphatic and Immune System)**   155

6.   **Nutritional Status and Pollutant Overload**   221

Glossary   481

Index   489

1   INTRODUCTION

Modern technology has given man many conveniences and the ability to explore the outer limits of knowledge. It has allowed us to travel to the moon and send probes into outer space. But this technology that has led us to uncover secrets of the universe has also brought into focus the severity of environmental pollution on earth. The Apollo astronauts emphasized the extent of this problem when viewing the earth from space. Although they initially called it the "blue planet," these astronauts saw at closer range pollution on all areas of the earth, which led them to state that "man has fouled his nest and this must be corrected."

The astronauts' observation reminds us of a simple truth: man's well-being is a function of his environment and living in polluted surroundings adversely affects health. Today, more than ever, substances that pollute the earth and pose health risks are many, including biological factors such as pollens, foods, water, bacteria, virus, fungus, and parasites, chemical factors such as inorganic and organic compounds, and physical forces such as heat, cold, weather, cyclic phenomenon, radon, light, sound, and electromagnetic fields.

As the number of dangerous environmental pollutants continues to multiply, so do reports of increasing numbers of people sensitive to these contaminants. And although identification of a causal link between an individual's pollutant exposure and any subsequent development of illness has been difficult to demonstrate through much of this century, recent technological advances have made possible scientific study of the effects of these many environmental contaminants upon individual health and have contributed directly to our understanding of environmentally triggered illness as a specific clinical entity. Further, our improved research methods and expanding knowledge of environmentally triggered illness have led to definition of the early parameters of pollutant injury and to specification of the evolving principles used for diagnosing and treating chemical sensitivity, which include total body load (total of all pollutants that are in the body at one time); adaptation (the increase of

1

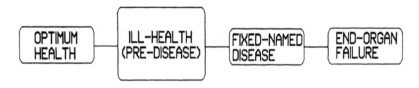

Figure 1.   The disease process.

the body's immune and nonimmune response systems to a new set point in order to accommodate environmental pollutants for acute survival); bipolarity (stimulation, withdrawal and depressive reactions to environmental pollutants); biochemical individuality (the uniqueness of individual responses to environmental pollution; the nature of these responses depends on each individual's genetic make-up, total body load, and state of nutrition); switch phenomenon (a patient's symptoms may totally change, i.e., from sinusitis to phlebitis to hemorrhoids), and spreading phenomenon (the patient may become sensitive to additional incitants, and new organs may become involved) (Figure 1).

Not only have technological advances made possible scientific study of pollutant effects upon individual health and improved our understanding of environmentally triggered illnesses as a particular entity; these advances have also made possible the development of more precise methods of diagnosing and treating these disease processes. The less polluted environmental control unit (ECU) developed by Randolph and perfected by Rea, for instance, has become the standard in its management. The ECU is used to aid precise definition of triggering agents of chemical sensitivity after a patient's total pollutant load has been reduced and deadaptation completed. This unit has further been used to carry out oral, injected, and inhaled challenges under controlled conditions. It has made possible the controlled evaluation of blood parameters, immune and nonimmune enzyme biochemical detoxification systems, levels of toxic chemicals, and nutrients found in the various body compartments. Safety features designed for use in the ECU have further been adapted by the Environmental Health Center-Dallas (EHC-Dallas) to control pollution levels in homes and public buildings, and we have used them to construct or modify over 17,000 less polluted environments for chemically sensitive patients.

This book reflects the strides made in defining and understanding environmentally triggered diseases. In detailing one aspect of this complex field — environmental sensitivity, and more specifically, chemical sensitivity — it pulls together both scientific data and clinical studies and makes available to the reader concepts, principles, and methods of diagnosis and treatment that are essential to health management of the chemically sensitive. To demonstrate the changes that are presently occurring in the diagnosis and treatment of chemical sensitivity, we present the following case report.

Case study — A 35-year-old white female was in perfect health for the first 30 years of her life. She then developed recurrent sinusitis with odor sensitivity that was treated with symptom-suppressing medication. She did well for 2 years when she developed severe premenstrual tension with recurrent head-aches, which were treated with symptom suppressing medication. She again did well for about 2 years until she developed back and leg pain caused by lumbar disc disease which required surgery. Postoperatively, she developed thrombophlebitis which responded to anticoagulants. After 6 months, this patient developed symptoms of anxiety, shakiness, weakness, and an inability to sleep. Her premenstrual tension increased including severe abdominal cramps. She was treated with tranquilizers but eventually became medication sensitive and developed phlebitis, which was refractory to medication. She became incapacitated, unable to walk, and developed a pulmonary embolus.

When, after 5 years from the onset of her initial symptoms, this patient had failed to regain her health through the usual course of treatment, she sought help at the EHC-Dallas where a new method of intervention was utilized. First, a careful environmental history was taken. It revealed that she had been working in an old clothing store when her sinusitis and odor sensitivity developed. This patient had noticed that when a severe rain would leak into the store followed by mold odor, she would have a stuffed up nose. She had also noticed that when boxes of new clothes were unpacked her nose would burn and run. She had related these facts to several physicians who, unaware of their significance, had dismissed them.

This patient next underwent challenge testing. She reacted with a severe stuffed up nose and sinus pain to intradermal challenge with mold extracts. Both intradermal and inhaled challenges of ambient doses of formaldehyde (<0.2 ppm) caused severe odor sensitivity, runny and burning nose, and sinus pain. Intradermal challenge of estrogen, an oral ingestion of milk, and an inhaled challenge of phenol (<0.002 ppm) reproduced her premenstrual tension and headaches. Oral and intradermal challenge of estrogen and intradermal and inhaled challenge of formaldehyde and phenol provoked her thrombophlebitis. After blood analysis showed deficiencies in vitamin C, magnesium, manganese, and molybdenum, supplementation of these nutrients eliminated her postdiscectomy pain. After both a 24 h urine analysis and an intravenous challenge of 25 meq of magnesium revealed a magnesium deficiency, supplementation of this mineral eliminated her premenstrual cramps, anxiety attacks, weakness, and inability to function.

This patient was diagnosed with environmentally triggered disease. As part of her treatment, she constructed an environmentally controlled oasis at home, where as many triggering agents as possible were removed. She has been active and vigorous for the past 15 years without the onset of new symptoms, the recurrence of old ones, or the need of medication.

This case illustrates an evolving health concept that acknowledges a causative link between environmental exposures to multiple toxic and nontoxic

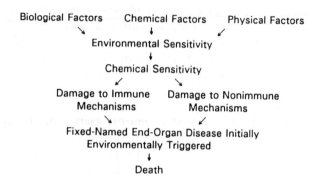

Figure 2.   The process of environmentally triggered disease.

agents, the eventual onset of chemical sensitivity, and the eventual development of named, clinical disease. This concept is based on four essential tenets (Figure 2). First, even in the presence of individual and nutritional variations, the development of chemical sensitivity is dependent on the relationship of the individual to his environment. Second, exposure to deleterious environmental factors if ignored and allowed to continue has potentially harmful effects. Third, a significant time span may exist between an initial toxic environmental exposure or repeated exposures and the development of chemical sensitivity which may lead to a specific fixed-named disease. It is critical to note that even after the onset of a named entity, chemical exposures may remain the prime factor in propagating this disease. Finally, from the onset of a toxic exposure to the development of end-stage clinical disease, there is a spectrum of multiple signs and symptoms which represent the biological markers of the effect of the exposure. They must be recognized and eliminated early if end-stage disease is to be prevented. Then for the chemically sensitive to achieve and maintain optimum health, their total load must be reduced, and they must remain deadapted in the alarm stage. In this state, their immune and enzyme detoxification systems respond more appropriately to pollutant insults in a finite manner. Their range of physiological adaptation then expands, and they can maintain symptom-free, optimum health even when they experience an isolated high-level pollutant exposure.

Environmentally triggered illnesses have long been recognized, but specific chemical sensitivity is a newly acknowledged and evolving entity of study. Its definition as well as development of unique methods of intervention in its process is the result of scientific and clinical study made possible by recent technological innovations. Presently chemical sensitivity is sufficiently well defined and its methods of diagnosis and treatment well enough understood so as to be useful to physicians and patients in identifying the meaning and significance of what they may otherwise assume are unrelated signs and symptoms of illness. But in order to be utilized, information about chemical sensitivity must first be made available. We therefore present in this book a comprehensive discussion of chemical sensitivity in order to assist physicians in better understanding and treating patients suffering from this condition.

2 DEFINITION OF CHEMICAL SENSITIVITY

CHEMICAL ENVIRONMENT

The rapidly accelerating rate of growth of modern technology has been accompanied by a proliferation of a wide variety of new toxic chemicals such as styrene, polyesters, polyethylene, etc. Recent studies[1-3] show that nearly 50% of the global pollutants isolated from natural products or synthesized which enter the atmosphere are generated by man. The pervasiveness of toxic chemical agents is well documented. In 1987, American industry poured 22 billion lb of toxic chemicals into the air, food, and water. Overall, Texas, ranking first in air and land releases,[4] dumped the most pollutants. Every day several million gallons of chemicals are emptied into Lake Erie, which is the source of drinking and bathing water for most cities from Toledo to Cleveland, OH, to Buffalo, NY.

Inorganic pollutants include ozone, carbon monoxide, nitrous oxides, sulfur dioxides, heavy metals,[5-10] and other metals (e.g., Al, Cu, etc.).[11,12] Organic pollutants include pesticides, formaldehyde,[13] solvents (e.g., toluene and xylene), drugs,[14] terpenes, cleaning chemicals, cigarette smoke, combustion products, consumer products (e.g., clothing, building materials, hygiene products, etc.),[15-17] and biological compounds (mold toxins).[18, 19] The most toxic organic pollutants are those classified as halogenated aromatic and aliphatic hydrocarbons.[20]

According to the EPA,[21] more than four million chemical compounds are currently recognized. Over 60,000 of these are produced commercially, and about 3 new compounds are introduced each day. The rampant widespread presence of hazardous chemicals in our environment has become critical.

7

Unfortunately, the link between chemical sensitivity and our individual well-being described by Randolph[22] over 35 years ago has been ignored until now. While celebrated instances of gross contamination through industrial waste have long been the object of professional attention, only recently have literally thousands of synthetic chemical products, heretofore believed innocuous, been incriminated as agents of homeostatic dysfunction. Throughout this book, we will emphasize understanding and interpreting the effects of this environmental load on individual malfunction.

In our opinion, the major stumbling block in recognizing the etiology of chemical sensitivity and many instances of resultant fixed-named diseases has been the general failure of the medical profession to appreciate the massive increase in and adverse effects of exposure to environmental pollution. Nonetheless, ten environmental centers and thousands of physicians and scientists on four continents acknowledge the effects of environmental pollution on human health and have contributed to the scientific and clinical evidence for the existence of chemical sensitivity.

CHEMICAL SENSITIVITY

Chemical sensitivity is one of the major manifestations of environmentally triggered disease and is the main focus of this book. It is an adverse reaction(s) to ambient levels of toxic chemical(s) contained in air, food, and water. The nature of these adverse reactions depends on the tissue(s) or organ(s) involved, the chemical and pharmacologic nature of the substance(s) involved (i.e., duration of time, concentration, and virulence of exposure), the individual susceptibility of the exposed person (i.e., nutritional state, genetic makeup, and toxic load at the time of exposure), and the length of time and amount and variety of other body stressors (i.e., total load), and synergism at the time of reaction(s).

ALLERGIC AND TOXIC RESPONSES

Chemical allergies are a small but significant part of the overall spectrum of chemical sensitivity. They may involve both allergic (immunologically mediated mechanisms including all of the four types of hypersensitivity reactions) and toxic (nonimmune mechanisms) responses. They involve the mechanisms of the IgE class of immunoglobulins. An example of chemical allergy is the IgE-mediated toluene diisocyanate antigen/antibody reaction which frequently manifests as asthma or some other form of respiratory or vascular dysfunction. Other immune mechanisms such as IgG, cytotoxic response, immune complex (IgG + complement), or T and B cell abnormalities are often involved in chemical sensitivity, although these reactions are frequently secondary responses following an initial enzyme detoxification response. Failure of enzyme detoxification appears to be the prime mechanism in chemical sensitivity. *Regardless of the mechanisms involved, clinical manifestations of chemical sensitivity may be the same.* For example, rhinitis may occur either as an IgE response to toluene diisocyanate, or it may be an enzyme detoxification system response to formaldehyde.

CAUSES OF CHEMICAL SENSITIVITY

Chemical sensitivities may arise in several ways. Individuals who survive near-fatal exposures to toxic substances often experience lowered resistance to disease as a result of the depletion of their nutrient pool brought on by exposure. They may then develop chronic symptoms of ill health. If these people are later exposed to ambient doses of toxic chemicals, they may experience additional and/or enhanced symptoms. "Spreading," which can involve both new organ systems and increased sensitivities to additional substances, may occur. For example, an individual working in a chemical plant may be exposed to high doses of xylene after an explosion. He immediately develops headaches and flu-like symptoms that become chronic. Weeks later, after ongoing ambient exposures in the workplace and at home, this person develops asthma and sensitivity to ambient doses of various toxic and nontoxic (e.g., perfume) substances. Of the chemically sensitive patients seen at the EHC-Dallas, 13% relate the onset of their sensitivity to a severe acute exposure.

Three major incidents have occurred in the 20th century which graphically illustrate that chemical sensitivity may be caused by a significant, acute exposure to toxic substances. The first occurred during World War I when many of the troops were gassed with mustard gas.[23,24] The clinical aftermath for many of the survivors was the development of chemical sensitivity. The second incident involved military personnel who were sprayed with defoliants while serving in Vietnam. The "agent orange" syndrome, as symptoms from this exposure were later dubbed, persisted for years after their initial contact. The third incident occurred in Bhopol, India, where an accidental atmospheric exposure to large quantities of cyanate left an estimated 86,000 people injured.[25] Several months later, many remained afflicted with recurrent symptoms that are today believed to be manifestations of chemical sensitivity.

Chemical sensitivity can occur subsequent to severe bacterial, viral, or parasitic infection. Of the patients treated for chemical sensitivity at the EHC-Dallas, 1% have traced the origin of their illness to such an event.

The onset of chemical sensitivity has also been attributed to exposure to ambient doses of toxic chemicals following massive trauma, childbirth, surgery, or immunizations. At the EHC-Dallas, 12% of our patients associated massive trauma with the start of their illness; 9% identified childbirth as the triggering event; 2% traced onset to surgery; and 1% linked their illness to other causes.

In contrast to those who develop chemical sensitivity as a result of acute exposure to toxic substances, there exists another group for whom a specific cause of illness is often difficult to discern. This group includes individuals who have become chemically sensitive following accumulative subacute toxic exposures over time. Of our patients at the EHC-Dallas, 60% fit into this category.

Only 2% of our chemically sensitive patients are unable to account for the specific events that precipitated onset of their illness.

Because we can link only 28% of our patients' illnesses to employment that involved exposure to high levels of toxic chemicals, such as work in pesticide plants or refineries or work involving pesticide spraying, we speculate that some cases of chemical sensitivity may evolve as the result of a series of events that occur with the passage of time. These events may include an individual's long-term exposure to ever-present, subtle levels of pollutants of which he is unaware. We believe that as these toxic, environmental exposures occur, an insidious breakdown in resistance mechanisms takes place. An individual's vulnerability increases and eventually chemical sensitivity ensues. Chapter 3, Principles, details how this phenomenon can occur.

The development of chemical sensitivity may be chronic and, therefore, insidious. Individuals are often unaware of their developing sensitivity until their chemical intolerance is such that only minuscule concentrations of chemicals are needed to trigger symptoms of illness. At this point, reactions to lower levels of toxic chemicals commonly occur. "Spreading," which involves both single-organ susceptibility to increasing numbers of toxic chemicals and increasing susceptibility of new organ systems to one or many toxic chemicals, may then follow. Fixed-named or end-stage disease may occur, and eventually its course may proceed autonomously. For example, an individual may develop arthralgia due to a sensitivity to formaldehyde. Symptoms may fluctuate from many months to years. Then the individual may develop sensitivities to more toxic chemicals, and arthritis may result. Eventually, if left environmentally untreated, this individual may develop fixed intractable arthritis that appears to be self-perpetuating.

"Low levels" of toxic chemicals have been implicated in the insidious onset of some chemical sensitivities. Because of this connection, use of the term "low level" itself needs to be reconsidered. We believe the term should not be used because it implies that such levels are harmless when, in fact, many chemicals are potentially lethal at low levels. The herbicide 2,4,5-T, for example, has been found to be harmful in the parts-per-trillion or even less. Furthermore, the levels of chemicals considered to be "low" by today's standards are based on levels found in the average population or environment and then termed "normal" and, hence, "low level." Unless a toxic chemical such as formaldehyde or pentane is generated by the body, the control levels should usually be nondetectable and not "low level." For example, mathematical calculations of a toxic substance reveal that the number of chlordane molecules per cell is between 700 and 1500 when serum levels are measured at one part per billion.

Too often, "safe" levels of toxic chemicals are assigned based on inferences from an unhealthy control population who are assumed healthy because they are able to function with a minimum of short-term illnesses such as a flu or cold, or they are functioning in seemingly good health with the assistance of symptom-masking medications. More commonly, levels are assigned from animal studies. We believe the control population from which "safe" levels of toxic chemicals ought to be derived should be those who are totally well and medication free. This population in western society, however, is hard to find. Until such a population can be isolated and appropriate "safe" levels of toxic chemicals determined from their examination, we advocate a reference point for blood and tissue levels of toxic chemicals that is nondetectable.

Because modern medical practitioners accept of occasional illness or medicated wellness as part of general good health, we often fail to detect what may be symptoms of the early stages of chemical sensitivity, and because we fail to recognize these stages for what they are, we miss the opportunity to intervene in time to reverse any ongoing damage or prevent the development of serious, fixed-named illnesses. Instead, we misdiagnose and mistreat symptoms, and as a result of our present attitudes and practices, we probably grossly underestimate the incidence of chemical sensitivity.

MANIFESTATION OF CHEMICAL SENSITIVITY

Symptoms of chemical sensitivity typically are multiple in nature. Usually, one main organ is affected with secondary symptoms occurring in others. End-organ responses are often in the smooth muscles of the cardiovascular, gastrointestinal, urogenital, respiratory systems, or in the nervous system. Also, common early responses may occur in the skin (such as nonjaundice yellowing or edema) or other body organs.

At their onset, symptoms of chemical sensitivity are almost always reversible. As end-organ involvement increases, however, responses are more difficult to decipher and reverse.

Pollutant damage can occur at the main site of pollutant entry or in the detoxifying organ or it can be random, affecting any end-organ. Usually, however, the weakest end-organ, that which has been genetically damaged or previously harmed by trauma or exposure, is the first affected.

The sicker the patient with chemical sensitivity, the more diverse and multiple are his responses to a large number of individual incitants, suggesting primary and secondary organ involvement. For example, a patient develops rhinitis on exposure to formaldehyde. Later in the course of his illness, symptoms and signs of cystitis and colitis develop in addition to the rhinitis.

Although these various illnesses involve multiple systems and organs, only one end-organ may ultimately be damaged as the result of repeated insults to the same resistance mechanism. After this damage occurs, however, end-organ failure follows and extreme fixed-named illness results. For example, a mechanic constantly exposed to car exhaust could develop general symptoms such as aches, pains, malaise, headaches, and fatigue. These symptoms might then continue for several months until finally renal failure or some other specific end-organ disease develops.

FACTORS INFLUENCING THE ONSET OF CHEMICAL SENSITIVITY

Onset of chemical sensitivity is influenced by multiple factors including total body load (burden), total toxic load, nutritional state, synergisms, competition for storage, bioaccumulation, and biological half-life of the chemicals themselves.

Total Toxic (Body) Load (Burden)

Total toxic (body) load is the sum of all pollutants in the body at one time. When this accumulation overloads the system, chemical sensitivity can occur.

Nutritional State

The nutritional state needed to maintain good health is depleted by toxic exposure. Overload of pollutants can increasingly tax the detoxification systems, eventually resulting in depletion of nutrients, system/organ malfunction, and susceptibility to illness.

Synergisms

Synergisms may be additive if the effects of the pollutants equal the sum of the individual pollutants involved. For example, a patient whose sensitivity to mold results in a runny nose and whose sensitivity to formaldehyde yields burning eyes might react to exposure to both with runny nose and burning eyes. Synergisms may also be potentiative where the effects of the potentially harmful substances exceed the sum of the individual substances involved. For example, mold toxin and formaldehyde combined may together give swollen eyes along with a swollen face and extremities in addition to or substituting for the burning eyes and runny nose. Occasionally, individual pollutants may have antagonistic effects. If introduced simultaneously, these substances may cancel each other's usual effects. For example, salicylic acid and acetophenomen introduced simultaneously reduce each others effects so that an individual exposed to these might exhibit slight or no symptoms.

Competition for Storage and Removal

Some chemicals may be competitive for both storage and removal. DDT, for example, increases and dieldrin decreases when they are introduced simultaneously. Both compete for the same enzyme sites for detoxification and metabolic function for detoxification; thus, one may circulate and even be deposited in a lipid membrane while the other is metabolized and used or cleared from the body. In our clearing studies, we have observed that certain toxic chemicals cannot be removed until others are mobilized and removed.

Bioaccumulation of Toxic Substances

The accumulation of a single agent in the body is dependent upon the dose level, interval and duration of exposure, and half-life and lipophilic nature of a chemical. Accumulation of a toxic substance also depends on an individual's quantity and quality of immune and enzyme detoxication responses along with his age and overall health. Accumulation may also occur with constant exposures that allow no time for clearing.

The factors influencing chemical accumulation are solubility quotient, the storage of chemicals in poorly perfused tissues, poor glomerular filtration, intensive tubular reabsorption, and slow biotransformation. This bioaccumulation may be likened to a layered sponge. Each layer fills due to the excess pollutants that are absorbed yet unable to be immediately metabolized. With each new contact, more layers fill with excess pollutants until the maximum load is exceeded and the disease process begun.

Biological Half-Life of Toxic Substances

The biological half-life of a chemical is one-half the time a chemical takes to disintegrate. It is usually calculated from animal exposures. It is also based on inadvertent acute exposures of healthy people to toxic substances. The half-life may have little relationship to the detoxification mechanisms available. Metabolism may be high for initial high-dose exposures, but very slow for low doses.[26] This latter response will leave residue in blood and tissues and may explain why low dose exposures can be a significant cause of disease. (See Chapter 4, Nonimmune Mechanisms).

Various investigators[27-29] have shown a relationship between the presence and the bioaccumulation of foreign chemicals in human tissue and the incidence of cancer. This correlation appears to exist for nonmalignant diseases as well. This finding should cause even greater concern because apparently some environmentally persistent halogenated hydrocarbons such as DDT or chlordane may have a significant negative effect on the human immune system.[30]

REFERENCES

1. National Research Council, Safe Drinking Water Committee. "Drinking Water and Health" (Washington, DC: National Academy of Sciences, 1977).
2. National Research Council. "Indoor Pollutants" (Washington, DC: National Academy Press, 1981), pp. 16–27.
3. Winslow, S. G. "The Effects of Environmental Chemicals on the Immune System: A Selected Bibliography with Abstracts" (Oak Ridge, TN: Toxicology Information Response Center, Oak Ridge National Laboratory, 1981).
4. "Is Your Drinking Water Safe?" (Washington, DC: U.S. Environmental Protection Agency, Office of Public Affairs, March 1977).
5. National Research Council, Safe Drinking Water Committee. "Drinking Water and Health" (Washington, DC: National Academy of Sciences, 1977).
6. Dooms-Goossens, A., A. Ceuterick, N. Vanmaete, and H. Degreef. "Follow-up study of patients with contact dermatitis caused by chromates, nickel, and cobalt," *Dermatologica* 160(4):249–260 (1980).
7. Freedman, B. J. "Sulphur dioxide in food and beverages: its use as a preservative and its effect on asthma," *Br. J. Dis. Chest* 74(2):128–134 (1980).
8. Mustafa, M. G., and D. F. Tierney. "Biochemical and metabolic changes in the lung with oxygen, ozone, and nitrogen dioxide toxicity," *Am. Rev. Respir. Dis.* 118(6):1061–90 (1978).
9. Vermeiden, I., A. P. Oranje, V. D. Vuzevski, and E. Stolz. "Mercury exanthem as occupational dermatitis," *Contact Dermatitis* 6(2):88–90 (1980).
10. Whittemore, A. S., and E. L. Korn. "Asthma and air pollution in the L. A. area," *Am. J. Public Health* 70(7):687–696 (1980).
11. National Research Council, Safe Drinking Water Committee. "Drinking Water and Health" (Washington, DC: National Academy of Sciences, 1977).
12. Clemmensen, O., and H. E. Knudsen. "Contact sensitivity to aluminum in a patient hyposensitized with aluminum precipitated grass pollen," *Contact Dermatitis* 6(2):305–308 (1980).

13. Fisher, A. A. "Dermatitis due to the presence of formaldehyde in certain sodium lauryl sulfate (SLS) solutions," *Cutis* 27(4):360–366 (1981).

14. Dahl, R. "Sodium salicylate and aspirin disease," *Allergy* 35(2):155–156 (1980).

15. Frigas, E., W. V. Filley, and C. E. Reed. "Asthma induced by dust from urea formaldehyde foam insulating material," *Chest* 79(6):706–707 (1981).

16. Imbeau, S. A., and C. E. Reed. "Nylon stocking dermatitis. An unusual case," *Contact Dermatitis* 5(3):163–164 (1979).

17. Larson, W. G. "Sanitary napkin dermatitis due to the perfume," *Arch. Dermatol.* 115(3):363 (1979).

18. Olson, K. R., S. M. Pond, J. Seward, K. Healey, O. F. Woo, and C. E. Becker. "Amanita phalloides-type mushroom poisoning," *West J. Med.* 37:282–289 (1982).

19. Wilson, C. W. M. "Hypersensitivity to Maine tap water in children: Its clinical features and treatment," *Nutr. Health* 2:51–63 (1983).

20. Laseter, J. L., I. R. DeLeon, W. J. Rea, and J. R. Butler. "Chlorinated hydrocarbon pesticides in environmentally sensitive patients" *Arch. Clin. Ecol.* 2(1):6 (1983).

21. Schnare, D. W., D. B. Katzin, and D. E. Root. "Diagnosis and Treatment of Patients Presenting Subclinical Signs and Symptoms of Exposure to Chemicals which Bioaccumulate in Human Tissue," P-150, Proceedings of the National Conference on Hazardous Wastes and Environmental Emergencies. Cincinnati, OH, May 14–16, 1985.

22. Randolph, T. G. "Sensitivity to petroleum: Including its derivatives and antecedents" *J. Lab. Clin. Med.* 40:931–932 (1952).

23. LeFebure, V. *The Riddle of the Rhine: Chemical Strategy in Peace and War* (New York: E. P. Dutton and Co., 1923).

24. Heller, C. E. "Chemical warfare in World War I: The American experience 1917–1918," Leavenworth papers, No. 10, Combat Studies Institute, Library of Congress Cataloging in Publication Data, Forth Leavenworth, KA (1984), pp. 24–25.

25. Shrivastava, P. *Bhopal: Anatomy of a Crisis* (Cambridge, MA: Ballinger Publishing Company, 1987).

26. Balazs, T. "Hepatic reactions to chemicals," in *Toxicology: Principles and Practices*, Vol. 1, A. L. Reeves, Ed. (New York: John Wiley & Sons, Inc., 1981), p. 93.

27. Williams, G. M., and J. H. Weisburger. "Chemical carcinogens," in *Caserett and Doull's Toxicology: The Basic Science of Poisons*, 3rd ed., C. D. Klassen, M. O. Amdur, and John Doull, Eds. (New York: Macmillan Publishing Co., Inc., 1986), pp. 124 and 153.

28. Unger, M., and V. Olsen. "Organochlorine compounds in the adipose tissue of deceased people with and without cancer," *Environ. Res.* 23:257–263 (1980).

29. Wasserman, N. M., D. P. Nogueira, S. Cucos, A. P. Mirra, H. Shibata, G. Arie, H. Miller, and D. Wasserman. "Organochlorine compounds in neoplastic and apparently normal gastric mucosa," *Bull. Environ. Contam. Toxicol.* 20:544–553 (1978).

30. Klotz, V. I., R. A. Bahayantz, V. G. Brysin, and A. Safarova. "Effect of pesticides on the immunological reactivity of the body of animals and man," *Gig. Sanit.* 9:35–36 (1978).

3   PRINCIPLES

INTRODUCTION

The diagnosis, treatment, and prevention of chemical sensitivity are based upon six basic principles: total body load, adaptation, bipolarity, spreading phenomenon, switch phenomenon, and biochemical individuality (Figure 1). The following case report illustrates the application of these principles.

Case study — A completely incapacitated 26-year-old white female presented with the complaints of odor intolerance, recurrent stuffed-up nose with episodes of rhinosinusitis, fatigue, weakness, bloating, and diarrhea. She was totally well until the age of 18 when she left home for the university. Upon moving into the dormitory, she noticed the gradual onset of fatigue and weakness. These symptoms occurred weekly and were traced to the routine spraying of pesticides in the dorm (increase in *total body load*). As her total body load increased, she noticed that she had problems digesting foods with additives and preservatives in them, although she could eat the same foods without additives and preservatives when cooked at home. She also noticed that as her total load increased she would develop rhinitis after showers and drinking chlorinated city water. Intake of the pollutants harbored in these further increased her *total load*. As she progressed through life, she seemed to get used to her environment, *adapting* to these exposures with a decrease in the symptoms related to her acute chemical exposures. However, she was left with severe chronic fatigue, weakness, and episodes of disorientation and memory loss (these latter two symptoms are examples of *switch phenomenon*, i.e., new organs are affected by or new symptoms manifested in response to chemical exposure). She then noticed that often during a 24 h period and again in monthly cycles, her energy level would vacillate from high energy to exhaustion. Parallel to these swings of energy, she developed mood swings (*stimulatory* and *withdrawal* phenomenon characteristic of *bipolarity*). She was placed on tranquilizers and most of her mood swings disappeared (*mask-*

17

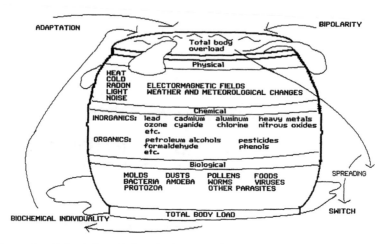

Figure 1.   Principles of chemical sensitivity.

ing phenomenon), but she developed recurrent cystitis (*switch phenomenon*) as her exposures continued. She gradually developed a severe sensitivity to odors of perfume, car exhausts, phenols, and formaldehyde (*spreading phenomenon*) as her total load increased and she became severely maladapted. After her total body load was severely exceeded during an exposure to a natural gas leak in her dormitory, she developed fulminant diarrhea and incapacity. Each of six other women who were simultaneously exposed developed different symptoms. One developed severe premenstrual tension, another asthma, another cardiac arrhythmia, another cystitis, another arthritis, and the final one migraine headaches, indicating *biochemical individuality of response* to a common exposure.

This patient was admitted to the Environmental Control Unit (ECU), where her total load was reduced by decreasing her intake of air, food, and water pollutants. After 5 days and near completion of her deadaptation process, her symptoms had dissipated and she was asymptomatic without medications for the first time in 5 years.

She was then challenged by inhaled and intradermal exposures to <0.002 ppm of phenol, <0.50 ppm of petroleum derived ethanol, <0.33 ppm of chlorine, <0.2 ppm of formaldehyde, and <0.0034 ppm of pesticide (2,4-DNP). These challenges reproduced all of her symptoms.

After going home, she continued to reduce her total pollutant load by constructing a less polluted room, drinking and bathing in less polluted water, and eating less polluted food. She maintained her health without symptom-suppressing medication by staying in the deadapted alarm state. So stabilized, she could perceive any exposures and take appropriate measures to minimize any damage they might cause; she could also avoid chronic exposure and a relapse. She has remained well without medication for the last 4 years.

Each principle demonstrated in this case report will now be discussed separately.

TOTAL BODY LOAD (BURDEN)

Total body load (burden) is the total of all pollutants in air, food, and water that the body incorporates and has then to process in order to maintain homeostasis.[1]

Pollutants that contribute to an individual's total body load may be biological (pollens, dusts, molds, foods, parasites, viruses, bacteria), chemical (organic or inorganic), or physical (heat, cold, electromagnetic radiation, light, radon, positive and negative ions, noise, weather changes)[2] (Figure 1).

Total body load increases as exposure to increased numbers of toxic chemical pollutants increases or as more tissue damage occurs, resulting in intolerance to an equal amount of chemicals. The accumulation of total body load pollutants involves two types of exposure. The first, a sudden, massive exposure, can be the result of a physical trauma such as an auto accident or a toxic injury such as an acute pesticide exposure or a massive viral or bacterial exposure. The second type of accumulation involves ongoing low-level toxic exposure to commonly occurring biological, chemical, and/or physical pollutants which then build up gradually.

Sublethal exposures to such common substances as pollens, dusts, molds, water contaminants, food, food contaminants, inhaled ambient doses of chemicals,[3-7] electromagnetic radiation, and positive air ions or electrical field changes[8] may individually contribute to increased total body load, or they may act in synergistic or additive fashion to cause insults as well as subsequent increased sensitivity to small doses of the aforementioned agents. Common examples of seemingly innocuous exposures of these kind include daily contact with such pollutants as sulfur dioxide from auto exhausts or refineries or formaldehyde fumes from new clothes or plywood. Exposures to radon or electromagnetic fields through contact with tight buildings that house computers provide another example. Even agents such as phenol, chlorine, formaldehyde, and various organic solvents used for wound cleansing and the prevention of infections can inadvertently be absorbed and consequently contribute to an individual's total body load. Bioaccumulation of toxic substances in the food chain can further increase this load. Since humans are at the end of the food chain, we tend to acquire higher levels of pollutants from this source.

As the facts of the massive pollution on earth are calculated, the causes of continuing increases in total body load become obvious. Four million distinct chemical compounds have been reported in the literature since 1965, with approximately 6,000 new compounds being added to the list each year. Of these, as many as 70,000 are in current commercial production.[9] Many of these chemicals are deliberately added to food, and over 700 have been identified in drinking water. When exposure to these substances is compounded by additional exposure from intake of pharmaceutical and/or over-the-counter medications, the direct exposure to individuals is considerable (Table 1).[12] A significant number of these toxic chemicals are lipid or fat soluble and tend to

Table 1. Chemicals with Toxicity Data

Source*	Number of Chemicals	Percent with Data
Commerce	49,000	20
Food additives	8,600	20
Cosmetics	3,400	26
Pesticides	3,400	36
Drugs	1,800	39
Water pollutants	1,800	

Adapted from: Schnare, David W. Examining the toxicology of low-level exposures: the approach and the literature, 1-10. Presented at the 151st National Meeting of the American Association for the Advancement of Science, May 30, 1985.
National Research Council—National Academy of Sciences. Toxicity testing: Strategies to determine needs and priorities, N.A.S. Washington D.C. 1984.

* Exposure to these will increase total body load.

accumulate in the fatty tissues, especially cell membranes, throughout the body, further increasing total load.

Excess total body load tends to disturb many of the body's homeostatic mechanisms, as evidenced by our studies of over 20,000 chemically sensitive patients who became ill after they were overtly exposed to pollutants. For people with known hereditary or acquired limitations, this overload becomes especially difficult to handle, as seen in individuals exposed to a large amount of contaminants such as those that might be released from a chemical explosion.[13,14] This overload may be too much for even a normal, healthy person to process. Disturbances in biological detoxification systems, such as changes in conjugation pathways, changes in cell receptor sensitivity, and/or depletion of nutrient fuels, may occur. Consequently, an individual may become susceptible with onset of generalized inflammatory disease or a specific change in one end-organ. Investigators have shown this phenomenon in studies of NO_2, SO_2, or O_3 overexposure, which produced damage to the bronchial mucosa.[15-17] This overload phenomenon also has been shown to occur in animals who ate foods containing pesticides and then developed disturbances of estrogen and progesterone levels.[18]

Although psychological stressors such as the death of a spouse, divorce, loss of a job, etc. add to the total body burden and can hasten the onset of disease,[19] an exaggerated psychological response is frequently secondary to a malfunctioning system.

In order to prevent disease, the body must either utilize, compartmentalize, or eliminate its total pollutant load. If this load becomes excessive and the body is unable to process it adequately, metabolic changes and symptoms may occur and not clear until this load is reduced. In our studies at the EHC-Dallas, improvement in energy level and in overall health usually begins as soon as the total load starts to diminish. Because time is needed to eliminate the total load, however, new symptoms may also occur during reduction due to mobilization of buried pollutants and inadvertent, new exposures.

This principle of reduction of total body load has been well documented and is commonly understood in relation to bacteria and body function. Reduction of bacterial load is practiced in nearly every facet of modern civilization by eliminating agents, including dust, garbage, vermin, and human and animal excrement, that are known to foster infectious diseases. Also, no physician today would consider treating a wound with antibiotics alone. He would first eliminate the overload of bacteria by vigorously cleansing the wound and applying a sterile bandage, thus reducing the total body burden of microbes. In environmental practice, an analogous situation occurs when a pollutant, e.g., a gas stove, is removed from a patient's home. The subsequent reduction in this patient's total body load allows for the metabolic systems to function better with more efficient overall detoxification.

The ECU was developed to aid in the reduction of total body burden. From our experience in using the ECU to reduce the total load in 3000 of our patients and in manipulating the outpatient environment of an additional 17,000 patients, we have concluded that the principle of reducing total body load is paramount to diagnosing and treating chemical sensitivity.

ADAPTATION

Adaptation is an acute survival mechanism which apparently allows an individual to "get used to" an acute toxic exposure in order initially to survive it. Adaptation involves a change in homeostasis (steady state) brought on by exposure to pollutants in the internal or external environment. Body function accommodates this exposure by adjusting to a new set point with induction and increased output of enzyme detoxification systems and immune system enhancement within a physiologic range. Adaptation can occur in any organ or tissue that has been affected by pollutant exposure.[20] Further, pollutant load may increase in all organs or just one (Figure 2).

Over time, adaptation that accompanies continued exposure to toxic substances can result in a long-term decrease in efficient functioning that can then lead to diminished longevity. Because an individual is unable to recognize the acute effects of toxic exposure during adaptation (masking — acute toxicological tolerance), he may inadvertently allow repeated exposures during which pollutants continue to enter and accumulate in his body. These substances may gradually contribute to an increased total body load and depletion of nutrient fuels as his body tries to counteract this build-up. Finally, depressed function occurs followed by end-organ failure.

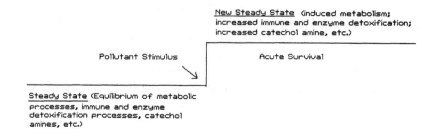

Figure 2.   Adaptation to an acute toxic exposure for survival.

Variations in metabolic changes during adaptation are dependent on the level, concentration, and virulence of pollutants as well as the volume of offending substances, exposure time, nutritional state of the organism, total body load, and the presence of other disease.[21-25] For example, an individual briefly exposed to cigarette smoke may develop a minor problem such as a runny nose. Constant exposure year in and year out for 30 years, however, increases the likelihood that he will develop lung cancer, lung failure, cardio-vascular disease, skin wrinkling, or a host of any other smoking-related conditions. Occasionally, no disease will occur. The aforementioned factors of total body load and an adequate nutritional state will finally determine the condition, if any, that results.

Studies of common inorganic pollutants such as nitrogen dioxide and ozone provide evidence that soundly supports the concept of adaptation.[26] For example, Stokinger and Coffin,[27,28] Bennett,[29] and the National Research Council[30] have pointed out that although daily exposures to pollutants may initially decrease pulmonary function 15 to 20%, by the fourth day of exposure, the pulmonary functions return to control levels. This type of activity demonstrates the adaptation phenomenon, but does not emphasize the required metabolic changes or the increased need for nutrient fuels in the adaptation process (Table 2).

Rinkel[31] described the adaptation concept in relation to foods and demonstrated that masking occurs with cyclic food sensitivity. Once a person becomes sensitive to a food, he usually adapts to it if he eats it daily. When he finally begins to develop symptoms of illness, he does not recognize the causal relationship between his food intake and the onset of his illness (Table 3).

Randolph[32] presented clinical demonstrations that specific adaptation is active in chemical sensitivities. His findings have since been confirmed by over 5,000 specialists in environmental medicine over the last 25 years; studies at the EHC-Dallas using environmentally controlled conditions have further confirmed the occurrence of adaptation in over 20,000 patients. Adaptation has also been observed in welders, cotton, grain, and wood workers,[33-35] and nitroglycerin workers and their families.[36]

Table 2. Clinical Adaptation to Ozone

Day	FEV$_1$ C - O (%)	Challenge
1	11	Ozone 0.3 ppm for 1 h
2	23	
3	4	
4	0	
5–11	0	No ozone
12	11	Ozone 0.3 ppm for 1 h
13	23	
14	4	
15	0	
16–22	0	No ozone
23	11	Ozone 0.3 ppm for 1 h
24	23	
25	4	
26	0	

Ozone challenge daily for 4 days showed FEV$_1$ to return to control levels, thus demonstrating adaptation. Avoidance for 4 days deadapts and allows the cycle to recur. This same phenomenon has been demonstrated for foods,[31] food pollutants, water pollutants, and other air pollutants.[1,27,28,75-77]

Table 3. Adaptation — Acute Toxicological Tolerance (Masking)

	Organism	Year
Rinkel,[31] food	Man	1944
Selye,[78] general adaptation	Animals	1946
Adolph,[79] specification adaptation	Animals	1956
Randolph,[32] chemical specific adaptation	Man	1962
Stokinger[27] and others,[80] pulmonary function	Man	1954–1964; 1968
Evans[81] and others,[82,83] cellular changes	Man	1962, 1973, 1975
Mustafa and Tierney,[84] metabolic changes	Man	1978

Misunderstanding of the adaptation phenomenon has led some to claim that chronic adaptation is beneficial. This misinterpretation of the value of adaptation has led some to argue that pollution, particularly ozone, is good for individuals because they can become used to it and, thus, build up tolerance. Continued exposure to pollutant stimuli may result in cellular and metabolic changes which are initially beneficial for protection, but eventually deplete nutrient fuels through overstimulation and overuse (see Chapter 4, Nonimmune Mechanisms). The seriousness of these changes provides evidence that defeats this specious argument, other than for acute survival.

Adaptation consists of three stages: alarm, masking, and end-organ failure. Each will be discussed separately.

Adaptation — Stage I (Alarm)

The first stage of adaptation is the alarm stage, in which an individual perceives a causal relationship between any exposures and the development of symptoms of ill health. If a stimulus is mild (i.e., sufficient enough to be cleared by the detoxification systems within a few hours and/or days), a pharmacologic effect will occur. If it is strong or prolonged, the response will be pathological with tissue changes (Figure 3).

Intervention during the alarm stage has the possibility of reversing any damage done. Without intervention, the adaptation process continues to the second stage, "masking." We have found that for optimum health, the individual is best kept in the alarm stage. Here, before masking has occurred, the individual can remain aware of environmental exposures and respond appropriately to them. Also, with intervention and prevention programs in place, he has the opportunity to expand the time and strength of his physiological adaptation without damaging his nutrient reserve. In other words, a person might be exposed to a pollutant while he or she has a high total load. With only a limited ability to combat this exposure, the individual's nutrient pool would likely be depleted and he or she would become vulnerable to further insults. With a decreased total load, however, the resources for sustained physiological adaptation and clearing without depletion of the nutrient pools are available. The individual has, therefore, the opportunity to maintain optimum health.

Adaptation — Stage II (Masking — Toxicological Tolerance)

Masking is the moving of the body's immune, metabolic, and detoxification systems to a new set point in order to accommodate an acute exposure (Table 3). The process of masking has two phases.

Phase 1 — The first phase of masking is a physiological adjustment through an induction of the immune and detoxification systems to combat an incitant.

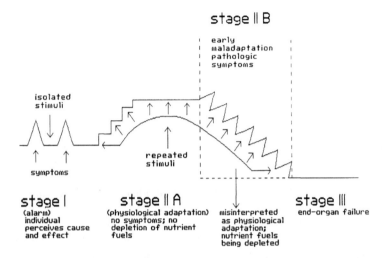

Figure 3.   **Adaptation phenomenon.**

This phase is probably defined by narrow limits that do not deplete nutrient fuels. It likely depends on the quantity of enzymes and the total load on immunodetoxification mechanisms as well as the nutrients available for fueling, induction, and response of these systems. In this phase, the system is minimally strained without chronic inflammation or severe metabolic or nutritional depletion occurring. For example, an individual might fill his car with gas. In response to gasoline fume exposures, his nose might begin to run. Then, even though the fumes remain in his body, his nose stops running as he moves away from the odor . Since little strain was placed on his metabolic pool or his immune and enzyme detoxification systems during this exposure, he continues about his business without the development of any additional problems.

Physicians rarely see patients in the first phase of masking unless a prevention program is being used for intervention with these patients. More commonly, most patients present in the second phase of masking or in stage III with early fixed-named disease including end-organ failure.

Phase 2 — Onset of the second phase of masking is signaled by the development of more severe difficulties and is really maladaptation (some opinion calls this phase early stage III maladaptation — the onset of end-organ failure). This phase occurs with prolonged exposure to or excess virulence of the incitants. This phase is pathologic, with tissue changes eventually occurring rather than simple physiologic adjustment. A series of metabolic events which strain the energy regulators, e.g., adenosine triphosphate (ATP),[37] metabolism of minerals, glucose,[38] carbohydrates, and fats,[39] occurs. Also, enzyme systems, such as the glucose-6 phosphate dehydrogenase,[40] glutathione peroxidase,[41] superoxide dismutase, monoamine oxidase,[42] aryl hydrocarbon hydroxylase, mixed function oxidase,[43] and cytochrome P-450 systems,[44] and many more are stimulated. Gradually these systems are overextended by continuing stress, which increases total body load by virtue of the body's gradual inability to detoxify substances. Gradual depletion of essential nutrients occurs. If an end-organ has become involved, it is more rapidly destroyed because of the concentrated overload of pollutants in this limited area. Also, the endocrine system may become involved with hormone deregulation and eventual deficiency. At this point, an individual may well remain unaware of the causal relationship between pollutant exposures and the onset of illness, and because the individual fails to recognize the ongoing effects of exposure, he may even continue to jeopardize his health by increasing his total burden as he inadvertently continues his exposures.

In this second stage of masking, an individual clinically acclimates to pollutants. This acclimation brings about metabolic alterations outside physiological parameters in that symptoms occur. For example, an individual who is sensitive to beef steak might eat a small portion and develop no symptoms. If, however, he eats a pound of steak at one meal and is unable to breakdown the toxins in the meat fast enough to reduce his load, he will overload his system and, unable to quickly reduce these pollutants, may experience severe symptoms of vascular spasm that result in Raynaud's phenomenon.

As apparent, correlated symptoms subside and the individual appears clinically to be no longer affected by exposure to a toxic substance, repeated exposures may, in fact, continue to damage immune and enzyme detoxification systems. Continued over time, this process can lead to further increases in the total body load.

In the second phase of masking, the stimulatory and depressive phases of bipolarity are accentuated.

Chronic exposure to toxic agents coupled with an inability to maintain the nutrient supply for detoxification lead to the third stage of adaptation, "end-organ failure."

Adaptation — Stage III (End-Organ Failure)

The process of maladaptation which leads to end-organ failure and is observed in the chemically sensitive individual may be one or both of two types (Figure 4). The first type occurs when an individual experiences frequent reexposure. Instead of completely clearing the pollutant load acquired from the initial exposure, an individual experiences only a short reaction. Continued subsequent exposures then lead to additional short reactions, none of which are sufficient to clear the expanding total load. Thus, as the load grows and the body responds increasingly less efficiently, reactions heighten and trigger more easily. Finally, if this process continues uninterrupted, the pollutant load becomes overwhelming and end-organ failure is inevitable.

The second type of maladaptation that leads to end-organ failure results from a minimal number of exposures over an extended time period. In this

A) Even with an adequate response system, the body is bombarded by pollutants and simply cannot protect itself because the timing of exposures is repeated and too frequent. This type of exposure occurs before the end of the body's short reactions and each of these responses becomes heightened and triggers more easily as reexposure continues.

B) Few exposures but long reaction time due to inadequate response system. Gradual build-up occurs because recovery time is so slow.

Figure 4.   Pollutant response leading to end-organ failure.

process, the reaction time is of an extended duration because the detoxifying mechanisms are inadequate to the clearing task. If a subsequent exposure occurs before complete clearing takes place, the defense system remains weakened. Continued inappropriate responses occur, leading inevitably to end-organ failure and fixed-named disease (Figure 4).

In the third stage of adaptation, fixed-named disease occurs with eventual end-organ failure or maladaptation. Diseases involving the heart, lung, blood vessel, gastrointestinal, genitourinary, or any of a host of other systems or tissues are easily recognized and given fixed names and usually are fixed and irreversible, e.g., coronary heart disease and lung failure.

Early chemical sensitivity may manifest itself clinically in late pathological adaptation (maladaptation) of phase 2, stage II masking, which actually may be early maladaptation of stage III. With early identification, diagnosis, and elimination of triggers and correction of nutritional deficits, end-organ failure often can be halted and reversed.

Individuals with good adaptive mechanisms are initially comfortable with a toxic load. They may, therefore, be more at risk for the sudden development of end-organ disease (which may have been developing over 20 to 30 years) than the sensitive individual, who is usually uncomfortable and perceives his polluted environment as he becomes aware of triggering agents. Initially, this perception of pollutants as causing symptoms works to the disadvantage of the chemically sensitive individual because of the discomfort that comes from the constant pollutant bombardment, but as time passes, he uses his awareness to clean up his environment. In so doing, he tends ultimately to put less strain on his immune and enzyme detoxication mechanisms, thus slowing the lifelong process toward end-organ failure.

Challenge tests performed on a patient in an adapted state are frequently negative, probably due to the increased activity of the induced immune and enzyme detoxification systems which can accommodate the pollutant without obvious clinical symptoms (Figure 5). Therefore, deadaptation by reduction of total body load must take place before causal relationships can be identified by challenge. At the EHC-Dallas, deadaptation has been accomplished in 20,000 patients whose subsequent participation in a total of 32,000 double-blind challenges revealed the causal relationship of the patients' signs and symptoms to specific environmental factors.

Avoidance of a suspected harmful substance for 3 to 4 days reduces total load and allows deadaptation to occur. A much more precise, immediate definition of a health problem can then be obtained when challenges are performed. In the majority of our studies presented in this book, we allowed a minimum 3 to 4 days of total load reduction and deadaptation to occur before challenges were performed. Failure to consider this adaptation principle has rendered many of the present published negative-challenge tests invalid.[45,46] For optimum results with challenge tests, a prolonged period of avoidance should not occur. If an individual avoids a substance for an extended time, he may repair tolerance to the substance, and the first challenge test will be negative. However, repeated challenges usually demonstrate the sensitivity.

BIPOLARITY

Bipolarity is a two-part response of the immune and enzyme detoxification and metabolic systems to exposure to a toxic substance(s) (Figure 6). The first phase involves a stimulatory/withdrawal reaction in which the stimulatory response is dominant. The second phase is a depressive reaction in which immune and enzyme detoxification and metabolic systems are unable to adequately process their total load. This inability leads to pathology. Each phase will be discussed separately.

Phase I

The first phase of bipolarity is a stimulatory/withdrawal reaction. Over a period of a few hours to a day or two, a finite stimulus induces enzyme detoxication and immune systems as well as biological amplification and mediator substances. A finite pharmacologic- or super-pharmacologic-type response then occurs in which the patient appears "high" as if under the initial stimulatory influence of alcohol. This phenomenon is not usually clinically observable unless a simultaneous brain reaction occurs. However, it can be measured physiologically through evaluation of enzyme and immune induction. The following hypothetical case is an example of an unmeasurable stimulatory reaction followed by observable withdrawal reactions.

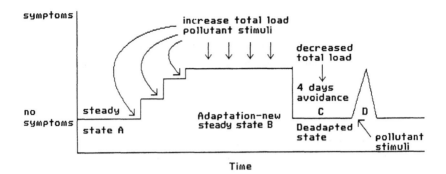

A — no symptoms

B — symptoms but no related cause and effect recognized

C — initially symptoms from withdrawal; then no symptoms

D — symptoms related to cause and effect even though
response was not as great as the no cause and
effect symptom state of new steady state adaptation

Figure 5.   Effect of total load on adaptation.

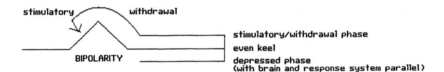

Figure 6.   Bipolarity.

An individual is exposed to pesticide fumes while at work. He perceives no effect until he returns home to an atmosphere free of pesticide. Here he develops withdrawal symptoms of headache, flu-like pains, malaise, and vomiting. After this sequence of events occurs several times, he consults his physician who compiles a detailed medical history and suspects chemical sensitivity. This person is told to avoid reexposure for 4 days, after which time he is challenged with ambient doses of pesticide. Typically, initial exposure to pesticide produces no reaction because the pollutants have been cleared and the individual is temporarily able to fend off an isolated reexposure. With each of three additional challenges, however, signs and symptoms are reproduced as the defense systems begin to rapidly malfunction. This scenario strongly suggests a stimulatory phase of bipolarity is occurring. Serial serum blood complements and eosinophil counts taken during challenge testing reveal alteration during the asymptomatic stimulatory phase. That they do not return to normal until completion of the symptomatic withdrawal phase suggests that asymptomatic induction of immune and metabolic systems has taken place.

If a stimulus is acutely removed and if the system has been affected adversely enough to change metabolism, a symptomatic withdrawal period may be experienced. The symptoms in this period are due to the slow turn-off of the response mechanisms involved in the immune and detoxification systems. *This withdrawal period is not to be confused with the depression (part of pathological and chronic phase 2 of bipolarity) which occurs with continued excessive stimulation* (Figure 7).

The clinically symptomatic withdrawal period of the finite pharmacologic-type phase is probably due to a sudden removal of an incitant with a slow 3- to 4-day turn off of the response systems, whereas the depressed part of the pathological phase of bipolarity is due to depletion of the response nutrient fuels with structural cellular damage. A prime example of withdrawal is seen in people who drink alcohol on a Saturday night in order to get "high" (stimulatory). The withdrawal period is expressed as a "hangover" on Sunday morning. If this habit continues, the individual will eventually move out of phase 1 into the second pathologic phase of bipolarity, where the system may be transiently stimulated but is mainly depressed due to the inability of the enzyme detoxification and immune systems to effectively combat the pollutant. Symptoms in this pathologic phase are the result of pollutant overload. Pathologic reactions to pollutants become autonomous and are virtually infinite, leading eventually to end-stage diseases such as alcoholic psychosis and/ or cirrhosis. In our experience at the EHC-Dallas, bipolarity can occur with exposure to most environmental pollutants.

Another example of withdrawal occurs in people who have hangovers on Saturday morning at home following exposure to toxic substances in the work place during the week. They feel well during the work week, but they experience withdrawal headaches, muscle aches, shakiness, and impaired ability to function through Saturday and early Sunday. By Sunday afternoon, as their systems rid themselves of the toxic substances and induced detoxification systems return to prestimulatory levels, the workers are again able to function well and feel fit enough to work on Monday, even though they do not feel totally well. This cycle may be repeated weekly for months or years before a person develops an end-stage disease. During this time, the symptoms continue to accentuate. It can easily be seen how in order to stave off withdrawal symptoms an individual might consciously or unconsciously seek reexposure in order to maintain a stimulatory phase and feel "high." Eventually, this behavior can lead to an addictive phenomenon which results in an increase in pollutant intake and, thus, increased total body load.

Our observations under environmentally controlled conditions have suggested that this stimulatory/withdrawal reaction leads to the addictive phenomenon which can spill over into cross-reactivity and the spreading phenomenon. For example, an individual working in a plastic factory may develop pathological thirst or hunger due to stimulation by ambient chemicals. He may then drink sugar phosphate-containing beverages or eat "junk foods" in order to prevent

Bipolarity of response to pollutants during the adaptation process may occur at any stage.

A. Pharmacologic - Super Pharmacologic Type Phase - Reversible - Acute - Limited Exposure

    1. Stimulatory part
        a. Stimulus intact
        b. Enzyme induction
        c. Immune induction
    2. Withdrawal part
        a. Stimulus removed acutely
        b. Enzyme system slow turn off up to 3 to 4 days
        c. Immune system slow turn off up to 3 to 4 days

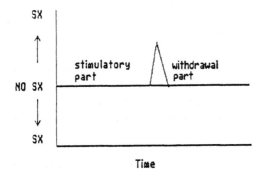

B. Pathologic Phase - Irreversible - Chronic exposure and response

    1. Stimulatory part
        a. Stimulus intact
        b. Enzyme induction continues
        c. Immune induction continues
    2. Depressed part - fixed end-organ disease
        a. Stimulus intact
        b. Enzyme induction depleted
        c. Immune induction depleted
        d. Nutrient fuels for induction and physiologic healing depleted
        e. Tissue destruction and scar formation occurs

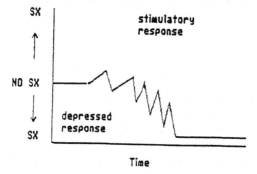

Figure 7.   Bipolarity.

uncomfortable withdrawal symptoms until he sets up more addictive patterns. Because he continues this intake, more harmful substances enter his body, thereby increasing his total load and decreasing his nutrient fuels. He may also take symptom-suppressing medications or drugs to maintain his precarious homeostatic balance thereby increasing, however inadvertently, his total load. Eventually, he may develop a spreading phenomenon and become sensitive to more unrelated incitants such as formaldehyde, TCE, etc. at increasingly lower doses. Antigen recognition sites, blood-organ barriers, and other mechanisms malfunction, allowing this spreading to occur as the response system defenses deteriorate. If the neurovascular system (i.e., autonomic nervous system that connects to the vascular system) is involved in pollution stimulation, as it is in 95% of observed reactions, the stimulatory withdrawal phenomenon of bipolarity occurs. As an individual's load increases, more damage occurs, causing additional sensitivity to new foods and chemicals which further deranges body metabolism. Periods of vulnerability ensue due to the depletion of response nutrient fuels, and fixed-named disease is allowed to set in.

We have seen this acute stimulatory/withdrawal pharmacologic-type phase of bipolarity occur in our environmental unit patients in over 20,000 individual oral and inhaled challenges of ambient doses of food and toxic chemicals. This stimulatory/withdrawal reaction can continue unabated until it finally moves into the pathologic phase II, where end-organ failure (fixed-named disease) develops.

Phase II

The second phase of bipolarity is predominantly depressive. When enzyme detoxification and immune systems are stimulated by a pollutant, a transient stimulatory response (similar to that which occurs in phase I) follows. The dominant depressive response then occurs as the various metabolic, enzyme, and immune systems become depleted (even temporarily) or sustain sufficient damage to inhibit their ability to respond effectively to the stimuli. If an individual continues daily intake of or exposure to a harmful substance, the damaging toxic long-term load increases. This damage occurs due to overuse of the nutrient fuels of induction and detoxification. After a period of time, be it minutes, months, or years, the body's defenses breakdown from overuse. Nutrient depletion and disabling symptoms of fixed end-organ disease develop. During the stimulation phase, enzyme and immune induction have occurred initially with increases in energy production, ATP,[47] glucose,[48] protein,[49] lipid metabolism,[50] and antipollutant enzyme response.[51] While in the depressive phase, these are inadequate or nonexistent, resulting in disturbed pathology. These stimulation and depression/exhaustion phenomena have been observed with many pollutant exposures including ozone, oxygen, and nitrous oxide.[52-57] Fixed pathological cellular changes then occur with abnormal

healing or even scar formation[58] followed by fixed-named end-organ disease. Increased tissue damage results in increased vulnerability to the same or a lower level of pollutants.

Phase I bipolarity occurs in all stages of adaptation. In stage I (alarm), bipolarity results in symptoms. In stage IIA (masking), no cause and effect symptoms are perceivable, but metabolic and physiologic processes are occurring subclinically. Exposure that exceeds the range of physiological response in stage IIA leads to stage IIB (masking), where symptoms are most severely manifested. Because of the discrepancy between the time of exposure and the onset of symptoms, however, causative links are not recognized (see the case report at the start of this chapter).

Phase I bipolarity of response to a pollutant in environmentally triggered adaptation is another factor which contributes to total body load. Often a brain reaction during which a person feels "high" accompanies other organ responses to pollutant stimuli. He is, therefore, inclined to misinterpret pollutant exposure as beneficial or not harmful and to continue to take in the pollutants, inadvertently increasing his total body load.

Phase II bipolarity is the depressive reaction in which immune and enzyme detoxification and metabolic systems are unable to adequately process their total load. This inability leads to pathology. Phase II bipolarity occurs only in stage III of adaptation (end-organ failure).

SPREADING

Spreading is a secondary response to pollutants that can involve new incitants or new target organs. Spreading that involves new incitants occurs when the body has developed increased sensitivity to increasing numbers of biological inhalants, toxic chemicals, and foods at increasingly smaller doses. At this time, overload becomes so taxing that a minute toxic exposure of any substance may be sufficient to trigger a response or autonomous triggering may occur. For example, a person initially may be damaged by a pesticide and then eventually have his disease process triggered by exposure to a myriad of toxic chemicals and foods, such as phenol, formaldehyde, perfume, beef, lettuce, etc.

Spreading may occur for many reasons. It may be due to a failure of the detoxification mechanisms — oxidation, reduction, degradation, and conjugation — brought about by pollutant overload, or it may occur because of depletion of the nutrient fuels of the enzyme or coenzyme, nutrient fuels, such as zinc, magnesium, all B vitamins, amino acid, or fatty acid. This depletion may account for the increasing inability of the body to detoxify and respond appropriately. The blood brain barrier or peripheral cellular membranes of the skin, lung, nasal mucosa, and gastrointestinal or genitourinary systems may be damaged, allowing previously excluded toxic and nontoxic substances to penetrate to areas that increase the risk of harm. Physiologic parameters including immune or pharmacologic releasing mechanisms, such as serotonin, kinin, and other vasoactive amines, may become so damaged that they are triggered by many toxic, then nontoxic (e.g., food) substances in addition to the specific one to which they initially reacted. It is well substantiated that antigen recognition sites may be disturbed or destroyed by pollutant overload. Hormone deregulation (feedback mechanisms) may occur, allowing for still greater dysfunction and sensitivity.

In contrast to patients who experience increased sensitivity to multiple triggering agents, some chemically sensitive patients may have one isolated organ involved in their disease process for years only to have dysfunction spread to other organs as their resistance mechanisms breakdown. This kind of spreading from one to another or multiple end-organs enables the progression of hypersensitivity and the eventual onset of fixed-named disease.

SWITCH PHENOMENON

The switch phenomenon is the changing of pollutant-stimulated responses from one end-organ response to another. This change usually occurs acutely, but it may occur over a much longer period of time. This phenomenon was first described by Savage in the 1800s.[59] He observed that when mental patients were at their worst they usually had a remission of their asthma or sinusitis. When they were better mentally and they were seen in the outpatient clinic, they had a much higher incidence of sinus and asthma problems. Randolph and most other environmentally oriented physicians have also observed this phenomenon. At the EHC-Dallas, we have observed similar occurrences in our patients and, in fact, take cognizance of this phenomenon when evaluating therapy outcome (Table 4).

In observing thousands of controlled challenges in the environmental unit, we have seen the target organ responses of many of our patients switch to several different ones during a long (i.e., 24 h) reaction. Often, we have seen, for example, transient brain dysfunction followed by arthralgia, followed by diarrhea, followed by arrhythmia.

The switch phenomenon has also been seen following unsuspected or unrecognized pollutant exposures. For example, an individual sprays his home with pesticides and subsequently visits a neurologist with complaints of headaches and a rheumatologist with symptoms of arthritis. Never noticing or suspecting a connection between his exposure and the onset of his symptoms, he fails to disclose to either doctor symptoms unrelated to their specialty or the fact of his exposure. Instead, he submits to symptomatic treatment by both physicians. His health may temporarily improve, but in all likelihood, his total body load will remain elevated and he will become increasingly vulnerable to additional exposures that result in a still greater variety of symptoms.

Table 4. Switch Phenomenon

Severity of Reaction[a]	Type of Reaction
4+	Manic
3+	Hypomanic, toxic
2+	Agitated
1+	High
0	Even keel
1-	Localized responses, e.g., rhinosinusitis, headache, PMS
2-	Systemic responses, vasculitis, phlebitis
3-	Aphasia, brain fag
4-	Depression

[a] During the course of pollutant challenge and injury, symptoms may switch from no reaction (0) to severe (4+) to decreased (−) reactions of some symptoms with simultaneous increase of others.

Even when therapy for pollutant injury appears to have been effective, the switch phenomenon may be disguising the fact that the body is still harboring a pollutant. In this case, a new set of symptoms may begin indicating that a pollutant response has simply switched to another end-organ. This phenomenon occurs frequently when symptom-suppressing medication therapy or inadequate environmental manipulation is used over a period of time. For example, a patient may have his sinusitis cleared by medication (e.g., cortisone), but later since the cause has not been eliminated, he may develop arthralgia and eventually arthritis, or his colitis may have cleared only later to have cystitis develop. Because the occurrence of switch phenomenon is both common and insidious, it is essential that physicians monitor their patients for the onset of any new symptoms or problems.

The switch phenomenon with its cluster of disparate symptoms signals a problem that is a part of a larger pattern needing further investigation. If physicians were cognizant of this phenomenon during initial patient evaluation, they could help curtail a life-long progression of illness through better diagnosis and treatment.

BIOCHEMICAL INDIVIDUALITY

The final principle necessary to understanding environmental aspects of health and disease and especially chemical sensitivity is that of biochemical

individuality. Biochemical individuality of response is an individual's uniqueness of response to pollutant exposure. This uniqueness depends on the differ-
ing quantities of carbohydrates, fats, proteins, enzymes, vitamins, minerals,
and immune and enzyme detoxification parameters with which an individual
is equipped to handle pollutant insults (Figure 8). These variations determine
an individual's ability to process the noxious substances he encounters. They
further contribute to the intensity of his reaction to toxic exposures and to his
susceptibility to chemicals. Thus, while a group of individuals may be exposed
to the same pollutant, one person may develop arthritis, one sinusitis, one
diarrhea, one cystitis, one asthma, and one may remain apparently unaffected
(Table 5).

The biochemical individuality of response of each individual is dependent
on at least three factors: genetics, the state of a fetus's nutritional health and
toxic body burden during gestation, and an individual's toxic body burden in
relation to his nutritional state at the time of exposure. Each will be discussed
separately (Figure 9).

Genetic Susceptibility

Biochemical individuality of response is determined in part by "genetic
susceptibility," a deficiency in or absence of genes able to respond appropriately to pollutant exposure and defend the body from pollutant effects.

An example of genetic susceptibility is that some individuals are born with
significantly less quantities (maybe 25, 50, or even 75%) of a specific enzyme.
While an individual with less than 100% of the quantity of an enzyme may to
some degree be able to counteract an environmental pollutant, his response is
often considerably less adequate than that of another individual born with
100% of the same detoxifying enzyme and appropriate immune parameters.
Once this former individual is overly stressed with toxic environmental chemicals, disturbed function occurs.

Genetic deficiency of some kind is found in virtually every person, and odds
are that any given individual may have one or more genetic defects.

Greater than 2,000 genetic metabolic defects are already described in the
literature and appear to be "time bombs" awaiting the appropriate combination
of environmental triggers to elicit their expression.[60] A well-known example
of the "time bomb" effect is seen in children with phenylketonuria who do well
as long as they do not take in phenylalanine, which they are genetically unable
to process. If their system becomes overloaded with phenylalanine, brain
damage occurs.[61] If excessive amounts of phenylalanine are not absorbed, they
grow to normal adulthood.

Smith[62] has shown that a group of individuals exists in the general population (approximately 20%) who are slow sulfonators. When they are exposed to
an excessive amount of a sulfur-containing substance such as s-carboxymethyl-

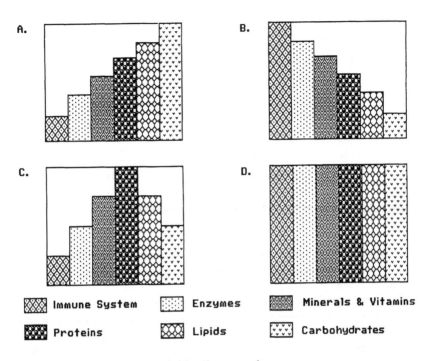

A.

B.

C.

D.

| Immune System | Enzymes | Minerals & Vitamins |
| Proteins | Lipids | Carbohydrates |

Figure 8.   Biochemical individuality — uniqueness.

Table 5. Effects of Xenobiotics on Detoxification Enzymes of the Chemically Sensitive Illustrating Biochemical Individuality

1. None
2. Induces
3. Depletes
4. Switches to other pathways
5. Destroys completely
6. Loss of enzymes or their nutrient fuels
7. Block of active site
   a.   Reversible
   b.   Irreversible
8. Loss of components
9. Death

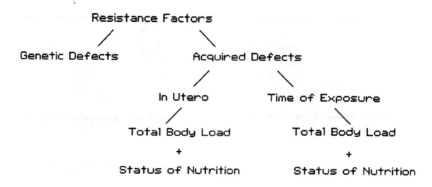

Figure 9.   Components of biochemical individuality.

L-cysteine, they will metabolize it slowly and become ill during the process. This defect may partially explain some individual's susceptibility to sulfur additives in foods and beverages and to sulfur-emitting refineries or sour gas fields.

Another example of this genetic time bomb effect was shown by Shear et al.,[63] who found that by exposing human lymphocytes to metabolites generated by a hepatic microsomal system, genetic differences in susceptibility can be demonstrated. Patients who have sulfonamide sensitivity due to genetic defects in N-acetylation are susceptible to aromatic epoxide metabolites.[64]

Idle[65] identified a group (about 10%) of the general population with genetic deficiency through the cytochrome P-450 system who could not tolerate debrisoquine compounds. In a select group of chemically sensitive people, Monro et al.[66] found similar results, showing as many as 60% of the patients studied had defects in these systems.

Debrisoquine challenge has been used to define a genetic defect in proper oxidative metabolism. The debrisoquine hydroxylation genetic locus has been shown to regulate the metabolism of drugs such as phenformin,[67] nortriptyline,[68] and phenacetin,[69] or it can produce exaggerated or adverse response to drugs and toxic chemicals. Thus, genetically determined oxidation status may be an important factor in individual susceptibility to certain diseases brought on by exposure to toxic environmental chemicals. Some people may have genetically very low cholinesterase levels that cause susceptibility to O-P pesticides.

State of Nutrition and Total Toxic Load In Utero

A person's biochemical individuality of response is not only affected by genetic susceptibility; it is also influenced by acquired nongenetic susceptibility brought on by environmental exposures that cause damage to sperm, ovum, or fetuses.[70-72] (See Chapter 28.)

Many different toxic chemicals, additives, preservatives, pesticides, and chlorinated hydrocarbons may cause vitamin, amino acid, lipid, enzyme, and mineral depletion that results in a selectively (nongenetically regulated), nutritionally depleted individual. This damage can occur at any stage of sperm, ovarian, or intrauterine development. This nutritional imbalance and increased total toxic load may be due to enzyme depletion in the fetus which can be accompanied and caused by bioconcentration of toxic chemicals. Dowty et al.[73] observed a bioconcentration of toxic chemicals in newborns when he studied blood levels of mothers and newborns simultaneously. This phenomenon has also been observed in babies of smokers and drug addicts.[74]

Direct toxic effects of chemicals on nutrients, competitive inhibition (e.g., some drugs such as isoniazid and hydralazine are selectively absorbed over B_6), overuse of the detoxification systems with depletion of their fuel sources (vitamins and minerals, etc.), or selective nutrient malabsorption due to toxic damage to intestinal or other membrane walls or flora can result in decreased resistance to environmental pollutants in the fetus. The fetus may then either experience susceptibility to chemicals or illness.

Detoxification systems in the liver and kidney of a fetus are not as well developed as its mother's. Thus, the mother's toxic load delivered to her fetus via the blood supply will not be processed well by the fetus. This inability on the part of the fetus to clear pollutants effectively ultimately leads to total load buildup and a potential for damage to developing cells and organs. Newborns may then emerge with depleted or malfunctioning immune and enzyme detoxification systems along with an increased total body load. If they have endured repeated exposures and/or excessive overload, they may experience end-organ damage in utero as well. Such in utero experience may lead to a newborn's inability to clear additional pollutants with which he may come in contact in routine living. Well-known conditions such as increased susceptibility to infection or inability to eat are seen in newborns of drug or cigarette addicts. Although infants whose problems result from maternal drug or cigarette use are routinely acknowledged to be at risk for health problems, less recognized are their counterparts, whose in utero pollutant exposure and consequent overload via their parents' pollutant exposures may have, if not causing gross abnormalities in their early development, predisposed them to a lifetime of potential health problems, including a risk for environmentally triggered illnesses and their associated end-organ diseases.

Bioaccumulation of pollutants in the food chain with accumulation in parents makes possible an increase in load with each generation. This accumulation also increases a fetus's chance of being overloaded. These findings help explain observations of generational increases of allergies and chemical sensitivity in babies and youth. Bioaccumulation of pollutants also may explain increased incidences of learning and behavioral problems and degenerative diseases seen with each successive generation.

State of Nutrition and Total Body Load at the Time of Exposure

Along with genetic susceptibility and state of nutrition and total toxic load in utero, biochemical individuality of response to pollutants is, finally, affected by an individual's state of nutrition and total load at the time of pollutant exposure. For example, an individual who is living in the city and therefore has an increased total body load, may not be able to eat any food to which he is sensitive without reacting to it. At the seashore, however, where his total load is reduced, he may be able to eat these same foods without experiencing any problems.

While an individual's total toxic load and nutritional state are throughout life influenced by genetic and acquired congenital factors, they are not fixed. They vary daily depending on the total number of pollutants an individual absorbs from the air he breathes, the food he eats, and the water he drinks and cleans with. They further vary depending on an individual's dietary habits which affect his nutritional state. Also, efficient functioning of biochemical and immune detoxification systems affects these variations. Since an individual's ability to defend himself against the effects of pollutant exposure is determined by his total load and nutritional state at the time of exposure, daily fluctuation of these often seems to determine an individual's potential for deleterious, long-term health risks and particularly susceptibility to chemicals, which could lead to sensitivity and, if left untreated, resultant fixed-named disease.

Daily variations in total body load are the result of both controllable (food, water, home environment) and many uncontrollable (weather cycles, outdoor air, public building pollutants) variables. The interaction of these variables makes possible many ways in which the total load at the time of exposure can vary (Figure 10).

An increased total load and depleted nutritional state account for the variability of responses under different test conditions on different days. These changes also account for periods of vulnerability that occur in each individual's life and promote the onset of disease. Manipulation of total body load within the individual's biochemical individuality of response is the basis for diagnosing chemical sensitivity and determining therapy for environmentally triggered responses.

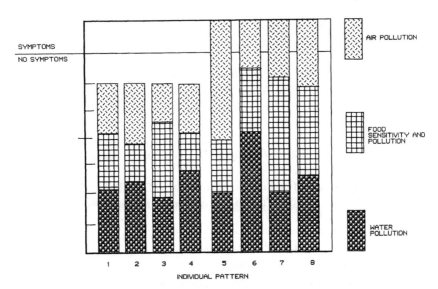

Figure 10. Variability schemes for components of total body load (other variations of components may be seen) due to biochemical individuality as a function of pollutant intake.

REFERENCES

1. Rea, W. J. "Review of cardiovascular disease in allergy," in *Bi-Annual Review of Allergy*, C. A. Frazier, Ed. (Springfield, IL: Charles C. Thomas, 1980), pp. 282–347.
2. Rea, W. J., and O. D. Brown. "Cardiovascular disease in response to chemicals and foods," in *Food Allergy and Intolerance*, J. Brostoff and S. J. Challacombe, Eds. (London: Bailliere Tindall, 1987), 737–783.
3. Dickey, L. D. "History and documentation of coseasonal antigen therapy intracutaneous serial dilution titrations, optimal dosage, and provocative testing," in *Clinical Ecology*, L. D. Dickey, Ed. (Springfield, IL: Charles C. Thomas, 1976), pp. 18–25.
4. Miller, J. B. *Food Allergy: Provocative Testing and Injection Therapy* (Springfield, IL: Charles C. Thomas, 1972).
5. Speer, F. *Migraine* (Chicago, IL: Nelson-Hall Publishers, 1977).
6. Randolph, T. G. "Food susceptibility (food allergy)," in *Current Therapy*, H. Conn, Ed. (Philadelphia: W.B. Saunders, 1960), pp. 418–423.
7. Rinkel, H. J., T. G. Randolph, and M. Zeller. *Food Allergy* (Springfield, IL: Charles C. Thomas, 1950).
8. Becker, R. O., and A. A. Marino. *Electromagnetism and Life* (Albany, NY: State University of New York Press, 1982).

9. Schnare, D. W., G. Denk, M. Shields, and S. Brunton. "Evaluation of a detoxification regimen for fat stored xenobiotics," *Med. Hypoth.* 9(3):265–282 (1982).

10. Adapted from Schnare, David W. "Examining the Toxicology of Low-Level Exposures: The Approach and the Literature," paper presented at the 151st National Meeting of the American Association for the Advancement of Science, Cincinnati, OH, May 30, 1985.

11. National Research Council. "Toxicity Testing: Strategies to Determine Needs and Priorities (Washington DC; National Academy of Sciences, 1984).

12. "Chemicals Identified in Biological Media, A Data-Base," EPA-560/13-80-036B, PB, 81-161-176; U.S. Environmental Protection Agency, Washington DC (1980).

13. Stanbury, J. B., J. B. Wyndgaarden, D. S. Fredrickson, J. L.Goldstein, and M. S. Brown, Eds. *The Metabolic Basis of Inherited Disease*, 5th ed. (New York: McGraw-Hill, 1983).

14. Williams, R. J. *Biochemical Individuality: The Basis for the Genotrophic Concept* (New York: John Wiley & Sons, 1956).

15. Nadal, A., and L. Y. Lee. "Airway hyperirritability induced by ozone," in *Biochemical Effects of Environmental Pollutants*, S. D. Lee, Ed. (Ann Arbor, MI: Ann Arbor Science Publishers, 1977).

16. Matsumura, Y, K. Mizuno, D. Miyamoto, T. Suzuki, and Y. Oshima. "The effect of ozone, nitrogen dioxide and sulfur dioxide on experimentally induced allergic respiratory disorder in guinea pigs," *Am. Rev. Respir. Dis.* 105:262 (1972).

17. Lee, L.Y., E. R. Bleecher, and J. Nadel. "Effects of ozone on bronchomotor response to inhaled histamine aerosol in dogs," *J. Appl. Physiol.* 43:626 (1977).

18. Hinsdill, R. D., and P. T. Thomas. "Effect of polychlorinated biphenyls on the immune responses of rhesus monkeys and mice," *Toxicol. Appl. Pharmacol.* 44:41–51 (1978).

19. Holmes, T. H., and R. H. Rahe. "The social adjustment rating scale," *J. Psychosom. Med.* 11:213–218 (1967).

20. Selye, H. "The general adaptation syndrome and the diseases of adaptation," *J. Allergy* 17:231–247 (1946).

21. Crapo, J. D., J. Marsh-Salin, P. Ingram, and P.C. Pratt. "Tolerance and cross-tolerance using NO_2 and O_2. II. Pulmonary morphology and morphometry" *J. Appl. Physiol.* 44:370 (1978).

22. Tierney, D. F., L. Ayers, S. Herzog, and J. Yang. "Pentose pathway and production of reduced nicotinamide adenin dinucleotide phosphate" *Am. Rev. Respir. Dis.* 108:1348 (1973).

23. Tierney, D. F., L. Ayers, and R. S. Kasuyama. "Altered sensitivity to oxygen toxicity," *Am. Rev. Respir. Dis. Suppl.* 2:59 (1977).

24. Young, S. L., and J. N. Knelsen. "Increased glucose uptake by rat lung with onset of edema," *Physiologist* 16:494 (1973).

25. Currie, W. D., P. C. Pratt, and A. P. Sanders. "Hyperoxia and lung metabolism," *Chest* 66(Suppl.):195 (1974).
26. Mustafa, M. G., and D. F. Tierney. "Biochemical and metabolic changes in the lung with oxygen, ozone, and nitrogen dioxide toxicity," *Am. Rev. Respir. Dis.* 118:1061–1090 (1978).
27. Stokinger, H. E. "Ozone toxicology: A review of research and industrial experience (1954–1964)," *Arch. Environ. Health* 10:719 (1965).
28. Stokinger, H. E., and D. L. Coffin. "Biological effects of air pollution," in *Air Pollution*, Vol. I, 2nd ed., A. C. Stern, Ed. (New York: Academic Press, 1968), p. 445.
29. Bennett, G. "Ozone contamination of high altitude aircraft cabins," *Aerosp. Med.* 33:969 (1962).
30. National Research Council. "Atmospheric Ozone Studies" (Washington DC: National Academy of Sciences, 1962).
31. Rinkel, H. J. "The role of food allergy in internal medicine," *Ann. Allergy* 2:115 (1944).
32. Randolph, T. G. *Human Ecology and Susceptibility to the Chemical Environment* (Springfield, IL: Charles C. Thomas, 1962).
33. Williams, N., A. Skonlas, and E. Merriman. "Exposure to grain dust. I. A survey of effects," in *Pulmonary Disease. Focus on Grain, Dust, and Health*, J. A. Dosman and D. S. Cotton, Eds. (New York: Academic Press, 1980).
34. Howie, A. D., G. Boyd, and F. Moran. "Pulmonary hypersensitivity to ramin (Gonystylus bancanus)," *Thorax* 31:585 (1976).
35. Hunter, D. *The Diseases of Occupations* (London: The English Universities Press Ltd., 1975), p. 1029.
36. Doum, S. "Nitroglycerin and alkyl nitrates," in *Environmental and Occupation Medicine*, W. Rom, Ed. (Boston: Little, Brown, 1983), pp. 639–648.
37. Young, S. L., and J. N. Knelson. "Increased glucose uptake by rat lung with onset of edema," *Physiologist* 16:1065 (1973).
38. Ospital, J. J., N. Elsayed, A. D. Hacher, M. G. Mustafa, and D. F. Tierney. "Altered glucose metabolism in lungs of rats exposed to nitrogen," *Am. Rev. Respir. Dis.* 113:107 (1976).
39. Young, S. L., and J. N. Knelson. "Increased glucose uptake by rat lung with onset of edema," *Physiologist* 16:1069 (1973).
40. Abe, M., and D. F. Tierney. "Lipid metabolism of rat lung during recovery from lung injury," *Fed. Proc.* 35:479 (1976).
41. Young, S. L., and J. N. Knelson. "Increased glucose uptake by rat lung with onset of edema," *Physiologist* 16:1070 (1973).
42. Said, S. I. "The lung as a metabolic organ," *N. Engl. J. Med.* 279:1330 (1968).
43. Palmer, N. S., D. N. Swanson, and D. L. Coffin. "Effect of ozone on benzpyrene hydroxylase activity in the sorian golden hamster," *Cancer Res.* 31:730 (1971).

44. Goldstain, B. D., S. Soloma, B. S. Pasternack, and D. R. Bickers. "Decrease in rabbit lung microsomal cytochrome P-450 levels following ozone exposure," *Res. Commun. Chem. Pathol. Pharmacol.* 10:759 (1975).

45. Bock, S. A., W.Y. Lee, L.K. Remigio, A. Hoist, and C. D. May. "Appraisal of skin tests with food extracts for diagnosis of food hypersensitivity," *Clin. Allergy* 8:559 (1978).

46. Blade, L. M. "Occupational exposure to formaldehyde recent NIOSH involvement," in *Formaldehyde-Toxicology-Epidemiology-Mechanisms*, J. J. Clary, J. E. Gibson, and R. S. Waritzs, Eds. (New York: Marcel Dekker, 1983), 1.

47. Hunter, D. *The Diseases of Occupations* (London: The English Universities Press Ltd., 1975), p. 1029.

48. Doum, S. "Nitroglycerin and alkyl nitrates," in *Environmental and Occupation Medicine*, W. Rom, Ed. (Boston: Little, Brown, 1983), pp. 639–648.

49. Young, S. L., and J. N. Knelsen. "Increased glucose uptake by rat lung with onset of edema," *Physiologist* 16:1065 (1973).

50. Ospital, J. J., N. Elsayed, A. D. Hacher, M. G. Mustafa, and D. F. Tierney. "Altered glucose metabolism in lungs of rats exposed to nitrogen," *Am. Rev. Respir. Dis.* 113:107 (1976).

51. Pelkonen, O. "Biotransformation of xenobiotics in the fetus," *Pharm. Ther.* 10:261–281 (1980).

52. Tierney, D. F., L. Ayers, S. Herzog, and J. Yang. "Pentose pathway and production of reduced nicotinamide adenin dinucleotide phosphate" *Am. Rev. Respir. Dis.* 108:1348 (1973).

53. Crapo, J. D., J. Marsh-Salin, P. Ingram, and P.C. Pratt. "Tolerance and cross-tolerance using NO_2 and O_2. II. Pulmonary morphology and morphometry" *J. Appl. Physiol.* 44:370 (1978).

54. Tierney, D. F., L. Ayers, S. Herzog, and J. Yang. "Pentose pathway and production of reduced nicotinamide adenin dinucleotide phosphate" *Am. Rev. Respir. Dis.* 108:1348 (1973).

55. Tierney, D. F., L. Ayers, and R. S. Kasuyama. "Altered sensitivity to oxygen toxicity," *Am. Rev. Respir. Dis. Suppl.* 2:59 (1977).

56. Young, S. L., and J. N. Knelsen. "Increased glucose uptake by rat lung with onset of edema," *Physiologist* 16:494 (1973).

57. Stokinger, H. E., and D. L. Coffin. "Biological effects of air pollution," in *Air Pollution*, Vol. I, 2nd ed., A. C. Stern, Ed. (New York: Academic Press, 1968), p. 445.

58. Zeek, P. M. "Periarteritis nodosa and other forms of necrotizing angitis," *N. Engl. J. Med.* 248:764 (1953).

59. Savage, G. M. *Insanity and Allied Neuroses: Practical and Clinical* (Philadelphia: Henry C. Lea's Son and Co., 1884).

60. National Research Council. "Toxicity Testing: Strategies to Determine Needs and Priorities (Washington DC; National Academy of Sciences, 1984), pp. 18 and 270.

61. National Research Council. "Toxicity Testing: Strategies to Determine Needs and Priorities (Washington DC; National Academy of Sciences, 1984), pp. 18 and 270.

62. Smith, R. L. "Inborn Errors of Metabolism of Drugs and Toxic Substances," report at the 4th Annual International Symposium on Man and his Environment in Health and Disease, Dallas, TX, 1986.

63. Shear, N. H., S. P. Spielbury, D. M. Grant, B. K. Tang, and W. Kalow. "Differences in metabolism of sulfonamides predisposing to idiosyncratic toxicity," *Ann. Intern. Med.* 105:179–184 (1986).

64. Motulsky, A. G. "Pharmacogenetics," *Prog. Med. Genet.* 3:49–74 (1964).

65. Idle, S. R., A. Mahgoub, R. Lancaster, and R. L. Smith. "Hypotensive response to debrisoquine and hydroxylation phenotype," *Life Sci.* 22:979–984 (1978).

66. Monro, J. Personal communication, Breakspear Hospital, London (1986).

67. Oats, N. S., R. R. Shah, J. R. Idle, and Smith, R.L. "Genetic polymorphism of phenformin 4-hydroxylation," *Clin. Pharmacol. Ther.* 32:81–89 (1982).

68. Mellstrom, B., L. Bertilsson, J. Säwe, H. U. Schulz, and F. Sjöquist. "E- and Z- 10- Hydroxylation of nortriptyline: Relationship to polymorphic debrisoquine hydroxylation," *Clin. Pharmacol. Ther.* 30(2):189–193 (1981).

69. Stein, M., S. E. Keller, and S. J. Schleifer. "Stress and immunomodulation. The role of depression and neuroendocrine function," *J. Immunol.* 135:8275 (1984).

70. Young, S. L., and J. N. Knelson. "Increased glucose uptake by rat lung with onset of edema," *Physiologist* 16:1065 (1973).

71. Soyka, L., and J. Joffe. "Male mediated drug effects on offspring," in *Drug and Chemical Risks to the Fetus and Newborn*, Progress in Clinical and Biological Research, Vol. 36, R. Schwarz and S. Yoffe, Eds. (New York: Alan R. Liss, Inc., 1980), pp. 49–66.

72. Khera, K. S. "Maternal toxicity of drugs and metabolic disorders — A possible etiologic factor in intrauterine death and congenital malformation: A critique on human data," *CRC Crit. Rev. Toxicol.* 17:345–375 (1987).

73. Dowty, B. J., A. L. Laseter, and J. Storer. "The transphental migration and accumulation in blood of volatile organic constituents," *Pediatr. Res.* 10:696–701 (1976).

74. Krauer, B., F. Krauer, F. Hyhen, and E. del Pozo, Eds. *Drugs and Pregnancy, Maternal Drug Handling, Fetal Drug Exposure* (London: Academic Press, Inc., 1989).

75. Randolph, T. G. "The specific adaptation syndrome," *J. Lab. Clin. Med.* 48:934 (1956).

76. Hackney, J., W. Linn, S. Mohler, and C. Collier. "Adaptation to short term effects of ozone on men exposed repeatedly," *J. Appl. Physiol.* 43(1):82–85 (1977).

77. Horvath, S. M., J. A. Gliner, and L. J. Folinsbee. "Adaptation to ozone: Duration effect," *Am. Rev. Respir. Dis.* 123:496–499 (1981).
78. Seyle, H. "The general adaptation syndrome and the diseases of adaptation," *J. Allergy* 17:289–358 (1946).
79. Adolph, E. F. "General and specific characteristics of physiological adaptations," *Am. J. Physiol.* 184:18 (1956).
80. Hackney, J., W. Linn, S. Karuza, R. Buckley, D. Law, D. Bates, M. Hazache, L. Pengelly, and F. Silverman. "Effects of ozone exposure in Canadians and Southern Californians. Evidence for adaptation?" *Arch. Environ. Health* 32(3):110–116 (1976).
81. Evans, M. J., L. J Cabral, R. J. Stephens, and G. Freeman. "Renewal of alveolar epithelium in the rat following exposure to nitrogen dioxide," *Am. J. Pathol.* 70:175 (1973).
82. Adamson, I. Y. R., and D. H. Bowden. "The type II cell as a progenitor of alveolar epithelial regeneration: A cytodynamic study on mice after exposure to oxygen," *Lab. Invest.* 30:35 (1974).
83. Evans, M. S., L. J. Cabral, R. S. Stephens, and G. Freeman. "Transformation of alveolar type II cells to type I cells following exposure to nitrogen dioxide," *Exp. Mol. Pathol.* 22:142 (1975).
84. Mustafa, M. G., and D. F. Tierney. "Biochemical and metabolic changes in the lung with oxygen, ozone, and nitrogen dioxide toxicity," *Am. Rev. Respir. Dis.* 118:6 (1978).

4   NONIMMUNE MECHANISMS

INTRODUCTION

Some researchers and physicians have long refused to acknowledge chemical sensitivity as a clinical entity both because its symptoms are diverse and because they have not taken the opportunity to study its identifiable patterns or its progression under controlled conditions. Because of this lack of interest in exploring the parameters of chemical sensitivity, sufficient clinical and experimental evidence has been unavailable to demarcate its underlying mechanisms. A significant body of research is finally emerging, however, that demonstrates the occurrence of alterations of detoxification systems by pollutant overload. As chemical sensitivity becomes more specifically identifiable, our ability to study and evaluate its symptoms and physiological and biochemical mechanisms of operation also increases. Today we are able to recognize symptomatic expression of chemical sensitivity and intervene with appropriate prevention, diagnosis, and treatment. Further study and increased understanding of the mechanisms that underlie this entity will facilitate our ability to manage it. This chapter, therefore, intends to introduce the enzymatic and nonenzymatic aspects of pollutant transformation that are influential in the onset and continuation of chemical sensitivity. Although we do not know at this time the initial mechanism by which good health gives way to chemical sensitivity, we do know that once this change has occurred, dysfunction of these biochemical mechanisms becomes relevant.

In this chapter, we identify the connection between the clinical aspects of chemical sensitivity and the biochemical processes that are involved in the initiation and propagation of this illness. We discuss how pollutants penetrate and affect surface barriers, which function to protect the body from foreign substances. We trace how, as penetration occurs, pollutants continue to damage systems, organs, cells, subcellular elements, molecular targets, and finally damage metabolism and the body's nutrient fuels.

We further show in this chapter how metabolic derangements seen in chemical sensitivity may affect one or any combination of the oxidation, reduction, degradation, or conjugation processes. Evidence that free radicals do occur in chemical sensitivity leads us to consider the part their creation may play in the initial change from good health to chemical sensitivity. Because enzymes such as superoxide dismutase, glutathione peroxidase, and catalase are often induced or suppressed in the chemically sensitive, we speculate that these contribute to the initial move from good health to the onset of chemical sensitivity. We discuss these systems and their involvement in chemical sensitivity at the end of this chapter, fully realizing that malfunction of any of these systems may be the result of damage to surface barriers and not the initial causative link between good health and the onset of chemical sensitivity.

Nonimmune detoxification mechanisms are those that clear pollutants from the body through the enzyme and nonenzymatic biochemical transformation systems. They are important to the study of chemical sensitivity in that their function or malfunction often affects the severity of the clinical aspects of chemical sensitivity.

Often the severity of chemical sensitivity depends on three factors: the ease with which pollutants are able to penetrate the various barriers, the amount of damage that is done at the point of entry, or the adverse chain of events set in motion by exposure, including altered metabolism, system(s) dysfunction, and/ or interrupted cellular function.

The following case demonstrates the route of pollutant injury from its source to its entry into the body, to its transport through the blood vessels to the detoxification mechanisms in the liver, to the sequestration in fat cells, to damage to the end-organ, and finally, to recovery due to repair systems with reduction of total body pollutant load and restoration of health.

Case report — A 35-year-old white male worked in a mixing room for solvents. The room was poorly ventilated and the vat was open to the air. He had no protection and often would stick his arm in the vat to mix the solution. At times, various types of solvents such as 1,1,1-trichlorethane, trichloroethylene, tetrachloroethylene, xylene, and toluene were used.

Initially, this patient developed signs of neurovascular dysfunction characterized by his hands and feet turning blue and then intermittently blanching white. He developed petechiae, spontaneous bruising, cold sensitivity, imbalance, breathlessness, and recurrent periorbital, hand and feet nonpitting edema. He could not stand on his toes with his eyes closed. Further objective measurements using the Iris Corder® showed autonomic nervous system dysfunction with cholinergic and sympatholytic expression. Balance studies using the computerized balance machine showed that he had a central nervous system defect in his brainstem including the cerebellum. This patient showed signs of liver stress with an increase in alkaline phosphatase, SGOT, SGPT, SPPT, CPK, and lactic dehydrogenase.

Analysis of his blood revealed elevated levels of toxic chemicals including toluene, xylene, 1,1,1-trichloroethane, trichloroethylene, tetrachloroethylene, and chloroform. The presence of these chemicals in the blood indicated that they probably entered through both the lungs and skin.

This patient developed odor sensitivity. He suddenly became intolerant of perfumes, car exhausts, fumes from pesticide, formaldehyde, foam rubber, fabric stores, and chlorine. Onset of odor sensitivity is compatible with the traveling of solvents up the olfactory nerve to the smell centers in the brain, through the thalamus, and to the hypothalamus. Triggering of odor sensitivity and autonomic nervous system dysfunction then results in blood vessel deregulation. Direct pollutant injury to the cell membranes along with autonomic dysfunction could have caused this patient's edema. Also, this cell wall injury might have caused protein denaturation and the formation of haptens, from which hypersensitivity could have resulted. However, a more likely metabolic scenario is that the conjugation systems that are used for detoxification were overloaded causing increased sensitivity to minute doses of new toxic chemicals.

This patient had to stop work. He gradually improved to a state of chronic illness and dysfunction. Fat biopsies at this time showed increased levels of the originally identified chemicals plus the presence of chloroform and a concomitant decrease in blood levels of these chemicals. His decreased blood levels suggest that his intake of pollutants had decreased, probably due in part to his having been off work and in part to more efficient liver function in that he was metabolizing these chemicals better. His fat biopsy indicated that some chemicals were being sequestered. The presence of chloroform, which was not initially identified, suggested that some of the chlorinated compounds were breaking down and thus his detoxification systems were to some extent working. The total effect of these processes was a decrease in his total body load.

This patient was placed in the ECU where he fasted for 4 days. His blood toxic chemical levels increased during this time, suggesting that he was mobilizing some of those that had been sequestered in fat. His clinical course exacerbated during this period with resultant edema, bruising, increased imbalance, petechiae, and peripheral cyanosis. Biopsy of the bruises showed a perivascular lymphocytic infiltrate, which is compatible with immune dysfunction.

His peripheral T lymphocytes, especially his suppressor T cells, were low, which is compatible with the presence of increased toxic chemicals in the blood and lymphocytes in the skin biopsy. His liver battery, which had returned to normal after he had stayed away from work during this period, increased. During this patient's stay in the controlled environment, his total pollutant load decreased, as was evidenced by a decrease in blood pollutant levels to below those levels prior to entering the ECU. Fat biopsy showed less pollutants in the fat than previously, increased T lymphocytes, an increase in the anti-pollutant enzymes superoxide dismutase, glutathione peroxidase, and catalase (that had been depressed prior to entering the ECU), and a decrease in the liver enzymes.

At this time, his brain function, which was measured by the Harrel-Butler profile and was grossly abnormal for toxic brain damage upon admission to the ECU, returned to normal as did his balance test and Iris Corder® measurement. This patient was challenged with individual biological inhalants (pollen, dusts, and mold), foods, and ambient doses of toxic chemicals (formaldehyde, <0.2 ppm; phenol, <0.0024 ppm; petroleum-derived alcohol, <0.5 ppm; pesticide [2,4DNP], <0.0034 ppm; and chlorine, <0.33 ppm). A wide variety of these substances reproduced all of his signs and symptoms, confirming his chemical sensitivity.

His reactions to so many substances suggest that antigen recognition sites on cells were damaged and that the metabolic pathways of oxidation, degradation, and conjugation were overloaded. This overload was confirmed by measuring blood and intracellular levels of vitamins, amino acids, and minerals. There were many abnormalities of the B vitamins (low), minerals (excessively high or low), and amino acids, which were high in the urine and low in the blood. Specifically, methionine, glutathione, and acetyl radicals were low in the blood, while mercapturate and glucaric acid were high in the urine. These findings support the notion that the conjugation systems and their cofactors, the vitamins and minerals, were abnormal, which made pollutant elimination difficult. The additional finding of low antipollutant enzymes with high levels of toxic chemicals in the fat and blood supports the idea that the toxics circulated or created free radicals, either of which overutilized or directly damaged the superoxide dismutase, glutathione, peroxidase, and catalase.

Widespread food sensitivities tend to be immune mediated or nonimmune. When they are nonimmune (oxidation, reduction, degradation, and conjugation), as some were in this case, they result in metabolic dysfunction of amino acid metabolism which can then lead to food sensitivity. For example, this patient could have had low taurine due to pollutant overload which would have caused poor bile conjugation and then gastrointestinal upset with food intolerance.

After a period of treatment with total reduction of pollutant load by avoidance of pollutants in his air, food, and water as well as nutrient supplementation, this patient improved. Because he remained supersensitive to odors, he was given 8 weeks of heat depuration-physical therapy. He went through an exacerbation of signs, symptoms, and laboratory parameters similar to those experienced in the ECU. This time all laboratory tests returned to normal, and after a period of 4 more months of environmental control, he became totally healthy. He was able to return to a job where he was exposed to decreased levels of chemicals, and he has remained well.

Pollutants can enter the body through the respiratory tree, gastrointestinal tract, genitourinary system, and the skin. Local factors and principles of pollutant injury in the chemically sensitive are discussed in detail in the individual chapters. Therefore, only general factors will be presented here (Figure 1).

Though not fully delineated, environmental pollutants, once they have entered the body, can trigger responses in numerous ways while at the same time disturbing immune and enzyme detoxification systems. Most compounds entering the body are subject to metabolic transformation irrespective of whether they have nutritive value. The body has to either utilize, compartmentalize, or eliminate foreign substances. The chemically sensitive appear to have difficulty in eliminating foreign substances because these change to other chemicals that may also be toxic to them but not necessarily to a normal individual. Responses to foreign materials require energy and, therefore, metabolism, with its nutrient requirements, is involved. The chemically sensitive appear to have more energy requirements because of their inefficiently functioning systems. Although there is a separate category of detoxification pathways for foreign compounds (mainly conjugation), they as well as nutrients may otherwise be metabolized by identical processes (oxidation, reduction, and degradation), even though the result of that process differs with the molecule or substance in question.[1]

Xenobiotics (foreign chemicals) are metabolized in order to reduce toxicity (though sometimes toxicity may be increased in certain steps). In contrast, nutrients are metabolized to maintain energy building blocks (see Chapter 6, Nutrition). Disturbance of these building block pathways as seen in the chemically sensitive may occur with an overload of toxic chemicals that interfere with the orderly digestion, transportation, incorporation, and utilization of nutrients. This fact partially explains why food, vitamin, and/or mineral intolerance occurs in the chemically sensitive. At times when the detoxication pathways may be damaged by an overload of toxic chemicals, disturbances may appear to be permanent. In our experience treating 20,000 patients, however, normal function usually returns after a period of nutrient treatment and removal of the incitants (reduction of total body load). The pattern here can be confusing and open to interpretation because biochemists focus on the mechanisms of detoxification and specific enzyme system involvement. For example, they might see a mixed enzyme reaction that is triggered by one chemical and inhibited by another. While this immediate response looks like chemical sensitivity to the clinician and, in fact, probably is, biochemists would not necessarily view it as such, calling it instead a mixed enzyme reaction. The clinician, in contrast, treats these mixed responses that may be triggered by many separate chemicals as chemical sensitivity without knowing which chemical was the initiator and which the inhibitor because they produce the same clinical response. *The complexity of pollutant injury and detoxication clearly emphasizes the multifaceted nature of chemical sensitivity.*

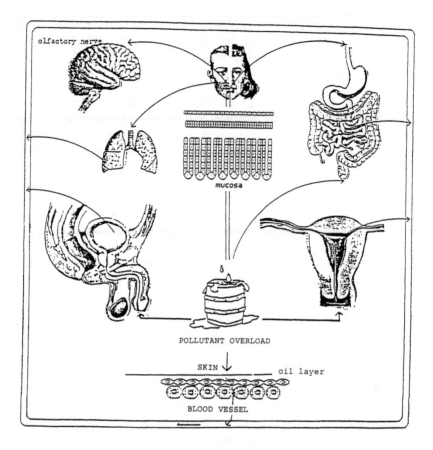

Figure 1.   Routes of pollutant entry leading to overload in the chemically sensitive.

SURFACE BARRIERS TO ENVIRONMENTAL POLLUTANTS

The surface barriers are the body's first defense against pollutants (toxic chemicals [synthetic and natural, organic and inorganic], radiation, dust, electromagnetic fields) that can trigger chemical sensitivity. These include the skin and the mucous layer of the respiratory, gastrointestinal, and genitourinary systems. Oily layers of the skin, as well as hair, tend to trap pollutants as does nasal hair and cilia of the respiratory tree. In addition, the stimulated mucous membranes secrete large amounts of mucous in their attempt to bind and trap pollutants. Obviously, damage to these barriers may allow a larger volume of pollutants to more rapidly penetrate the body. Further damage to the underlying tissues or vascular tree may then occur increasing the body's vulnerability to chemical sensitivity.

The chemically sensitive individual often has odor sensitivity, which occurs rapidly to minute exposures. For example, one whiff of perfume in the normal person has no effect. In the chemically sensitive, however, incapacitating neurovascular responses such as cold hands and feet, inability to remember, and flu-like symptoms may result. This type reaction suggests an inadequate barrier to a pollutant with penetration to the neurovascular system. Similar responses to minute amounts of pollutants entering through the gastrointestinal tract, bronchial tree, or genitourinary system may occur. For example, water used for bathing or drinking contains many solvents which enhance absorption of toxins. Thus, engaging in either of these activities could enable toxins to enter the body via the skin or orally. Inhalation of toxic fumes and aerosols from ambient air may further allow toxins to enter through the respiratory tract. Studies with workers exposed to magnesium, titanium, and chlorine have shown that chemicals in work air and food can break down local mucosal resistance. Toxins may also enter the body through food and water ingested into the gastrointestinal tract, or they may enter through the genitourinary tract as a result of sexual contact or genital manipulation involving birth control gels and apparatus.

Once pollutants penetrate the mucosal barrier, free radicals may be created or direct toxic damage to organs may occur.[3] The free radicals may be O_2^-, O_3, OH^-, lipid peroxide (ROO^- and RO^-), or many others which may damage systems, organs, cells, or subcellular components, such as mitochondrial and cellular membranes, or even molecules. Macrophages and inflammatory cells occur and act as barriers in response to penetration. Both the nonimmune and immune systems may be involved.

Skin and mucous membrane and other membrane resistance can be broken down by many solvents such as trichloroethane, tetra- and trichloroethylene, chloroform (all found in the blood of chemically sensitive patients), as well as other toxic chemicals. The attraction of toxins to skin and mucous membranes is usually due to the lipophilic nature of toxins. They are attracted to the lipid structural layers. Once solvents are incorporated or attached to the lipid part of the cell membrane or its oily layer, dissolution or dysfunction may occur. Solvents as well as other toxic chemicals enter with the eventual result of an increase in total body pollutant load, creation and damage by free radicals, disordered metabolism, and eventually chemical sensitivity with impaired enzyme detoxification and immune responses if the body cannot deal properly with them (see Chapter 25).

DAMAGE TO SYSTEMS AND ORGANS ONCE POLLUTANT PENETRATION OCCURS

In chemical sensitivity, pollutant damage to major anatomical systems such as the cardiovascular, genitourinary, gastrointestinal, respiratory, dermal, neurological, musculoskeletal, and hematological often occurs. Resultant patho-physiology will vary depending on the system involved and the intensity of exposure and response. For example, if the cardiovascular system of a chemi-cally sensitive patient were damaged, early pollutant injury would result in edema with swelling around the eyes, hands, feet, and ankles. Severe injury, in contrast, would result in bruising, petechiae, and purpura. Examples of damage to other major systems are discussed in their respective chapters of this book and will not be discussed further here.

In chemical sensitivity, pollutant injury to specific organs (e.g., heart, liver, brain) rather than systems may also occur, resulting in a more concentrated set of signs and symptoms. For example, an environmentally triggered organ response would be seen in an individual who had nonarteriosclerotic spastic vessel-induced arrhythmias. Examples of damage to other organs are discussed within subsections of other volumes on clinical responses of systems to pollut-ants and will not be discussed further here.

After pollutants penetrate the surface barrier, they move into the blood stream and produce a generalized response, which is expressed as an imbalance of the autonomic nervous system and its attachments to blood vessels. This response, which is often seen in the chemically sensitive, is manifested by vessel deregulation (spasm or dilation), edema, bruising, purpura, petechiae, acne, and cold sensitivity. Although the neurovascular tree is the site of an initial pollutant injury or trigger after surface barrier penetration, the primary detoxification process usually occurs in the liver and respiratory system. How-ever, the kidney and other organs will perform detoxification to a lesser degree (Figure 2).

With damage to these different components throughout the body, dysfunc-tion occurs, resulting in chemical sensitivity. The vascular tree seems to be where the most common early end-organ responses to pollutant challenge occur in the chemically sensitive (before fixed end-organ failure occurs), with the spread of dysfunction often moving through its autonomic nervous system attachments, including the hypothalamic system and its limbic system involve-ment in the brain (see Vol. III, Chapter 26). This dysfunction results in symptoms of autonomic nervous system and vascular system deregulation. These signs and symptoms may include an inability to sweat, bloating, nausea, inappropriate digestion, urinary urgency and frequency, cold sensitivity, jitteriness, as well as the previously mentioned vascular symptoms. It appears that if the triggering agents of autonomic nervous system dysfunction can be perceived, diagnosed, and removed, environmentally triggered disease can be halted before it becomes fixed-named and irreversible. Early recognition of pollutant-triggered autonomic-vascular dysfunction and action toward the re-duction of the total individual pollutant load allows for cost-effective preven-tion of disease. With neurovascular deregulation, nutrient delivery and waste elimination may be hampered (see Vol. III, Chapter 20 and Chapter 5 for more information).

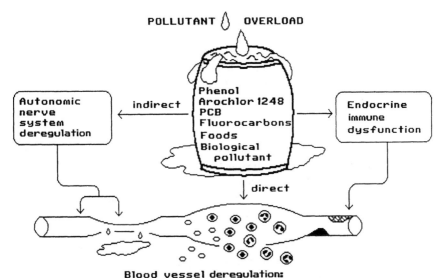

Figure 2. **Potential pollutant damage to blood vessels in the chemically sensitive.**

Once pollutant penetration and overload is beyond this stage, vessel damage may occur. According to Zeek, vessel wall damage may be mild with fluid leakage (generalized edema or localized hives), and as it progresses the leaks get larger, allowing red blood cells to go out through the vessel wall (bruising, purpura, petechia).[4] With severe damage to the wall, clotting may occur, resulting in distal peripheral tissue damage. Attempts at healing may occur in various ways such as granulomatous or fibrous scar formations (see Vol. III, Chapter 20 for more information). All of these stages have been observed in some chemically sensitive patients.

Either local, direct toxic chemical exposure or indirect free radical generation can damage the vascular tree. Nour-Eldon has shown that some substances such as phenol have an affinity for the cardiovascular system.[5] Yevic demonstrated cardiovascular changes in sea animals exposed to oil spills, further emphasizing impaired metabolism due to pollutants.[6] Perivascular lymphocytic chloracne lesions have been produced in monkeys fed a pesticide, Arochlor-1248 PCB product.[7] (For further details, see Vol. III, Chapter 20).

This effect has been seen in humans exposed to agent orange[8] as well as in other chemically sensitive individuals who were exposed to chlorinated hydrocarbons.[9] Free radicals can disrupt the vascular cell membranes and the mitochondrial membranes, resulting in vascular deregulation. Even with the presence of toxic chemicals, some free radicals can be neutralized or prevented from forming by various antioxidant enzymes, vitamins, and minerals[10] (see

Chapter 6, Nutrition). The detoxification enzymes include superoxide dismutase,[11] glutathione peroxidase,[12] lipid peroxidase,[13] aryl hydrocarbon hydroxylase, glucose 6-phosphate dehydrogenase, elastase,[14] cytochrome P-450 monooxygenase, and many others (see Enzymes in Chapter 6). With pollutant injury these enzymes may be induced, with overstimulation occurring, which may result in clearing the toxic chemicals as well as the free radicals (see Free Radicals, this chapter). Eventually, toxic overload with depression of the production of these enzymes may yield an insufficient quantity of equalizing substances to neutralize pollutant effects. This overload then results in neurovascular dysfunction (see Adaptation in Chapter 3 and Lung Enzymes in Vol. III, Chapter 18).

After pollutant entry, the intrinsic mechanisms by which blood vessels or any system in the body is triggered can basically be divided into two categories: those which involve the immune system and those which involve the nonimmune enzyme detoxification systems. Though there is overlap, each is discussed separately. (For Immune, see Chapter 5).

Nonimmune triggering of the vessel or other cell walls by toxic chemicals or their by-products may occur. Activation of these nonimmune mechanisms often appears to be primary, with the activation of the immune system being secondary. At times, however, immune activation appears to be first. Some mechanisms such as the alternate complement pathway may be triggered immunologically or nonimmunologically directly by molds, foods, or toxic chemicals.[15] Both mechanisms may be triggered simultaneously. Of the chemically sensitive patients sick enough to enter our hospital wing, 40% have abnormal complement levels. Also, mediators like kinins,[16] prostaglandins,[17] etc. may be directly triggered (see Vol. III, Chapter 25 for further discussion of the lipoxygenase system). These reactions may act as or enhance amplification systems, causing vascular spasm with resultant hypoxia, which in turn causes the release of lysozymes that further accelerate the cycle with more spasm, hypoxia, inflammation, etc. Eventually, after days, months, or years of excessive total body load with repetitive tissue triggering, end-organ disease occurs (Figure 3). The vascular spasm can be recognized by the finding of decreased peripheral pulses with cold hands and feet, blanching or blue skin, disordered brain function, and low energy in the chemically sensitive patient. Eventually the inflammation clinically manifests as aching and tenderness.

Immune Triggers

Immune triggering of the vascular tree is similar to that shown in the section on the immune response to toxic chemicals. This involves Gell and Coomb, types I, II, III, and IV. See Chapter 5.

Triggering of the enzyme detoxication systems also may occur, generally in the liver and respiratory mucosa. However, detoxification occurs to a lesser

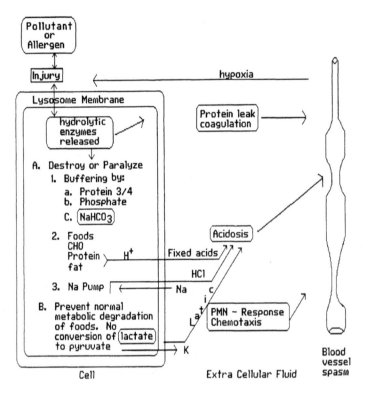

Figure 3.   Cellular response.

extent in all systems. Biotransformation of foreign compounds varies depending on genetic and environmental factors, age, sex, nutrition, health status, and
the size and virulence of the dose. Many pathways may be involved in the
detoxifying process and will be described on the following pages. As pollutants
enter the body, different levels of injury and stages of detoxication processes
are seen. This varied level of injury and stages of detoxication allow for the
many different facets and types of symptoms seen in different chemically
sensitive individuals. Even though the areas of injury in the chemically sensitive may occur simultaneously or closely in sequence, each will now be
discussed separately. As suggested by Brabec and Bernstein, these levels are
arbitrarily divided into cellular, subcellular, and molecular targets of foreign
compounds.[18]

POLLUTANT DAMAGE TO CELLS

Pollutant damage to cells can cause or exacerbate chemical sensitivity.
Cellular responses to pollutants vary because of unique metabolic and physical
composition of cells.

When cell injury occurs, manifestation is by either acute response or accommodation reaction (adaptation). Each type of response is biochemically damaging but clinically different in the chemically sensitive. The acute response manifests clinically as a given set of symptoms, whereas the accommodation response initially has no identifiable symptoms that suggest cause/effect. For acute survival, adaptation usually occurs. As this process begins, the cells accomodate foreign substances. Accommodation depends on three factors and may result in the individual initially being unaware of his chemical sensitivity.

The reactions responsible for masking (adaptation, accommodation) are (1) *sufficient inducible detoxification systems*, e.g., *inducible microsomal enzyme systems* such as superoxide dismutase must have enough RNA, protein, and minerals to sustain them for the sufficient period of time that the stimulus is present. The chemically sensitive individual often has low levels of antipollutant enzymes, poorly inducible enzyme systems, and low pools of nutrients. The individual may therefore have a problem with continuing to detoxify if his load is too high or virulent (see Chapter 6 and Vol. IV, Chapter 30). (2) *Enforcement of the permeability barrier or sequestration.* The integrity of the cell membrane may be enhanced as long as the nutrient fuel lasts. Solvents frequently seen in the chemically sensitive dissolve parts of the membrane, allowing holes for pollutant penetration to occur. Once penetration occurs, sequestration may be the next line of pollutant defense. Sequestration can occur in any organ, particularly the fat (as evidenced in biopsies of the chemically sensitive; see Vol. IV, Chapter 30) or liver, but a price is often paid. Nutrients may become lost or displaced and as a result the body's defense mechanisms may be weakened. The body then becomes vulnerable to new or increased pollutant damage. For example, the liver synthesizes and stores thionen for zinc metabolism; during cadmium exposure, thionen binds and sequesters large amounts of the cadmium in the liver. This sequestration of cadmium tends to reduce its toxicity, but zinc metabolism becomes deranged in the process. Then, because the detoxification system is zinc dependent, chemical sensitivity is exacerbated.

This sequestering of many xenobiotics may account for some of the metabolic impairments seen in the chemically sensitive. Sequestration in lipid membranes may create free radicals, cell leaks, and autoantibodies and render the body open for autotransfusion of toxins back into the bloodstream in times of stress. (3) *Mitosis of cells adjacent to necrotic cell* (liver cells frequently multiply, while in brain cells multiplication is nearly impossible). Enough nutrient fuel must be present for proper mitosis or repair to occur, and this process is inadequate in some chemically sensitive. One may often develop necrosis immediately adjacent to the pollutant if it is virulent enough. However, cell damage is not always fatal because subcellular repair via biosynthetic routes is often accelerated by trauma (e.g., carbon tetrachloride exposure shows a loss of ability to conduct substrate oxidation due to mitochondrial damage; this is coupled with synthesis of ATP, which intensifies cell

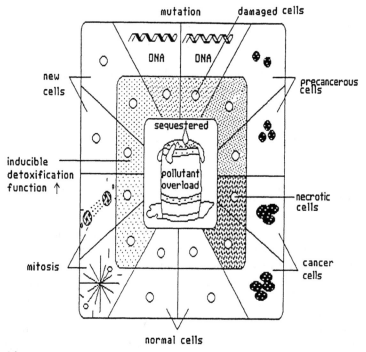

mutation   damaged cells

DNA   DNA

new
cells

precancerous
cells

sequestered

inducible
detoxification
function ↑

pollutant
overload

necrotic
cells

mitosis

cancer
cells

normal cells

must have:
1. sufficient inducible enzymes, DNA, nutrients
2. sufficient substances to enforce the permeability barrier
3. sufficient substances to allow for sequestration of the pollutant or chemical
sensitivity may occur

Figure 4.   Pollutant damage to cells.

metabolism with more rapid healing occurring). The cellular process can restore the damaged mitochondria and cell to activity without necrosis. This phenomena also can happen with any organelle.[19] Cells may also mutate and form malignant or nonmalignant tumors or become hypofunctional. This action may be the mechanism by which organochlorine pesticides induce malignant changes in breast tissue.[20] If the stimulus continues, as happens in most chemically sensitive, nutrient depletion with an excess or lack of responses occurs (Figure 4).

Various types of agents can trigger different types of responses in cells. These triggering agents of cellular dysfunction may be (1) *infectious* (bacteria, virus, fungus, parasitic), (2) *chemical* (SO_2, phenol, hydrocarbons),[21-23] (3) *nutritional*,[24] or (4) *traumatic* (physical environmental agents [which can allow cells with attached chemicals to concentrate at damaged tissue area], vibrations [EMF], heat, cold, pressure, light, noise, air ions, or radiations). Specific cellular responses to the initiating agents are varied, probably due in part to the properties of the specific agent. For example, (1) staphylococcal infections usually give a neutrophilic infiltration (leukocytoclastic); (2) intracellular pathogens or toxic chemicals usually give a lymphocytic infiltration; (3) certain parasitic infections give an eosinophilic[25] infiltration; and (4) most infectious organisms give mixed infiltration.[26] This varied cellular response pattern also applies to the biochemical response patterns of the individual. Therefore, even though the clinical responses may appear to be the same, different pathologic patterns in different chemically sensitive individuals occur.

Susceptibility of a cell to a toxin depends on three factors: (1) *inherent specialization* of the cell; some cells are more susceptible to one type of pollutant than others. For example, myocardial cells are especially susceptible to hypoxia, or rapidly dividing intestinal crypt cells are especially susceptible to DNA synthesis inhibitors such as nitrogen mustards; thus, the chemically sensitive may have different symptoms and responses (see Chapter 29). (2) *Unequal distribution* of the toxin within the body; toxins may be common in one area over another due to their point of entry and subsequent distribution. For example, toxins in the stomach go to the liver, while in the lung direct absorption sends toxins throughout the body as well as causing local lung membrane and blood vessel damage. The cells nearest to the point of entry are usually the first to be damaged; therefore, when damage occurs in different organs, different types of chemical sensitivity occur in different people. (3) *What the cell does with the toxin*. Different cells respond differently to toxins. For example, liver containing microsomal enzymes destroy toxins quickly, while other tissues with fewer enzymes destroy more slowly.[27] These responses to the toxins may be total or partial depending upon the total load and the quantity and quality of nutrient fuels present. This involvement of different cells would again allow each chemically sensitive individual to have different symptoms, which may be as varied as the textbook of medicine (see Chapter 29).

In the chemically sensitive, an increase in the total body burden of incitants, either due to excess intake or severe virulence of the agents, overloads these cellular detoxifying mechanisms, damaging other systems and triggering mediators and lysozymes, which causes more vascular and end-organ dysfunction. Even biological substances such as pollens, terpenes, molds, and foods can produce chemicals that may be harmful. The spreading phenomenon (see Chapter 1) of chemical sensitivity occurs and gives varied cellular responses. For example, pollens have been shown to have toxic substances contained within them that will trigger hemolysis and cell membrane swelling.[28] In addition to the usual allergic reaction caused by plants, the odors, colors, and flavors of terpenes from various plants will also trigger various types of cellular responses. It is well known that some molds and bacteria produce toxic substances such as cancer-causing aflatoxin[29] and clostridia toxins.[30] Many foods, especially legumes, contain lectins which contain phytohemagglutinins.[31] These may cause reactions that lower the body's resistance and trigger

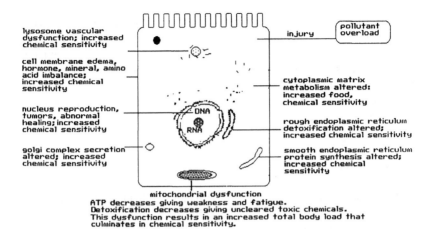

lysosome vascular
dysfunction; increased
chemical sensitivity

cell membrane edema,
hormone, mineral, amino
acid imbalance;
increased chemical
sensitivity

nucleus reproduction,
tumors, abnormal
healing; increased
chemical sensitivity

golgi complex secretion
altered; increased
chemical sensitivity

injury

pollutant
overload

cytoplasmic matrix
metabolism altered:
increased food,
chemical sensitivity

rough endoplasmic reticulum
detoxification altered;
increased chemical sensitivity

smooth endoplasmic reticulum
protein synthesis altered;
increased chemical
sensitivity

mitochondrial dysfunction
ATP decreases giving weakness and fatigue.
Detoxification decreases giving uncleared toxic chemicals.
This dysfunction results in an increased total body load that
culminates in chemical sensitivity.

**Figure 5.   Pollutant damage to subcellular elements which may account
for some different types or various responses of the chemically
sensitive.**

chemical sensitivity reactions with various cellular responses. We have seen a
case of dizziness, the result of vascular deregulation of the vestibular apparatus
and cerebellum, and pancreatitis apparently repeatedly triggered by foods
high in phytohemagglutinins (see Pancreatitis in Chapter 21 and see Chapter
8) with various types of cellular responses resulting. The responses may be
eosinophilic, neutrophilic, or lymphocytic. The most common vascular lesion
is the lymphocytic infiltrates around the small blood vessel.

POLLUTANT DAMAGE TO SUBCELLULAR ELEMENTS

If the pollutants penetrate the cell membrane and are not sequestered, many
contents of the cell can be damaged. The chemically sensitive individual has
symptoms that have been attributed to malfunction of these components, any
of which may be damaged. More frequently, selective involvement occurs,
potentially resulting in the varied responses of the chemically sensitive. Each
component part is discussed separately (Figure 5).

Nuclei

Nuclei containing DNA and RNA in the chromatin can be damaged in two
ways. Fidelity of replication[32] can be disturbed. Because of this failure, aberrant cell lines may then be produced. Also, the quantity of a gene can be
disturbed.[33] The amount of DNA and RNA synthesized can be affected by
altering the activity of the synthetase involved without altering the fidelity of

the replicative processes, e.g., alpha amanitine (mushroom toxin) inhibits RNA polymerase II; cycloheximide blocks DNA synthesis as well as cytoplasmic protein synthesis.[34] These changes cannot be recognized clinically, but laboratory studies using cell cultures looking for aberrant changes or chromatin fragments will aid in the diagnosis.

Mitosis is inhibited by some toxic chemicals, e.g., by colchicine (short term is not fatal) and some plant alkaloids.[35]

Mitochondria

Mitochondria dysfunction due to pollutant overload occurs yielding low ATP levels and a disturbance of the redox state of the cell[36] (e.g., there is decreased mitochondrial function due to copper exposure with resulting riboflavin deficiency). One gets proliferation of mitochondria, yielding functionally inadequate megamitochondria. Supplementation of nutrients returns the cell to normal.[37] These dysfunctions, which can result in weakness, muscular and nonmuscular fatigue, and an inability to cope with new exposures, are commonly seen in chemically sensitive individuals.

Energy-driven ATP synthesis is disturbed by many substances such as dinitrophenol; mitochondrial biogenesis is inhibited by chloromycetin and oxytetracycline since mitochondria have a genetic system similar to bacteria.[38] *Pollutant-driven switching from aerobic to anaerobic metabolism reduces energy generation 36 times,[39] again emphasizing the reasons for the severe loss of energy seen in some chemically sensitive individuals.*

Mitochondrial enzymes contain sulfhydryl groups sensitive to heavy metals, whose damage results in a propensity to accumulate calcium, which destroys mitochondrial activity.[39] Many patients with chemical sensitivity studied at the EHC-Dallas have been observed to have high intracellular calcium. These dysfunctions probably account for the low energy frequently observed in the chemically sensitive as well as the impaired calcium and magnesium metabolism seen in some chemically sensitive individuals. Fortunately, overall, mitochondria are very resilient to pollutant damage and usually repair themselves if the total pollutant load is lifted.

The electron transport area is disturbed by cyanide, azide, rotenone, and dicumarol, accounting further for the impaired mineral and energy metabolism found in the chemically sensitive individual (see Chapters 6 and 37).

Lysosomes

It is clinically relevant that these intracellular vesicles sequester the hydrolytic enzymes (lysosomes), e.g., nucleases, phosphatases, and peptidases, from the cytoplasm. Lysosomes may be disrupted by pyrogenic steroids, polyfene antibiotics, other chemicals, and hypoxia, exacerbating chemical sensitivity.

Glucocorticoids, chloroquine, and chlorpromazine stabilize the lysosome membrane, thereby temporarily preventing the enzyme release until the pollutant overload becomes too overwhelming; then breakthrough occurs with an exacerbation of chemical sensitivity.

Other Cellular Bodies

Membrane-bound structures are susceptible to solvents. When the membrane comes in contact with a solvent, it leaks and chemical sensitivity may be exacerbated. For example, microbodies (peroxisomes) containing oxidative enzymes and crystals of uricase will leak and disrupt oxidative metabolic function. Peroxisomes are labile to exposure to ultraviolet light.

Endoplasmic Reticulum (ER)

This organelle contains rough (with ribosomes microsomal enzymes) and smooth units (protein generation). ER is 15% of the total cell volume and 8.5% greater than mitochondria by surface area and 37.5% of the surface area greater than the cell membrane. This size alone explains why so many chemicals can be detoxified. This large size also explains why pollutant damage to the ER exacerbates chemical sensitivity since clearing capability is diminished by this damage.

Microsomal enzymes are the most important contents of the rough ER. ER contain electron transfer enzymes responsible for the oxidation of various lipophilic compounds including steroids, long-chain fatty acids, and xenobiotics. Glucose-6-phosphate dehydrogenase is a sensitive marker of ER membranes. Injury of the ER results in membrane inhibition or induction (with subsequent ER proliferation). (If excess induction occurs, increased lipid peroxidation occurs with degranulation and dilation of the ER and with marked decrease in protein synthesis and glucose-6-phosphatase and cytochrome P-450 being inhibited). (See Pathways and Toxins in the Liver, Chapter 21.) Since the enzymes of the ER metabolize primarily endogenous substances, natural proliferation may result in abnormal levels of hormones, bile salts, and other normal metabolites. These abnormalities may then result in various symptomatology, such as gastrointestinal upset with bloating, diarrhea and food intolerance, premenstrual syndrome, etc., all of which are seen in the chemically sensitive.

Plasma Membrane

The plasma membrane has microheterogenicity (Figure 6). The cell membrane has antigen recognition sites. The disturbance of these sites may be one

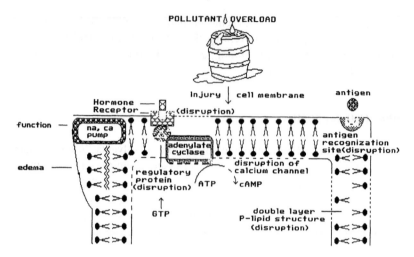

Figure 6. Pollutant injury to cell membrane showing many ways chemical sensitivity is exacerbated.

explanation for the widespread reaction seen in the severely damaged chemically sensitive patient. The severely chemically sensitive individual, particularly, develops problems with antigens to which he has never been exposed. Many severely sensitive patients will react to foods they have never eaten such as poi, llama, amaranth, etc. As their total load is reduced (sometimes acutely when in the environmental unit) and as antigen recognition is apparently restored, they are able to differentiate between the foods to which they are sensitive and those to which they have never been exposed. Sometimes, clinically, this reduction of total load must be coupled with intradermal injection therapy (see Chapter 36) in order to restore antigen recognition.

Membranes transport enzymes. For example, the sodium pump is in the membranes. Pollutants disturb this pump with resulting edema, as seen in the chemically sensitive. Generalized and periorbital edema appears to be one of the most common reversible clinical responses seen in medicine today. Most chemically sensitive people commonly have problems with mild recurrent edema. Many toxins attack the membranes, producing changes in permeability. Calcium channels are also disrupted.[40]

Hormone receptor sites are often damaged by pollutants with resultant premenstrual and thyroid dysfunction syndromes that are common in severely affected chemically sensitive patients.

Messenger activators are at the membrane and may be rendered dysfunctional. Therefore, they may not regulate properly and a whole array of symptoms, as seen in some chemically sensitive patients, may occur.

Membrane inhibition or proliferation with free radical generation may occur with pollutant injury, causing a further exacerbation of symptoms in the chemically sensitive.

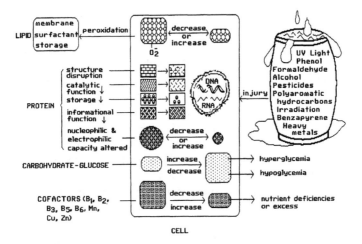

Figure 7.   Pollutant effect to the molecular targets in chemical sensitivity.

POLLUTANT DAMAGE TO MOLECULAR TARGETS

The molecules that are frequently the target of foreign compounds in chemical sensitivity are those concerned with the dynamic execution of the living process (Figure 7). These are involved in production, transfer, and compartmentalization of energy and information. All of these molecules have been noted to be damaged in some chemically sensitive patients. Often only one or two areas are involved.

Protein (Amino Acids)

Proteins (amino acids) are the genetic vehicles by which a cell expresses its potential. Pollutant injury can be disruptive to the functions of enzymes, nucleic acid, structural protein, carrier or storage proteins, and regulatory proteins, aggravating chemical sensitivity.

Enzymes

Pollutant injury to enzymes can be either widespread or limited to a specific enzyme, resulting in induction or exacerbation of chemical sensitivity. For example, catalytic enzymes suppressed by pollutant exposure will function at a suboptimum level, leaving an individual susceptible to other pollutants and thus cause increased chemical sensitivity.

Examples of catalytic enzymes affected by pollutant injury are lactic dehydrogenase and cytochrome P-450. The effect of toxic chemicals on enzyme function can be studied by observing the alteration in the kinetic behavior of the modified enzyme. This alteration of kinetics may in part explain the fragility of the chemically sensitive. The reversibility depends on the metal and enzyme, e.g., lead at 10^{-3} to 10^{-4} M inhibits a wide range of enzymes. Aminolevulinic acid dehydrase is blocked at 10^{-5} M, while lipoamide dehydrogenase (part of pyruvate and α-ketoglutaric acid dehydrogenases) is inhibited by 10^{-6} M. All are reversible by administration of EDTA or dithiothreitol. The heavy metal ions will also form inactive ligands with histadyl and tryptophanyl residues, which are frequently involved in catalysis. Another example is a nerve gas, diisopropylfluorphosphate, which blocks the activity of acetylcholinesterase, which hydrolyzes acetylcholine.[41] Inhibition of the alteration in cholinesterase is one response to exposure to organophosphate and carbamyl insecticides. In one series of chemically sensitive patients at the EHC-Dallas, superoxide dismutase and glutathione peroxidase were found to be depressed. This finding suggests overutilization or suppression of catalytic action due to pollutant damage (see Chapters 6 and 30). Cytochrome P-450 is discussed later in this chapter.

Noncatalytic nucleophilic and electrophilic capacity designed to neutralize xenobiotics nonenzymatically may be disturbed in the chemically sensitive by pollutant injury. This change also will then alter much detoxication. This phenomenon is discussed later in this chapter.

Nucleic Acid

DNA can be disrupted irreversibly by irradiation and aromatic polycyclic hydrocarbons such as benzopyrene.[42] Actinomycin D will interfere with RNA or protein synthesis.[43] Some compounds such as Acridine orange and ethidium bromide can give reversible DNA reactions.[44] The alterations could continue to propagate chemical sensitivity.

Structural Protein

Muscle mass may be wasted and collagen may not be available for healing in the chemically sensitive due to alteration of protein metabolism. The EHC-Dallas has seen 50 patients who arrived with massive weight loss of 20 to 50 lb, severely malnourished, weighing 60 to 90 lb, requiring intravenous and tube feeding in order to replace structural deficiency as well as remedy the other functional problems. These patients apparently had experienced pollutant damage to these proteins and their metabolism (see Chapter 6).

In normal physiology, sulfhydryl linkage of cysteinyl residues may form to stabilize the suprastructure of proteins. Pathologically, the ions of heavy metals form mercaptides with cysteinyl residues, thus disturbing that stability; due to this damage, future responses may be fragile, brittle, and inappropriate. Brittleness is the trademark of the chemically sensitive.

Carrier or Storage

Transferrin and red blood cell mass are frequently injured in the chemically sensitive. Heme proteins are affected by CO, NO_2, analine, CN, and H_2S.[45]

Informational or Regulatory Proteins

Informational or regulatory proteins may be injured in the chemically sensitive, causing mood swings and abnormal neurological responses.

Coenzymes

Coenzymes are present in limited number and are in part dependent upon the dietary supply of minerals and vitamins and absorption as well as the speed of utilization in detoxification (see Chapter 6). A chemical of similar structure that is toxic and cannot act as a building block or be a direct enzyme poison (e.g., heavy metals, ultraviolet light, hydrazine neutralizing pyridoxa-1-5-phosphate) may be preferentially absorbed over nutrients, thus causing nutritional deficiencies. Free radical attack on NADPH leads to its rapid destruction with exacerbation of chemical sensitivity.

Lipids

The integrity of lipid membranes seems to be paramount to chemical sensitivity. Chemical damage to lipid membranes results in loss of nutrients and edema. In the chemically sensitive, lipid membranes may be affected by pollutants in a variety of ways. For example, lipid membranes may be stimulated or inhibited by pollutants, as may lipid surfactants and lipid storage or mobilization. Lipid peroxidation by halogenated hydrocarbons and ethanols stimulates free radical production that can lead to more disturbance or even necrosis. Normally free radical reactions seem to be an inherent but limited part of cell metabolism. Membrane dissolution or inactivation may occur. Organic solvents and amphoteric detergents destroy membranes; so do bile salts. The heavy metal ions Hg^{2+} and Cd^{2+} join with the phospholipid base ethanolamine and choline. This expands the surface area of the membrane, decreasing its fluidity and flexibility. Lead increases red blood cell fragility. Disturbances in steroid metabolism occur with some insecticides and birth control medications (estrogen and progestin). Diethylstilbestrol also gives malignant changes. Cholesterol metabolism may be disordered by diet, smoking, stress, and probably several other toxic chemicals.[46] Surfactants of the lungs may be destroyed or disturbed by pollutants. Pollutant damage may result in an increase in blood cholesterol and other lipids with resultant arteriosclerosis (see Lipids in Chapter 6).

Carbohydrates

While the bulk of opinion concurs that the mass and redundancy of CHO molecule is a diffuse target for toxic chemicals, some think that they are not frequent targets. However, toxic chemicals seem to disturb CHO metabolism severely (see Chapter 19 and Pancreas in Chapter 21). Many people seen in our center with chemical overload are intolerant of carbohydrates, particularly refined sugars (especially cane and beet).

POLLUTANT DAMAGE TO THE METABOLISM OF FOREIGN COMPOUNDS

Detoxication mechanisms are often damaged by an increase in total body pollutant load, thus making the chemically sensitive more vulnerable to other pollutants. Several mechanisms participate in the removal of foreign substances. One is the enzymatic aspect of detoxication with *catalytic* conversion of foreign compounds, while the other is *noncatalytic*. Both mechanisms appear to be effected in some chemically sensitive, as is evidenced by the positive response to the administration of nutrient components and measurable antipollutant enzyme deficiencies seen in the series at the EHC-Dallas (see Chapters 30 and 37). For example, some chemically sensitive individuals respond to catalytic supplementation such as glutathione peroxidase and superoxide dismutase, while many also respond to noncatalytic albumin, niacin, or soda bicarbonate. Each mechanism will be discussed separately.

Disturbed Noncatalytic Reactions in the Chemically Sensitive

Noncatalytic reactions occur between nucleophiles (an electron donor in chemical reactions in which the donated electrons bond other chemical groups) and electrophiles (having an affinity for electrons) in which one is of exogenous and the other of intracellular origin (Table 1). These reactions are similar to an acid and base combination, which uses no catalysts where soda bicarbonate neutralizes HCl. For example, *reduced glutathione*, present in most tissues at concentrations of about 5 mM, is a major resource of nucleophilic capacity

**Table 1. Pollutant Overload Can Disturb the Following Type of
Noncatalytic Reactions between Nucleophils and Electrophils in the
Chemically Sensitive**

Electrophil		Nucleophil		Less Toxic End Product
1,2 disubstituted ethanes	+	Glutathione	→	Ethylene
Aryl nitroso compounds	+	Glutathione	→	Aryl amines Aryl hydroxyl amines
Nitro oxide Sodium nitrite	+	Cysteine	→	S-Nitrocysteine
Nitro benzene	+	NADPH	→	Phenylhydroxyl-amine
Benza pyrene	+	NADH		
Phenyl esters Phenyl acetate	+	Albumin	→	Bound + out
Biogenic aldehydes Biogenetic ketones	+	Amino deriv-atives of Schiff's bases	→	Alcohols → CO_2 + H_2O
Hydrazines Hydrazides	+	Carbonyl group	→	Hydrazones
Cocaine Some benzopyrenes	+	Neutral aqueous solutions (e.g., plasma)	→	Less toxic substances
Xenobiotics	+	Acid media (e. g., stomach)	→	Less toxic substances
Xenobiotics		NaH CO_3	→	Less toxic substances
Reaction with each other				Less toxic substances
Salicyclic acid	+	Acetophenomen		Less toxic substances

and is capable of reacting with toxic chemicals to make them less toxic even
without the intervention of protein catalysts. A lack of availability of sufficient
reduced glutathione or other of its supply nutrients by toxic overload will then
hamper orderly detoxication (the process of normal xenobiotic metabolism)
and detoxification (correction of a toxic state), making the chemically sensitive
individual much worse by increasing his total body load. These nutrient reduc-

tions may even allow chemical sensitivity to occur. Supplementation with reduced glutathione has increased the clearing of chemicals and decreased reactions of the chemically sensitive at the EHC-Dallas.

Most nonenzymatic biotransformations involving endogenous nucleophiles other than water are often excellent neutralizers of toxins and are mediated by thiols particularly glutathione (GSH). An example of this change is the conversion of 1,2-disubstituted ethanes to less toxic ethylene by GSH.[47] Another biotransformation is the change of aryl nitroso compounds by GSH to yield arylhydroxylamines or arylamines.[47,48]

Another example of noncatalytic enzyme neutralization of toxins is cysteine, which has been shown to react with nitric oxide (NO) or sodium nitrite (NaNO$_2$) to yield less toxic S-nitrosocysteine.[49] This compound and other S-nitrosothiols are pollutants that are strong activators of coronary artery guanylate cyclase, which gives coronary vasodilation.[50]

The smaller nucleophilic compounds are noncatalytic chemical neutralizers such as B$_3$-dependent NADPH and NADH and will reduce nonenzymatically environmental pollutants such as nitrobenzene or benzapyrene. Nitrosobenzene is converted to phenylhydroxylamine, a less toxic substance.[51]

Macromolecules such as serum albumin may also aid in the biotransformation of substances such as benzopyrene. Albumin will also nonenzymatically detoxify phenyl esters and acetates such as acetyl salicylic acid.

Biogenic aldehydes and ketones such as formaldehyde, aceto acid aldehydes, acetone, pyruvates, ketoglucaric acids, and acetoacetic acids form a group of endogenous electrophils which are able to react nonenzymatically with a variety of amino derivatives for many Schiffs bases. These are reactions of amines such as hydrazine and hydrazides with carbonyl groups.[52] They yield less toxic hydrazones.

Many xenobiotics are nonenzymatically transformed into less toxic substances by neutral aqueous media such as plasma. Some cocaines and benzapyrenes are broken down this way. Acid media of the stomach have a strong effect on some xenobiotics, breaking them down. Since over 50% of the chemically sensitive are hypo- or achlorohydric, one essential component used to counteract ingested chemicals is impaired. Therefore, these states tend to increase total load and exacerbate sensitivity.

Some xenobiotics will break down each other, such as the reaction between acetyl salicylic acid and acetaminophen. Lack of noncatalytic reactions may in part explain the synergistic and potentiative effects seen with some chemical exposures in the chemically sensitive. Certainly well-functioning noncatalytic reactions will explain the antagonistic effects of some chemicals when they enter the body of the chemically sensitive.

Pollutant overload and/or nutritional deficiency may disturb any or all of these reactions, allowing for continued or new toxic products to enter and thus cause further damage, increasing the total body load with continued propagation of the chemical sensitivity.

The detoxication role of enzymes during pollutant exposure is made possible by the availability of means for partitioning substrates between phases. This partitioning may not occur as effectively in the chemically sensitive.

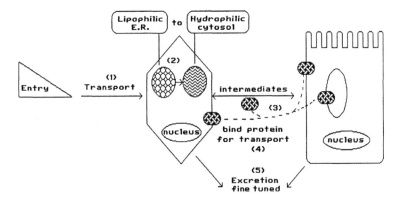

Figure 8.   Areas of potential pollutant damage to the catalytic conversion of foreign compounds.

Catalytic Conversion of Foreign Compounds

The detoxication role of enzymes in the chemically sensitive is made possible by the availability of means for partitioning substrates (toxic chemicals) between phases (extracellular cytoplasm, nucleus, mitochondria, endoplasmic reticulum), e.g., (1) the transport of xenobiotics from their port of entrance to the initial site of conversion; (2) the lipophile environment of the endoplasmic reticulum to the more hydrophilic cytosol; (3) the subsequent shuttling of intermediates from one organ to another; (4) the activity of binding proteins that serve as a vehicle for transport both in the circulation and through permeability barriers; and (5) the exquisitely tuned systems for excretion (Figure 8). If any of these malfunction due to pollutant injury, as seen in the chemically sensitive, function is impaired and chemical sensitivity intensifies. Damage to these functions is clearly part of the problem in the chemically sensitive patient, allowing build-up of toxic chemicals in the blood and lipid membranes, as evidenced by studies of the toxic levels in blood and fat at the EHC-Dallas. (See Chapters 30 and 35.)

Present knowledge suggests that the major share of biotransformation is by the enzyme systems, but the nonenzymatic neutralization of chemicals should not be discounted in the chemically sensitive since evidence of their significance is growing. Pollutant injury to any of the aforementioned steps may easily occur with an increase in total body load or virulence of the chemicals involved.

The metabolism of foreign compounds usually occurs in the microsomal fraction (rough ER) of liver cells.[53] Cytochrome P-450 is the main system for detoxifying foreign compounds, and if this fraction is disturbed by pollutant overload, dysfunction occurs. This microsomal monooxygenase system has three components: (1) cytochrome P-450; (2) NADPH (derivative of niacin B_3) cytochrome C reductase or cytochrome P-450 reductase; and (3) phosphotidylcholine (derivative of B_6).[54] The reaction requires both NADPH and molecular oxygen. If any of these parts are not present in sufficient quantities or are disturbed by pollutant overload, appropriate detoxication may not occur, resulting in chemical sensitivity.

The types of chemical transformations attributed to cytochrome P-450 in hepatic microsomes include: (1) oxidative reactions in which an atom of molecular oxygen is inserted into an organic molecule, such as aliphatic and aromatic hydroxylation, N-oxidation, sulfoxidation, epoxidation, N-S-O-dealkylations, desulfurations, deaminations; (2) reductive reactions involving direct transfer such as reduction of azo, nitro, N-oxide, and epoxide groups and dehalogenations (Table 2). These and aryl (phenol)-substituted aliphatic amines,

$$\left(-\overset{|}{\underset{\underset{\displaystyle CH_2}{|}}{C}}H - CH - NH_2\right)$$

e.g., tranylopromine (parnate), such as amphetamine, acetylcholine, and 4-aminodiphenyl are deaminated by oxidation; (3) degradations with water; and (4) conjugations with peptide, acetyl, sulfurs, methyls, and glucuronic acid. Disturbances of parts or all of these steps have been observed in selected chemically sensitive patients, rendering them intolerant to the aforementioned chemicals.

The protein cytochrome P-450 is the liver microsomal oxidase responsible for the metabolism of drugs and various foreign compounds. The level of microsomal cytochrome P-450 in the liver can be increased by the administration of inducers, e.g., phenobarbital, 3-methylcholanthrene, benzopyrene, 5,6-benzoflavone, pregnenolone 16 carbonitrile, and 2,3,7,8-tetrachlorodibenzo-p-dioxin, and perhaps many more substances, and decreased by treatment of animals with inhibitors, e.g., cobalt chloride and carbon tetrachloride, SKF525-A, metyrapone, 7,8-benzoflavone, 2,4-dichloro-6-phenylphenol oxyethylamine (DBPA), carbon monoxide, and perhaps many more[55-61] (Figure 9).

Individuals are temporarily protected from a given chemical if induction of the enzyme has commenced previously. This increase in cytochrome P-450 may explain the stimulatory phase in chemical sensitivity. This enzyme induction provides protection for acute survival and is probably one of the processes by which adaptation occurs. With chronic long-term exposure, as seen in the chemically sensitive, adaptation breaks down and the person becomes ill.

If the cytochrome systems are excessively stimulated and overly utilized, they are eventually depleted or immediately suppressed by toxic chemicals to the point of poor response. Then chemical sensitivity may occur, eventually followed by nutrient depletion, and finally fixed-named disease.[62] The most

**Table 2. Potential Pollutant Injury to the Categories of
Biotransformation Reactions for Metabolism of Food Stuffs and
Elimination of Toxins in the Chemically Sensitive**

Oxidation	Lose electrons	O_2
		$O_2 + H_2O$
		NADPH
		Enzymes
Reduction	Gain electrons need	Lack of O_2
		Carbon monoxide
		NADPH
		Enzymes
Degradation	$+H_2O$ + carbosyltransferase	
Conjugation	Acetyl	
	Acyl (peptide)	
	Sulfur	Enzymes
	Methyl	
	Glucuronic acid	

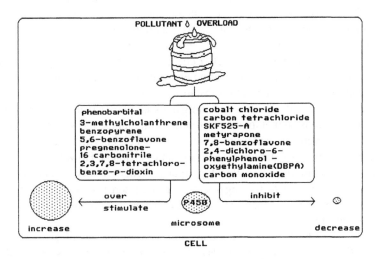

**Figure 9.   Effects of pollutants on microsomes and thus the cytochrome
P-450 system in the chemically sensitive.**

severe chemically sensitive patients in our series at the EHC-Dallas have abnormally low antipollutant enzymes in addition to toxic suppression and nutrient depletion. In some instances, antibodies are also produced against cytochrome P-450 and thus may inhibit or decrease its effectiveness.[62]

The mechanism by which cytochrome P-450 catalyzes is probably the following, involving iron peroxide:[63]

$$Fe^{3+} - OOH \xrightarrow[Cu^{++}]{^-OH} Fe^{5+} = O \xrightarrow{RH} Fe^{4+} - OH + R \rightarrow Fe^{3+} + ROH$$

Replenishment of iron stores with intravenous infusion to correct deficiencies may explain why the chemically sensitive patient improves dramatically in their detoxification processes, as seen in many patients at the EHC-Dallas. Leiberman[64] sees massive loss of Hb in his detoxification patients undergoing heat depuration physical therapy. By increasing iron to 54 to 81 mg/day, loss is checked. Copper also helps catalyze the cytochrome systems (see Chapter 6). Although the liver microsomal system is the primary site for oxidation of xenobiotics, the cytochrome P-450 system is found in other tissues that are exposed to environmental compounds like the skin, lungs, gastrointestinal tract, the kidneys, placenta, corpus luteum, lymphocytes, monocytes, pulmonary alveolar macrophages, adrenal, testis, and brain. The cytochrome P-450 is not only localized in the mitochondria of the corpus luteum, kidneys, adrenal, and liver, but in the nuclear membrane. These isozymes are affected differently by dietary components as well as pollutant overload and are unevenly distributed among the sexes.[65] This uneven distribution may explain the predominance of women with earlier chemical sensitivity seeking help from physicians. It also may explain varied end-organ damage seen in a group of individuals exposed to the same substance.

The discussion of organics and inorganics is interspersed throughout the following sections on metabolism of foreign compounds. Inorganic chemicals such as lead, mercury, cadmium, aluminum, nickel, beryllium, etc. can also disturb foreign body catabolism (see Inorganics chapter). However, they may be just as toxic, more or less, than the organics. Their effects on metabolisms will be interwoven with the organic chemicals, often severely affecting the chemically sensitive adversely and certainly always contributing to an increase in total body load.

There are 13 basic toxic compound classes of organic chemicals including acids (RCOOH), aldehydes (R–$\overset{\text{O}}{\overset{\|}{\text{C}}}$–H), alcohols (R–O–H), ketones (R$_1$–$\overset{\text{O}}{\overset{\|}{\text{C}}}$–R$_2$),

esters (R$_1$–$\overset{\text{O}}{\overset{\|}{\text{C}}}$–O–R$_2$), ethers (R$_1$–O–R$_2$), amines (R$_1$–NH$_2$), nitro compounds

(R–C–NH$_2$), thio compounds (R–SH), aromatic hydrocarbons (◯–), phenols (◯–OH), and alkyl (R–Cl) and aryl halides (◯–Cl), seen responding to detoxication and thus are extremely important in understanding the problems the chemically sensitive have upon exposure to these compounds. All 13 are usually initially changed into 6 types of less toxic substances by oxidation, reduction, or hydrolysis before or after final conjugation generally occurs.[66] Breakdown of each will be discussed in the following sections. It is necessary to understand some facts concerning these changes of compounds in order to better understand some of the underlying mechanisms that are disturbed in chemical sensitivity (Figure 10).

Of the essentially four types of biotransformation reactions — oxidation, reduction, degradation, and conjugation — conjugation is predominant.

Conjugation is almost exclusively observed for foreign compounds, with the other three being for breakdown and incorporation of nutrients. There is much overlap between all four systems. More than one pathway is involved in most of the detoxification reactions of the chemically sensitive, which in turn are strongly influenced by many nutrients including vitamins and minerals. More than one pathway may be involved with an individual toxic chemical; e.g., simple compounds like benzene or phenol may be excreted by multiple pathways: (1) phenylglucuronide (50%); (2) potassium phenylsulfate (40%); (3) quinol (10%); and (4) catechol (1%)[67] (Figure 11). Breakdown of the main pathways may occur with pollutant overload, resulting in shunting to lesser pathways, overloading them, and disturbing orderly nutrient metabolism, thus exacerbating chemical sensitivity.

Ironically, though most foreign compound biotransformation is to reduce the toxicity of a substance, occasionally more toxic intermediates may be formed. A table of many of these substances is presented for emphasis (Table 3). These may exacerbate chemical sensitivity. For example, the pesticide heptachlor epoxide is found in many chemically sensitive individuals. This compound may be a more toxic substance than the parent heptachlor compounds, thereby aggravating the chemical sensitivity.

It is becoming increasingly evident that xenobiotics and their metabolizing enzymes function in an ambivalent manner. On one hand, lipid-soluble chemicals are converted to water-soluble forms that are readily excreted via the kidney or bile duct; on the other hand, the same enzymes may generate reactive electrophilic molecules that modify essential cellular macromolecules with resulting toxicity or chemical sensitivity or carcinogenic effects.

A few biotransformations are predominantly nonmicrosomal-dependent redox reactions involving alcohols, aldehydes, and ketones.[68] These will be discussed first. A discussion of microsomal enzymes follows.

Oxidation (Loss of Electrons)

Oxidation loses an electron. $O_2 \rightarrow O^-$ is blocked by CO, excess toxic chemicals, lack of enzymes, lack of nutrients and/or loss of O_2. Blocking of this

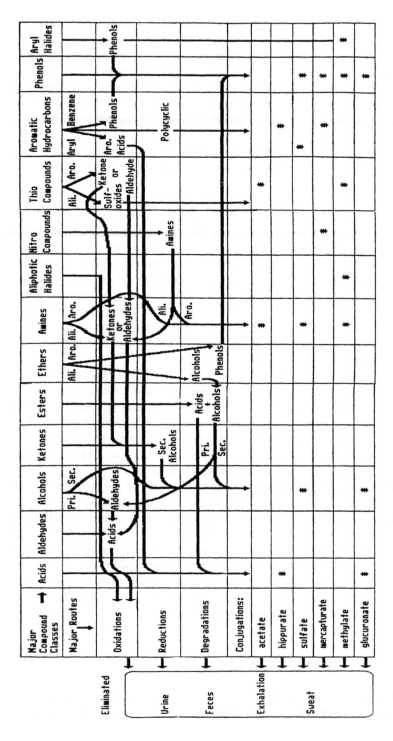

Figure 10.   Typical routes of foreign compound metabolism potentially alterable in the chemically sensitive.

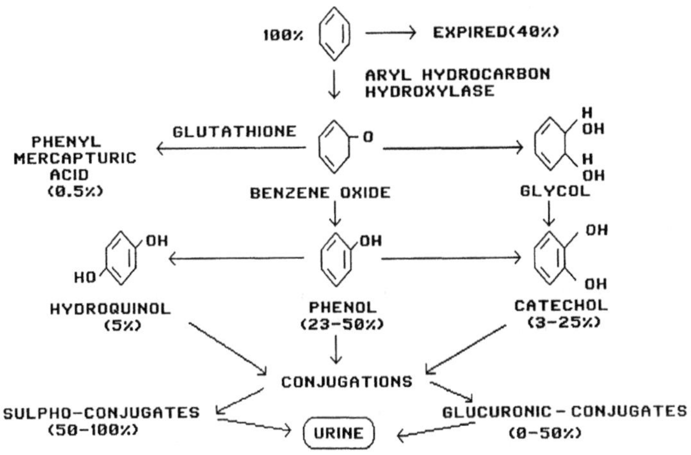

Figure 11.   Pathway of benzene metabolism and elimination.

Table 3. Mechanisms of Metabolic Activation of Carcinogenic Foreign Compound Breakdown Where the Product Becomes More Toxic to the Chemically Sensitive

Types of Metabolic Activation	Theoretical Active Metabolite	Specific Examples
Epoxidation	Arene oxide	Benzo[a]pyrene
C-Hydroxylation	Carbonium ion and 1,2-dimethyl-hydrazine	Dimethylnitrosamine
N-Hydroxylation	Nitrenium ion	Naphthylamine and N-acetylaminofluorene
Free radical formation and redox cycle	Carbon-centered radical Oxygen-centered	Carbon tetrachloride Quinones and nitrocompounds
S-oxidation	Sulfinic acid	Thioacetamide

reaction will result in increased chemical sensitivity. Another way to look at this is that alcohol loses two hydrogens. This reaction is, therefore, called a dehydrogenase reaction [ROH → RO⁻]. If this process is damaged by excess pollutants, the usually orderly process may result in a build-up of more toxic substances in the target tissue instead of a change to less toxic ones. Damage to this oxidation process has been observed in most chemically sensitive patients at the EHC-Dallas with a resultant increase in total body load. For example, build-up and supersensitivities to formaldehyde may occur, resulting in the spreading phenomenon occurring with sensitivity to more chemicals such as ketones and alcohols (Figure 12). This process has been seen in chemically sensitive patients.

Nonmicrosomal Enzymatic or Nonenzymatic Oxidation of Alcohols, Ketones, and Aldehydes

Nonmicrosomal enzymatic or nonenzymatic oxidation of *alcohols* (R–O–H), *ketones* (R$_1$–COR$_2$), and *aldehydes* (R–C–H) does not usually use

$$\overset{\|}{O}$$

microsomal enzymes for their breakdown (Figure 12). Primary alcohols (some secondary and tertiary) are oxidized by strong oxidizing agents. They may or may not use nonmicrosomal enzymes. Some may be just nucleophilic-electrophilic reactions being oxidized to ketones to aldehydes and acids. Tertiary alcohols are oxidized without rupture of the C–C bond. The metabolic elimination of alcohols in humans occurs first by oxygenation to the aldehyde, then to the carboxylic acid, then to CO_2 and water. This pathway is the primary one for breakdown. Alcohols broken down into aldehydes require slow-acting NAD (B$_3$)-dependent alcohol dehydrogenases, which are in turn broken down into acids by rapid-acting flavin (B$_2$)-requiring aldehyde oxidases, and then broken down to CO_2 and water[69] (Figure 12). Pollutant overload may result in damage to B$_2$ or B$_3$ or the aldehyde dehyrodgenases or oxidases. This damage renders the chemically sensitive individual intolerant to formaldehyde compounds, causing severe tissue crosslinking due to the aldehyde buildup. Vasculitis has been seen with this buildup, resulting in seizures and brain damage (see Chapter 26).

Microsomal Enzyme Used — Alcohols

Two lesser pathways using microsomal enzyme mechanisms occur. These pathways for the elimination of alcohol with a type of peroxidase used may occur by conjugation with glucuronic acid using microsomal energy transfers. Here, alcohol can also be oxidized by peroxide in a catalase-dependent reaction or by a high-K$_m$, NADPH-dependent microsomal oxidizing system.[70,71] The pathway involved in breakdown depends on the structure of the alcohol. Xanthine oxidase, a complex protein [Fe-S-Mb-FAD(B$_2$)], plays an important role in the oxidation of purines. These enzymes may be damaged in chemical sensitivity, resulting in an inability to tolerate xanthine-containing compounds such as coffee and chocolate.

Aldehydes are generally more toxic than their precursor alcohols and will interact readily with cellular material. They are cross-linkers, especially of protein. If crosslinking occurs, it is usually permanent. Aldehydes are not only exogenous, they are also produced in vivo from many biological reactions as intermediates, e.g., the monoamine oxidase-catalyzed oxidation of biogenic amines, the metabolism of plasmalogen in brain, and the peroxidation of polyunsaturated lipids (Figure 12).

Though most aldehydes occur as oxidative intermediates, some can be brought into the body from outside sources, such as ethylene glycol and methanol breakdown products as well as direct formaldehyde and glutaraldehyde exposure. The ultimate metabolic products of methyl and ethylene glycol, formate and oxalate, are more toxic than their precursors. This phenomena suggests why the chemically sensitive have more problems with these precursor products, since their breakdown products cause brain and other organ dysfunction.

Excess aldehydes may overload the detoxication system, reversing the reaction and giving an increase in alcohols. This reversal may account for some of the symptomatology such as foggy thinking and loss of fine tuning and judgment, which is frequently seen in the formaldehyde-sensitive individual who has brain reactions (see Chapter 14).

The various pharmacologic, addictive, and pathologic consequences of ethanol consumption are directly related to their biochemical properties, rate of elimination, and formation of metabolic byproducts. Basically, the availability of detoxification enzymes and NAD are the dominant factors in ethanol detoxification. When deficient in these substances, the individual may become extremely intolerant of ethanol, as is seen in the chemically sensitive.

Other external sources such as benzaldehyde and acid aldehyde can cause discomfort, while others can be lethal. Mutagens and carcinogens include glicidaldehyde,[72] macrosamine,[73] 3,4,5-trimethoxycinnamaldehyde,[74] formaldehyde,[75] and a metabolite of safrole.[76] Exposure to drugs, i.e., phenobarbital and herbicides[77-84] including 2,3,7,8-tetrachlorodibenzo-p-dioxin,[83-85] can cause an increased induction of aldehyde dehydrogenase in rats. One enzyme is free and the other is complexed with GSH.[88-90] Chronic exposure to the aforementioned chemicals could damage induction of aldehyde dehydrogenase to the point that formaldehyde sensitivity, which is observed in chemical sensitivity, is propagated.

Fructose and vitamin B_3 (niacin) also have been shown to accelerate clearance of aldehydes in some chemically sensitive patients.

In addition to individual chemical overload in food, water, and home, the sheer weight of the total outdoor environmental load that an individual takes in daily may overwhelm these systems causing intoxication symptoms. For example, in addition to the large amounts of manmade alcohols and aldehydes that the individual takes in, acetone is inhaled from city air constantly. As discussed in Chapter 9, a typical city like Dallas, TX, has over 200 tons of acetone in its air in 1 year. In order to function daily, an individual living under these conditions has to overcome the ambient intake. If a chemically sensitive individual is near his limit or has already exceeded his total body load, his symptoms will be aggravated by overloading these detoxication systems.

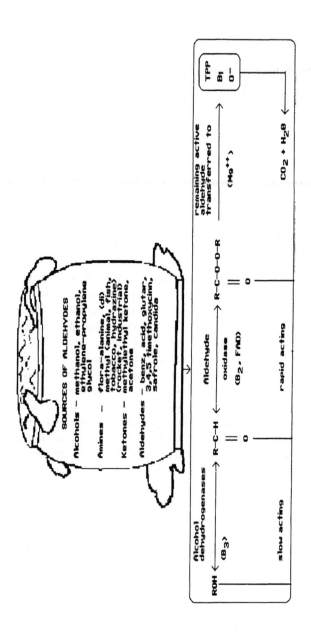

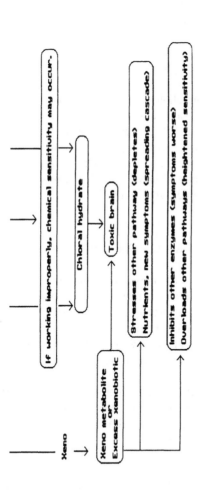

Figure 12. Effects of pollutant overload on the aldehyde, acid, and alcohol oxidation mechanisms influencing chemical sensitivity.

Hydroxylation

Hydroxylation requires water, oxygen, and the microsomal enzyme (Figure 13). This is the mechanism of action for detoxifying (1) straight-chain aliphatics with or without amines, (2) aromatics (aryl-s-phenols) with or without amines (3) and amines alone. Often the chemically sensitive seen at the EHC-Dallas have trouble handling these compounds, as evidenced by reported adverse reactions upon inadvertant exposure, blood levels, and intentional controlled injection and inhalation exposures.

The simplest molecules of nonamine aliphatic hydrocarbons are metabolized by hydroxylation and include N-butane, N-pentane, N-hexane, N-heptane, octane, decane, etc. All are oxidized to alcohols. Many chemically sensitive individuals have high levels of pentanes, hexanes, and/or heptanes in their blood. These products decrease as the patient detoxifies. They remain elevated if the detoxification process is damaged as seen in many chemically sensitive patients.

Examples of the simple aliphatic amines are glycine, serine, and valine (Figure 13). Toxic amines are substances such as methylamine, dimethylamine (from animal waste, fish processing, tobacco smoke), or hydrazine (H_2N-NH_2) (rocket industry, industrial plants) may also be hydroxylated (see Chapter 12, Organic Chemical Pollutants). The hydrazine becomes reactive when its breakdown product 920 dye is oxidized, resulting in free radicals, which can exacerbate chemical sensitivity.[91]

$$R - \overset{\overset{\textstyle O}{\|}}{C} - O - R' + H_2O \rightarrow R - \overset{\overset{\textstyle O}{\|}}{C} - OH + HOR' \text{ (carboxylester hydrolysis)}$$

In most cases the hydrolysis of an ester or amide bond in a toxic compound means detoxication because of the increased hydrophilicity and accelerated excretion. However, there are instances of hydrolysis resulting in an increase in toxicity such as the action of liver carboxylesterases upon phenacetin,[92] the herbicide propanil,[93] and the food flavoring allyl esters, which produce toxic metabolites.[94] These substances often exacerbate chemical sensitivity in our experience.

The mixed-function oxidase system, including cytochrome P-450, C reductase, and phosphatidylcholine, is the microsomal monooxygenase system which includes flavin adeninedinucleotide (B_2)-containing monoamine oxidases (MAO) in the detoxification of amines.

$$R - CH_2 + O_2 + H_2O \xrightarrow[\text{Oxidase FAD}]{\text{(mitochondrial) monoamine}} R - \underset{\underset{\textstyle O}{\|}}{C} - H + \boxed{NH_3} + H_2O_2$$

1. The amines are oxidized into aldehydes, with a loss of ammonia, by mitochondrial MAO. If this enzyme is blocked (e.g., MAO inhibitors) certain amines will not be metabolized and chemical sensitivity will be exacerbated. For example, tyramine is formed in the bowel by bacterial action on tyrosine found in many cheeses. Subsequent hypertensive reaction will follow if the tyramine builds up, and an exacerbation of chemical sensitivity will follow. Microsomal MAO is responsible for the biotransformation of amphetamine to phenylacetone by the removal of ammonia. Inability to perform this function due to pollutant damage will result in an exacerbation of chemical sensitivity due to the continued presence of amphetamines.

2. Mitochondrial enzyme diamine oxidase metabolizes the diamines such as histamine, cadaverine, and putrescine by removal of one ammonia. Pollutant damage to this enzyme will allow for continued presence of these substances and probably explains the frequent flushing seen in a small subset of the chemically sensitive.

3. A special case of aliphatic amine biotransformation occurring by C-hydroxylation is diethylamine plus nitrite plus hydrochloric acid (in the stomach), causing nitrosamine, a very potent carcinogen. Vitamin E and C will block or neutralize this reaction.[95] Slow cooking and microwave cooking will prevent or minimize this biotransformation. Mostly, fat is involved.

The EHC-Dallas and Bionostics have shown that pollutant stimulation speeds up deamination in the chemically sensitive. Acceleration of normal deamination leads to an increase in blood ammonia, resulting in exacerbation of the symptoms of chemical sensitivity. Ammonia and the symptoms it produces ameliorate with the administration of nutrients such as α-ketoglutaric acid, B_1, B_2, B_3, administration of oxygen, or by decreasing the total body load.

Aromatic Side Chain

Aromatic Side Chains (C–O–R). An example of aromatic side chain break-down by oxidation is alkyl-substituted aromatic compounds such as phenylalkane (typically split by the microsomal enzyme between the C^1 and C^2 side). These will change irrespective of the length and structure of the side chain, and they are oxidized to benzoic acid. A hydroxyl group can be added to an aromatic system by two means. One is by direct insertion into the C–H bond, and the other is by oxygen addition to the carbon-carbon double bond (bond) of the aromatic system and subsequent rearrangement to form a hydroxyl group. Pollutant injury to the microsomal enzymes responsible for this process will cause an increase in aromatics with exacerbation of the chemical sensitivity.

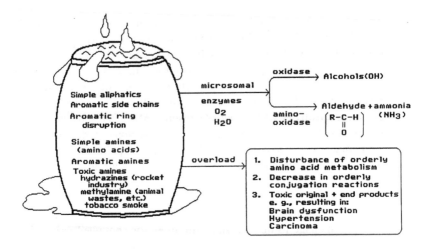

Figure 13.   Pollutant damage to hydroxylation reactions.

Hydroxylation of Ring System. Aromatic and alicyclic compounds such as benzene, xylene, ethylbenzene, or cyclohexane, all seen in some chemically sensitive individuals (without oxidizable side chains), may be hydroxylated (OH) on the ring to substances such as benzyl alcohol and methyl phenylcarbinol. This hydroxylation is more readily achieved with cyclic ring-like cyclohexane to cyclohexanol than with the benzene-ring of aromatics. Slower breakdown may explain the high toxicity of benzene since the microsomal aryl hydroxylase has relative difficulty in its oxidation to phenol. If aromatic and alicyclic amines are condensed and the alicyclic ring becomes oxidized first, carcinogenic metabolites develop.[96] Benzene compounds are seen in over 80% of the chemically sensitive patients examined at the EHC-Dallas and cause severe problems upon challenge. This increase in benzene in the chemically sensitive in part appears to be due to damage of the aryl hydroxylase.

Aromatic Amines (◯–NH₂) Such as Tyrosine, Phenylalanine or Amino-pyridine-4-Aminodiphenyl. These may be N-hydrolated (oxidized) by the hydroxylase enzyme, resulting in less reactive derivatives. The lesser toxic substances such as amino acids are predominantly broken down by this pathway.

Some toxic chemicals, such as 4-amino biphenyl and others industrially generated, are thought to play a role in carcinogenesis[97] as well as chemical sensitivity. Aromatic primary amines (RNH_2) are believed to induce tumors as a consequence of their N-oxidation followed by a second metabolic transformation to derivatives that are capable of reaction with nucleic acids.[98] It is clear that overloading this breakdown system with industrially generated amines may distort normal amino acid metabolism and cause an exacerbation of the chemical sensitivity. Often nutrient depletion may result.

In animal species capable of N-acetylation, arylhydroxamic acids appear to play key roles in the carcinogenic process. The enzymatic conjugation of

N-acyloxyarylamines derived from arylhydroxamic acids or by enzymatic N,O-acyl transfer may be critically involved in the formation of liver and extrahepatic tumors in rats and other species.[99] Aromatic primary amines (e.g., arylhydroxamines) decompose under acidic conditions to yield a very reactive amino cation (nitrenium ion $ArNH^+$) that may be responsible for their toxicity. Aromatic hydroxylamines have been implicated in a number of toxicological responses. Their ability to interact with cellular nucleophils such as proteins, thiols, and nucleic acids is believed to be the cause of a number of toxic reactions. N-Hydroxyamides reactivity is similar to the hydroxyamines and less toxic.

N-Oxidation

N-Oxidation occurs naturally. For example, trimethylamine is excreted as trimethylamine N-oxide. Some N-oxidations like to detoxify the naturally occurring alkaloids and synthetic medical amines to less toxic substances. However, N-oxidatives of arylamines are more toxic, even leading to carcinogenic compounds.

N-Oxidation of tertiary amines (R_3N) yields readily excreted more polar less toxic amine oxides of low molecular weight. Many secondary (R_2NH) and tertiary amines are N-oxidated to yield primary and secondary amines (R_2NH) together with aldehydes as products. Demethylation, N-dethylation, and N-debutylation are known; the products from the breaking of the carbon and nitrogen bond, in addition to the amines are formaldehyde, acetaldehyde, and buteraldehyde, respectively.

Aromatization of Alicyclic Compounds

All cyclic compounds may be aromatized and oxidized by the liver or kidney mitochondrial enzymes into benzoic acid or ring cleavage products. Here benzene ring compounds are attached to the alicyclics.

Epoxidations $(-\overset{|}{C}-\overset{|}{\underset{\diagdown \; O \; \diagup}{C}}-)$

Epoxides are cyclic ethers found in nature. They are also man-made. Most are metabolically derived from unsaturated aliphatic and aromatic hydrocarbons and organochlorine pesticide precursor molecules.[100] Many reactions with the microsomal epoxide hydrolase (epoxide hydrase or hydratase) may result in a toxicant producing or accelerating chemical sensitivities. Many are responsible for acute or chronic toxicity and sensitivity including necrosis, mutagenesis, carcinogenesis, and teratogenesis. The hydrolase enzymes catalyze the biotransformation of many chemically reactive and highly toxic epoxides to dehydro forms which in many instances are devoid of these properties, but in other instances are precursors of more reactive dihydrodiol epoxides or phenol.[101]

Because of ring polarization, epoxides are a chemically reactive species.[102] There are three major types of epoxides: (1) alkene oxides derived from nonaromatic olefins; (2) arene epoxides formed from aromatic hydrocarbons; (3) diols formed from dihydrodiol derivatives of polyaromatic hydrocarbons. Production of these may result in exacerbation of the chemically sensitive (Figure 14).

The mechanism of the enzymatic reaction appears to be a nucleophilic (DNA-RNA) attack by H_2O_2 or OH^- from the side of the molecule opposite to the epoxide ring;[103] consequently, the resulting diols have a transconfiguration.[104-106] If a product diol is derived from a large hydrophobic molecule, it may serve as a substrate for the monooxygenases, resulting in a diol epoxide.[107] If the epoxide moiety is situated in the bay region, an especially vigorous reaction may result.[108] The scope of this reaction is complex,[109,110] although the majority of epoxides are inactivated. The enzyme hydrolases play a dual role with the large polycyclic hydrocarbons possessing a bay region. They inactivate monofunctional epoxides and produce a precursor molecule for the dihydrodiol-bay region-epoxide formation. The relative importance of microsomal epoxide hydrolases will also be dictated by the presence of competing reactions.

Many epoxides, being good substrates for certain of the GSH transferases,[111,112] readily react with GSH.[113,114] Epoxides derived from large lipophilic compounds severely limit the effectiveness of GSH and GSH transferase,[115] thus exacerbating chemical sensitivity.

The induction of the hydrolase by several foreign compounds is similar to the monooxygenases. Phenobarbital, 3-methylcholanthrene, and pregnenolone 16x-carbonitrile induce both.[116-120] *trans*-Stilbene oxide, 7-ethoxycoumarin O-deethylation, and benzo[a]pyrene also induce epoxide hydrolase.[121,122] Again, overload of these enzyme systems by stimulation with these compounds will eventually lead to a build-up of toxic pollutants with an exacerbation or induction of chemical sensitivity.

Stable epoxides which are toxic, mutagenic, or carcinogenic, occur as major metabolites of the insecticides heptachlor and aldrin, becoming heptachlor epoxide and dieldrin.[123] (See Chapter 13.) These compounds are frequently present in chemically sensitive patients. Heptachlor epoxide is frequently seen in chemically sensitive patients whose home was termite proofed with chlordane.[124] Often, these substances can be removed from the blood with good environmental control, thereby decreasing the chemical sensitivity (see Chapter 31). Heat depuration therapy appears to enhance their removal also. Intermediately stable epoxides are thought to be the leukemogenic agents of benzene degradation. Epoxide hydrolase has a broad specificity for substrates including arene and alkene oxides.[125,126]

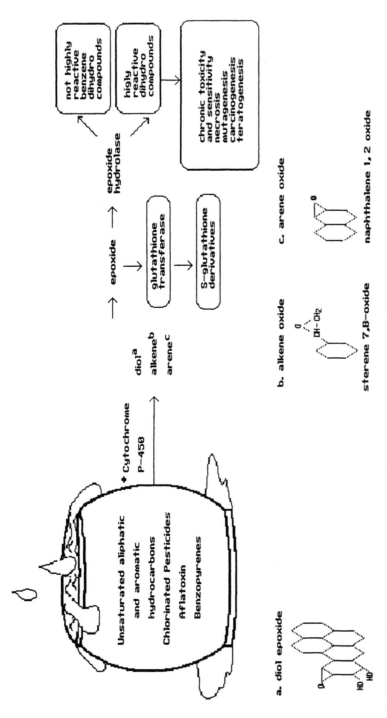

Figure 14.   Excess total body load of xenobiotics producing more toxic breakdown compounds that exacerbate chemical sensitivity.

88   **Chemical Sensitivity: Principles and Mechanisms**

Alkene Epoxidation. The cytochrome P-450 is capable of oxidizing C=C double bonds that are not part of the aromatic system, leading to epoxides as products. Although the diol which is formed from the epoxide upon addition of water in alkene epoxidation is more polar than the parent compound, the epoxide is a reactive intermediate. The epoxide of an alkene group usually cannot be considered a detoxification reaction.

Among the compounds that are activated to toxic metabolites by alkene or other epoxidation are some of the most important carcinogens as well as inducers and propogators of chemical sensitivity. These include (1) aflatoxin B (from *Aspergillus flavis* AFB, the most important carcinogen known to cause liver tumors); (2) 7,8-dihydrodiol of benzopyrene; (3) vinyl chloride; (4) heptachlor; (5) aldrin; (6) trichloroethylene; (7) cyclohexene; (8) styrene; and (9) safrole. It has been observed that the asymmetrically substituted chlorinated ethylenes (vinyl chloride, vinylidine chloride, and trichloroethylene) are much more mutagenic than the symmetrically substituted members (tetrachloroethylene + E^- + 2-1,2-dichloroethylene). Because of the electron-drawing properties of chlorine, this property would be expected to increase the reactivity and electrophilicity of chlorinated ethylenes and their epoxide analogues; metabolism of these, again, undergoes oxidation to aldehydes on hydrolysis of the acylchlorides to acids. Many of these compounds are found in the blood of chemically sensitive patients and exacerbate their symptoms.

Oxidation of Organic Sulfur

Oxidation of organic sulfur is necessary for normal body function, and pollutant injury will produce symptoms by altering desulfuration and sulfoxidation function. Understanding these functions is important to understanding some aspects of chemical sensitivity, since many chemically sensitive patients cannot tolerate inorganic or organic sulfur exposure often presented to them in minute doses.

Desulfuration. Desulfuration of compounds containing C=S or P=S functional groups involves the substitution of oxygen for sulfur and is the mixed-function oxidase type mediated by cytochrome P-450 (Figure 15). FAD is also needed for this reaction to occur. Desulfuration does not necessarily lead to detoxification, but is rather a mechanism leading to activation and resulting in products capable of covalent binding to tissue macromolecules.

Also, amine oxidase-catalyzed desulfuration reactions occur.[127] The sulfur in aliphatic combinations or in aromatic side chains is sometimes replaced with oxygen by means of microsomal enzymes, and the compounds are converted to their oxygen analogues. For example, parathion is metabolized to paraoxon, but this reaction is inhibited by the presence of carbon monoxide.[128] Therefore, chemically sensitive individuals will have problems with this process. Also, it can be inhibited by antibodies to cytochrome P-450.[129] Thiobarbital transforms

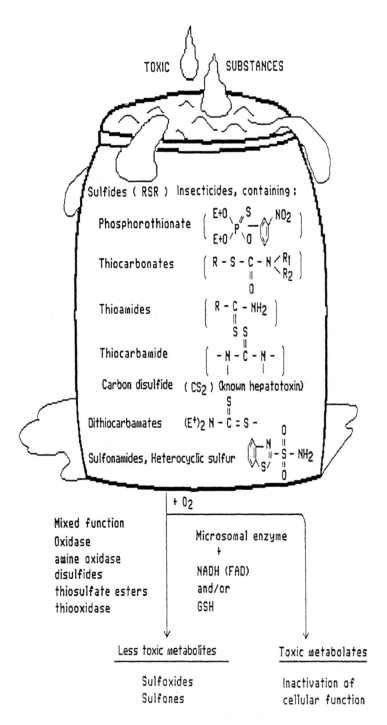

Figure 15.   Pollutant injury to the oxidation of organic sulfur.

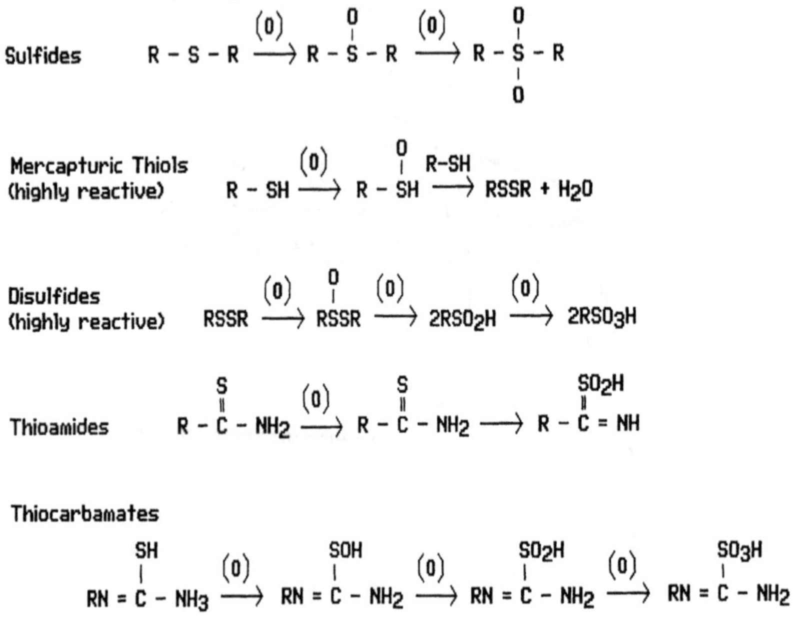

Sulfides $R - S - R \xrightarrow{(0)} R - \overset{O}{\underset{}{S}} - R \xrightarrow{(0)} R - \overset{O}{\underset{O}{S}} - R$

Mercapturic Thiols
(highly reactive) $R - SH \xrightarrow{(0)} R - \overset{O}{\underset{}{S}} H \xrightarrow{R-SH} RSSR + H_2O$

Disulfides
(highly reactive) $RSSR \xrightarrow{(0)} R\overset{O}{\underset{}{S}}SR \xrightarrow{(0)} 2RSO_2H \xrightarrow{(0)} 2RSO_3H$

Thioamides $R - \overset{S}{\underset{}{C}} - NH_2 \xrightarrow{(0)} R - \overset{S}{\underset{}{C}} - NH_2 \longrightarrow R - \overset{SO_2H}{\underset{}{C}} = NH$

Thiocarbamates

$RN = \overset{SH}{\underset{}{C}} - NH_3 \xrightarrow{(0)} RN = \overset{SOH}{\underset{}{C}} - NH_2 \xrightarrow{(0)} RN = \overset{SO_2H}{\underset{}{C}} - NH_2 \xrightarrow{(0)} RN = \overset{SO_3H}{\underset{}{C}} - NH_2$

Figure 16.   S-Oxidation of organic sulfur compounds.

into barbital this way. Performic acid cleaves each cysteine residue in insulin to cysteic acid.[130]

Sulfoxidation (–S–Oxidation). Sulfoxidation occurs when oxygen is added to sulfur. Both disulfides (GSSG) and thiosulfate esters (RSSO3) are mild oxidizing agents that can convert sulfhydryl groups to disulfides in low-molecular-weight substances as well as in such macromolecules as proteins. Heterocyclic sulfur, for example, methylene blue

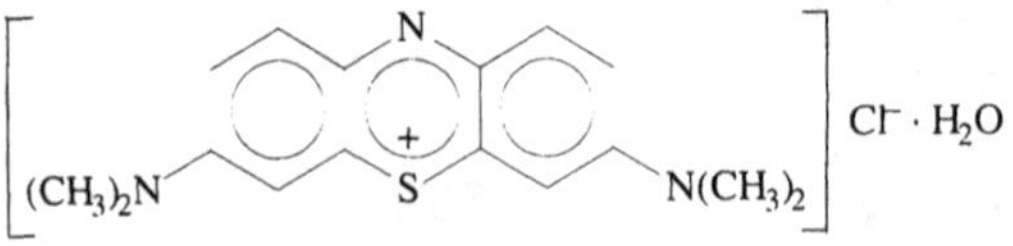

$Cl^- \cdot H_2O$

chlorpromazine dimethyl sulfoxide (C_2H_6OS), and other nonphenolated compounds are oxidized into sulfoxides[131] or disulfones[132] (Figure 16).

Oxidation of the sulfhydryl groups often modifies the biological activity of the parent molecules and in many cases results in inactivation of an essential cellular function. Oxidation of nucleophilic divalent sulfur atoms such as sulfides (RSR), thiols (mercaptans RSH), disulfides (RSSR), and thiones, thioamides ($RCNH_2$) or thiocarbamates ($RN=CNH_2$) produce an extraordinarily large number of products. Alkyl and aryl sulfides are readily oxygenated and eliminated in the urine as sulfoxide ($R=SO$) and sulfones ($R–SO_2–R$).

Oxidation of aryl and alkyl yields extremely reactive sulfenic acids (SO_2H) that react instantly with thiols, yielding disulfides and, in turn, can be further oxidized to sulfonic acids through thiosulfenic and sulfinic acid intermediates. While binary complex formation of metallo enzymes may explain their toxicity in part, the ability of sulfinic acid to react with free intracellular thiols, particularly GSH, may be more critical in toxicity and sensitivity.

Alkyl and aryl sulfenic acids react with GSH mixed disulfides. Oxidation of certain thiols and disulfides can yield highly reactive electrophiles capable of eliciting toxic responses. Classic studies, mostly with enzymes, demonstrate that proteins require intact sulfhydryl groups for their functions,[133] and the reversible oxidation of such groups has been suggested as a physiological control mechanism.[134] Heavy metals like lead or mercury may interfere with the GSH replenishing cycle, allowing excess oxidation to occur due to the absence of sufficient reduced GSH and thereby disturb this control mechanism. This disturbance results in increased chemical sensitivity.

Thiol oxidases in the extracellular fluid and di- and monooxygenases in cytosol plus oxygen react with a broad specificity of thiols in order to reduce toxicity. Low-molecular-weight mercaptans (RSH) such as 2-mercaptoethanol, cystine, cysteine, GSH, dithiothreitol, and benzene thiol, as well as the thiol (–SH) groups of reduced ribonuclease serve as substrates. Thio oxidase has a higher catalytic efficiency for endogenous mercaptans, cysteine, and GSH than 2-mercaptoethanol. At least two other types of enzymes catalyze the oxidations at the endogenous thiol (HSH) (GSH) to a disulfide (GSSG). One is the selenium-containing GSH peroxidase, and the other is the nonselenium-dependent GSH transferase B (Ligandin). The importance of reduction of exogenous disulfides is further attested by the observation of "disulfide poisoning," as is found in glycolysis in various cells.[135] Mitosis is accompanied by a rise in the titer of intracellular sulfhydryl group,[136] and dormant bacterial spores[137] and dormant fungal conidia[138] have high disulfide contents that decrease during germination. This decrease of disulfide during germination may well explain why the severely chemically sensitive patient who is extremely intolerant of foods can eat fresh, germinated plants with a reduction in problems, but may still be intolerant of a mature plant.

Thioltransferases (Thiol Disulfide Oxidoreductases). Thioltransferases may also catalyze the thiol disulfide reaction. Thiols in pesticides may be cholinesterase inhibitors and excess loads or direct damage to the enzyme may exacerbate the chemical sensitivity. (See Chapter 13.)

Even proteins may act as substrates, which implies that mercaptans (SH) may react with protein disulfide groups and disulfides with protein thiol groups to form mixed disulfides of proteins and mercaptans. In this way, xenobiotic mercaptans and their corresponding disulfides may effect the biological activities of proteins,[139] perhaps rendering the chemically sensitive more vulnerable.

Thus, very complex cellular processes seem to be regulated by the "thiol (RSH)/disulfide (RSSR) status"[140,141] of the cell, and increasing the disulfide content may be expected to inhibit cell division which would tend to increase sensitivity to chemicals. In general, intact active cells have a high thiol/disulfide ratio, and when subjected to oxidative stress by various processes, this ratio is maintained by reduction of disulfides and, at least in some cells, by transport of GSH (a mercaptan) disulfide (G-SS-R) from the interior of the cell to the surrounding medium.[142-144] Consequently, it appears important for the cell to have the capability of reducing sulfur-sulfur bonds for both the biotransformation of xenobiotics and the regeneration of endogenous thiols that have been oxidized by oxygen or by compounds activated after reaction with oxygen, e.g., H_2O_2 and hydroperoxides. The latter type of compound may be formed by oxidative metabolism of various xenobiotics[145] (pesticides, phenols) or by irradiation. The formation of the hydroperoxides due to overload by these xenobiotics often exacerbates the chemically sensitive.

Oxidative Dehalogenation

The cytochrome P-450 system is one of several capable of dehalogenating organic compounds by oxidation.[146] Halogenated pesticides, chlorinated drinking water, anesthetics, cleaning agents, propellants, and chemical intermediates all challenge the detoxification systems of humans, especially the chemically sensitive. As stated previously, many of these products are found in the chemically sensitive (Figure 17). The final breakdown product of the haloforms is to carbon monoxide. Carbon monoxide has been found to inhibit the oxidative metabolism of halogenated hydrocarbons including 1,1,2-trichloroethane, chloroform, and fluorexene.[147] Therefore, when the chemically sensitive are exposed to these halogenated substances, a self-cycling downhill course occurs with increased sensitivity.

Chlorine is preferentially removed from that carbon atom, bearing two chlorine atoms and a hydrogen atom. Dihalobromines and methanes requiring NADPH (B_3) for detoxification are converted to formylhalide. This detoxification process explains why the chemically sensitive, who frequently lack B vitamins, need both these and oxygen when they are exposed to toxic halogens. The metabolism of chloroform ($CHCl_3$) is oxidized by the cytochrome P-450 system to trichloromethanol which is spontaneously converted to phosgene by dehydrochlorination. CCl_4 is broken down into hexachloroethane. *The products formed by oxidative dehalogenation, e.g., carbon monoxide, formylhalide, and phosgene are toxic.* The same is true for halogen anesthetics. This breakdown with those final-end products explains why chemically sensitive patients (whose detoxification systems are already damaged) have so much trouble with halogenated compounds, with each new exposure inevitably making them worse. Many female patients report the onset of their chemical sensitivity after a delivery in which halogenated hydrocarbons were used as anesthetics.

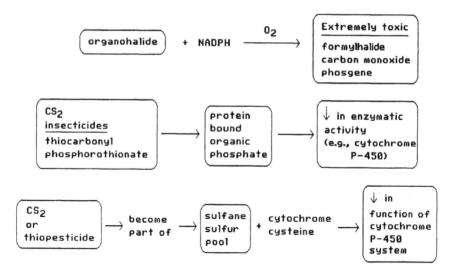

The end result of these detoxication processes is exacerbation of chemical sensitivity.   At times, these processes may precipitate it.

Figure 17.   Pollutant injury by organohalides, organophosphates, carbamates, and CS-2 during detoxication.

Oxidative Denitrification

2-Dinitropropane may be oxidatively denitrated by the microsomal enzyme to form acetone and nitrate. O_2 and NADPH are needed. The reaction can be induced by phenobarbital and inhibited by carbon monoxide. This detoxification process will again be disturbed by pollutant overload and adversely affect the chemically sensitive.

REDUCTIONS (GAIN ELECTRONS)

Reductions in foreign compound metabolism are much less common than oxidations because they run counter to the general trend of biochemical reactions in living tissue. Generally, they require a lack of oxygen. All enzymatic reactions are, however, fundamentally reversible. Cytochrome P-450 (a mixed-function oxidase) may catalyze reductases under anaerobic conditions. The direction of a reaction depends upon the chemical equilibrium. If the reduced form of a redox equilibrium produces products that are more easily excreted than the oxidized form, the law of mass action will push the reaction in a direction of easiest elimination (Table 4). Reversal may change the status of chemical sensitivity.

Table 4. Reductions

Require Lack of O_2

	Microsomal enzymes or reductase	
Toxic	$\xrightarrow{\hspace{4cm}}$	More easily excretable
substances	NADPH	substances

Law of mass actions pushes toward easiest elimination of toxins

Aldehydes and Ketones

Aldehyde and ketones may be reduced to the corresponding alcohols by reverse action of alcohol dehydrogenese. These reactions are usually nonmicrosomal and can be either enzymatic or nonenzymatic. This reverse action may partially explain the cerebral symptoms in the aldehyde-overloaded chemically sensitive patients. Chloral hydrate reverses to trichloroethanol. The final product of some aldehydes may be chloral hydrate resulting from trichloroethanols. Many of the chemically sensitive patients who have brain dysfunction develop sleepy states in which they can hardly stay awake. They often do fall asleep, especially when exposed to carbon monoxide and freeway fumes. Some may have higher levels of chloral hydrate in their brain due to this type of reaction. This sequence does not apply to acetone since it can directly enter the cycle of aerobic metabolism through aceto acetate and acetyl CoA.

Aldehyde (glucuronic, mevalonate, lactaldehyde, daunorubicin) reductase reduces such carbonyl compounds to alcohols. Reductases are low molecular weight, have cytoplasmic localization, broad substrate specification (aromatic, aliphatic and sugar aldehydes), and depend upon B_3 containing NADPH as the coenzyme. NADPH and cytochrome P-450 reductase is needed, while O_2 and CO inhibit tertiary amine and arene oxide. Highest levels of reductase are in the kidney followed by liver and brain. Special attention has been paid to the enzymes of the brain because of their preference for aldehydes derived from the neurotransmitters containing a B-hydroxyl group on the side chain, e.g., octopamine and noradrenaline.[148-156] Inducers of reductases are benzaldehydes, octopamine, and noradrenaline. Inhibitors are $NADP^+$ or P-nitrobenzyl alcohol by competition, anticonvulsant drugs such as barbiturate, glutethamide, succinamide, hydantins, and oxazolinedese compounds, tetramethyleneglutaric acid, flavanoids (quercetin, quercitin phenothiazine, chlorpromazine),[157-161] diuretics,[162] as well as biogenic acids (results from biogenic aldehyde of the oxidative pathway) such as 5-hydroxyindolacetic acid, 4-hydroxyphenylacetic acid, or homovanillic acid. If these reductases are depleted in the brain, excess aldehyde build-up may occur, explaining the cerebral reactions in some chemically sensitive patients.

Carbonyl-containing substances not biologically oxidized are reduced in the detoxification process. They are frequently hydrophobic and may be retained in the tissues. Their reduction to hydrophilic alcohols and subsequent conjugation is critical.[163] The in vivo metabolism of xenobiotic ketones to free alcohol or conjugated alcohols by ketone reductases has been demonstrated for aromatic,[164-166] aliphatic,[167] alicyclic,[168] or unsaturated ketones.[169] Warfarin is metabolized by these. These enzymes prefer NADPH rather than NADH as cosubstrate.

Aromatic Nitro and Azo Groups

Aromatic nitro and azo groups may be reduced by microsomal nitro reductase and azo reductase to analines, phenols, and picramic acid. Nitrophenols under certain conditions may inhibit reductases, which might prevent breakdown to less toxic substances. Azo compounds consisting of two aromatic rings joined by an azo (–N=N–) bond are synthetic and not found in nature. They are broken down by reductases. Overload of this system in the chemically sensitive will result in exacerbation of symptoms with an inability to detoxify. Some of the variations of activity as a function of structure stem from the susceptibility of the specific molecule to detoxification enzyme systems or in reverse to biochemical activation systems. The carcinogenic azo compounds typically have 4-amino substitution. This same type of configuration appears to adversely affect the chemically sensitive.

Disulfides (R–SS–R)

Disulfides split by the reductase enzyme into free thiols, analogously to the cystine-cystine reaction. Diethyl disulfide becomes ethyl mercaptan.

Double Bonds

Double bonds of certain aliphatic (e.g., alkenes) or alicyclic compounds such as cyclohexane and unsaturated monocyclic terpenes are reduced by the reductase enzymes and thus metabolized. Most chemically sensitive patients have trouble with terpenes, suggesting that reductase may be functioning inappropriately. Alkenes and alkynes (e.g., ethylene, propylene, butadiene, and acetylene) are frequently excreted unchanged.

Dehydroxylations

Hydroxylations will reverse in certain instances. Substituted catechols are in part reduced to the corresponding substituted monophenols by the hepatic and bacterial hydroxylases. Dehydrogenation may occur in the side chain as does noradrenaline to M-hydroxyphenylacetic acid.

Valence Reduction

Valence reduction of inorganic elements in the course of their metabolism may be accompanied by increased toxicity and sensitivity (e.g., pentavalent arsenic [As^{5+}] may be reduced to trivalent arsenic [As^{3+}], as is hexavalent selenium [Se^{6+}] to tetravalent selenium [Se^{4+}], by the reductase enzymes).

Reductive Dehalogenation

Under anaerobic conditions, carbon tetrachloride (CCl_4) is metabolized by liver microsomes to chloroform ($CHCl_3$) and to a species which binds to microsomal protein.[170] This process again requires the absence of O_2 and the presence of NADPH. Both are inhibited by CO and metyrapone. Halothane undergoes a similar reaction and may explain the toxicity of this anesthetic. Cobaltus chloride reduces toxicity in rats.[171] Of the chemically sensitive patients studied at EHC-Dallas, 37% have chloroform in their blood.

DEGRADATIONS

Certain foreign compounds require degradation before they can be further metabolized. This degradation is usually accomplished by combining them with water. These reactions are driven by hydralase enzymes (Figure 18).

Hydrolysis of esters ($R_1-C(=O)-O-R_2$), amides ($R_1-C(=O)-N(H)(H)$), hydrazides ($H_2N \cdot NH_2$), and nitriles ($-C::N-$) occurs. The hydrolysis of esters is catalyzed by a number of enzymes of generally low specificity and wide distribution such as carboxylesterase, monoglycerol lipase, cholinesterases and pseudocholinesterase, and ali and aryl amide esterases.

Carboxylesterases are proteins capable of catalyzing hydrolytic reactions of the following types:[172-174] carboxylester, carboxyamide, and carboxy thioester. Proper function of these are advantageous in the chemically sensitive.

Metrione[175] demonstrated a 50-fold increase in the activity of leucine aminopeptidase if the peptide NH group in good substrates was replaced by S. Often amides are more stable to enzymatic hydrolysis than corresponding esters of a similar structure, e.g., phenyl acetate is hydrolyzed faster than acetanilide.[176] B-Esterase is a well-defined group of carboxylesterases which

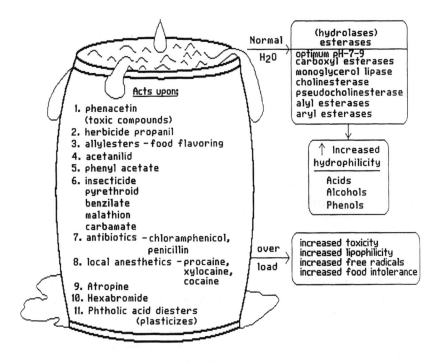

Normal → (hydrolases) esterases

H_2O → optimum pH-7-9
carboxyl esterases
monoglycerol lipase
cholinesterase
pseudocholinesterase
alyl esterases
aryl esterases

↑ Increased
hydrophilicity

Acids
Alcohols
Phenols

Acts upon:
1. phenacetin
   (toxic compounds)
2. herbicide propanil
3. allylesters – food flavoring
4. acetanilid
5. phenyl acetate
6. insecticide
   pyrethroid
   benzilate
   malathion
   carbamate
7. antibiotics – chloramphenicol,
                 penicillin
8. local anesthetics – procaine,
                        xylocaine,
                        cocaine
9. Atropine
10. Hexabromide
11. Phtholic acid diesters
    (plasticizes)

over
load → increased toxicity
increased lipophilicity
increased free radicals
increased food intolerance

Figure 18. Pollutant overload to degradation.

seems to be the most relevant for detoxification hydrolysis. In most cases, the hydrolysis of an ester or amide bound in a toxic compound means detoxification because of increased hydrophilicity and accelerated excretion. However, the action of carboxylesterases upon phenacetin,[177] the herbicide propanil,[178] or the food flavoring allyl esters[179] produces toxic metabolites. This observation suggests one of the reasons that the chemically sensitive patient often cannot tolerate artificial flavors. These artificial flavors will often result in exacerbation or production of gastrointestinal upset, asthma, vasculitis or other symptomatology. Much cross-reactivity, which emphasizes the importance of keeping these enzymes vigorous, is seen in this type of chemically sensitive patient.

Since many carboxylesterases hydrolyze carboxyl ester bonds in pyrethroids,[180-182] benzilate insecticides,[183] and malathion-type insecticides, they are sensitive to the inhibitors of certain organophosphates which have no antiacetylcholinesterase activity (see Chapter 13). Such organophosphates can potentiate the toxicity of malathion and related compounds by reducing the rate of esterolytic detoxification,[184,185] thus increasing chemical sensitivity. The carbamate[186,187] and many organophosphate insecticides inhibit B-type carboxylesterases sites by rapid esterification of a serine residue in the active site.[188] This reaction is followed by a hydrolysis of the new ester bond.[188] The carbamates and organophosphates are poor substitutes for B-esterases (serine hydrolase) carboxylesterases that are not inhibited by toxic organophosphorus triesters, but hydrolyze them instead as A-esterases (arylesterases).[189]

Some herbicides,[190] antibiotics[191-193] (e.g., chloramphenicol, penicillin),[194] and phthalic acid diesters (plasticizers) are also detoxified by the carboxylesterases/amidases. Again, with excess chemicals such as these in the total body load, this system may become impaired or destroyed in the chemically sensitive patient, thus prolonging or initiating illness. B-Esterases occur in liver,[195] kidney,[196-199] brain,[200,201] muscle,[202] serum,[203,204] and saliva[205] (Figure 18).

Substances that are detoxified by carboxyl esters and/or amidases are stable at pH 7.4[206, 207] and unstable at pH 5.0. Optimum pH is 7 to 9.[208] The hydrolysis of esters yields the corresponding acid ion, alcohol, or phenol. Numerous drugs such as aspirin, procaine, atropine, cocaine, lidocaine, and hexabendine are degraded in this manner.

Hydrolysis of amides, hydrazides, and nitriles occurs similarly to hydrolysis of esters. These reactions sometimes occur as minor pathways.

Besides other biotransformations, purified isoenzymes of the aliesterase type show a preference for simple aliphatic esters, e.g., methyl butyrate, relative to phenol esters, e.g., 4-nitrophenyl acetate. Fluor-2,4-dinitrophenyl will cleave amino acids from polypeptides,[209] as can trypsin and chymotrypsin.[210] Phenylisocyanate is also changed by polypeptides. Most B-type esters appear to have a preference for the more lipophilic esters, as compared to polar or charged substrates.

Dealkylations

$(CH_3 + O_2$ or N or S$) \rightarrow$ phenol, amine, thiol. Alkyl groups bound to oxygen, nitrogen, or sulfur may be removed by microsomal enzymes to yield the corresponding phenol, amine, or thiol. Carboxylase also hydrolyzes estrone acetate and propanodid,[211,212] phenol esters of short- and medium-chain fatty acids.[213-215]

The brain contains arylamides which are inhibited by serotonin and related compounds,[216,217] and related to acetylcholine esterase.[218] These facts again emphasize why the chemically sensitive with brain reactions may not function well upon exposure to these toxic chemicals if their system is already overloaded.

Ring Scissions

Ring scissions may occur with alicyclic or heterocyclic compounds such as hydantoin and coumarin where there is hydrolytic scission of the ring.

Obviously, an excess of total body load can disturb the mechanisms of oxidation, reduction, and degradation to such an extent that food intolerance (seen in 80% of the chemically sensitive) can occur. This intolerance could be due to direct metabolic enzyme damage or damage to nutrient fuels used for detoxification or damage to transport mechanisms in the gut or imbalance of the gut flora. Clearly, acute and chronic reduction of total body load has usually stopped this food sensitivity as evidenced in the chapters on ECU, Laboratory, and Avoidance of Food Pollutants (Chapters 31, 30, and 34). In addition to directly overloading these mechanisms, conjugation systems may become overloaded with the need for detoxification of certain chemicals spilling over into the other oxidative, reductive, and degradation systems occurring.

CONJUGATIONS (DETOXICATION REACTIONS IN THE CHEMICALLY SENSITIVE)

The biotransformations discussed heretofore are adaptations and modifications of biochemical reactions originally developed for the metabolism of nutrients. As shown previously, many toxic compounds are broken down as well as generated by these processes, the result of which may be exacerbation of chemical sensitivity. Conjugation seems to have developed chiefly for the metabolism of foreign compounds and, therefore, is extremely important in understanding the metabolism of chemical sensitivity.

There are basically five general conjugation categories. These include acetylations, acylation (peptide conjugation with amino acids), sulfur conjugations (with sulfates, GSH, and rhodenase), methylations, and conjugation with glucuronic acid. Pollutant overload, which may initially stimulate these processes, may overstrain or even stop them. This alteration results in a build-up of toxic chemicals in the body or an increase in tissue vulnerability to future insult with either movement of the chemicals to other pathways usually involving nutrition metabolism (thus, disordered metabolism) (see "amino acids" in Chapter 6) or creating various types of antigen antibody reactions. Either may result in or exacerbate chemical sensitivity. Though the process of conjugations occurs simultaneously, each facet will be discussed separately.

Acetylations

Acetylations through acetyl coenzyme A using N-acetyl transferase (found in the liver and gut) are general reactions of intermediary metabolism (e.g., acetylation of the hydroxyl group in choline to acetyl choline). This process needs B_5 to function. Acetylation, usually acting with a bimodal activity curve at a range of pH from 5.6 to 7.2, is the chief degradation pathway for the foreign compounds containing aromatic (aryl) amines such as P-amino salicylic acid, procaine amide, para-aminobenzoic acid, benzidine, histamine, serotonin, 4-aminodiphenyl, aniline, amino fluorine, methylane-bis-o-chloro analine, furfurylamine, P-phenetidin, and others (see "Table of Aromatic Amines" in

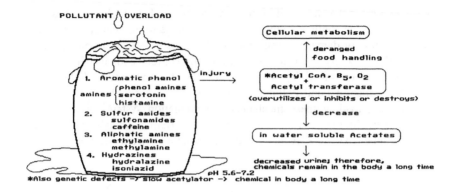

Figure 19.   Pollutant injury to acetylation.

Chapter 12). Acetylation is also a pathway for sulfur amides, such as sulfuramide, caffeine, sulfamethazine, sulfamerizine sulfanilamide, and sulfadiazine.

A third group of amines that is metabolized by acetylation includes the aliphatic amines, such as ethylamine, and methylamine. However, most of the less toxic simple amines, like the basic amino acids, are oxidized. Complex hydrazines are usually metabolized by acetylation. They contain N-N bonds, which contain aromatic rings (like isoniazid and hydralazine). (For lists of toxic bicyclic and polycyclic aromatic and aliphatic amines see tables in Chapter 12.) Disturbance of these processes by pollutant overload of these xenobiotics will often exacerbate chemical sensitivity and disturb the food handling mechanisms.

Acetylation can be influenced by pollutant overload and/or enzyme deficiency whether acquired or hereditary (Figure 19). Some genetic defects have been described, which explain some aspects of certain chemically sensitive patients. According to Evans et al.,[219] there is a group of patients in the general population who are slow acetylators. (See Enzymes of Biotransformation in Chapter 6.) This deficiency occurs in perhaps up to 50% of the European population. Presumably the N-acetyltransferase activity is reduced, thus prolonging the action of drugs and other toxic chemicals, and thereby enhancing their toxicity and human sensitivity. Since adverse reactions to sulfonamides occur in much less than 50% of the population, other factors must also be involved.[220] The acetylation polymorphism has been associated with differences in human drug toxicity between the two acetylation phenotypes. Slow acetylators accumulate high concentrations of isoniazid with prolonged intake of ordinary doses of the drug and are more inclined to peripheral neuropathy,[221] which is seen in some chemically sensitive patients. They are also predisposed to drug-induced lupus erythematosus from hydralazine[222,223] and procainamide.[224]

The incidence of antinuclear antibodies in tuberculosis patients treated with isoniazid is also significantly higher among slow acetylator phenotypes[225,226] (see Biochemical Individuality in Chapter 3). Monro[227] has shown that an extremely high percentage of chemically sensitive patients (up to 80%) are slow acetylators. Recently, sulfonamides have also been shown to convert to metabolites capable of covalently binding to macromolecules and killing human cells in vitro.

The metabolites are generated by cytochrome P-450 mixed-function oxidases and are detoxified in part by conjugation with glutathiones.[228] These hereditary differences are attributed primarily to polymorphism of the acetyl-CoA-dependent N-acetyl-transferase (EC 2.3.1.5). Since arylamine carcinogens, such as aminofluorene and benzidine, are acetylated by the N-acetyltransferase, the possibility of genetic differences in acetylating capacity suggests differences in susceptibility to the carcinogenicity of arylamines and to the susceptibility to chemical sensitivity.[229] These genetic acetylation differences may also partly explain why some patients develop various symptoms in certain types of chemical sensitivity. However, nutritional depletion and chemical overload *may* also trigger susceptibility in the healthiest individual who has no genetic weakness (see Chapter 3).

Acylation (CO–R)

Acylation is by acyl CO-A with glycine, glutamine, and taurine. Conjugation with amino acids results in the formation of a peptide bond joining a carboxylase group of the acid to the amino nitrogen of the amino acid. Peptide (two amino acids) conjugation is the characteristic metabolic reaction of aromatic carboxylic acids (monocyclic and polycyclic), such as benzoic acid 8-glycine minus H_2O goes to hippuric acid. The larger and more complex cyclic acids tend to be excreted as glycosides, while the simpler acids, e.g., benzoic, arylacetic, and bile acids, are more commonly eliminated in the form of amino acid conjugates.[230] These are water soluble, usually less toxic than the precursor, and are excreted in the urine. Intrahepatic conjugation of bile acids with glycine or taurine is a prerequisite to efficient removal of these potentially toxic amphipathic compounds from the liver. Animals probably evolved these reactions to facilitate the removal of a variety of plant acids which they could not degrade or utilize completely. Taurine is the most common amino acid conjugated with bile.[230] Disturbed acylation by pollutant overload decreases proper bile in the gastrointestinal tract, resulting in inappropriate lipid assimilation and disturbed cholesterol metabolism, as well as insufficient absorption of fat-soluble vitamins A and E seen in some chemically sensitive patients.

The bioavailability of the amino acids used in acylation comes from endogenous and dietary sources. This is apparently why some chemically sensitive patients seen at the EHC-Dallas need specific amino acid supplementation in order to clear their symptoms resulting from reactions to various toxic chemicals. Certain organ systems become amino acid depleted and patients are

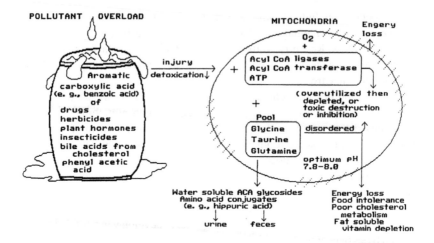

Figure 20.   Pollutant injury to acylation (peptide conjugation) in the chemically sensitive.

intolerant of food, particularly proteins, and therefore continue in a downward spiral of nutritional depletion, developing an inability to use peptide conjugation efficiently (Figure 20).

Peptide bonding is an energy-requiring reaction which is accomplished by activation of the acid to a thioester derivative of coenzyme A at the expense of ATP, which is converted to adenosine monophosphate (AMP) and pyrophosphate (Figure 21). Enzymes catalyzing this reversible reaction are ATP-dependent acid, CoA ligases (AMP) synonymous to acyl-CoA synthetases (ligases), or acid activating enzymes. In a subsequent separate reaction, the coenzyme A thioester then transfers its acyl moiety to the amino group of the acceptor amino acids; the amide derivative of the acid thus becomes synthesized and the coenzyme A carrier regenerated. This second reaction is irreversible and is catalyzed by acyl-CoA (amino acid N-acyl-transferases [amino acid acylases]).

Both the acid-activating and acyltransferase enzymes responsible for conjugating benzoates and acylacetates are found in the matrix of mitochondria (liver, kidney). Pollutant damage to the mitochondria may disturb this process, and overload of xenobiotics will then cause an increased utilization of ATP, which may finally result in weakness and fatigue in the chemically sensitive.

If enough pollutant damage occurs to the mitochondrial membrane, difficulty conjugating toxic chemicals through this pathway might result and chemical sensitivity could then occur or be accentuated. The enzymes responsible for activating benzoates and arylacetates are intermediate chain-length fatty acids (acyl-CoA synthetases), which also activate aromatic carboxylic acid of medium chain-length fatty acids. It is evident that the presence of a toxic chemical, e.g., heptachlor, due to its lipophilic nature, could be attracted to the fatty acid and destroy or limit the peptide conjugation pathway in that area.

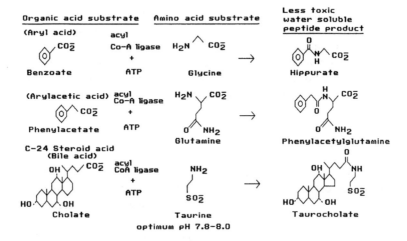

Figure 21. Examples of peptide conjugation of organic acids with amino acids in the chemically sensitive.

The oxidative degradation of cholesterol in mammalian liver results in the formation of certain hydroxylated C_{24} steroid acids (bile salts). Enzymatic conjugation of bile acids proceeds by the sequential activation acylation mechanism, similar to those of the benzoates and arylacetates. Regardless of the final definition of similar or different systems, it is clear that toxic chemical overload can damage cholesterol metabolism. In fact, 25% of our chemically sensitive patients have abnormal cholesterols or triglycerides, which are remedial to reduction of total body load. (For information on transition of liver cells into fat by toxic damage, see the section on fat, Chapter 6.)

Although glycine is generally available, the capacity to synthesize taurine may be limited[231] (see Chapter 6). In humans, low intrahepatic specific activity of cysteine-sulfinic acid decarboxylase is the rate-limiting step in taurine biosynthesis,[232] thus decreasing the pool of taurine available for bile acid conjugation. Pollutant damage to this enzyme directly or by overload resulting in depletion by overuse may accentuate taurine deficiency, exacerbating chemical sensitivity. Taurine and its precursor, cysteine, are available from animal protein in the diet, perhaps explaining the prevalence of taurine conjugates in carnivores relative to herbivores.[233] Human neonates fed breast milk (which is relatively high in taurine content) conjugate almost exclusively with taurine. After weaning, however, they predominantly conjugate with glycine.[233]

The optimum pH for taurine- and glycine-dependant reactions is 7.8 to 8.0. One of the reasons for alkalinization of chemically sensitive patients with sodium bicarbonate, calcium carbonate, and potassium bicarbonate is to allow optimal detoxification enzyme activity through this as well as other pathways.

Pollutant injury to this process may occur by direct toxic injury to the acyl CoA, ATP, acyl CoA synthetase, cysteine-sulfonic acid decarboxylase, or to the glycine, glutamine, taurine. Also, excess total pollutant load may overstress the system, resulting in overutilization of the components of the system with their subsequent depletion due to an inability to keep up with nutrient demand. In addition, damage to transport systems may occur and result in poor absorption or depletion of nutrients or generation of amino acids due to the pollutant damage. Hypochlorite reactions will decrease with a resultant increase in chemical sensitivity (see Methylation and Amino Acids in Chapter 6).

Sulfur Conjugation

Inorganic sulfur compounds are not usually found in humans. They are known enhancers of polyaromatic hydrocarbons as carcinogens and appear to be the same for chemical sensitivity (Table 5). Nonphysiologic organic sulfur compounds are widely distributed in the environment. (See Chapters 9 and 11.) Sulfur pollutants come from crude oil and other fossil fuels, sulfur-containing pesticides, other industrial compounds, food preservatives, and certain foods and drugs. Toxic sulfurs also can originate endogenously from cells, which can cause problems. The toxicity of sulfur in the chemically sensitive patient is well recognized, but the molecular basis for the toxicity and sensitivity is poorly understood. It is clear that the balance of sulfur-containing compounds is important for detoxifying many other toxic chemicals as well as endogenous ones. The balance between function and malfunction, intoxication and detoxication in one or more of the various types of sulfur conjugations depends upon defects, which can be genetic or acquired from pollutant overload. These detoxication conjugation functions are in the sulfation systems using 3'-phosphoadenosine-5'-phosphosulfate (PAPS) and sulfotransferases for detoxifying toxic chemicals with hydroxy groups such as phenols resulting in etheryl (ethereal) sulfates (RSO_3^-), those utilizing GSH with GSH transferases for detoxifying such toxins as anthracene to mercapturic acids (acetylcysteines), those using inorganic sulfates with the enzyme rhodenase to detoxify inorganic cyanides and sulfides to thiocyanates, and those used for the conversions of heavy metals to insoluble sulfides. Each and their variations will be discussed separately.

Sulfonation

The primary function of sulfonation (Figure 22) is to transfer inorganic sulfate to the hydroxyl group present in phenols including naphthalol, estrone, aliphatic alcohols, hydroxylamines, arylamines, and alicyclic hydroxysteroids (dihydro epiandrosterone, bile salts [lithocolic acid]). Sulfonation is done through the use of sulfotransferases. The resulting products are sulfate esters (etherial sulfate). These products are ionized sulfates (RSO_3^-), which are less toxic and more readily excreted than the parent compound. The major sulfate donor in this conjugation reaction is PAPS, which with organelle-attached

Table 5. Sulfur Conjugation Systems

			Sulfotransferase	
ROH Xenobiotics	+	PAPS	$\xrightarrow{\hspace{2cm}}$ O_2	Ethyl sulfates
Xenobiotics or toxic intermediates of cyto P-450	+	Glutathione	Glutathione transferase $\xrightarrow{\hspace{2cm}}$ O_2	Mercapturic acids
Inorganic cyanides or sulfides	+	Inorganic sulfates	Rhodanase $\xrightarrow{\hspace{2cm}}$ O_2	Thiocyanates
Heavy metals	+	Sulfur from	Cysteines $\xrightarrow{\hspace{2cm}}$ O_2	Insoluble sufides

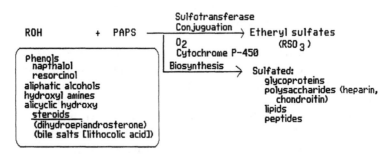

Sulfonation conjugates low doses of phenol because it has a low capacity but a high affinity for it. Gluconation will detoxify the high doses of phenol.

Figure 22.   Pollutant injury to sulfonation results in disorders of conjugation and biosynthesis in the chemically sensitive.

enzymes also performs biosynthetic roles in the formation of sulfated glycoproteins, polysaccharides (heparin and chondroitin), lipids, and peptides. Sulfonation is an important step in steroid synthesis as well as conjugation.[234,235] Pollutant injury may severely disturb the functions of production of these aforementioned substances, causing disturbed metabolism, steroid output, and immune function as seen in some chemically sensitive patients.

PAPS is synthesized in a multistep process from inorganic sulfate and ATP, with the major sources of the sulfate coming from the oxidation of cysteine.[236] The enzymes involved are ATP sulfurase and a derivative of 5-phosphosulfate kinase.[237,238] In reactions catalyzed by the sulfotransferases, the SO_3 group of PAPS is transferred in a reaction involving a nucleophilic attack of the phenolic oxygen or the amine nitrogen on the sulfur atom with the subsequent displacement of adenosine-3′,5′-diphosphate. The overall reactions are enhanced by pyrophosphatases that catalyze the exergonic hydrolysis of inorganic pyrophosphstate. An increased demand on this enzyme would tend to deplete ATP, resulting in loss of energy as seen in the chemically sensitive.

Sulfonation has a high affinity but low capacity for conjugation of phenols. Therefore, low doses of phenol will be detoxified as etherial sulfates while high doses will go through glucuronation pathway.

As the dose of phenol is increased, the percent of the dose and occasionally the absolute amount excreted as a sulfate may decrease with a proportionate increase in the amount excreted as glucuronide. With chemical overload, depletion of the available supply of organic sulfate may occur, indicating depletion or inadequate release of ATP or cysteine (or nutritional depletion) or a depleted sulfur pool or damage to sulfotransferase production or assembly. All of these conditions have been observed in some chemically sensitive patients at the EHC-Dallas. Most chemically sensitive patients have difficulty handling low doses of phenol (e.g., as a preservative in allergy injections), probably as a result of the sulfonation process being disturbed.

Though most of the sulfoconjugates are excreted in the urine, some are broken down enzymatically by aryl sulfates in gut microflora. Some of the organic sulfates are bound to plasma proteins[239,240] or in hydrophobic lipid-soluble complexes. These have a role in the transport of sulfoconjugate through lipid membrane permeability barriers. Some breakdown products are more toxic than the substrate such as the N-O sulfate of the carcinogen N-hydroxy-2-acetylaminofluorene and exacerbate chemical sensitivity.

The sulfate transfer by PAPS is accepted by a variety of substances such as vitamin C, hydroquinone, and ommatin D.[241] With rats, parenterol administration of L-ascorbic acid $2[35_S]$ sulfate was shown to result in fecal excretion of cholesterol $[35_S]$ sulfate.[242] We have treated thousands of chemically sensitive patients with intravenous vitamin C with immediate results of resolution of reactions. Part of the response undoubtedly is through this acceptor mechanism.

Regulation of Conjugation of Sulfate. Because of the previously mentioned facts, regulation of the conjugation of sulfates is extremely important in the chemically sensitive.

Sulfate conjugation is regulated by three factors: the supply of inorganic sulfate, the capacity for synthesis of PAPS, and the control of the level and distribution of the individual sulfotransferases. Each is discussed separately (Table 6).

Table 6. Regulation of Conjugation of Sulfate in the
Chemically Sensitive

Is Due To
The supply of inorganic sulfate
Capacity of synthesis of PAPS
Vitamin A deficiency
Niacin
Cysteine
Hydrocortisone
Estrone, testosterone
Level and distribution of sulfo transferases
High concentration of phenol inhibit

Supply of Inorganic Sulfate. Sulfation of phenol is a function of sulfate ion concentration. Not only is sulfate readily permeable to cells in perfusion or tissue culture systems or by intravenous or intraperitoneal administration; the oral route is also effective for rapid absorption in the gastrointestinal tract.[243] Since the chemically sensitive have poor absorption, they may not absorb sulfate or organic sulfur precursors properly; therefore, this may be one of the reasons for sulfur deficiency. This deficiency of sulfur-containing amino acids may also be due to inadequate intake due to intolerance of sulfur-containing foods such as beef and beans.

Capacity of Synthesis of PAPS. The primary source of inorganic sulfate required for PAPS synthesis appears to be derived from the oxidation of cysteine by cysteine dioxygenase. This increase in synthesis of PAPS may be enhanced by hydrocortisone, nicotinamide, or cysteine.[244] B_3 deficiency seen in 20% of the chemically sensitive may disturb this synthesis.

Vitamin A deficiency seen in 15% of the chemically sensitive reduces the production of PAPS because of the resultant low level of PAPS-sulfurylase. This effect does not appear to involve direct interaction of the vitamin and the enzyme.[245] The need for sulfate may explain why intravenous $MgSO_4$ rather than $MgCl_2$ will boost some patients with chemical sensitivity. This sulfate radical in $MgSO_4$ probably aids in phenol detoxification. Both the aryl (phenol) and hydroxysteroid sulfotransferase actively appear to be under hormonal regulation. Estradiol will decrease the level of activity when combined with N-hydroxy-2-acetylaminofluorene, while increased activity occurs with the administration of testosterone. Although castration and/or adrenalectomy have little effect in the male, a marked decrease in enzyme concentration is noted 3 to 9 weeks after hypophysectomy or thyroidectomy. Studies with the hydroxysteroid sulfotransferases indicate a dependence on both gonadal[246] and adrenal[247] hormones.

Level and Distribution of the Individual Sulfotransferases. Direct regulation of the enzymes by substrate is also possible, since both types of aryl (phenol) sulfotransferases are inhibited at higher phenol concentrations.[248] This phenomenon is another reason why patients who are highly exposed to phenols and contain them in their blood (as seen in the chemically sensitive patients) do not function well. Their sulfur conjugation along with other detoxification enzymes is overloaded; both orderly biosynthetic and conjugation function are hampered. Potassium phenylsulfate, the breakdown product of phenol metabolism, is much less toxic than phenol; 40% of phenol is broken down this way.

Sulfotransferases are necessary for phenols,[249,250] aliphatic and aromatic steroids,[250-253] and certain aryl amines,[253-256] alcohols,[257] tyrosine methyl esters,[258] and hydroxylamine.[259] At least three homogeneous aryl sulfotransferases[260,261] and three homogeneous steroid sulfotransferases[262-264] have to be isolated.

Glutathione Conjugation

Glutathione (L-y-glutamyl cysteinylglycine) conjugation (Figure 23) is used for the transformation of xenobiotics such as aromatic disulfides, naphthalene, anthracene, phenanthracin compounds, aliphatic disulfides, thiamine s-sulf cystamine, thiosulfate esters, as well as regeneration of endogenous thiols (RSH) from disulfides (RSSR'). Reduced GSH has a nucleophilic sulfhydryl group (SH) which is catalyzed either by cytosolic (5 to 40 times more activity) or microsomal GSH thiotransferases. The reduced GSH combines with these xenobiotics, which contain electrophilic carbon atoms, to form less toxic mercapturic acids (N-acetyl cysteine).[265,266]

Mercapturic acids are S-conjugates of N-acetylcysteine; the sulfur substituent may be alkyl, aryl, or heterocyclic group with or without halogen, nitro, amino, or other groups. These include halogenated and nitrogenated compounds such as benzyl chloride, 3,4-dichloronitro benzene, diethylmalente, methyl iodide, and many others. The glutathione S-conjugates are converted to mercapturic acids in stages by three enzyme-catalyzed reactions: hydrolysis, autotranspeptidetion, and tranpeptidation.

Sulfur-sulfur bonds occur in exogenous compounds in addition to those naturally found in cells. The most important classes of these sulfur-containing compounds are disulfides (RSSR'), and thiosulfate esters (Bunte salts [RSSO$_3$]). Both are mild oxidizing agents that modify biological activity and can result in inactivation of cellular function (see Oxidation, this chapter). Because proteins require intact sulfhydryl groups for their functions, the reversible oxidation of such groups may be a physiological control mechanism.

The xenobiotic sulfur compounds of interest in this detoxication reaction include various aliphatic and aromatic disulfides (RSSR') and thiosulfate esters (RSSO$_3$) originating outside the body. In addition to exogenous compounds,

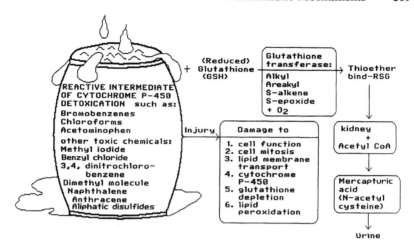

Figure 23. **Pollutant injury to glutathione conjugation resulting in increased chemical sensitivity.**

some may be chemically constructed intracellularly from corresponding thiols (RSH). Some examples of practical interest of the first category are disulfide derivatives of thiamine[265] and the radioprotector S-sulfocysteamine.[268] Examples of the thiosulfate ester category are the mixed disulfides of penicillamine, which can be formed with GSH or cysteine during treatment of cystinosis.[269] Furthermore, the endogenously present thiol(RSH)-bearing compounds may form disulfides (RSSR') from natural pathways. In addition, the thiosulfate esters S-sulfocysteine and S-sulfoglutathione are known to occur naturally.[270] The mechanism of reduction of sulfur-to-sulfur bonds occurs basically through thiol transferases and reductases. There are several enzymes involved with the metabolism of these compounds.

Several amino acids (the neutral L-isomers of cysteine, glutamine, methionine, alanine, and serine) can act as acceptors of the glutamyl moiety.[271,272] This enzyme is inhibited by borate.[273-275] The metabolic turnover of GSH is closely geared to the activity of y-c-glutamyl transpeptidase, one of the thiotransferases which has been used as marker for neoplasm.

Glutathione transferases (GTS) are catalysts for (1) reactions in which GSH participates; (2) binding proteins serving a storage function for toxic compounds such as bilirubin in the liver; and (3) scavengers for alkylating agents.[276] The glutamyl and glycine parts of the conjugates are subsequently lost by hydrolysis, and the foreign compounds end up as their acetylcysteyl derivatives (mercapturic [sulfur] acids).

GTS have a capacity to bind an enormous amount of certain compounds that have the following features: (1) must be hydrophobic; (2) must contain an electrophilic carbon atom; and (3) must react nonenzymatically with GSH at some measurable rate. The aforementioned xenobiotics have this property. The GTS may also serve as ligands for a number of compounds and the lipophilic

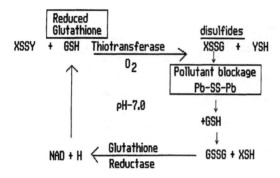

* Heavy metals, e. g., lead, will stop the cycle exacerbating chemical sensitivity.

Figure 24.   Pollutant* injury to glutathione replenishing cycle.

remains serve as the binding sites of chemicals. The GTS enzymes provide a means of reacting to the diverse array of electrophilic xenobiotics with the nucleophilic GSH, thus preventing, to a degree, the reactions of these compounds with essential constituents of the cells. GSH S-transferases act to detoxify the reactive intermediates produced by cytochrome P-450 detoxication of substances such as bromobenzenes, chloroforms, and acetaminophans. Processes involving thiodisulfide and thiosulfate ester reactions take place spontaneously at a pH of 7.0, but the thio transferases speed up this reaction.[277-285] Once reduced, GSH is oxidized. There is a cycle of replenishment (Figure 24). If a heavy metal is caught in the cycle, replenishment will not occur and chemical sensitivity will be exacerbated.

Probably the y-glutamyl moiety serves as a carrier for amino acids. The glutamamylamino acid formed via the action of transpeptidase is metabolized further by y-glutamyl cyclotransferase, a cytosolic enzyme, the products of which are free amino acid and 5-oxoproline (pyroglutamate or pyrrolidone 5-carboxylate). The latter is cleaved by an ATP-dependent enzyme to glutamate which, together with cysteine and glycine (arising from hydrolysis of Cys-Gly), can be utilized intracellularly to regenerate GSH, thereby completing a cyclic series of enzymatic reactions termed the "y-glutamyl cycle." Since a major portion of transpeptidase is located on the cell surface with its active site apparently exposed to the extracellular environment, the proposed role of the enzyme in transport requires either a transmembrane catalysis involving intracellular GSH and incoming amino acid or cell-surface catalysis between extracellular GSH and amino acid followed by translocation of y-glutamylamino acid. The latter possibility is important because intracellular GSH is normally translocated to cell surface supporting the interorgan transport of GSH.[286-288] Pollutant injury to the cell membrane may disturb this orderly balance of transfer across the membrane, resulting in exacerbation of chemical sensitivity.

Other possible physiological functions for transpeptidase could be related to its ability to produce ammonia from glutamine and to its potential for conversion from GSH to GSH disulfide.

There is a delicate balance within cells between the rate of formation of the reactive metabolites and their inactivation by GSH. An excess of these reactive intermediates can deplete GSH. Since GSH is a cofactor for GSH peroxidase, its lack of availability can lead to lipid peroxidation. Over 200 chemically sensitive patients who were undergoing detoxification programs at the EHC-Dallas received intravenous GSH after their blood levels had been found to be deficient. The patients got dramatic response of clearing of their symptoms 70% of the time. Presumably, this conjugation system as well as other functions of GSH were enhanced.

In aerobic cells the thiol (RSH)/disulfide (RSSR) ratio tends to decrease, owing to oxidative events, and the proper balance is secured by reductive process which effect the cleavage of sulfur to sulfur bonds.

This endogenous conjugation involves the mercaptans (RSH) which contain the highly reactive thiol –SH group to which the toxic substance is transferred. The chemistry of the thiol group is somewhat similar to that of the hydroxyl group. However, the higher acidity and the higher polarizability of the thiol make possible many reactions that cannot be realized under physiological conditions by the hydroxyl group. Sulfur is a particularly powerful nucleophilic because the electron shells can be readily distorted to fulfill the requirements of a transition state of the reaction even at relatively large distances between reacting molecules.[289-291]

Because of their reactivity, free thiol groups are not abundant outside living cells. In the presence of O_2, mercaptans (RSH) are oxidized in the cells to disulfides (RSSR), a process that limits the amounts entering from the environment of the organism. Excess environmental mercaptans, even though small in quantity, like odorants in natural gas, have been enough to trigger symptoms in some chemically sensitive individuals. Many mercaptans are present in the gastrointestinal tract produced from metabolism of intestinal bacteria. Some aromatic xenobiotics do have free thiol groups which undergo chemical modifications when entering the body. Small disulfide-containing molecules that can penetrate the inner mitochondrial membrane may be reduced by interaction with protein-bound dihydrolipoly groups as well,[292] but the physiological importance of this pathway is uncertain.

Chemically sensitive patients often have organochlorine pesticides at the parts per billion ranges in their blood. Often there are high fat stores. Clearly an exogenous or endogenous infusion of these substances might detoxify slowly. However, bursts of overexposure or excess mobilization overload this type of breakdown. Exogenous administration of GSH intravenously or orally helps blunt reactions and speed up clearing.

In summary, the mercapturic acid pathway appears to serve as a detoxication mechanism for a large number of xenobiotic compounds and may have

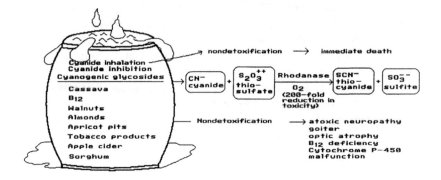

Figure 25. Pollutant injury of cyanide detoxification.

evolved as a protective mechanism against such compounds in the same manner as the immunological defense mechanism of the animal. The transpeptidase enzyme may perform other physiological functions including that proposed for transport of amino acids and peptides.[293-296] Other roles for GSH, most involving the thiol group, have been reviewed elsewhere.[297-301]

Reduction of Cyanides and Inorganic Sulfides Using Thiotransferase and Rhodanese

Reduction of cyanides and inorganic sulfides using thiotransferases (rhodanase [3-mercuptopyruvate sulfase transferase y-cystathionase]) to thio-cyanates (Figure 25) is another factor in chemical sensitivity. Rhodanese (thiosulfate cyanide sulfurtransferase — $SSO_3{}^{2-}$–CN–S–transferase) is consid-ered the main mitochondrial enzyme to catalyze cyanide and inorganic sulfide. This function and thiosulfate reduction can produce sulfane sulfur as well as detoxify cyanide. Cyanide is one of the major problems in human health on a world scale. This problem has to do with the prevalence of cyanogenic glyco-sides (in plant materials) commonly used for food (see Chapter 8). Sulfur catalyzed by rhodanese is responsible for the conversion of inorganic cyanides (CN^-) to thiocyanates (SCN^-), which are partially secreted through saliva and result in a 200-fold reduction in toxicity. Rhodanese will detoxify the cyanides of plants including those present in cassava, vitamin B_{12}, walnuts, almonds, apricot pits (amygdalin), and tobacco products. All cassava-dependent diets (Africa) contain substantial amounts of cyanogenic materials, and despite the various processing regimens in common use, many entire populations are afflicted with chronic cyanide poisoning from this source. The result is widespread ataxic neuropathy and, indirectly, goiter (see Thyroid in Chapter 38) and a variety of other ills. The etiology of these disease states is complex, and a number of nutritional interactions can be implicated.[302] Nevertheless, understanding this rhodanase reaction is important in understanding one aspect of chemical sensitivity.

Even in areas where the population is not strongly dependent on food stuffs heavily laced with cyanogenic glycosides, the importance of cyanide detoxification mechanisms is evident by the occasional triggering of the hereditary disease, Leber's optic atrophy. These patients' responses, including extraordinary sensitivity to the relative traces of cyanide found in tobacco smoke, apple cider, sorghum, etc., are similar to those observed in the chemically sensitive. Such patients display both a diffuse neuropathology similar to that seen in cassava-consuming populations. Sudden, permanent blindness on first exposure to cyanide in small amounts may occur when they begin to smoke cigarettes.[303] This defect is another example of a genetic time-bomb awaiting an environmental trigger. To a lesser degree, many chemically sensitive patients cannot tolerate sulfur or cyanide compounds. Probably the fault lies in this mechanism with the deficiency being either a hereditary weakness or an acquired one.

The reduction of inorganic cyanides and sulfides involves three thiotransferase enzymes. These include (1) thiosulfide cyanide sulfurtransferase (rhodanese); (2) mercaptan sulfur-transferase; and (3) y-cystathionase. The thiosulfate and mercaptothiosulfurtransferase use cyanide as a sulfur acceptor and thereby detoxify the cyanide, and these enzymes, along with thiosulfate reductase and cystasthionase, can produce persulfides that spontaneously produce inorganic sulfide.[304]

Inorganic sulfide (RSH) is practically as toxic to mammals as is cyanide. Sulfide ($-S$, e.g., H_2S, CS_2) is abundant in both mineral and biological form (see Chapter 11). In mineral form as metallic sulfides, this material appears to pose little threat. There is an annual toll of inorganic sulfide-related illness and death from industrial contact with hydrogen sulfide gas.[305] These are usually from refineries and sour gas fields (see Chapter 11). The current rate of release of H_2S into the atmosphere by microbiological processes is of the order of 10^8 tons of sulfur annually on a worldwide basis.[306] Mammalian tissues contain enzymes that produce persulfides ($R-S-S^-$) that can release H_2S spontaneously. Somehow, we avoid poisoning by endogenous inorganic sulfide, probably due to rhodanese and other enzymes. Humans who are overcome by accidental heavy exposure to H_2S usually recover. However, they may develop chemical sensitivity as a result of the exposure, probably due to the damage of these enzymes (see Chapter 11).

Nutritional interrelationships of cyanide and inorganic sulfide are now becoming established. Their damage is exacerbated in the presence of nutritional deficiencies as seen in most chemically sensitive patients. The nonactive form of B_{12} is cyanocobalamin. Cyanide thus has the potential for creating an effective B_{12} deficiency because of this chemical union.[307] A less obvious relationship has to do with B_2 (riboflavin) and cyanide. West African diets may be riboflavin deficient, contributing to the complexity of the characteristics of chronic cyanide poisoning. In addition, the diet of persons subsisting principally on cassava is typically a protein-deficient diet. This deficiency is important since the ultimate source of most of the sulfur for cyanide detoxication is ingested cysteine and methionine from protein. A deficient sulfane pool might not only fail to detoxify all of the cyanide but will fail in its other normal physiological roles as well. **We have found that almost 35% of chemically sensitive patients are deficient in intracellular sulfur** (see Chapter 6). This lack of sulfur may reflect a deficiency in the sulfane pool and partially explain the inability of many chemically sensitive to detoxify some sulfur-containing and other toxic chemicals.

An additional set of cyanide-related problems besetting cassava-dependent populations has to do with thyroid function; goiter is common (see Chapter 38). As the sulfane pool components are converted to thiocyanate, which is not excreted well by the kidneys, accumulation occurs. Although thiocyanate passes the glomerulus well, it is reabsorbed well in the tubule, very much similar to chloride. Unfortunately, the pseudohalogen character of thiocyanate also extends to its behavior in thyroid tissue, where it competes with iodine, resulting in thyroid insufficiency. Moreover, although thyroid peroxidase,[308] like other peroxidases,[307-310] is capable of oxidizing thiocyanate, the products are sulfate and cyanide, thus effectively undoing the detoxication that had been achieved at the expense of the sulfane pool (see Thyroid in Chapter 38) (Figure 26). Hypothyroidism is often seen in chemical sensitivity.

Detoxication for cyanide and inorganic sulfide causes a reversible inhibition of cytochrome oxidase (Figure 27). This inhibition is based on the common trait of high affinity of the substrates for the metal ions in this terminal electron carrier of the cytochrome oxidase. The catalyst for this process is rhodanese[312] (thiosulfate:cyanide sulfurtransferase). Rhodanese prevents endogenous release of free sulfide (hydrosulfide [HS⁻]). Rhodanese is found mainly in the liver and may not be involved in the CNS reactions of cyanide. Liver biopsy of patients with hereditary optic neuropathy shows normal rhodanese, but an inappropriate pool of the right kind of sulfane sulfur to effect conversion.[313] Again, this temporary inhibition of the cytochrome oxidase systems may occur in the chemically sensitive allowing more dysfunction.

Treatment of cyanide excess involves the administration of nitrates and aminophenols, which allow some of the cyanide or sulfide to be stored temporarily as methemoglobin cyanide or sulfide. This treatment is to be discouraged because it is the generator of the methemoglobin. In the case of cyanide poisoning, injection of thiosulfate and thiosulfonate as well as cobalt salts also aids recovery. No comparable stratagems appear to have been developed in sulfide poisoning. Hume has shown that α-ketoglutaric plus thiosulfate appears to be up to 80 times more detoxifying than amylnitrate.[314]

Two other enzymes can cleave carbon-sulfur bonds and produce new sulfane sulfur. These are 3-mercaptopyruvate sulfurtransferase and y-cystathionase. Each will be discussed separately.

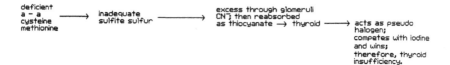

Figure 26.   The pseudohalogen effect of cyanide upon the thyroid.

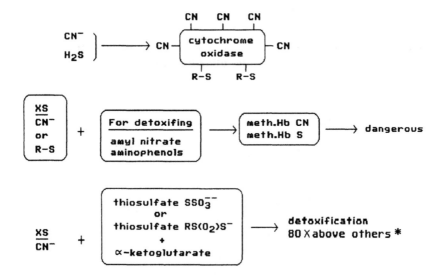

Figure 27.   Treatment of the inhibition of cytochrome oxidase by cyanide or hydrogen sulfite.* Moore, Norris, Ho, and Hume.[314]

3-Mercaptopyruvate Sulfurtransferase (3MPS Thiotransferase). 3-MPS thiotransferase (Figure 28), equal in quantity to rhodanese, is believed to arise metabolically from cysteine by transamination.[315] In addition to detoxication, 3MPS thiotransferase can also produce sulfane sulfur. Its principal pathway of further metabolism involves cleavage to pyruvate with transfer of the sulfur as atom to sulfite-forming thiosulfate or to a thiol forming the corresponding persulfide or to cyanide-forming thiocyanate.

The complicating observation is that less thiocyanate than pyruvate is formed in the enzyme-catalyzed reaction.

3MPS thiotransferase catalyzes the transfer to all of these acceptors, but 3-mercaptopyruvate is its only known donor substrate. Clearly, this enzyme might be as much involved in cyanide and sulfide metabolism as is rhodanese. In addition, it may be involved in the thiolation of pyrimidine bases in RNA. Whether 3-mercaptopyruvate or cysteine directly is the better source of sulfur for this purpose in vivo is still a matter of dispute.[316-321]

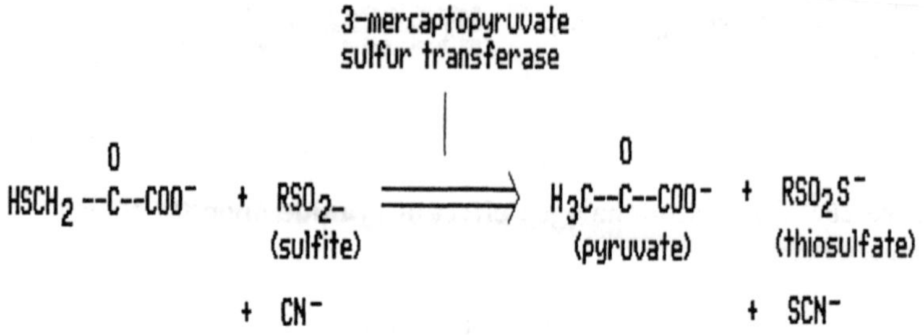

3-mercaptopyruvate
sulfur transferase

$$HSCH_2 \text{--}\overset{O}{\underset{}{C}}\text{--}COO^- + RSO_2{}_- \rightleftharpoons H_3C\text{--}\overset{O}{\underset{}{C}}\text{--}COO^- + RSO_2S^-$$

(sulfite)   (pyruvate)   (thiosulfate)

+ CN⁻ ... + SCN⁻

Figure 28. Detoxication of sulfide and cyanide through the enzyme 3-mercapto-pyruvate sulfur transferase.

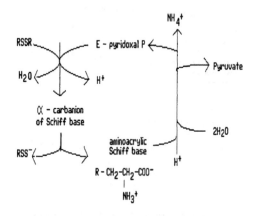

Figure 29. Mechanism of y-cystathionase action on cystine.

y-Cystathionase. One enzymatic activity of animal tissues is known to produce sulfane sulfur. y-Cystathionase cleaves cystine asymmetrically to yield cysteine persulfide, pyruvate, and ammonia[322,323] (Figure 29).

3MPS thiotransferase, the cleavage of a carbon-sulfur bond, gives rise to a sulfane sulfur atom. The difference is in the donor substrate, which is a thiol for the sulfurtransferase and a disulfide for cystathionase.

Conjugation Using Sulfur for the Conversions of Heavy Metal to Sulfides

Here there is conjugation with cysteine sulfur of a heavy metal to sulfide. It is not sure if this occurs in humans. If so, the sulfur pool would be necessary as in all other conjugation reactions.

Sulfur Pool. The completeness of the sulfur pool is essential in the chemically sensitive. In studies at the EHC-Dallas, the sulfur pool appears to be depleted in over 33% of our chemically sensitive patients' cases.

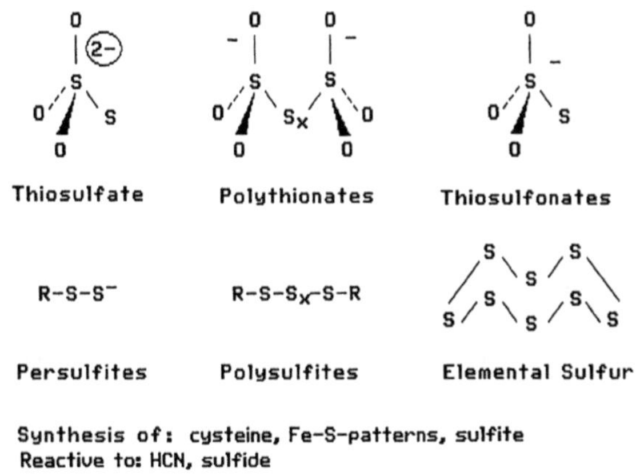

Thiosulfate   Polythionates   Thiosulfonates

R–S–S⁻   R–S–S$_x$–S–R

Persulfites   Polysulfites   Elemental Sulfur

Synthesis of: cysteine, Fe–S–patterns, sulfite
Reactive to: HCN, sulfide

Figure 30.   Natural sulfane sulfur atoms.

Thiosulfonate   Thiosulfate

$$RS(O_2)S^- + SO_3^{2-} \; \rule[0.5ex]{1.5em}{0.4pt}\rule[0.5ex]{1.5em}{0.4pt} \; RSO_2^- + SSO_3^{2-}$$

Sulfite

Figure 31.   Detoxication of sulfite.

The sulfanes are inorganic compounds consisting of chains of divalent atoms or in compounds, e.g., outer sulfur atom of thiosulfate ion S SO_3^{2-}, the internal chain of organic polysulfides (R–S–Sx–S–R), and the terminal atoms of persulfides (RSS⁻) (Figure 30). In normal individuals, there appears to be a significant pool of sulfur at the oxidation level that will react with cyanide or other toxic substances. In the chemically sensitive, often inadequate amounts are present. This pool of sulfane sulfur is very reactive with cyanide.[324] Sulfane sulfur is used for the synthesis of cysteine and iron-sulfur proteins, which are often deficient in the chemically sensitive and very important in detoxifying the hypochlorite ion.

Injection of 35S-labeled sulfane in the sulfane form has been shown to give rise to labeled thiosulfate,[325] polythionates,[326] and protein-associated elemental sulfur.[327] The persulfides (RSS⁻) will also react with thiols (R′SH) to yield disulfides (R–S–S–R′) and inorganic sulfide: RSS⁻ + R′SH=RSSR′ + HS⁻.

Reaction with Sulfite (SO_3^{2-}). By use of a sulfane-doner substrate alternative to thiosulfate $RSSO_3^{2-}$, e.g., a thiosulfonate of the form $RS(O_2)S^-$, it is easy to show that sulfite can serve as a sulfur-acceptor substrate alternative to cyanide[326] (Figure 31).

When persulfides (RSS) are used as donor substrates, the use of sulfite (SO_3^{2-}) as acceptor provides an enzyme-catalyzed alternative to the generation of sulfide (SH) from persulfides since persulfides (R–S–S⁻) can be produced

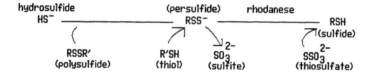

Figure 32.   Detoxication of endogenous sulfide.

by several enzymes with the resultant thiosulfate being nontoxic[329] (Figure 32). This reaction has been proposed as the basis for what is in effect the detoxication of endogenous sulfide.[330] With dihydrolipoate as acceptor substrate, rhodanese itself also produces sulfide (–S) by way of the particularly labile lipoate persulfide.[331,332]

As a detoxication process, sulfation is considered to be effective by decreasing the activity of pharmacologically active compounds as well as by giving rise to the formation of highly ionized organic sulfates (e.g., phenyl sulfate

$$SO_4,\ \overset{\overset{\displaystyle O}{\parallel}}{\text{sulfane R–S–R}})$$

that are more soluble in aqueous solution than the parent compound and thereby facilitate excretion. Once a reaction is turned on, it is difficult to turn off in some chemically sensitive patients. Failure to detoxify these pharmacologically active amplifiers may be one of the reasons for toxic exposure exacerbation of chemical sensitivity. Some of the organic sulfates are avidly bound to plasma proteins[333,334] and some can interact with large hydrophobic molecules to form lipid-soluble complexes.[335] Either type of binding may reflect a functional role in the transport of sulfoconjugates[336] through membrane permeability barriers and into the circulation. The sulfate ion is also a common conjugating agent forming sulfate esters[337] (Figure 33).

Pesticide Metabolism and Sulfane Sulfurs. A number of studies during the middle and late 1970s showed that the sulfur of carbon disulfide and of various thiocarbonyl and phosphorothionate pesticides is retained in sulfane form when these compounds are oxidized metabolically.[338-344] The cellular system involved is the cytochrome P-450 system of the endoplasmic reticulum (Figure 34).

The oxidative desulfuration of these compounds results in protein-bound sulfur that is thought to be present as persulfide groups of the protein cysteinyl residues. One of the proteins so affected is the apoprotein of cytochrome P-450 itself, with consequent decrease in the activity of this system.[345] This explains another way that the chemically sensitive develop spreading phenomenon which eventually causes a near universal reactor. It is not known whether this sulfane sulfur equilibrates with that generated by the 3MPS thiotransferase and, possibly, the y-cystathionase reactions. Neither is it known how directly any of the sulfane-handling enzymes aid in the recovery of cytochrome P-450 activity in vivo.

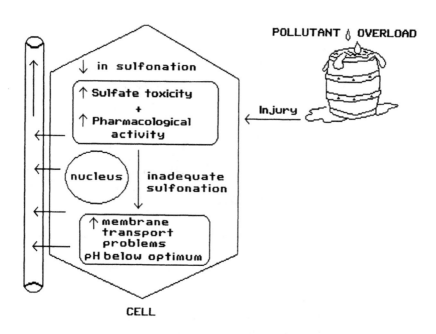

Figure 33.   Pollutants effects on sulfane sulfur sulfation.

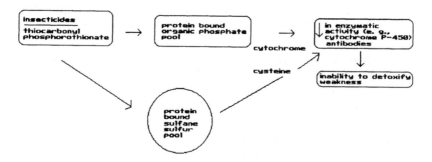

Figure 34.   The effects of sulfur pollutants on sulfane sulfur pool by carbon disulfide and thiocarbamyl pesticides during detoxication.

Alkylations by Methionine and Ethionine

The process most often disturbed in the chemically sensitive involves biological N- and O-methylation reactions catalyzed by S-adenosyl-L-methionine-dependent methyltransferases[346] (Figure 35). It also plays a role in modulating macromolecules (proteins, nucleic acids). Methionine is the chief methyl donor by group specific methylases to detoxify amines,

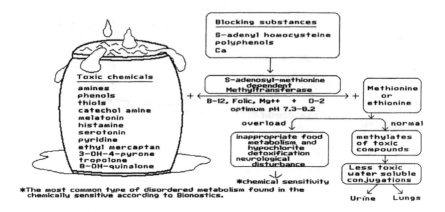

Figure 35.   Pollutant effects on methylation in the chemically sensitive.

phenols, thiols, noradrenaline, adrenaline, dopamine, melatonin, L-dopa, a-methyldopa, isoproterenol, histamine, serotonin, pyridine, pyrogallol, ethylmercaptin, sulfites, selenites, hypochlorites, and tellurites into colifile dimethyl compounds excreted through the lungs. The amino acid ethionine acts analogously as ethyl donor in transethylation reactions. Cyanogen bromide clears polypeptides at the methionine peptide bond.[347] Methionine is needed to detoxify the hypochlorite reaction, which is produced by respiratory bursts during phagocytes and xenobiotic overload. If methylation is impaired by pollutant injury, sensitivity to many other chemicals increases and chemical sensitivity is exacerbated.

Methyltransferase enzymes are mainly found in the liver, but also in the brain, placenta, breast, red blood cells, and lungs.[348-355] Activity of O-methyltransferase requires Mg^{2+}. Many chemically sensitive patients are observed to be deficient in magnesium. This is one of the reasons that magnesium supplementation helps in stabilizing some patients (see Mg in Chapter 6). Many have disordered methionine metabolism causing inadequate detoxification. Optimum pH is between 7.3 and 8.2. Methyltransferase is inhibited by Ca^{2+}, S-adenosylhomocysteine (AdoHcy), and poly-phenolic compounds such as tropolone, 8-hydroxyquinoline, and 3-hydroxy-4-pyrone.[356-358] This system may then become imbalanced by these toxic chemicals, resulting in inappropriate food metabolism and chemical sensitivity. Catechol without vitamin C can act as a substrate.[359] Catechol O-methyltransferase activity may be genetically less in quantity, or it may be increased by cold, benzpyrene, pregnancy, and testosterone. Changes also occur in depression, Parkinson's disease, hypertension, neuroblastoma, pheochromocytoma, duodenal ulcers, and vagotomy increases.[360-371] According-ing to Bionostics,[372] *impaired methionation is the most common disordered metabolism in the chemically sensitive.*

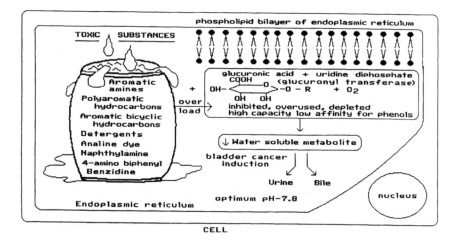

Figure 36. Pollutant injury to glucuronic conjugation.

Gluconation (Glucuronidation) with Glucuronic Acid

The last essential conjugation process that may be impaired in chemical sensitivity is gluconation (Figure 36). Gluconation is one of the major conjugation reactions involved in the metabolic conversion of xenobiotics and of numerous endogenous compounds to polar water-soluble metabolites. **There is a high capacity but low affinity for phenols in this reaction in contrast to sulfonation, which has a high affinity but low capacity.** These products are removed via the bile and urine using a two-stage process[373] involving the enzyme(s) uridine diphosphate and UDP-glucuronosyltransferase, located in the phospholipid bilayer of the microsomal fractions of the endoplasmic reticulum.[374,375] They are the key catalyst(s) in glucuronide formation working at an optimum pH of 7.8. Glucuronidation appears to be the only membrane-associated conjugation reaction except for microsomal S-glutathione transferase activity.[376] Pollutant injury will then disturb this process at the membrane level in chemical sensitivity. Bladder cancer induction by aromatic amines (analine dye) and arylamines (such as 1- and 2-naphthylamine, 4-aminobiphenyl, benzidine, and polycyclic aromatic hydrocarbon [benzopyrene]) may occur through this pathway.

Other carcinogeneses occur through conjugation also. In the liver, arylamines (phenol) are metabolized by the mixed-function oxidase systems to produce ring-hydroxylated intermediates[377] as well as N-hydroxy derivatives.[378-381] Benzo[a]pyrene is metabolized by the cytochrome P-450 system and epoxide hydrase to a variety of epoxides, phenols, quinones, dihydrodiols, and dihydrodiol oxides.[382-384] Metabolic activation is required for the expression of the mutagenic or carcinogenic effects of benzo[a]pyrene. This transferase catalyzes the translocation of glucuronic acid from UDP-a-D-glucuronic acid to an appropriate acceptor to form the B-D-glucuronide[385] and then to bilirubin.[386] Many benzene-containing compounds are degraded through the same pathway using the enzyme aryl hydrocarbon hydrolase to convert to benzene oxide then to phenol. After this, fulfunation, gluconation, and methylation will occur to render the substance harmless.[387] Detergents may induce the microsomes.[388] Liver, kidney, gastrointestinal tract, and the skin, in that order, are involved in glucuronic action.[389] Four general categories of glucuronides are the (1) O-, (2) N-, (3) S-, and (4) C-glucuronides.

Different pathways may occur for man and animals,[390] so animal data should be viewed with caution. Glucuronic acid has substantial advantages over glucose as a detoxifying agent because glucuronides are ionizable while glucosides are not. This difference is sometimes cited as the reason higher animals have better resistance to environmental chemicals than lower ones. However, after years of evaluating the chemically sensitive, we conclude that there is little, if any, more resistance in higher animals or humans. This idea of higher resistance appears to be just a matter of varied expression in the higher animals. It is much easier to rationalize out and negate the effects in humans. This rationalization is because of the adaptive mechanisms which allow modern medicine the luxury of using the term, "psychological origin" or "psychosomatic" (implying cause) rather than meticulously defining the cause. This misrepresented human resistance may be the basis of the reported selective action of pesticides on insects but not humans. Alcohols, phenols, enols, carboxylic acid, amines, hydroxyamines, carbamides, sulfonamides, and thiols conjugate with glucuronic acid.[391] Ignorance of these facts may allow the exacerbation of chemical sensitivity by overexposure or lack or proper nutrients.

POLLUTANT INJURY CREATING FREE RADICALS

Understanding pollutant generated free radicals is essential to understanding chemical sensitivity since they seem to be so harmful in that they alter metabolism and membranes as well as other tissues (Figure 37).

Pollutant injury may create free radicals such as the superoxide anion ($O_2^-\cdot$), singlet oxygen ($^1O_2$), hydrogen peroxide (H_2O_2), and the hydroxy radicals (OH^-) as well as oxygen centered radicals of polyunsaturated fatty acids of the general structures of alkoxy (RO^-) or peroxy (ROO^-) radicals and perhaps many others. These will then exacerbate chemical sensitivity.

The superoxide anion ($O_2^-\cdot$) is the product of an electron reduction of molecular oxygen. It is produced by autooxidation of a multitude of reduced biomolecules in enzymatic reactions, during electron flow in the respiratory chain. These can be extremely damaging to cell membranes and membranes of

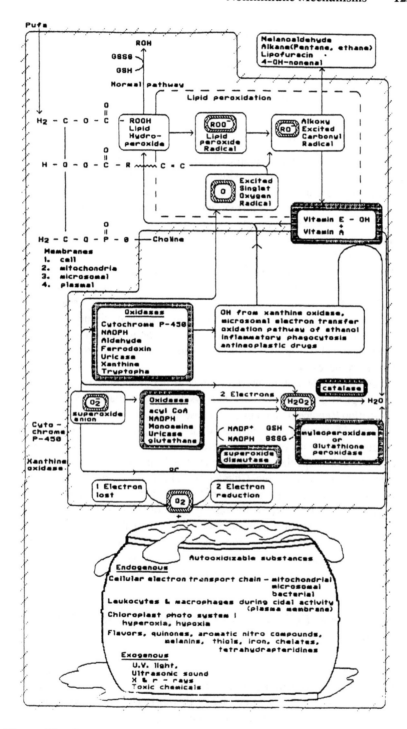

Figure 37.   Pollutant injury creating free radicals.

almost any organelle or other substances contained by membranes, thus exacerbating chemical sensitivity.

Defense mechanisms against these free radicals involve certain enzymes such as superoxide dismutases, catalase, GSH peroxidase, myleoperoxidase, vitamins A and E, but do not replace pollutant reduction. Administration of these will often ameliorate some aspects of chemical sensitivity. Superoxide dismutase (see Enzymes in Chapter 6) provides the primary defense against oxygen toxicity.[392-394] It will also protect against radiation damage to DNA,[395] viruses and mammalian cells in culture,[396] and bacteria.[397-399] GSH peroxidase protects against the oxidative breakdown of hemoglobulin. It also prevents the GSH-induced high-amplitude swelling of mitochondria that occurs with pollutant injury (see Treatment of Enzymes in Chapter 6). Vitamin E will hook onto the free radicals, thus neutralizing toxicity (see Vitamin E in Chapter 6). Vitamin A quenches oxidant species such as singlet oxygen nonchemically.

Factors influencing defense mechanisms against the production of oxygen free radicals include diet, level of oxygenation, integrity of membrane, amount of inflammation, status of the microcirculation, as well as compartmental aspects on the intraorgan and interorgan levels. Adequate tissue, vitamin, and mineral levels are necessary to allow detoxification to function normally and prevent chemical sensitivity. Also, adequate tissue levels of the antipollutant enzymes are necessary.

To counteract the pollutant effects on cells in chemical sensitivity, antioxidant biological detoxification defense systems are available. (See Chapter 6 and Nutrition Replacement chapters.) These are sophisticated systems utilizing nonenzymatic, nutrient-derived antioxidants acting in concert with nutrient-modulated antioxidant enzymes to reduce and thereby detoxify oxygen-derived free radical species that are generated from pollutants as well as the pollutants themselves[400] (see Chapter 6). The antioxidant systems overlap extensively with the immune system[400] in their nutrient requirements and present a complete battery of interacting physiological and biochemical defenses against attack from environmental agents. Once the cycle of oxidant and antioxidant weakening is set in motion by free radical release or direct toxic damage, an individual may eventually progress to an extreme degenerative state with fixed vascular or other chronic degenerative diseases.[401]

FACTORS AFFECTING RATES OF BIOTRANSFORMATION OF FOREIGN RADICALS

Several factors affect the rates and quality of foreign chemical biotransformation in chemical sensitivity. These are intrinsic factors related to the chemical and the biochemical individuality of the host. They are important in the creation and maintenance of chemical sensitivity.

Chemical Intrinsic Factors

Intrinsic factors related to the chemical and the concentration of the foreign compound at the active center of enzymes are involved in the biotransformation. If an individual is unfortunate enough to take in an extremely toxic chemical or one that is held in the body for a long time, like DDT or chlordane, he may develop chemical overload, thus creating the potential for chemical sensitivity. Lipophilic nature is another factor because this property results in readily better absorption than outer solubility. An increase in the lipophilic nature allows the substance to cross membranes faster to where enzymes are and to hook onto the fat part of the membranes. Compounds containing amine, carboxyl, phosphate, sulfate, and phenolic hydroxyl that are ionized at physiologic pH are generally more water soluble and are less readily transferred across cell membranes than compounds not containing these groups. Binding of the xenobiotics by proteins especially reduces the availability for detoxication, thus slowing clearance. Dose and concentration are important since some enzyme systems have a high affinity and low capacity for xenobiotics (e.g., sulfonation of phenol). Other systems have low affinity but high capacity (e.g., gluconation of phenol). Potentially one could decrease the percentage detoxified this way, exacerbating the chemical sensitivity.

Biochemical Individuality (Host Variables)

One host variable in chemical sensitivity appears to be enzyme induction. This is an event that requires *de novo* protein synthesis, which has been shown to be disturbed in the chemically sensitive. If the systems are already induced, chemicals will be more readily detoxified to a point, but constant overstimulation of induction by pollutant overload will result in a depletion of the system with exacerbation of chemical sensitivity. Inducing agents may make morphological changes, particularly in the hepatocytes, which enhance the biochemistry with resultant increase biotransformation. The patterns of induction are different for different chemicals, e.g., methyl anthrene vs. benz[a]pyrene, which may also influence clearing. Genes must be appropriately functioning for induction to occur with a proper binding protein to trigger a response. However, in the chemically sensitive, these genes may be damaged or deficient. Time of induction may be rapid or slow, thus influencing detoxication. Often slow induction as evidenced by the slow acetylators described by Monro and Smith[402] occurs in some chemically sensitive patients. Some inducing agents do not induce enzymes as well as in the liver, and some not at all in peripheral tissue, so peripheral detoxication may not be as good. Location of an inducer is important. For example, xenobiotics will induce an endoplasmic reticulum but may not in cytosol; therefore, more damage might occur in the cytosol. Insufficient cofactors and nutrients often seen in the chemically sensitive will alter detoxication as will competition for active sights. Decreased biosynthesis and increased breakdown of enzymes may occur, retarding detoxication. This decrease in biosynthesis can be generalized for proteins or specific for one enzyme system, e.g., cytochrome P-450. Pool depletion of antioxidants (seen in the chemically sensitive) may occur, thereby limiting detoxication. Pool contamination (seen in the chemically sensitive) can occur from sulfur- and phosphate-containing pesticides, thus damaging the vigor of the cytochrome P-450 action. Generally, the female is initially more effected. In humans, sex-dependent differences occur in the biotransformation of nicotine, acetyl salicylic, and heparin. The ratio of females to males with initial recognized chemical sensitivity is 7 to 1. However, end-stage mortality is 50:50, with the male living a shorter life.

In summary, there are numerous routes of foreign compound metabolism in man which can be damaged by pollutant overload and injury resulting in chemical sensitivity. Of the 13 major toxic compound classes, 6 are finally conjugated into acetates, hippurate, sulfates, mercapturate, methylates, and gluconates. All 13 are usually initially or eventually oxidized, reduced, and/or degraded. These include the acids, aldehydes, alcohols, ketones, esters, ethers, amines, nitro compounds, thio compounds, aromatic hydrocarbons, phenols, and alkyl and aryl halides. Environmental overload with increased total body load may disturb or even destroy the orderly catabolism of toxic pollutants, resulting in metabolic dysfunction and eventual disease production, especially chemical sensitivity.

REFERENCES

1. Reeves, A. L. "The metabolism of foreign compounds," in *Toxicology: Principles and Practice*, Vol. 1, A. L. Reeves, Ed. (New York: John Wiley & Sons, 1981), pp. 1–28.
2. Rakhimova, M. T. "Nonspecific immunological reactivity characteristics and sickness rate among workers at a titanium magnesium combined plant," *Gig. Tr. Prof. Zabol.* 21(10):29–32 (1977).
3. Levine, S. A., and Kidd, P. M. *Antioxidant Adaptation: Its Role in Free Radical Pathology*, (San Leandro, CA: Biocurrents Division, Allergy Research Group, 1986), p. 84.
4. Zeek, P. M. "Periarteritis nodosa and other forms of necrotizing angiitis," *N. Engl. J. Med.* 248:764–769 (1953).
5. Nour-Eldon, R. "Uptake of phenol by vascular and brain tissue," *Microvasc. Res.* 2:2245 (1970).
6. Yevick, P. "Oil Pollutants in Marine Life," Eighth Advanced Seminar, Society of Clinical Ecology, Instatape, Tape II (1975).
7. Street, J. C., and R. P. Sharma. "Quantitative aspects of immunosuppression by selected pesticides," *Toxicol. Appl. Pharmacol.* 29(1):135–136 (1974).

8. Stanton, S. L. "Scoring Methodology for Estimating Agent Orange Exposure Status U.S. Army Personnel in the Republic of Vietnam," in *U.S. Department of Health and Human Services Public Health Service: Comparison of serum levels 2,3,7,8-Tetrachlorodibenzo-p-Dioxin with Indirect Estimates of Agent Orange Exposure among Vietnam Veterans,* 33. Atlanta, GA:30333 (1989).

9. Rea, W. J., and O. D. Brown. "Mechanisms of environmental vascular triggering," *Clin. Ecol.* 3(3):122–128 (1985).

10. Levine, S. A., and P. M. Kidd. *Antioxidant Adaptation: Its Role in Free Radical Pathology,* (San Leandro, CA: Biocurrents Division, Allergy Research Group, 1986), pp. 291–293.

11. Levine, S. A., and P. M. Kidd. *Antioxidant Adaptation: Its Role in Free Radical Pathology,* (San Leandro, CA: Biocurrents Division, Allergy Research Group, 1986), pp. 47–49.

12. Jakoby, W. B. "Detoxification enzymes," in *Enzymatic Basis of Detoxification,* Vol. 1, W. B. Jakoby, Ed. (New York: Academic Press, 1983), pp. 1–6.

13. Sies, H., and E. Cadenas. "Biological basis of detoxication of oxygen free radicals," in *Biological Basis of Detoxication,* J. Caldwell and W. B. Jakoby, Eds. (New York: Academic Press 1983), 181–211.

14. Calabrese, E. J. *Pollutants and High Groups: The Biological Basis of Increased Human Susceptibility to Environmental and Occupational Pollutants,* (New York: John Wiley & Sons, 1978).

15. Rea, W. J., and C. W. Suits. "Cardiovascular disease triggered by foods and chemicals," in *Food Allergy: New Perspectives,* J. W. Gerrard, Ed. (Springfield, IL: Charles C. Thomas, 1980), pp. 99–143.

16. Bell, I. R. "A kinin model of mediation for food and chemical sensitivities: Biobehavioral implications," *Ann. Allergy* 35:206–215 (1975).

17. Buisseret, P. D., L. J. F. Youlten, D. I. Heinzelmann, and M. H. Lessof. "Prostaglandin-synthesis inhibitors in prophylaxis of food intolerance," *Lancet* 1:906–908 (1978).

18. Brabec, M. H., and I. A. Bernstein. "Cellular, subcellular, and molecular targets of foreign compounds," in *Toxicology: Principles and Practice,* Vol. I, A. L. Reeves, Ed. (New York: John Wiley & Sons, 1981), pp. 29–47.

19. Brabec, M. H., and I. A. Bernstein. "Cellular, subcellular, and molecular targets of foreign compounds," in *Toxicology: Principles and Practice,* Vol. I, A. L. Reeves, Ed. (New York: John Wiley & Sons, 1981), pp. 29–47.

20. Unger, M., H. Kiaer, M. Blichert-Toft, J. Olsen, and J. Clausen. "Organochlorine compounds in human breast fat from deceased with and without breast cancer and in biopsy material from merely diagnosed patients undergoing breast surgery," *Environ. Res.* 34:24–28 (1984).

21. Reeves, Andrew L. "The metabolism of foreign compounds" in *Toxicology: Principles and Practice*, Vol. 1, Andrew L. Reeves, Ed. (New York: John Wiley & Sons, 1981), p. 24.
22. Alarie, Y., L. Kane, and C. Barrow. "Sensory irritation: The use of an animal model to establish acceptable exposure to airborne chemical irritants," in *Toxicology: Principles and Practices*, Vol. 1., A. L. Reeves, Ed. (New York: John Wiley & Sons, 1981), pp. 86–88.
23. Clayson, D. B. "Occupational cancer," in *Toxicology: Principles and Practice*, Vol. I, A. L. Reeves, Ed. (New York: John Wiley & Sons, 1981), p. 173.
24. Bunn, H. F., G. R. Lee, and M. M. Wintrobe. "Pernicious anemia and other megaloblastic anemias," in *Harrison's Principles of Internal Medicine*, 8th ed., G. W. Thorn, R. D. Adams, E. Braunwald, K. J. Isselbacher, and R. G. Petersdorf, Eds. (New York: McGraw-Hill 1977), p. 1658.
25. Weinstein, L., and M. N. Swartz. "Host responses to infection," in *Pathologic Physiology Mechanisms of Disease*, 7th ed., W. A. Sodeman, Jr. and T. M. Sodeman, Eds. (Philadelphia: W. B. Saunders, 1985), p. 552.
26. Weinstein, L., and M. N. Swartz. "Host responses to infection," in *Pathologic Physiology Mechanisms of Disease*, 7th ed., W. A. Sodeman, Jr. and T. M. Sodeman, Eds. (Philadelphia: W. B. Saunders, 1985), pp. 548–549.
27. Brabec, M. H., and I. A. Bernstein. "Cellular, subcellular, and molecular targets of foreign compounds," in *Toxicology: Principles and Practice*, Vol. I, A. L. Reeves, Ed. (New York: John Wiley & Sons, 1981), p. 40.
28. Freed, D. L., J. C. H. Buckely, Y. Tsivion, H. Sharon, and D. H. Katz. "Nonallergic haemoloysis in grass pollens and housedust mites," *Allergy* 38:477–486 (1983).
29. Ames, B. N. "Six common errors relating to environmental pollution," *Regulat. Toxicol. Pharmacol.* 7:379–383 (1987).
30. Blumenthal, D. S. *Introduction to Environmental Health*, (New York: Springer-Verlag, 1985).
31. Freed, D. L. J. "Dietary lectins and disease," in *Food Allergy and Intolerance*, J. Brostoff and S. J. Challacombe, Ed. (London: Bailliere Tindall, 1987), pp. 375–400.
32. Brabec, M. H., and I. A. Bernstein. "Cellular, subcellular, and molecular targets of foreign compounds," in *Toxicology: Principles and Practice*, Vol. I, A. L. Reeves, Ed. (New York: John Wiley & Sons, 1981), p. 39.
33. Brabec, M. H., and I. A. Bernstein. "Cellular, subcellular, and molecular targets of foreign compounds," in *Toxicology: Principles and Practice*, Vol. I, A. L. Reeves, Ed. (New York: John Wiley & Sons, 1981), p. 39.
34. Brabec, M. H., and I. A. Bernstein. "Cellular, subcellular, and molecular targets of foreign compounds," in *Toxicology: Principles and Practice*, Vol. I, A. L. Reeves, Ed. (New York: John Wiley & Sons, 1981), p. 40.

35. Brabec, M. H., and I. A. Bernstein. "Cellular, subcellular, and molecular targets of foreign compounds," in *Toxicology: Principles and Practice*, Vol. I, A. L. Reeves, Ed. (New York: John Wiley & Sons, 1981), p. 39.

36. Brabec, M. H., and I. A. Bernstein. "Cellular, subcellular, and molecular targets of foreign compounds," in *Toxicology: Principles and Practice*, Vol. I, A. L. Reeves, Ed. (New York: John Wiley & Sons, 1981), pp. 39–40.

37. Brabec, M. H., and I. A. Bernstein. "Cellular, subcellular, and molecular targets of foreign compounds," in *Toxicology: Principles and Practice*, Vol. I, A. L. Reeves, Ed. (New York: John Wiley & Sons, 1981), p. 46.

38. Brabec, M. H., and I. A. Bernstein. "Cellular, subcellular, and molecular targets of foreign compounds," in *Toxicology: Principles and Practice*, Vol. I, A. L. Reeves, Ed. (New York: John Wiley & Sons, 1981), p. 40.

39. Brabec, M. H., and I. A. Bernstein. "Cellular, subcellular, and molecular targets of foreign compounds," in *Toxicology: Principles and Practice*, Vol. I, A. L. Reeves, Ed. (New York: John Wiley & Sons, 1981), p.40.

40. Rasmussen, H. "The cycling of calcium as an intracellular messenger," *Sci. Am.* 261:66–73 (1989).

41. Brabec, M. H., and I. A. Bernstein. "Cellular, subcellular, and molecular targets of foreign compounds," in *Toxicology: Principles and Practice*, Vol. I, A. L. Reeves, Ed. (New York: John Wiley & Sons, 1981), p. 32.

42. Cook, J. W., C. L. Hewett, and I. Hieger. "The isolation of a cancer producing hydrocarbon from coal tar. Parts I, II, and IV," *J. Chem. Soc.* 395 (1933).

43. Brabec, M. H., and I. A. Bernstein. "Cellular, subcellular, and molecular targets of foreign compounds," in *Toxicology: Principles and Practice*, Vol. I, A. L. Reeves, Ed. (New York: John Wiley & Sons, 1981), p. 37.

44. Brabec, M. H., and I. A. Bernstein. "Cellular, subcellular, and molecular targets of foreign compounds," in *Toxicology: Principles and Practice*, Vol. I, A. L. Reeves, Ed. (New York: John Wiley & Sons, 1981), pp. 44–47.

45. Brabec, M. H., and I. A. Bernstein. "Cellular, subcellular, and molecular targets of foreign compounds," in *Toxicology: Principles and Practice*, Vol. I, A. L. Reeves, Ed. (New York: John Wiley & Sons, 1981), p. 34.

46. Brabec, M. H., and I. A. Bernstein. "Cellular, subcellular, and molecular targets of foreign compounds," in *Toxicology: Principles and Practice*, Vol. I, A. L. Reeves, Ed. (New York: John Wiley & Sons, 1981), p. 36.

47. Testa, Bernard. "Nonenzymatic biotransformation" in *Biological Basis of Detoxication*, J. Caldwell and W. B. Jakoby, Eds. (New York: Academic Press, 1983), pp. 137–150.

48. Dolle, B., W. Topner, and H. G. Neumann. "Reaction of aryl nitroso compounds with mercaptans," *Xenobiotica* 10:527–536 (1980).

49. Ignarro, L. J., J. Lippton, J. C. Edwards, W. H. Baricos, A. L. Hymon, P. J. Kadowitz, and C. A. Gruetter. "Mechanism of vascular smooth muscle relaxation by organic nitrates, nitrites, nitroprusside and nitric oxide: Evidence for the involvement of S-nitrosothiols as active intermediates," *J. Pharmacol. Exp. Ther.* 218:739–749 (1981).

50. Brabec, M. H., and I. A. Bernstein. "Cellular, subcellular, and molecular targets of foreign compounds," in *Toxicology: Principles and Practice*, Vol. I, A. L. Reeves, Ed. (New York: John Wiley & Sons, 1981), p. 34.

51. Brabec, M. H., and I. A. Bernstein. "Cellular, subcellular, and molecular targets of foreign compounds," in *Toxicology: Principles and Practice*, Vol. I, A. L. Reeves, Ed. (New York: John Wiley & Sons, 1981), p. 34.

52. Brabec, M. H., and I. A. Bernstein. "Cellular, subcellular, and molecular targets of foreign compounds," in *Toxicology: Principles and Practice*, Vol. I, A. L. Reeves, Ed. (New York: John Wiley & Sons, 1981), p. 34.

53. Brabec, M. H., and I. A. Bernstein. "Cellular, subcellular, and molecular targets of foreign compounds," in *Toxicology: Principles and Practice*, Vol. I, A. L. Reeves, Ed. (New York: John Wiley & Sons, 1981), pp. 41–42.

54. Lu, A. Y. H., D. M. Jerina, and W. Levin. "Resolution of the cytochrome P-450-containing w-hydroxylation system of liver microsomes into three components," *J. Biol. Chem.* 244:3714–3721 (1969).

55. Reeves, A. L. "The metabolism of foreign compounds," in *Toxicology: Principles and Practice*, Vol. 1, A. L. Reeves, Ed. (New York: John Wiley & Sons, 1981), pp. 43 and 95.

56. Coon, M. J., J. L. Vermilion, K. P. Vatsis, J. S. French, W. L. Dean, and D. A. Haugen. "Biochemical studies on drug metabolism: Isolation of multiple forms of liver microsomal cytochrome P-450," in *Drug Metabolism Concepts*, D. M. Jerina, Ed. (Washington, DC: American Chemical Society, 1977), pp. 46–71.

57. Mannering, G. J. "Properties of cytochrome P-450 as affected by environmental factors: Qualitative changes due to administration of polycyclic hydrocarbons," *Metab. Clin. Exp.* 20:228–245 (1971).

58. Levin, W. "Purification of liver microsomal cytochrome P-450: Hopes and promises," in *Microsomes and Drug Oxidations*, V. Ullrich, I. Roots, A. Hildebrandt, R.W. Estabrook, and A.H. Conney, Eds. (Pergamon: Oxford, 1977), pp. 735–747.

59. Lu, A. Y. H. "Liver microsomal drug-metabolizing enzyme system: Functional components and their properties," *Fed. Proc. Fed. Am. Soc. Exp. Biol.* 35:2460–2463 (1976).

60. Johnson, E. F. "Multiple forms of cytochrome P-450: Criteria and significance," *Rev. Biochem. Toxicol.* 1:1–26 (1979).

61. Poland, A., E. Glover, and A. S. Kede. "Stereospecific, high affinity binding of 2,3,7,8-tetrachlorodibenzo-p-dioxin by hepatic cytosol. Evidence that the binding species is the receptor for the induction of aryl hydrocarbon hydroxylase," *J. Biol. Chem.* 251:4936–4946 (1976).

62. Kamataki, T., D. H. Belcher, and R. A. Neal. "Studies of the metabolism of diethyl *p*-nitrophenyl phosphorothionate (parathion) and benzphetamine using an apparently homogeneous preparation of rat liver cytochrome *P*-450 antibody preparation," *Mol. Pharmacol.* 12:921–932 (1976).

63. Mueller, G. C., and J. A. Miller. "The metabolism of 4-dimethylaminoazobenzene by rat liver homogenates," *J. Biol. Chem.* 176:535–544 (1948).

64. Leiberman, Allan. Personal communication (1990).

65. Katz, R. Sex-related differences in drug metabolism, *Drug Metab. Rev.* 3:1–32 (1974).

66. Reeves, A. L. "The metabolism of foreign compounds," in *Toxicology: Principles and Practice*, Vol. 1, A. L. Reeves, Ed. (New York: John Wiley & Sons, 1981), p. 27.

67. Reeves, A. L. "The metabolism of foreign compounds," in *Toxicology: Principles and Practice*, Vol. 1, A. L. Reeves, Ed. (New York: John Wiley & Sons, 1981), p. 24.

68. Reeves, A. L. "The metabolism of foreign compounds," in *Toxicology: Principles and Practice*, Vol. 1, A. L. Reeves, Ed. (New York: John Wiley & Sons, 1981), pp. 10–11.

69. Reeves, A. L. "The metabolism of foreign compounds," in *Toxicology: Principles and Practice*, Vol. 1, A. L. Reeves, Ed. (New York: John Wiley & Sons, 1981), pp. 4–6

70. Teschke, R., Y. Hasumurn, and C. S. Lieber. "Hepatic microsomal alcohol-oxidizing system affinity for methanol, ethanol, propanol, and butanol," *J. Biol. Chem.* 250:7393–7404 (1975).

71. Teschke, R., S. Matsuzaki, K. Ohnishi, L. DeCarl, and C.S. Lieber. "Microsomal ethanol oxidizing system (MEOS): Current status of its characterization and its role," *Alcohol: Clin. Exp. Res.* 1:7–15 (1977).

72. Corbett, T. H., C. Heidelberger, and W. F. Dove. "Determination of the mutagenic activity to bacteriophage T_4 of carcinogenic and noncarcinogenic compounds," *Mol. Pharmacol.* 6: 667–679 (1970).

73. Schoental, R. "Carcinogens in plants and microorganisms," in *Chemical Carcinogens*, C. E. Searle, Ed. (Washington, DC: American Chemical Society, 1976), 173:626–689.

74. Schoental, R., and S. Gibbard. "Nasal and other tumors in rats given 3,4,5-trimethyoxy-cinnamaldehyde, a derivative of sinapaldehyde and of other α, β-unsaturated aldehyde wood lignin constituents," *Br. J. Cancer* 26:504–505 (1972).

75. Levine, S. A., and Kidd, P. M. *Antioxidant Adaptation: Its Role in Free Radical Pathology*, (San Leandro, CA: Biocurrents Division, Allergy Research Group, 1986), pp. 71–73.
76. Schoental, R., and S. Gibbard. "Nasal and other tumors in rats given 3,4,5-trimethyoxy-cinnamaldehyde, a derivative of sinapaldehyde and of other α, β-unsaturated aldehyde wood lignin constituents," *Br. J. Cancer* 26:504–505 (1972).
77. Deitrich, R. A. "Genetic aspects of increase in rat liver aldehyde dehydrogenase induced by phenobarbital," *Science* 173:334–336 (1971).
78. Deitrich, R. A., Collins, A. C., and V. G. Erwin. "Genetic influence upon phenobarbital in rat liver supernatant aldehyde dehydrogenase activity," *J. Biol. Chem.* 247:7232–7236 (1972).
79. Deitrich, R. A., P. A. Troxell, and V. G. Erwin. "Characteristics of the induction of aldehyde dehyrogenase in rat liver," *Arch. Biochem. Biophys.* 166:543–548 (1975).
80. Koivula, T., and M. Koivusalo. "Partial purification and properties of a phenobarbital-induced aldehyde dehydrogenase of rat liver," *Biochim. Biophys. Acta* 410:1–11 (1975).
81. Eriksson, S., P. Askelöf, K. Axelsson, K. Carlberg, C. Guthenberg, and B. Mannervik. "Resolution of glutathione-linked enzymes in rat liver and evaluation of their contribution to disulfide reduction via thiol-disulfide interchange," *Acta Chem. Scand. Ser. B.* 28:922–930 (1974).
82. Peterson, D. R., A. C. Collins, and R. A. Deitrich. "Role of liver cytosolic aldehyde dehydrogenase isozymes in control of blood acetaldehyde concentrations," *J. Pharmacol. Exp. Ther.* 201:471–481 (1977).
83. Torronen, R., U. Nousiainen, and M. Marselos. "Inducible aldehyde dehydrogenases in the hepatic cytosol of the rat," *Acta Pharmacol. Toxicol.* 41:263–272 (1977).
84. Nakanishi, S., E. Shiohara, and M. Tsukada. "Rat liver aldehyde dehydrogenases: Strain differences in the response of the enzymes to phenobarbital treatment," *Jpn. J. Pharmacol.* 14: 1853–1865 (1965).
85. Deitrich, R. A., P. Bludeau, T. Stock, and M. Roper. "Induction of different rat liver supernatant aldehyde dehydrogenases by phenobarbital and tetrachloro-dibenzo-p-dioxin," *J. Biol. Chem.* 252:6169–6176 (1977).
86. Marselos, M., and R. Torronen. "Increase of hepatic and serum aldehyde dehydrogenase activity after TCDD treatment," *Arch. Toxicol. (Suppl.)* 1:271–273 (1978).
87. Lindahl, R., M. Roper, and R. A. Deitrich. "Rat liver aldehyde dehydrogenase immunochemical identity of 2,3,7,8-tetrachlorodibenzo-p-dioxin inducible normal liver and 2-acetylaminofluorene inducible hepatoma isozymes," *Biochem. Pharmacol.* 27:2463–2465 (1978).
88. Strittmater, P., and E. G. Ball. "Formaldehyde dehydrogenase, a glutathione-dependent enzyme system," *J. Biol. Chem.* 213:445–461 (1955).

89. Goodman, J. I., and T. R. Tephly. "A comparison of rat and human liver formaldehyde dehydrogenase," *Biochim. Biophys. Acta.* 252:489–505 (1971).

90. Uotila, L., and M. Koivusalo. "Formaldehyde dehydrogenase from human liver. Purification, properties, and evidence for the formation of glutathione thiol esters by the enzyme," *J. Biol. Chem.* 249:7653–7663 (1974).

91. Heymann, E. "Carboxylesterases and amidases," in *Enzymatic Basis of Detoxication*, Vol. II, W. B. Jakoby, Ed. (New York: Academic Press, 1980), pp. 291–323.

92. Heymann, E., K. Krisch, H. Buech, and W. Buzello. "Inhibition of phenacetin- and acetanilide-induced methemoglobinemia in the rat by the carboxylesterase inhibitor bis(*p*-nitrophenyl)phosphate," *Biochem. Pharmacol.* 18:801–811 (1969).

93. Singleton, S. D., and S. D. Murphy. "Propanil (3,4-dichloropropionanilide)-induced methemoglobin formation in mice in relation to acylamidase activity," *Toxicol. Appl. Pharmacol.* 25:20–29 (1973).

94. Silver, E. H., and S. D. Murphy. "Effect of carboxylesterase inhibitors on the acute hepato-toxicity of esters of allyl alcohol," *Toxicol. Appl. Pharmacol.* 45:377–389 (1978).

95. Calabrese, E. J., and M. J. Dorsey. *Healthy Living in an Unhealthy World* (New York: John Wiley & Sons, 1984).

96. Reeves, A. L. "The metabolism of foreign compounds," in *Toxicology: Principles and Practice*, Vol. 1, A. L. Reeves, Ed. (New York: John Wiley & Sons, 1981), p. 8.

97. Clayson, D. B. "Occupational cancer," in *Toxicology: Principles and Practice*, Vol. I, A. L. Reeves, Ed. (New York: John Wiley & Sons, 1981), pp. 155–189.

98. Clayson, D. B., and R. C. Garner. "Carcinogenic aromatic amines and related compounds," in *Chemical Carcinogens*, C.E. Searle, Ed. (Washington, DC: American Chemical Society, 1976), pp. 366–461.

99. King, C. M., and W. T. Allaben. "The role of arylhdyroxamic acid N-O-acyltransferase in the carcinogenicity of aromatic amines," in *Conjugation Reactions in Drug Biotransformation*, A. Aitio, Ed. (Amsterdam: Elsevier North-Holland, Inc., 1978), pp. 431–441.

100. Reeves, A. L. "The metabolism of foreign compounds" in *Toxicology: Principles and Practice*, Vol. 1, A. L. Reeves, Ed. (New York: John Wiley & Sons, 1981), pp. 8–9.

101. Oesch, F. "Microsomal epoxidide hydrolase," in *Enzymatic Basis of Detoxication*, Vol. II, W. B. Jakoby, Ed. (New York: Academic Press, 1980), pp. 277–290.

102. Jerina, D. M., H. Yagi, and J. W. Daly. "Arene oxides-oxepins," *Heterocycles* 1:267–299 (1973).

103. DuBois, G. C., E. Appella, W. Levin, A. Lu, and D. M. Jerina. "Hepatic microsomal epoxide hydrase involvement of a histidine at the active site suggests a nucleophilic mechanism," *J. Biol. Chem.* 253:2932–2939 (1978).

104. Oesch, F. "Mammalian epoxide hydrases: Inducible enzymes catalyzing the inactivation of carcinogenic and cytotoxic metabolites derived from aromatic and olefinic compounds," *Xenobiotica* 3:305–340 (1973).

105. Matthews, H. B., and J. D. McKinney. "Dieldrin metabolism to cis-dihydroaldrindiol and epimerization of cis- to trans-dihydroaldrindiol by rat liver microsomes," *Drug Metab. Dispos.* 2:333–340 (1974).

106. Forrest, T. J., C. H. Walker, and K. A. Hassall. "The metabolism of a chlorinated epoxide (MME): Diol formation by pig liver microsomes," *Biochem. Pharmacol.* 28:859–865 (1979).

107. Sims, P., P. L. Grover, A. Swaisland, K. Pal, and A. Hewer. "Metabolic activation of benzo[a]pyrene by a diol-epoxide," *Nature London,* 252:326–328 (1974).

108. Jerina, D. M., R. Lehr, M. Schaefer-Ridder, H. Yagi, J. M. Karle, D. R. Thakker, A. W. Wood, A. Y. H. Lu, D. Ryan, S. West, W. Levin, and A. H. Conney. "Bay-region epoxides of dihyrodiols: A concept explaining the mutagenic and carcinogenic activity of benzo[a]pyrene and benzo[a]anthracene," in *Origins of Human Cancer*, H. H. Hiatt, J. D. Watson, and J. A. Winsten, Ed. (Cold Spring Harbor, New York: Cold Spring Harbor Laboratory, 1977), pp. 639–658.

109. Wood, A. W., W. Levin, A. Y. H. Lu, O. Hernandez, D. M. Jerina, and A. H. Conney. "Metabolism of benzo[a]pyrene and benzo[a]pyrene derivatives to mutagenic products by highly purified hepatic microsomal enzymes," *J. Biol. Chem.* 251:4882–4890 (1976).

110. Bentley, P., F. Oesch, and H. R. Glatt. "Dual role of epoxide hydratase in both activation and inactivation," *Arch. Toxicol.* 39:65–75 (1977).

111. Fjellstedt, T. A., R. H. Allen, B. K. Duncan, and W. B. Jakoby. "Enzymatic conjugation of epoxides with glutathione," *J. Biol. Chem.* 248:3702–3707 (1972).

112. Bend, J. R., Z. Ben-Zvi, J. Van Anda, P. M. Dansette, and D. M. Jerina. "Hepatic and extrahepatic glutathione S-transferase activity toward several arene oxides and epoxides in the rat," in *Polynuclear Aromatic Hydrocarbons*, R. Freudentahl and P. W. Jones, Eds. (New York: Raven Press, 1976), pp. 63–75.

113. Jakoby, W. B. "The glutathione S-transferases: A group of multi-functional detoxification proteins," *Adv. Enzymol.* 46:383–414 (1978).

114. Oesch, F., and J. Daly. "Conversion of naphthalene to trans-naphthalene dihydrodiol: Evidence for the presence of a coupled aryl monooxygenase-epoxide hydrase system in hepatic microsomes," *Biochem. Biophys. Res. Commun.* 46:1713–1720 (1972).

115. Glatt, H. R., and F. Oesch. "Inactivation of electrophilic metabolites by glutathione transferases and limitation of the system due to subcellular localization," *Arch. Toxicol.* 39:87–96 (1977).

116. Nebert, D. W., and H. V. Gelboin. "Substrate-inducible microsomal aryl hydroxylase in mammalian cell culture," *J. Biol. Chem.* 243:6242–6249 (1968).

117. Conney, A. H. "Pharmacological implications of microsomal induction," *Pharmacol. Rev.* 19:317–366 (1967).

118. Gielen, J. E., and D. W. Nebert. "Aryl hydrocarbon hydroxylase induction in mammalian liver cell culture. I. Stimulation of enzyme activity in nonhepatic cells and in hepatic cells by phenobarbital, polycyclic hydrocarbons, and 2,2-bis(p-chlorophenyl)-1,1,1-trichloroethane," *J. Biol. Chem.* 246:5189–5198 (1971).

119. Remmer, H. "Induction of drug metabolizing enzyme system in the liver," *Eur. J. Clin. Pharmacol.* 5:116–136 (1972).

120. Lu, A. Y. H., A. Somogyi, S. West, R. Kuntzman, and A. H. Conney. "Pregnenolone-16a-carbonitrile: A new type of inducer of drug-metabolizing enzymes," *Arch. Biochem. Biophys.* 152:457–462 (1972).

121. Oesch, F. "Differential control of rat microsomal 'aryl hydrocarbon' monooxygenase and epoxide hydratase," *J. Biol. Chem.* 251:79–87 (1976).

122. Bucker, M., M. Golan, H. U. Schmassmann, H. R. Glatt, P. Stasiecki, and F. Oesch. "The epoxide hydratase inducer trans-stilbene oxide shifts the metabolic epoxidation of benzo[a]pyrene from the bay to the K-region and reduces its mutagenicity," *Mol. Pharmacol.* 16: 656–666 (1979).

123. Nakatsugawa, T., M. Ishida, and P. A. Dahm. "Microsomal epoxidation of cyclodiene insecticides," *Biochem. Pharmacol.* 14:1853–1865 (1965).

124. Laseter, J. L., I. R. Deleon, W. J. Rea, and J. Butler, Jr. "Chlorinated hydrocarbon pesticides in environmentally sensitive patients," *Clin. Ecol.* 2(1):3–12 (1983).

125. Bentley, P., H. Schmassmann, P. Sims, and F. Oesch. "Epoxides derived from various polycyclic hydrocarbons as substrates of homogeneous and microsome-bound epoxide hydratase," *Eur. J. Biochem.* 69:97–103 (1952).

126. Lu, A. Y. H., D. M. Jerina, and W. Levin. "Liver microsomal epoxide hydrase hydration of alkene and arene oxides by membrane-bound and purified enzymes," *J. Biol. Chem.* 252:3715–3723 (1977).

127. Poulsen, L. L., R. M. Hyslop, and D. M. Ziegler. "S-oxidation of thioureylene catalyzed by a microsomal flavoprotein mixed function oxidase," *Biochem. Pharmacol.* 23:3431–3440 (1974).

128. Norman, B. J., R. E. Poore, and R. A. Neal. "Studies of the binding of sulfur released in the mixed-function oxidase-catalyzed metabolism of diethyl P-nitrophenylphosphorothionate (parathion) to diethyl P-nitrophenyl phosphate (paraoxon)," *Biochem. Pharmacol.* 23:1733–1744 (1974).

136   Chemical Sensitivity: Principles and Mechanisms

129. Kamataki, T., D. H. Belcher, and R. A. Neal. "Studies of the metabolism of diethyl p-Nitrophenyl phosphorothionate (parathion) and benzphetamine using an apparently homogenous preparation of rat liver cytochrome *P*-450: Effect of a cytochrome P-450 antibody preparation," *Mol. Pharmacol.* 12:921–932 (1976).
130. Reeves, A. L. "The metabolism of foreign compounds" in *Toxicology: Principles and Practice*, Vol. 1, Andrew L. Reeves, Ed. (New York: John Wiley & Sons, 1981), pp. 18–20.
131. Reeves, A. L. "The metabolism of foreign compounds," in *Toxicology: Principles and Practice*, Vol. 1, A. L. Reeves, Ed. (New York: John Wiley & Sons, 1981), pp. 9–10.
132. Masters, B. S. S. "The role of NADPH-cytochrome c (P-450) reductase in detoxication," in *Enzymatic Basis of Detoxication*, Vol. 1, W. B. Jakoby, Ed. (Orlando, FL: Academic Press, 1980), pp. 183–200.
133. Boyer, P. D. "Sulfhydryl and disulfide groups of enzymes," in *The Enzymes*, Vol. 1, 2nd ed., P.D. Boyer, H. Lardy, and K. Myrback, Eds. (New York: Academic Press, 1969), pp. 511–580.
134. Barron, E.S.G. "Thiol groups of biological importance," *Adv. Enzymol.* 11:201–266 (1951).
135. Nesbakken, R., and L. Eldjarn. "The inhibition of hexokinase by disulphides," *Biochem. J.* 87:526–532 (1963).
136. Mazia, D. "SH and growth," in *Glutathione*, S. Colowick, A. Lazarow, E. Racker, D.R. Schwarz, E. Stadtman, and H. Waelsch, Eds. (New York: Academic Press, 1954), pp. 209–223.
137. Setlow, B., and P. Setlow. "Levels of acetyl coenzyme A, reduced and oxidized coenzyme A, and coenzyme A in disulfide linkage to protein in dormant and germinated spores and growing and sporulating cells of bacillus megaterium," *J. Bacteriol.* 132:444–452 (1977).
138. Fahey, R.C., S. Brody, and S.D. Mikolajczyk. "Changes in the glutathione thiol disulfide status of neurospora crassa conidia during germination and aging," *J. Bacteriol.* 121:144–151 (1975).
139. Mannervik, Bengt. "Mercaptans," in *Metabolic Basis of Detoxication: Metabolism of Functional Groups*, William B. Jakoby, John R. Bend, and John Caldwell, Eds. (New York: Academic Press, 1982), pp. 185–206.
140. Kosower, E.M., and N.S. Kosower. "Manifestations of changes in the GSH-GSSG status of biological systems," in *Glutathionine*, L. Flohe, H.C., Benohr, H. Sies, H.D. Waller, and A. Wendel, Eds. (Stuttgart: Georg Thieme, 1974), pp. 287–295.
141. Kosower, E.M., and N.S. Kosower. "Chemical basis of the perturbation of glutathione-glutathione disulfide status of biological systems by diazenes," in *Glutathione: Metabolism and Function*, I.M. Arias and W.B. Jakoby, Eds. (New York: Raven Press, 1976), pp. 139–152.
142. Srivastava, S. K., and E. Beutler. "Cataract produced by tyrosinase and tyrosine systems in rabbit lens in-vitro," *Biochem. J.* 112:421–425 (1969).

143. Srivastava, S. K., and E. Beutler. "The transport of oxidized glutathione from human erythrocytes," *J. Biol. Chem.* 244:9–16 (1969).

144. Bartoli, G.M., and M. Sies. "Reduced and oxidized glutathione efflux from liver," *FEBS Lett.* 86:89–91 (1978).

145. Mannervik, Bengt. "Thioltransferases," in *Enzymatic Basis of Detoxication,* Vol. II, William B. Jakoby, Ed. (New York: Academic Press, 1980), pp. 229–244.

146. Van Dyke, R. A., and A. J. Gandolfi. "Characteristics of a microsomal dechlorination system," *Mol. Pharmacol.* 11:809–817 (1975).

147. Van Dyke, R. A., and C. G. Wineman. "Enzymatic dechlorination. Dechlorination of chloroethanes and propanes in vitro," *Biochem. Pharmacol.* 20:463–470 (1971).

148. Ris, M. M., and J. P. von Wartburg. "Heterogeneity of NADPH-dependent aldehyde reductase from human and rat brain," *Eur. J. Biochem.* 37:69–77(1973).

149. Erwin, V.G. 1975. Oxidative-reductive pathways for metabolism of biogenic aldehydes," *Biochem. J.* 152:709–712 (1975).

150. Bronaugh, R.L., and V.G. Erwin. "Partial purification and characterization of NADPH-linked aldlehyde reductase from monkey brain," *J. Neurochem.* 21:809–815 (1973).

151. Tabakoff, B., and V. G. Erwin. "Purification and characterization of a reduced nicotinamide adenine dinucleotide phosphate-linked aldehyde reductase from brain," *J. Biol. Chem.* 245:3263–3268 (1970).

152. Turner, A. J., and K. F. Tipton. "The purification and properties of an NADPH-linked aldehyde reductase from pig brain," *Eur. J. Biochem.* 30:361–368 (1972).

153. Tabakoff, B., Anderson, R., and G. A. Alivisatos. "Enzymatic reduction of biogenic aldehydes in brain," *Mol. Pharmacol* 9:428–437 (1973).

154. Anderson, R.A., Meyerson, L.R., and Tabakoff, B. "Characteristics of enzymes forming 3-methoxy-4-hydroxy-phenylethyleneglycol (MOEPG) in brain," *Neurochem. Res.* 1:525–540 (1976).

155. Reyes, E., and V. G. Erwin. "Distribution and properties of NADPH-linked aldehyde reductases from rat brain synaptosomes," *Neurochem. Res.* 2:87–97 (1977).

156. Huffman, D. H., and N. R. Bachur. "Daunorubiein metabolism by human hematological components," *Cancer Res.* 32:600–605 (1972).

157. Ris, M. M., and J. P. von Wartburg. "Heterogeneity of NADPH-dependent aldehyde reductase from human and rat brain," *Eur. J. Biochem.* 37:69–77(1973).

158. Bronaugh, R.L., and V.G. Erwin. "Partial purification and characterization of NADPH-linked aldlehyde reductase from monkey brain," *J. Neurochem.* 21:809–815 (1973).

159. Erwin, V.G., W.D.W. Heston, and B. Tabakoff. "Purification and characterization of an NADH-linked aldehyde reductase from bovine brain," *J. Neurochem.* 19:2269–2278 (1972).

160. Tabakoff, B., and V. G. Erwin. "Purification and characterization of a reduced nicotinamide adenine dinucleotide phosphate-linked aldehyde reductase from brain," *J. Biol. Chem.* 245:3263–3268 (1970).

161. Bronaugh, R.L., and G. Erwin. "Further characterization of a reduced nicotinaminde-adenine dinucleotide phosphate-dependent aldehyde reductase from bovine brain," *Biochem. Pharmacol.* 21:1457–1464 (1972).

162. Smolen, A., and A. D. Anderson. "Partial purification and charcterization of a reduced nicotinamide adenine dinucleotide phosphate-linked aldehyde reductase from heart," *Biochem. Pharmacol.* 25:317–323 (1976).

163. Felsted, R. L., and N. R. Bachur. "Ketone reductases," in *Enzymatic Basis of Detoxication,* Vol. I, W. B. Jakoby, Ed. (Orlando, FL: Academic Press, 1980), p. 281.

164. Leibman, K. C. "Reduction of ketones in liver cytosol," *Xenobiotica* 1:97–104 (1971).

165. Culp, H. W., and R. E. McMahon. "Reductase for aromatic aldehydes and ketones. The partial purification and properties of a reduced triphosphopyridine nucleotide-dependent reductase from rabbit kidney cortex," *J. Biol. Chem.* 243:848–852 (1968).

166. Ahmed, N. K., R. L. Felsted, and N. R. Bachur. "Comparison and characterization of mammalian xenobiotic ketone reductases," *J. Pharmacol. Exp. Ther.* 209:12–19 (1979).

167. Leibman, K. C. "Reduction of ketones in liver cytosol," *Xenobiotica* 1:97–104 (1971).

168. Leibman, K. C., and E. Ortiz. "Mammalian metabolism of terpenoids. I. Reduction and hydroxylation of camphor and related compounds," *Drug Metab. Dispos.* 1:543–551 (1973).

169. Fraser, I. M., M. A. Peters, and M. G. Hardinge. "Purification and some characteristics of an α,β-unsaturated ketone reductase from dog erythrocytes and human liver," *Mol. Pharmacol.* 3:233–247 (1967).

170. Uehleke, H., K. H. Hellmer, and S. Tabarelli. "Binding of ^{14}C-carbon tetrachloride to microsomal proteins in vitro and formation of $CHCL^3$ by reduced liver microsomes," *Xenobiotica* 3:1–11 (1973).

171. Suarez, K. A., and P. Bhonsle. "The relationship of cobaltous chloride-induced alterations of hepatic microsomal enzymes to altered carbon tetrachloride hepatotoxicity," *Toxicol. Appl. Pharmacol.* 37:23–27 (1976).

172. Junge, W., and E. Heymann. "Characterization of the isoenzymes of pig liver carboxylesterase. II. Kinetic studies," *Eur. J. Biochem.* 95:519–525 (1979).

173. Greenzaid, P., and W. P. Jencks. "Pig liver esterase. Reactions with alcohols, structure-reactivity correlations and the ccyl-enzyme inter-mediate," *Biochemistry* 10:1210–1222 (1971).

174. Kurooka, S., M. Hashimoto, M. Tomita, A. Maki, and Y. Yoshimura. "Relationship between the structures of S-acyl thiol compounds and their rates of hydrolysis by pancreatic lipase and hepatic carboxylic esterase," *J. Biochem. (Tokyo)* 79:533–541 (1976).

175. Metrione, R.M. "The thiolesterase activity of leucine aminopeptidase," *Biochim. Biophys. Acta* 268:518–522 (1972).

176. Junge, W., and E. Heymann. "Characterization of the isoenzymes of pig liver carboxylesterase. II. Kinetic studies," *Eur. J. Biochem.* 95:519–525 (1979).

177. Heymann, E., K. Krisch, H. Buech, and W. Buzello. "Inhibition of phenacetin- and acetanilide-induced methemoglobinemia in the rat by the carboxylesterase inhibitor bis(*p*-nitrophenyl)phosphate," *Biochem. Pharmacol.* 18:801–811 (1969).

178. Singleton, S. D., and S. D. Murphy. "Propanil (3,4-dichloropropionanilide)-induced methemoglobin formation in mice in relation to acylamidase activity," *Toxicol. Appl. Pharmacol.* 25:20–29 (1973).

179. Silver, E. H., and S. D. Murphy. "Effects of carboxylesterase inhibitors on the acute hepato-toxicity of esters of allyl alcohol," *Toxicol. Appl. Pharmacol.* 45:377–389 (1978).

180. Ecobichon, D. J. "Hydrolytic transformation of environmental pollutants," in *Handbook of Physiology*, D. H. K. Lee, Ed. (Baltimore, MD: Williams & Wilkins, 1977), pp. 441–454.

181. Ahmad, S., and A. J. Forgash. "Nonoxidative enzymes in the metabolism of insecticides," *Drug Metab. Rev.* 5:141–164 (1976).

182. Heymann, E. "Hydrolysis of carboxylic esters and amides," in *Metabolic Basis of Detoxication*, W. B. Jakoby, J. R. Bend, and J. Caldwell, Eds. (New York: Academic Press, 1982), pp. 229–245.

183. Murphy, S. D., K. L. Cheever, A. Y. K. Chow, and Brewster, M. "Organophosphate insecticide potentiation by carboxylesterase inhibitors," *Proc. Eur. Soc. Toxicol.* 17:292–300 (1976).

184. Ecobichon, D. J. "Hydrolytic transformation of environmental pollutants," in *Handbook of Physiology*, D. H. K. Lee, Ed. (Baltimore, MD: Williams & Wilkins, 1977), pp. 441–454.

185. Murphy, S. D., K. L. Cheever, A. Y. K. Chow, and Brewster, M. "Organophosphate insecticide potentiation by carboxylesterase inhibitors," *Proc. Eur. Soc. Toxicol.* 17:292–300 (1976).

186. Ecobichon, D. J. "Hydrolytic transformation of environmental pollutants," in *Handbook of Physiology*, D. H. K. Lee, Ed. (Baltimore, MD: Williams & Wilkins, 1977), pp. 441–454.

187. Ryan, A. J. "The metabolism of pesticidal carbamates," *CRC Crit. Rev. Toxicol.* 1:33–54 (1971).

188. Aldridge, W. N., and E. Reiner. *Enzyme Inhibitors as Substrates. Interaction of Esterases with Esters of Organophosphors and Carbamic Acids* (Amsterdam: North-Holland Publishing Co., 1972).

189. Krisch, K. "Carboxylic ester hydrolase," in *The Enzymes*, Vol. 5, 3rd ed., P.D. Boyer, Eds. (New York: Academic Press, 1971), pp. 43–69.
190. Ecobichon, D. J. "Hydrolytic transformation of environmental pollutants," in *Handbook of Physiology*, D. H. K. Lee, Ed. (Baltimore, MD: Williams & Wilkins, 1977), pp. 441–454.
191. Kuhn, D. *"Charakterisierung Einiger Carboxylesterase/amidase-Isoenzyme aus Meerschweinchen-Leber und Identifizierung des Chloramphenicol-spaltenden Enzyms,"* Dissertation, University of Kiel, West Germany (1980).
192. Satoh, T., and K. Moroi. "Solubilization, purification and properties of isocarboxazid hydrolase from guinea pig liver," *Biochem. Pharmacol.* 21:3111–3120 (1972).
193. Vining, L. C. "Chloramphenicol hydrolase," in *Methods in Enzymology*, Vol. 43, J. H. Hash, Ed. (New York: Academic Press, 1976), pp. 734–737.
194. Satoh, T., and K. Moroi. "Solubilization, purification and properties of isocarboxazid hydrolase from guinea pig liver," *Biochem. Pharmacol.* 21:3111–3120 (1972).
195. Junge, W., and K. Krisch. "The carboxylesterases/amidase of mammalian liver and their possible significance," *CRC Crit. Rev. Toxicol.* 3:371–434 (1975).
196. Krisch, K. "Carboxylic ester hydrolase," in *The Enzymes*, Vol. 5, 3rd ed., P.D. Boyer, Eds. (New York: Academic Press, 1971), pp. 43–69.
197. Chow, A. Y. K., and D. J. Ecobichon. "Characterization of the esterases of guinea pig liver and kidney," *Biochem. Pharmacol.* 22:689–701 (1973).
198. Kleine, R. "Charakterisierung der nierenesterasen verschiedener saugetierspezies," *Acta Biol. Med. Ger.* 28:283–297 (1972).
199. Kleine, R., and P. Meisel. "Occurrence and some properties of a mercury-activatable amino acid esterase from rat kidney microsomes," *FEBS Lett.* 37:120–123 (1973).
200. Rumsby, M. G., H. M. Getliffe, and P. J. Riekkinene. "Association of nonspecific esterase activity with central nerve myelin preparations," *J. Neurochem.* 21:959–967 (1973).
201. Eto, Y., and K. Suzuki. "Cholesterol ester metabolism in rat brain. A cholesterol ester hydrolase specifically localized in the myelin sheath," *J. Biol. Chem.* 248:1986–1991 (1973).
202. Ecobichon, D. J., and W. Kalow. "Properties and classification of the soluble esterase of human skeletal and smooth muscle," *Can. J. Biochem.* 43:73–79 (1965).
203. Junge, W. "Human microsomal carboxylesterase (EC 3.1.1.1.). Distribution in several tissues and some preliminary observations on its appearance in serum," *Enzymes Health Dis. Inaug. Sci. Meet. Int. Soc. Clin. Enzymol.* pp. 54–58 (1978).

204. Brogren, C. H., and T. C. Boeg-Hansen. "Enzyme characterization in quantitative immunoelectrophoresis. An enzymological study of human serum esterases," *Scand. J. Immunol.* 4(2):37–51 (1975).

205. Dudman, N. P. B., and B. Zerner. "Carboxylesterases from pig and ox liver," in *Methods in Enzymology*, Vol. 35, J. M. Lowenstein, Ed. (New York: Academic Press, 1975), pp. 190–208.

206. Krisch, K. "Carboxylic ester hydrolase," in *The Enzymes*, Vol. 5, 3rd ed., P.D. Boyer, Eds. (New York: Academic Press, 1971), pp. 43–69.

207. Rumsby, M. G., H. M. Getliffe, and P. J. Riekkinene. "Association of nonspecific esterase activity with central nerve myelin preparations," *J. Neurochem.* 21:959–967 (1973).

208. Junge, W., and K. Krisch. "The carboxylesterases/amidase of mammalian liver and their possible significance," *CRC Crit. Rev. Toxicol.* 3:371–434 (1975).

209. Lehninger, A. L., Ed. *Principles of Biochemistry*, (New York: Worth Publishers, 1982).

210. Harper, H. A., V. W. Roedwell, and P. A. Mayers. *Review of Physiological Chemistry*, 17th ed., (Los Altos, CA: Lange Medical Publications, 1977), p. 248.

211. Junge, W., E. Heymann, K. Krisch, and H. Hollandt. "Human liver carboxylesterase. Purification and molecular properties," *Arch. Biochem. Biophys.* 165:749–763 (1974).

212. Junge, W., and K. Krisch. "The carboxylesterases/amidase of mammalian liver and their possible significance," *CRC Crit. Rev. Toxicol.* 3:371–434 (1975).

213. Hashinotsume, M., K. Higashino, T. Hada, and Y. Yamamura. "Purification and enzymatic properties of rat serum carboxylesterase," *J. Biochem. (Tokyo)* 84:1325–1333 (1978).

214. Choudhury, S.R. "Substrate hydrolysis by an esterase isoenzyme of rat serum," *Biochim. Biophys. Acta* 350:484–490 (1974).

215. Ecobichon, D.J. "Characterization of the esterases of feline serum," *Can. J. Biochem.* 52:1073–1078 (1974).

216. Oommen, A., and A. S. Balasubramanian. "Aryl acylamidase of monkey brain and liver: Response to inhibitors and relation to acetylcholinesterase," *Biochem. Pharmacol.* 27:891–895 (1978).

217. Paul, S. M., and A. Halaris. "Rat brain de-acetylating activity: Stereospecific inhibition by LSD and serotonin-related compounds," *Biochem. Biophys. Res. Commun.* 70:207–211 (1976).

218. Ecobichon, D.J. "Characterization of the esterases of feline serum," *Can. J. Biochem.* 52:1073–1078 (1974).

219. Evans, D. A. P., M. F. Bullen, J. Houston, C. A. Hopkins, and J. M. Vetters. "Antinuclear factor in rapid and slow acetylator patients treated with isoniazid," *J. Med. Genet.* 9: 53–56 (1972).

220. Shear, N. H., S. P. Spielburg, D. M. Grant, B. K. Tang, and W. Kalow. "Differences in metabolism of sulfonamides predisposing to idiosyncratic toxicity," *Ann. Intern. Med.* 105:179–184 (1986).

221. Devadatta, S., P. R. J. Gangadharam, R. H. Andrews, W. Fox, C. V. Ramakrishnan, J. B. Selkon, and S. Velu. "Peripheral neuritis due to isoniazid," *Bull. World Health Org.* 3:85–87 (1974).

222. Perry, H. M., Jr., E. M. Tan, S. Carmody, and A. Sakamoto. "Relationship of acetyltransferase activity to antinuclear antibodies and toxic symptoms in hypertensive patients treated with hydralazine," *J. Lab. Clin. Med.* 76:114–125 (1970).

223. Strandberg, I., G. Bowman, L. Hassler, and F. Sjoqvist. "Acetylator phenotype in patients with hydralazine-induced lupoid syndrome," *Acta. Med. Scand.* 200:367–371 (1976).

224. Wollsely, R. L., D. E. Drayer, M. M. Reidenberg, A. S. Nies, K. Carr, and J. A. Oates. "Effect of acetylator phenotype on the rate at which procainamide induces antinuclear antibodies and the lupus syndrome," *N. Engl. J. Med.* 298:1157–1160 (1978).

225. Ecobichon, D.J. "Characterization of the esterases of feline serum," *Can. J. Biochem.* 52:1073–1078 (1974).

226. Alarcon-Segovia, D., E. Fishbein, and H. Alcala. "Isoniazid acetylation rate and development of antinuclear antibodies upon isoniazid treatment," *Arthritis Rheum.* 14:748–752 (1971).

227. Monro, J. "The relationship of enzyme deficiencies to multiple sensitivity states," paper presented at the 4th Annu. Int. Symp. Man and His Environment in Health and Disease, Dallas, TX (1986).

228. Evans, D. A. P., M. F. Bullen, J. Houston, C. A. Hopkins, and J. M. Vetters. "Antinuclear factor in rapid and slow acetylator patients treated with isoniazid," *J. Med. Genet.* 9: 53–56 (1972).

229. Glowinski, I.B., Radtke, H.E., and Weber, W.W. "Genetic variation in N-acetylation of carcinogenic arylamines by human and rabbit liver," *Mol. Pharmacol.* 14:940–949 (1978).

230. Killenberg, P. G., and L. T. Webster, Jr. "Conjugation by peptide bond formation," in *Enzymatic Basis of Detoxication*, W. B. Jakoby, Ed. (New York: Academic Press, 1980), pp. 154–163.

231. Spaeth, D. G., and D. L. Schneider. "Taurine synthesis, concentration and bile salt conjugation in rat, guinea pig, and rabbit," *Proc. Soc. Exp. Biol. Med.* 147:855–858 (1974).

232. Awapara, J. "The metabolism of taurine in the animal," in *Taurine*, R. Huxtable and A. Barbeau, Ed. (New York: Raven Press, 1976), 1–19.

233. Brueton, M. J., H. M. Berger, G. A. Brown, L. Ablitt, N. Iyngkaran, and B. A. Wharton. "Duodenal bile acid conjugation patterns and dietary sulphur amino acids in the newborn," *Gut* 19:95–98 (1978).

234. Hadd, H. E., and R. T. Blickenstaff. *Conjugates of Steroid Hormones*, (New York: Academic Press, 1969).

235. Roberts, K. D., and Lieberman, S. "The biochemistry of the 3-B-hydroxy-5-steroid sulfates," in *Chemical and Biological Aspects of Steroid Conjugation*, S. Bernstein and S. Soloman, Eds. (New York: Springer-Verlag New York, 1970), pp. 219–220.

236. Bernstein, S., and R. W. McGilvery. "The enzymatic conjugation of M-aminophenyl," *J. Biol. Chem.* 198:195–203 (1952).

237. Bandurski, R. S., L. G. Wilson, and C. L. Squires. "The mechanism of 'active sulfate' formation," *J. Am. Chem. Soc.* 78:6408–6409 (1956).

238. Robbins, P. W., and F. Lipmann. "Enzymatic synthesis of adenosine-5'-phosphosulfate," *J. Biol. Chem.* 233:686–690 (1958).

239. Hearse, D. J., G. M. Powell, A. H. Olavesen, and K. S. Dodgson. "The influence of some physico-chemical factors on the biliary excretion of a series of structurally related aryl sulphate esters," *Biochem. Pharmacol.* 18:181–195 (1969).

240. Wang, D. Y., and R. D. Bulbrook. "Binding of the sulfate esters of dehydroepiandrosterone, testosterone, 17-acetoxypregnenolone, and pregnenolone in the plasma of man, rabbit and rat," *J. Endocrinol.* 39:405–413 (1967).

241. Butenandt, A., E. Biekart, N. Koga, and P. Traub. "Uber ommochrome. 21. Konstitution und synthese des ommatins. D," *Hoppe-Seyler's Z. Physiol. Chem.* 321:258–275 (1960).

242. Verlangieri, A. J., and R. O. Mumma. "*In vivo* sulfation of cholesterol by ascorbic acid 2-sulfate," *Atherosclerosis* 17:37–48 (1973).

243. Krijgsheld, K.R., H. Frankena, E. Scholtens, J. Zweens, and G.J. Mulder. "Absorption, serum levels and urinary excretion of inorganic sulfate after oral administration of sodium sulfate in the conscious rat," *Biochim. Biophys. Acta* 586:492–500 (1979).

244. Yamaguchi, K., S. Sakakibara, K. Kaga, and I. Veda. "Induction and activation of cysteine oxidase of rat liver," *Biochim. Biophys. Acta* 237:502–512 (1971).

245. Levi, A. S., and G. Wolf. "Purification and properties of the enzyme ATP-sulfurylase and its relation to vitamin A," *Biochim. Biophys. Acta* 178:262–282 (1968).

246. Singer, S. S., and S. Sylvester. "Enzymatic sulfation of steroids. 2. The control of the hepatic cortisol sulfotransferase activity and of the individual hepatic steroid sulfotransferases of rats by gonads and gonadal hormones," *Endocrinology* 99:1346–1352 (1976).

247. Singer, S. S. "Enzymatic sulfation of steroids. 4. Control of the hepatic glucocorticoid sulfotransferase activity and the individual glucocorticoid sulfotransferases from male and female rats by adrenal glands and corticosteroids," *Endocrinology* 103:66–73 (1978).

248. Jakoby, W. B., R. D. Sekura, E. S. Lyon, C. J. Marcus, and J. L. Wang. "Sulfotransferases," in *Enzymatic Basis of Detoxication*, Vol. II, W. B. Jakoby, Ed. (New York: Academic Press, 1980), p. 209.

249. Gregory, J. D., and F. Lipmann. "The transfer of sulfate among phenolic compounds with 3′,5′-diphosphoadenosine as coenzyme," *J. Biol. Chem.* 229:1081–1090 (1957).

250. Banerjee, R. K., and A. B. Roy. "The formation of cholesteryl sulfate by androstenolone sulfotransferase," *Biochim. Biophys. Acta* 137:211–213 (1967).

251. Nose, Y., and F. Lipmann. "Separation of steroid sulfokinases," *J. Biol. Chem.* 233:1348–1351 (1958).

252. Banerjee R. K., and A. B. Roy. "The sulfotransferases of guinea pig liver," *Mol. Pharmacol.* 2:56–66 (1966).

253. Adams, J. B., and A. Poulos. "Enzymatic synthesis of steroid sulfates. III. Isolation and properties of estrogen sulphotransferase from bovine adrenal gland," *Biochim. Biophys. Acta* 146:493–508 (1967).

254. Nose, Y., and F. Lipmann. "Separation of steroid sulfokinases," *J. Biol. Chem.* 233:1348–1351 (1958).

255. Roy, A. B. "The enzymatic synthesis of aryl sulphomates," *Biochem. J.* 74:49–56 (1960).

256. Roy, A. B. "The enzymatic synthesis of aryl suphomates. II. The effects of 3 beta-methodoxyandrost-5-en-17-one on arylamine suphokinase,"*Biochem J.* 79:253–261 (1961).

257. Vestermark, A., and H. Borstrom. "Studies on ester sulfates. V. On the enzymatic formation of ester sulfates of primary aliphatic alcohol," *Exp. Cell Res.* 18:174–177 (1959).

258. Jones, J. G., and J. M. Dodgson. "Biosynthesis of L-tyrosine *O*-sulphate from the methyl and ethyl ester of L-tyrosine," *Biochem. J.* 94:331–336 (1965).

259. Wu, S.-C.G., and K. D. Straub. "Purification and characterization of *N*-hydroxy-2-acetylaminoflurorene sulfotransferase from rat liver," *J. Biol. Chem.* 251:6529–6536 (1976).

260. Sekura, R. D., and W. B. Jakoby. "Phenol sulfotransferases," *J. Biol. Chem.* 254:5658–5663 (1979).

261. Sekura, R. D., M. W. Duffel, and W. B. Jakoby. "Aryl sufotransferases," *Methods Enzymol.* 77:197–206 (1981).

262. Marcus, C. J., R. D. Sekura, and W. B. Jakoby. "A hydroxy-steroid sulfotransferase from rat liver," *Anal. Biochem.* 107:296–304 (1980).

263. Lyon, E. S., and W. B. Jakoby. "The identity of alcohol sulfotransferase with hydroxysteroid sulfotransferase,"*Arch. Bichem. Biphys.* 202(2):474–481 (1980).

264. Lyon, E. S., C. J. Marcus, J. L. Wang, and W. B. Jakoby. "Hydroxysteroid sulfotransferase," *Methods Enzymol.* 77:206–213 (1981).

265. Wood, J. L. "Biochemistry of mercapturic acid formation," in *Metabolic Conjugation and Metabolic Hydrolysis*, Vol. II, W. H. Fishman, Ed. (New York: Academic Press, 1970), pp. 261–299.

266. Chasseaud, L. F. "Conjugation with glutathione and mercapturic acid excretion," in *Glutathione: Metabolism and Function*, I. M. Arias and W. B. Jakoby, Eds. (New York: Raven Press, 1976), pp. 77–114.

267. Nogami, H., J. Hasegawa, and K. Noda. "Thiamine derivatives of disulfide type. I. Formation of thiamine from propyl disulfide in rat intestine in vitro," *Chem. Pharm. Bull.* 17:219–227 (1969).

268. Kelley, J. J., N. F. Hamilton, and O. M. Friedman. "Studies on latent derivative of aminoethanethiols as potentially selective cytoprotectants. III. Reactions of cysteamine-S-sulfate in biological media," *Cancer Res.* 27:143–147 (1967).

269. Seegmiller, H. E., T. Friedmann, H.E. Harrison, V. Wong, and J. A. Schneider. "Cystinosis," *Ann. Intern. Med.* 68:883–905 (1968).

270. Mannervik, B., and S. A. Eriksson. "Enzymatic reduction of mixed disulfides and thiosulfate esters," in *Glutathione*, L. Flohe, H. C. Benohr, H. Sies, H. D. Waller, and A. Wendel, Eds. (Stuttgart: Georg Thieme, 1974), pp. 120–131.

271. Thompson, G. A., and A. Meister. "Interrelationships between the binding sites for amino acids, dipeptides, and Y-glutamyl donors in Y-glutamyl transpeptidase," *J. Biol. Chem.* 138:177–188 (1970).

272. Tate, S. S., and A. Meister. "Interaction of Y-glutamyl transpeptidase with amino acids, dipeptides, and derivatives and analogs of glutathione," *J. Biol. Chem.* 249:7593–7602 (1974).

273. Meister, A., and S. S. Tate. "Glutathione and related y-glutamyl compounds: Biosynthesis and utilization," *Annu. Rev. Biochem.* 45: 559–604 (1976).

274. Orlowski, M., and Meister, A. "The Y-glutamyl cycle: Possible transport system for amino acids," *Proc. Natl. Acad. Sci. U.S.A.* 67:1248–1255 (1970).

275. Meister, A. "On the enzymology of amino acid transport," *Science* 180:33–39 (1973).

276. Jakoby, W. B., and W. H. Habig. "Gluathione transferases," in *Enzymatic Basis of Detoxication*, Vol. II, W. B. Jakoby, Ed. (New York: Academic Press, 1980), p. 64.

277. Racker, E. "Glutathione-homocystine transhydrogenase," *J. Biol. Chem.* 217:867–874 (1955).

278. Chang, S. H., and D. R. Wilkin. "Participation of the unsymmetrical disulfide of coenzyme A and glutathione in an enzymatic sulfhydryl-disulfide interchange. I. Partial purification and properties of the bovine kidney enzyme," *J. Biol. Chem.* 241:1304–1312 (1966).

279. Eriksson, B., and S. A. Eriksson. "Synthesis and characterization of the L-cysteine-glutathione mixed disulfide," *Acta. Chem. Scand.* 21:1304–1312 (1967).

280. Nagai, S., and S. Black. "A thiol-disulfide transhydrogenase from yeast," *J. Biol. Chem.* 243:1942–1947 (1968).

281. Wendell, P. L. "Distribution of glutathione reductase and detection of glutathione-cystine transhydrogenase in rat tissues," *Biochim. Biophys. Acta* 159:179–181 (1968).
282. States, B., and S. Segal. "Distribution of glutathione-cystine transhydrogenase activity in subcellular fractions of rat intestinal mucosa," *Biochem. J.* 113:443–444 (1969).
283. Kohno, K., K. Noda, M. Mizobe, and I. Utsumi. "Enzymatic reduction of disulfide-type thiamine derivatives," *Biochem. Pharmacol.* 18:1685–1692 (1969).
284. Eriksson, S. A., and B. Mannervik. "The reduction of the L-cysteine-glutathione mixed disulfide in rat liver. Involvement of an enzyme catalyzing thiol-disulfide interchange," *FEBS Lett.* 7:26–28 (1970).
285. Tietze, F. "Disulfide reduction in rat liver. I. Evidence for the presence of nonspecific nucleotide-dependent disulfide reductase and GSH-disulfide transhydrogenase activities in the high-speed supernatant fraction," *Arch. Biochem. Biophys.* 138:177–188 (1970).
286. Griffith, O. W., and A. Meister. "Translocation of intracellular glutathione to membrane-bound y-glutamyl transpeptidase as a discrete step in the y-glutamyl cycle: Glutathionuria after inhibition of transpeptidase," *Proc. Natl. Acad. Sci. U.S.A.* 76:268–272 (1979).
287. Griffith, O. W., A. Novogrodsky, and A. Meister. "Translocation of intracellular glutathione to membrane-bound y-glutamyl transpeptidase as a discrete step in the y-glutamyl cycle: Glutathionuria after inhibition of transpeptidase," *Proc. Natl. Acad. Sci. U.S.A.* 76:2249–2252 (1979).
288. Bartoli, G.M., and M. Sies. "Reduced and oxidized glutathione efflux from liver," *FEBS Lett.* 86:89–91 (1978).
289. Seegmiller, H. E., T. Friedmann, H.E. Harrison, V. Wong, and J. A. Schneider. "Cystinosis," *Ann. Intern. Med.* 68:883–905 (1968).
290. Mannervik, B., G. Persson, and S. Eriksson. "Enzymatic catalysis of the reversible sulfitolysis of glutathione disulfide and the biological reduction of thiosulfate esters," *Arch. Biochem. Biophys.* 163:283–289 (1974).
291. Mannervik, B., and K. Axelsson. "Reduction of disulphide bonds in proteins and protein mixed disulphides catalysed by A thioltransferase in rat liver cytosol," *Biochem. J.* 149:785–788 (1975).
292. Eldjarn, L., and J. Bremer. "The disulphide-reducing capacity of liver mitochondria," *Acta. Chem. Scand.* 17(1):59–66 (1963).
293. Tate, S. S., and A. Meister. "Interaction of Y-glutamyl transpeptidase with amino acids, dipeptides, and derivatives and analogs of glutathione," *J. Biol. Chem.* 249:7593–7602 (1974).

294. Orlowski, M., and Meister, A. "The Y-glutamyl cycle: Possible transport system for amino acids," *Proc. Natl. Acad. Sci. U.S.A.* 67:1248–1255 (1970).

295. Tietze, F. "Disulfide reduction in rat liver. I. Evidence for the presence of nonspecific nucleotide-dependent disulfide reductase and GSH-disulfide transhydrogenase activities in the high-speed supernatant fraction," *Arch. Biochem. Biophys.* 138:177–188 (1970).

296. Meister, A. "Current status of y-glutamyl cycle," in *Functions of Glutathione in Liver and Kidney*, H. Sies and A. Wendel, Eds. (New York: Springer-Verlag, 1978), pp. 43–59.

297. Meister, A. "Biochemistry of glutathione," in *Metabolic Pathways*. Vol. 7, 3rd ed., D. M. Greenberg, Ed. (New York: Academic Press, 1975), pp. 43–59.

298. Tate, S. S., and A. Meister. "Interaction of Y-glutamyl transpeptidase with amino acids, dipeptides, and derivatives and analogs of glutathione," *J. Biol. Chem.* 249:7593–7602 (1974).

299. Flohe, L., H. C. Benohr, H. Sies, H. D. Waller, and A. Wendel, Eds. *Glutathione* (Stuttgart: Georg Thieme, 1974).

300. Arias, I. M., and W. B. Jakoby, Eds. *Glutathione: Metabolism and Function.* (New York: Raven Press, 1976).

301. Sies, H., and A. Wendel, Eds. *Functions of Glutathione in Liver and Kidney* (New York: Springer-Verlag, 1978).

302. Wilson, J. "Cyanide and human disease," in *Chronic Cassava Toxicity* (Monogr. IDRC-010e), (Ottawa, Canada: International Development Research Centre, 1973), pp. 121–125.

303. Wilson, J. "Leber's hereditary optic atrophy: A possible defect of cyanide metabolism," *Clin. Sci.* 31:1–7 (1966).

304. Westley, J. "Sulfane-transfer catalysis by enzymes," *Bioorg. Chem.* 1:371–390 (1977).

305. Burnett, W. W., E. G. King, M. Grace, and W. F. Hall. "Hydrogen sulfide poisoning: A review of five years' experience," *Can. Med. Assoc. J.* 117:1277–1280 (1977).

306. Kellog, W. W., R. D. Cadle, E. R. Allen, A. L. Lazrus, and E. A. Martell. "The sulfur cycle," *Science* 175:587–596 (1972).

307. Wilson, J., and D. M. Matthews. "Metabolic inter-relationships between cyanide, thiocyanate, and vitamin B_{12} in smokers and nonsmokers," *Clin. Sci.* 31:1–7 (1966).

308. Maloof, F., and M. J. Soodak. "The oxidation of thiocyanate by a cytoplasmic particulate fraction of thyroid tissue," *J. Biol. Chem.* 239:1995–2001 (1964).

309. Sorbo, B., and J. G. Ljunggren. "The catalytic effect of peroxidase on the reaction between hydrogen peroxide and certain sulfur compounds," *Acta. Chem. Scand.* 12:470–476 (1958).

310. Oram, J. D., and B. Reiter. "The inhibition of streptococci by lactoperoxidase, thiocyanate, and hydrogen peroxide," *Biochem. J.* 100:382–388 (1966).

311. Chung, J., and J. L. Wood. "Oxidation of thiocyanate to cyanide and sulfate by the lactoperoxidase-H_2O_2 system," *Arch. Biochem. Biophys.* 141:73–78 (1970).

312. Lang, K. "Die rhodanbildung im tierkorper," *Biochem. Z.* 259:243–256 (1933).

313. Wilson, J. "Cyanide and human disease," in *Chronic Cassava Toxicity* (Monogr. IDRC-010e), (Ottawa, Canada: International Development Research Centre, 1973), pp. 121–125.

314. Moore, S. J., J. C. Norris, I. K. Ho, and A. S. Hume. "The efficiency of α-ketoglutaric acid in the antagonism of cyanide intoxication," *Toxicol. Appl. Pharmacol.* 82:40–44 (1986).

315. Meister, A., P. E. Fraser, and S. V. Tice. "Enzymatic desulfuration of B-mercaptopyruvate to pyruvate," *J. Biol. Chem.* 206:561–575 (1954).

316. Wong, T. W., S. B. Weiss, G. L. Eliceiri, and J. Bryant. "TRNA sulfurtransferase from bacillus subtilis W168," *Biochemistry* 9:2376–2386 (1970).

317. Abrell, J. W., E. E. Kaufman, and M. N. Lipsett. "The biosynthesis of 4-thiouridylate," *J. Biol. Chem.* 246:294–301 (1971).

318. Wong, T. W., M. H. Harris, and C. A. Jankowicz. "tRNA sulfurtransferase isolated from rat cerebral hemisphere," *Biochemistry* 13:2805–2812 (1974).

319. Harris, C. L., F. T. Kerns, and W. St. Clair. "tRNA sulfurtransferase activity in rat liver and chemically induced hepatomas," *Cancer Res.* 35:3608–3610 (1975).

320. Wong, T. W., M. H. Harris, and H. P. Morris. "The presence of an inhibitor of tRNA sulfurtransferase in morris hepatomas," *Biochem. Biophys. Res. Commun.* 65:1137–1145 (1975).

321. Harris, C. L. "Mammalian tRNA sulfurtransferase," *Nucleic Acids Res.* 5:599–613 (1978).

322. Cavallini, D., B. Mondovi, C. DeMarco, and A. Scioscia-Santoro. "The mechanism of desulphydration of cysteine," *Enzymologia* 24:253–266 (1962).

323. Flavin, M. "Microbial transulfuration: The mechanism of an enzymatic disulfide elimination reaction," *J. Biol. Chem.* 237:768–777 (1962).

324. Schneider, J. F., and J. Westley. "Metabolic interrelations of sulfur in proteins, thiosulfate, and cystine," *J. Biol. Chem.* 244:5735–5744 (1969).

325. Sorbo, B. "Sulfite and complex-bound cyanide as sulfur acceptors for rhodanese," *Acta. Chem. Scand.* 11:628–633 (1957).

326. Harris, C. L., F. T. Kerns, and W. St. Clair. "tRNA sulfurtransferase activity in rat liver and chemically induced hepatomas," *Cancer Res.* 35:3608–3610 (1975).

327. Flavin, M. "Microbial transulfuration: The mechanism of an enzymatic disulfide elimination reaction," *J. Biol. Chem.* 237:768–777 (1962).

328. Schneider, J. F. and J. Weseley. "Metabolic interrelations of sulfur in proteins, thiosulfate, and cystine," *J. Biol. Chem.* 744:5735–5744 (1969).

329. Sorbo, B. "The pharmacology and toxicology of inorganic sulfur compounds," in *Sulfur in Organic and Inorganic Chemistry*, A. Senning, Ed. (New York: Marcel Dekker, 1972), pp. 143–169.

330. Koj, A., and J. Frendo. "The activity of cysteine desulfhydrase and rhodanase [sic] in animal tissues," *Acta. Biochim. Pol.* 9:373–379 (1962).

331. Villarejo, M., and J. Westley. "Mechanism of rhodanese catalysis of thiosulfate-lipoate oxidation-reduction," *J. Biol. Chem.* 238:4016–4020 (1963).

332. Villarejo, M., and J. Westley. "Rhodanese-catalyzed reduction of thiosulfate by reduced lipoic acid," *Biol. Chem.* 238:PC1185–PC1186 (1963).

333. Hearse, D. J., G. M. Powell, A. H. Olavesen, and K. S. Dodgson. "The influence of some physico-chemical factors on the biliary excretion of a series of structurally related aryl sulphate esters," *Biochem. Pharmacol.* 18:181–195 (1969).

334. Wang, D. Y., and R. D. Bulbrook. "Binding of the sulfate esters of dehydroepiandrosterone, testosterone, 17-acetoxypregnenolone, and pregnenolone in the plasma of man, rabbit and rat," *J. Endocrinol.* 39:405–413 (1967).

335. Burstein, S. 1962. "Interaction of androsternolone sulphate with cationic detergents and phosphatides," *Biochim. Biophys. Acta.* 62:576–578 (1962).

336. Jakoby, W. B., R. D. Sekura, E. S. Lyon, C. J. Marcus and J.-L. Wang. "Sulfotransferases," in *Enzymatic Basis of Detoxication.* Vol. II, W. B. Jakoby, Ed. (New York: Academic Press, 1980), p. 201.

337. Reeves, A. L. "The metabolism of foreign compounds" in *Toxicology: Principles and Practice*, Vol. 1, A. L. Reeves, Ed. (New York: John Wiley & Sons, 1981), p. 19.

338. DeMatteis, F. "Covalent binding of sulfur to microsomes and loss of cytochrome P-450 during the oxidative desulfuration of several chemicals," *Mol. Pharmacol.* 10:849–854 (1974).

339. Catignani, G. L., and R. A. Neal. "Evidence for the formation of a protein-bound hydrosulfide resulting from the microsomal mixed function oxidase catalyzed desulfuration of carbon disulfide," *Biochem. Biophys. Res. Commun.* 65:629–636 (1975).

340. Hunter, A. L., and R. A. Neal. "Inhibition of hepatic mixed-function oxidase activity in vitro and in vivo by various thiono-sulfur-containing compounds," *Biochem. Pharmacol.* 24:2199–2205 (1975).

341. Seawright, A. A., J. Hrdlicka, and F. Mattise. "The hepatotoxicity of O,O-diethyl-O-phenyl phosphorothionate (SV[1])for the rat," *Br. J. Exp. Pathol.* 57:16–22 (1976).

342. Kamataki, T., and R. A. Neal. "Metabolism for diethyl p-nitrophenyl phosphorothionate (parathion) by a reconstituted mixed function oxidase enzyme system," *Mol. Pharmacol.* 12:933–944 (1976).

343. Neal, R. A., T. Kamataki, and C. Catignani. "Monooxygenase catalyzed activation of thiono-sulfur containing compounds to reactive intermediates," *Hoppe-Seyler's Z. Physiol. Chem.* 357:1044 (1976).

344. Jarvisalo, J., A.H. Gibbs, and F. Matteis. "Accelerated conversion of heme to bile pigments caused in the liver by carbon disulfide and other sulfur-containing chemicals," *Mol. Pharmacol.* 14:1099–1106 (1978).

345. Westley, J. "Rhodanese and the sulfane pool," in *Enzymatic Basis of Detoxication.*, Vol. II, W. B. Jakoby, Ed. (New York: Academic Press, 1980), p. 257.

346. Usdin, E., R. T. Borchardt, and C. R. Creveling. *Transmethylations* (New York: Elsevier, 1979).

347. Lehninger, A. L., Ed. *Principles of Biochemistry*, (New York: Worth Publishers, 1982), pp. 131–132.

348. White, H. L., and J. C. Wu. "Properties of catechol-O-methyltransferase from brain and liver of rat and human," *Biochem. J.* 145: 135–143 (1975).

349. Gugler, R., R. Knuppen, and H. Breuer. "Purification and characterizaton of human placenta S-adenosylmethionine: catechol-O-methyltransferase," *Biochim. Biophys. Acta.* 220:10–21 (1970).

350. Darmenton, P., L. Cronenberger, and H. Pacheco. "Purification and properties of catechol-O-methyltransferase from human placenta," *Biochimie.* 58:1401–1403 (1976).

351. Assicot, M., G. Contesso, and C. Bohoun. "Catechol-O-methyltransferase in human breast cancers," *Eur. J. Cancer* 13:961–966 (1977).

352. Assicot, M., and C. Bohuon. "Presence of two distinct catechol-O-methyltransferase activities in red blood cells," *Biochimie.* 53:871–874 (1971).

353. Axelrod, J., and C. K. Cohn. "Methyltransferase enzyme in red blood cells," *J. Pharmacol. Exp. Ther.* 176:650–654 (1971).

354. Raymond, F. A., and R. M. Weinshilboum. "Microassay of human erythrocyte catechol-O-methyltransferase: Removal of inhibitory calcium ion with chelating resin," *Clin. Chim. Acta.* 58:185–194 (1975).

355. Weisiger, R. A., and W. B. Jakoby. "S-Methylation: Thiol S-Methyltransferase," in *Enzymatic Basis of Detoxication*, Vol. II, W. B. Jakoby, Ed. (New York: Academic Press, 1980), pp. 131–140.

356. Guldberg, H. C., and C. A. Marsden. "Catechol-O-methyltransferase: Pharmacological aspects and physiological role," *Pharmacol. Rev.* 27:135–206 (1975).

357. Axelrod, J., and C. K. Cohn. "Methyltransferase enzyme in red blood cells," *J. Pharmacol. Exp. Ther.* 176:650–654 (1971).

358. Coward, J. K., M. d'Urso-Scott, and W. D. Sweet. "Inhibition of catechol-O-methyl-transferase by S-adenosylhomocysteine and S-adenosoylhomocysteine sulfoxide, a potential transition-state analog," *Biochem. Pharmacol.* 21:1200–1203 (1972).

359. Blaschke, E., and G. Hertting. "Enzymatic methylation of L-ascorbic acid by catechol-O-methyltransferase," *Biochem. Pharmacol.* 20:1362–1370 (1971).

360. Inscoe, J. K., J. Daly, and J. Axelrod. "Factors affecting the enzymatic formation of O-methylated dihydroxy derivatives," *Biochem. Pharmacol.* 14:1257–1263 (1965).

361. Wurtman, R. J., J. Axelrod, and L. T. Potter. "The disposition of catecholamines in the rat uterus and effect of drugs and hormones," *J. Pharmacol. Exp. Ther.* 144:150–155 (1964).

362. Landsberg, L., J. DeChamplain, and J. Axelrod. "Increased biosynthesis of cardiac norepinephrine after hypophysectomy," *J. Pharmacol. Exp. Ther.* 165:102–107 (1969).

363. Cohn, C. K., D. L. Dunner, and J. Axelrod. "Reduced catechol-O-methyltransferase activity in red blood cells of women with primary affective disorder," *Science* 170:1323–1324 (1970).

364. Barass, B. C., D. B. Coult, and R. M. Pinder. "3-Hydroxy-4-methoxyphenylethylamine: The endogenous toxin of parkinsonism," *J. Pharm. Pharmacol.* 24:201–266 (1951).

365. Sjoerdsma, A. "Relationships between alterations in amine metabolism and blood pressure," *Circ. Res.* 9:734–743 (1961).

366. Comoy, E., and C. Bohuon. "Isohomovanillic acid determination in human urine," *Clin. Chim. Acta.* 35:369–375 (1971).

367. Crout, J. R. "Phenochromocytoma," *Pharmacol. Rev.* 18:651–657 (1966).

368. Weinshilboum, R. M. "Correlation of erythrocyte catechol-O-methyltransferase activity between siblings," *Nature (London)* 252:490–501 (1974).

369. Scanlong, P. D., F. A. Raymond, and R. M. Weinshilboum. "Catechol-O-methyltransferase: Thermolabile enzyme in erythrocytes of subjects homozygous for allele for low activity," *Science* 303:63–65 (1978).

370. Barth, H., H. Troidl, W. Lorenz, H. Rohde, and R. Glass. "Histamine and peptic ulcer disease: Histamine methyltransferase activity in gastric mucosa of control subjects and duodenal ulcer patients before and after surgical treatment," *Agents Actions* 7:75–79 (1977).

371. Snyder, S. H., and J. Axelrod. "Sex differences and hormonal control of histamine methyltransferase activity," *Biochim. Biophys. Acta* 111:416–421 (1965).

372. Bionostics, Inc. Technical Memorandum 2, Functions of Amino Acids (1986), p. 8.

373. Kasper, C. B., and D. Henton. "Glucuronidation," in *Enzymatic Basis of Detoxication,* Vol. II, W. B. Jakoby, Ed. (New York: Academic Press, 1980), p. 4.

374. Black, M., B. H. Billing, and K. P. M. Heirwegh. "Determination of bilirubin UDP-glucuronyl transferase activity in needle-biopsy specimens of human liver," *Clin. Chim. Acta.* 29:27–35 (1970).

375. Dutton, G.J. "The biosynthesis of glucuronides," in *Glucuronic Acid,* G. J. Dutton, Ed. (New York: Academic Press, 1966), chap. 3.

376. Morgenstern, R., J. W. DePierre, and L. Ernster. "Activation of microsomal glutathione S-transferase activity by sulfhydryl reagents," *Biochem. Biophys. Res. Commun.* 87:657–663 (1979).

377. Clayson, D. B., and R. C. Garner. "Carcinogenic aromatic amines and related compounds," *ACS Monogr.* 173:366–461 (1976).

378. Cramer, J. W., J. A. Miller, and E. C. Miller. "N-hydroxylation: A new metabolic reaction observed in the rat with the carcinogen 2-acetylaminofluorene," *J. Biol. Chem.* 235:885–888 (1960).

379. Booth, J., and E. Boyland. "The biochemistry of aromatic amines. 10. Enzymatic N-hydroxylation of arylamines and conversion of arylhydroxylamines into o-aminophenols," *Biochem. J.* 91:362–369 (1964).

380. Poulsen, L. L., B. S. S. Masters, and D. M. Ziegler. "Mechanism of 2-naphthylamine oxidation catalyzed by pig liver microsomes," *Xenobiotica* 6:481–498 (1976).

381. Uehleke, H. "N-hydroxylation of carcinogenic amines in vivo and in vitro with liver microsomes," *Biochem. Pharmacol.* 12:219–221 (1963).

382. Selkirk, J. K., R. G. Croy, P. P. Roller, and H. V. Gelboin. "High-pressure liquid chromatographic analysis of benzo[a]pyrene metabolism and covalent binding and the mechanism of action of 7,8-benzoflavone and 1,2-epoxy-3,3,3-trichlor-propane," *Cancer Res.* 34:3474–3480 (1974).

383. Holder, G., H. Yagi, P. Dansette, D. M. Jerina, W. Levin, A. Y. H. Lu, and A. H. Conney. "Effects of inducers and epoxide hydrase on the metabolism of benzo[a]pyrene by liver microsomes and a reconstitutes system: Analysis by high pressure liquid chromatography," *Proc. Natl. Acad. Sci. U.S.A.* 71:4356–4360 (1974).

384. Yang, S. K., D. W. McCourt, P. P. Roller, and H. V. Gelboin. "Enzymatic conversion of benzo[a]pyrene leading predominantly to the diol-epoxide r-7, t-8-dihydroxy-t-9, 10-oxy-7,8,9,10-tetrahydrobenzo(a)pyrene through a single enantiomer of r-7, t-8-dihydroxy-7,8-dihydro-benzo[a]pyrene," *Proc. Natl. Acad. Sci. U.S.A.* 73:2594–2598 (1976).

385. Kasper, C. B., and D. Henton. "Glucuronidation," in *Enzymatic Basis of Detoxication.* Vol. II, W. B. Jakoby, Ed. (New York: Academic Press, 1980), p. 5..

386. Bionostics, Inc. Technical Memorandum 2, Functions of Amino Acids (1986), p. 8.

387. Reeves, A. L. "The metabolism of foreign compounds" in *Toxicology: Principles and Practice,* Vol. 1, A. L. Reeves, Ed. (New York: John Wiley & Sons, 1981), p. 8.

388. Bock, K. W., G. van Ackeren, F. Lorch, and F. W. Birke. "Metabolism of naphthalene to naphthalene glucuronide in isolated dihydrodiol hepatocytes and liver microsomes," *Biochem. Pharmacol.* 25:2351–2356 (1976).

389. Yang, S. K., D. W. McCourt, P. P. Roller, and H. V. Gelboin. "Enzymatic conversion of benzo[a]pyrene leading predominantly to the diol-epoxide r-7, t-8-dihydroxy-t-9, 10-oxy-7,8,9,10-tetrahydrobenzo(a)pyrene through a single enantiomer of r-7, t-8-dihydroxy-7,8-dihydro-benzo[a]pyrene," *Proc. Natl. Acad. Sci. U.S.A.* 73:2594–2598 (1976).

390. Yang, S. K., D. W. McCourt, P. P. Roller, and H. V. Gelboin. "Enzymatic conversion of benzo[a]pyrene leading predominantly to the diol-epoxide r-7, t-8-dihydroxy-t-9, 10-oxy-7,8,9,10-tetrahydrobenzo(a)pyrene through a single enantiomer of r-7, t-8-dihydroxy-7,8-dihydro-benzo[a]pyrene," *Proc. Natl. Acad. Sci. U.S.A.* 73:2594–2598 (1976).

391. Reeves, A. L. "The metabolism of foreign compounds" in *Toxicology: Principles and Practice*, Vol. 1, A. L. Reeves, Ed. (New York: John Wiley & Sons, 1981), p. 23.

392. Fridovich, I. "Superoxide and superoxide dismutases," in *Advances in Inorganic Biochemistry,* G. L. Eichhorn and D. L. Marzilli, Eds. (New York: Elsevier, 1979), pp. 67–90.

393. McCord, J. M., B. B. Keele, Jr., and I. Fridovich. "An enzyme-based theory of obligate anaerobiosis: The physiological function of superoxide dismutase," *Proc. Natl. Acad. Sci. U.S.A.* 68:1024–1027 (1971).

394. Fridovich, I. "Oxygen radicals, hydrogen peroxide, and oxygen toxicity," in *Free Radicals in Biology*, Vol. 1, W. A. Pryor, Ed. (New York: Academic Press, 1976), pp. 239–277.

395. Brabec, M. H., and I. A. Bernstein. "Cellular, subcellular, and molecular targets of foreign compounds," in *Toxicology: Principles and Practice,* Vol. I, A. L. Reeves, Ed. (New York: John Wiley & Sons, 1981), pp. 36–37.

396. Clive, D., and JT-SY Spector. "Laboratory procedure for assessing specific locus mutations at the TK locus in cultured L5178Y mouse lymphoma cells," *Mutat. Res.* 31:17 (1975).

397. McCann, J. N., E. Spingarn, J. Kobori, and B. N. Ames. "Detection of carcinogens as mutagens: Bacterial strains with R factor plasmids," *Proc. Natl. Acad. Sci. U.S.A.* 72:979 (1973).

398. Ames, B. N., W. F. Durston, E. Yamasaki, and F. D. Lee. "Carcinogens are mutagens: A simple test system combining liver homogenates for activation and bacteria for detection," *Proc. Natl. Acad. Sci. U.S.A.* 70:2281 (1973).

399. Legator, M., and S. Zimmering. "Genetic toxicology," *Annu. Rev. Pharmacol.* 15:387 (1975).

400. Mustafa, M. G., and D. F. Tierney. "Biochemical and metabolic changes in the lung with oxygen, ozone, and nitrogen dioxide toxicity," *Am. Rev. Respir. Dis.* 118:6 (1978).
401. Levine, S. A., and Kidd, P. M. *Antioxidant Adaptation: Its Role in Free Radical Pathology* (San Leandro, CA: Biocurrents Division, Allergy Research Group, 1986), pp. viii–xiii.
402. Monro, J., and C. W. Smith. Personal communication (London: The Lister Hospital, 1986).

5 POLLUTANT EFFECTS ON THE BLOOD AND RETICULOENDOTHELIAL SYSTEM (LYMPHATIC AND IMMUNE SYSTEM)

INTRODUCTION

Effects of pollutants on the blood and reticuloendothelial system are now becoming well known. Nitrous oxide can cause bone marrow depression; pesticides can depress T cells. Chemotherapy using antimetabolites can suppress leukocytes, leaving an individual vulnerable to recurrent infections. Mold toxins such as cyclosporin can suppress the tissue transplant rejection phenomenon. Clearly, pollutant effects on these systems can exacerbate chemical sensitivity and in some cases may even induce it. Toluene diisocyanate, for example, may sensitize T lymphocytes, while formaldehyde may suppress T cells and cause asthma or vasculitis.

An example of hematological pollutant injury is a 19-year-old white female who entered the ECU because of recurrent hemorrhage. As a hairdresser, she worked in close proximity to many toxic chemicals, and over time her bone marrow became depressed. When she was away from her work environment, her bone marrow improved along with her overall health. Upon her return to work, however, her bone marrow became suppressed. This cycle continued unabated until this patient was eventually diagnosed and treated for chemical sensitivity. She subsequently came to realize that perception of toxic odors followed by symptoms of illness was the first stage of her chemical sensitivity, which in its most severe state resulted in bone marrow suppression.

155

Although it is difficult to isolate various pollutant effects in the study of blood and lymphatics, we attempt to do just this as we discuss each area of the blood and reticuloendothelial system.

HEMATOLOGICAL SYSTEM

Pollutant effects upon the hematological system have been recognized for a long time. Thirty years ago hair dressers were known to be prone to bone marrow depression due to exposure to toxic substances used in their profession. With the explosion of knowledge about the immune system, a multitude of effects related to pollutant exposure have now been observed. Observations clearly show that there can be a spectrum of changes from minimal shifts (e.g., mild anemia) at one end to complete cessation of function (e.g., aplastic anemia) at the other. Though the hematologic system is complex, we make an attempt in this chapter to describe the pollutant effects on each entity separately. It appears that some types of damage are the result of chemical overexposure, and the final outcome of these exposures may be chemical sensitivity.

Bone Marrow

Many substances able to suppress or stimulate bone marrow have been identified. Environmental agents readily associated with pancytopenia include ionizing radiation, benzene, antimetabolites, lindane, chlordane, mustards, arsenics, chloramphenicol, trinitrotoluene, gold, hydantoin derivatives, and phenylbutazone.[1] The hematopoetic system of workers in many professions is at risk for chemical overexposure and resultant chemical sensitivity. Refinery and petrochemical workers are especially at risk as are workers whose jobs involve the use of pesticides. We have seen hair dressers who, due to repeated and widespread use of toxic chemicals in their profession, have not only had a high incidence of aplastic anemia, but a wide range of environmentally triggered vascular dysfunction as well (Figure 1).

Many agents such as folic acid antagonist drugs (methotrexate)[2] or antimalarials (pyrfmethanine chlorguanide)[3] will also give megaloblastic anemia.

Under normal circumstances, the bone marrow in humans is the primary source of all blood cells with the possible exception of lymphocytes. Marrow contains a complex mixture of cell types, which are either the direct precursors of circulating blood cells or which act in a still uncertain manner to nurture these precursors. At least three of the common blood cell types — the red blood cell, the platelet, and granulocytic white blood cell — are ultimately derived from a single precursor cell, the pluripotential stem cell. These

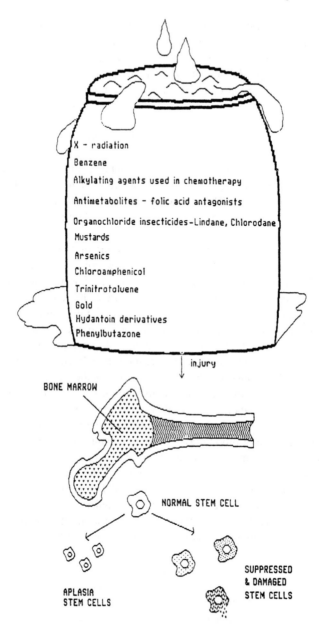

Figure 1.   Pollutants that damage bone marrow.

precursor cells can differentiate within the bone marrow into any one of the earliest identifiable forms of the more mature cells. The entire process is under the control of a feedback mechanism, whereby a deficiency in a cell type leads to increased production of that cell; for example, anemia leads to increased red blood cell production. The earliest identifiable precursors go through an orderly maturation process within the bone marrow, during which their function matures. This feedback system could be a prime target for pollutant overload because these are premature cells and more vulnerable to pollutants.

In the bone marrow, the toxic substances that affect the earliest stem cells, such as benzene or X-ray, tend to decrease the production of all three cell types and lead to pancytopenia. The following case is illustrative.

Case study— A 63-year-old white female was well until she noticed that she had become sensitive to the odors of many toxic chemicals after living in a gas-heated home for most of her life. She was short of breath and developed progressive weakness with a progressive yellow color to her skin. She was found to be anemic with a hemoglobin of 7.6 g%. She was admitted to the hospital. Her university physicians were unaware of her chemical sensitivity. Despite her repeated warnings that she could not tolerate substances contained in plastic bags, they gave her blood transfusions from plastic bags. After each transfusion, this patient developed a large area of gangrene on her abdominal wall (Figure 2). Each area of gangrene measured 10×10 cm and was extremely painful to touch. Biopsy of the area adjacent to the gangrene revealed a perivascular lymphocytic infiltrate. At this point, she was transferred to the ECU. Her admitting laboratory revealed a pancytopenia (Table 1).

After a week in the ECU when her lesions had begun to subside, this patient was given a half pint of blood drawn from and contained in a glass bottle. She experienced no problems. The unused portion of this blood was then stored in a plastic blood bag and given to her the following day. She immediately developed pain and edema at the periphery of her gangrene sites only. In addition, she developed a recurrence of her malaise, weakness, and yellow skin. The blood was stopped and the patient's symptoms cleared. Reinstitution increased her pain and edema. For several weeks in the ECU, the patient was treated with avoidance of pollutants in air, food, and water and with a rotary diet. She did well with relief of her pancytopenia.

Sufficient bone marrow damage markedly decreases total bone marrow cellularity, resulting in aplastic anemia. This illness can be fatal, particularly if exposure to an offending toxic substance continues unchecked. Death is usually due to hemorrhage because of lack of platelets or to infection due to lack of granulocytes. Bone marrow toxic substances including autoantibodies, which can be pollutant triggered, can also affect the development of just one cell type, leading to anemia, leukopenia, or thrombocytopenia.

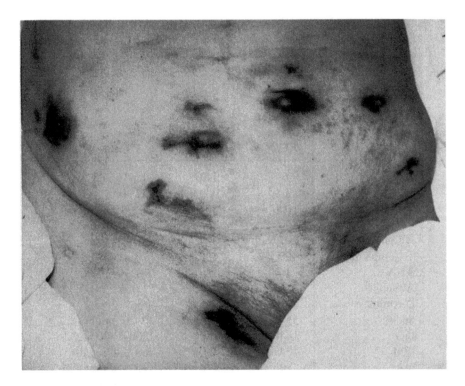

Figure 2.   Spontaneous bruising and gangrene from blood contained in plastic.

Cancer of bone marrow precursor cells can take a number of forms. Acute myelogenous leukemia is an uncontrolled proliferation of the earliest identifiable granulocytic cell precursor. Granulocytic cell types of greater maturity are found in chronic myelogenous leukemia. Treating tetanus with nitrous oxide, Gorsmen noted that on the fifth day of treatment granulocyte suppression occurred.[4] Lassen et al. studied 13 cases of tetanus, all of whom developed aplastic anemia after prolonged use of nitrous oxide.[5] Kieler et al., using various concentrations of nitrous oxide on mouse heart myoblasts, concluded that nitrous oxide was a mitotic poison.[6] In 1959, Lassen and Kristensen reported prolonged administration of nitrous oxide in myloid leukemia.[7] In two patients the total number of myloid white cells reduced after 5 to 15 days. The white blood cell count rose rapidly after withdrawal of nitrous oxide. In studies with rats, Okamoto showed a similar response.[8] Eastwood et al. confirmed these observations.[9]

There is a similar abnormal production of mature red blood cells (polycythemia vera), as well as other far less common neoplasms of red cells

Table 1. 63-Year-Old White Female in the ECU for 8 Weeks
(Pancytopenia with Severe Chemical Sensitivity)

	Laboratory		
	Initial Admission to the ECU	8 Weeks in the ECU	Control
Hb	7.6	11.0	12–16 g/dl
HCT	23	33	37–48%
WBC	1000	5000	$4.8–10.8 \times 10^3$
T Lymphocytes	250	1200	$1260–2650/mm^3$
B Lymphocytes	207	300	$82–477/mm^3$
T Complement	40%	100%	70–120%
ANA	+1:80	1:40	<1:40
Albumin	2.9	3.5	3.0–5.5 mg/dl
D combs	+	—	—
Sed	55	10	0–20 mm/hr
IgG	1440	1500	800–1800 mg/dl
IgM	254	260	60–280 mg/dl
IgA	282	270	90–450 mg/dl
IgE	3	5	1.0–180 Iu/dl
TP	5.1	5.2	6.8–8.5 gm/dl
T Eos	0		150–250
BUN	43	20	7–26 mg/dl
D_3	8 ng/dl	8	27–76 pg/ml
Platelets	50,000	200,000	$130–400 \times 10^3$

Transfusion—blood contained in glass: no problems.
Transfusion—blood contained in plastic bag: reproduction of gangrene and symptoms.
Long-term follow-up: patients did well.

and platelets. These disorders are to some extent clinically related, presumably reflecting the common nature of the stem cell. Certain of these cancers have been associated with exposure to radiation, benzene, and other solvents, and to alkylating agents used in cancer chemotherapy (e.g., a patient successfully treated for Hodgkin's disease with chemotherapy has a higher than usual risk of subsequently developing acute myelogenous leukemia).

According to the American Cancer Institute, approximately 85% of all cancer is due to environmental factors. The process by which chemicals produce cancer begins with a direct alteration of DNA that results in somatic mutation.[10] (See Chapter 4.)

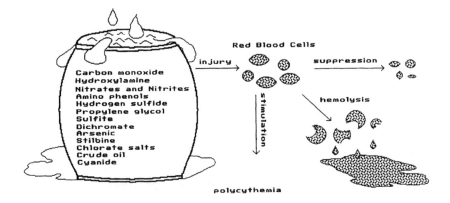

Figure 3.   Pollutant damage to red blood cells.

Red Blood Cells

Damage to the red cells can result in fragility, frank hemolysis, or stimulation giving polycythemia (Figure 3).

Carbon monoxide is well known to cause damage to the red cells. (For a complete discussion, see Chapter 11.) Levels of carbon monoxide have been found in smokers as well as people exposed to car exhausts.[11] In addition, levels are found due to endogenous catabolism of heme proteins (especially hemoglobulins), catalase, and the cytochromes.[12] The significance of these internally generated levels, coupled with the intake of the higher levels in the outdoor environment and passive smoking in the indoor environment, cannot be overlooked, since the cumulative effect may disturb many metabolic functions.

Methemoglobinemia can be generated by exposure to nitrates, especially in the young. Hydroxylamine will also trigger this condition as well as amyl nitrite, amino phenols, other aliphatic nitrites and nitrates, N-hydroxy-p-amino toluene, N-hydroxy-p-acetophenone, aromatic amines, arylnitro compounds, aniline, and nitro benzenes.[13] Hydrogen sulfide can produce hemolysis.

Heinz body hemolysis can be produced by the exposure of aromatic amines (aniline), nitro (nitro benzenes) compounds, phenols, propylene glycol, ascorbic acid, sulfites, dichromate, arsenic, stilbene, hydroxylamine, and chlorate salts.[14] Crude oil ingestion gives hemolytic anemia in birds.[15]

The following case represents hemolysis due to toxic chemical exposure that results in chemical sensitivity.

Case report — A 34-year-old white female entered with the chief complaint of hemolytic anemia. At 5 years prior to admission, this patient developed an unexplained episode of pneumonia which eventually cleared. At 4 years prior to admission, she developed colitis and recurrent staphlococceal vaginitis. At 2 years prior to admission to the ECU, she developed

Table 2. ECU: Hemolysis in a 34-Year-Old White Female

Medications	Date	Hb	Hct	Retic	LDH	(c-combs)
Prednisone (60 pp)	3/9/82	7.7	24.0	12.3	1000	+
None	3/23/82	10.5	33.0	4.2	600	–
None	4/5/82	10.8	34.0	4.2	446	–
None	4/9/82	11.3	35.0	0.9	197	–
None	4/11/82	11.9	36.8	0.5	175	–

Long-term Follow-Up at Home 1982–1983

Episodes of Hemolysis

1. 4 hr in front of TV camera
2. 1 day in city with papermill odor
3. Spoke in church — lots of odors
4. Malathion exposure

All cleared with only environmental control at home; prednisone was not used.

what was diagnosed as autoimmune hemolytic anemia of the IgG type. She was treated with 60 mg of prednisone per day, but due to a poor response, she had a splenectomy in May of the same year. After 1 month, she was able to stop taking prednisone; 9 months later, however, hemolysis began again. Prednisone (60 mg/day) was reintroduced, but it continued unabated, dropping to 7.7 g% with a reticulocyte count of 12.3. On admission to the ECU, her hematocrit was 24 and platelets were 716,000 mm^3. LDH was 1,000 units. She was combs positive. This patient was placed in the ECU and fasted for 6 days. A safe, less polluted water was found. During that week, she was gradually weaned off prednisone. Her symptoms cleared and her hemolysis stopped (Table 2).

The patient was then challenged by inhalation under controlled conditions in a double-blind manner with phenol, <0.0020 ppm, and chlorinated pesticides (2,4 DNP), <0.0034 ppm, to which she did hemolyze (Figure 4). The flames of an open gas stove also produced hemolysis. Formaldehyde, <0.2 ppm, and saline placebo did not trigger hemolysis. The patient created an oasis of less polluted materials at home. She ate less polluted food and drank less polluted water. She did not take any more prednisone. During the next year, exposure to pollutants resulted in four episodes of hemolysis, each of which cleared with her retreat into a less polluted environment. The environmentally induced hemolysis was due to odors from a TV camera, papermill, church, and malathion exposure (Table 2).

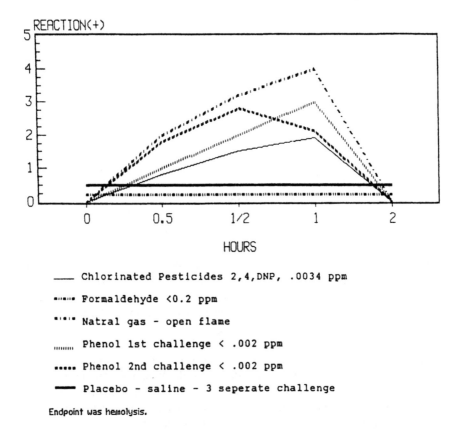

Figure 4.   Double-blind bronchial challenge exposure 15 min in the ECU after 4 days deadaptation with total load decreased.

Red blood cells can be damaged without hemolysis occurring, but with fragility resulting. This damage occurs only with exposure to certain, specific chemicals. Hydrogen cyanide and hydrogen sulfide may cause such a lesion. Red cells exposed to hydrogen cyanide and hydrogen sulfide are unable to utilize oxygen. These chemicals also interrupt the cytochrome oxidase electron transports as a result of inhibition of oxidative metabolism in which phosphorylation is compromised. Clinically, a flush occurs due to the high O_2 found in the venous return. Cyanide, which is found in cassava, amygdalin in fruit pits of peaches, apricots, and almonds, also directly stimulates the chemoreceptor of the carotid and aortic bodies with a resultant hyperpnea. Arrhythmias start, but the heart invariably outlasts the respirations. Death occurs due to respiratory arrest of central cellular origin. Nitro prusside has also caused this problem (for more information, see Chapter 4).

Thrombocytes (Platelets)

Thrombocytes are known to be adversely effected by pollutants, as seen in thrombocytopenia and in increased platelet fragility due to eminations from the plastics in a heart-lung machine.

Thrombocytopenia

A low platelet count may result from one or a combination of mechanisms. These include abnormal platelet production, disordered platelet distribution, and increased rate of destruction (Table 3).[16]

Defective production of platelets may result from (1) the administration of drugs such as the cytotoxic agents used in cancer chemotherapy (the most common cause of thrombocytopenia in a large medical center), gold, sulfonamides, and ethanol, and irradiation of the bone marrow; (2) generalized decrease in production of all marrow cells (aplastic anemia); (3) marrow replacement or infiltration, including marrow fibrosis (myelophthisis), leukemia, and metastatic carcinoma; and (4) congenital deficiency of a thrombopoietic factor.[17]

Defective maturation of platelets is associated with vitamin B_{12} or folate deficiency, the myeloproliferative disorders, paroxysmal nocturnal hemoglobinuria, and two hereditary disorders, the Wiskott-Aldrich syndrome and the May-Hegglin anomaly. In these disorders the bone marrow shows a normal or increased number of megakaryocytes, but thrombopoiesis is ineffective in a manner analogous to ineffective erythropoiesis.[18]

Thrombocytopenia may be caused by disordered distribution of platelets. In normal individuals about 30% of the circulating platelets are present in the spleen, whereas with massive splenomegaly, up to 80% of the total number of circulating platelets may be in the spleen, resulting in thrombocytopenia in the peripheral blood. Thrombocytopenia may occur in any patient with a significantly enlarged spleen. However, it is rarely severe enough to produce hemorrhagic disease, and splenectomy, though not usually required, will restore the platelet count to normal levels.[19]

Accelerated destruction of platelets is a frequent cause of thrombocytopenia. The survival time of platelets is markedly decreased from the normal of 10 days to less than 1 day. The most common causes of accelerated platelet destruction are antibody-mediated platelet injury, increased platelet utilization in disseminated intravascular coagulation, and massive blood transfusion.[20]

Antibody-mediated thrombocytopenia may be due to (1) autoantibodies in idiopathic thrombocytopenic purpura, systemic lupus erythematosus, chronic lymphocytic leukemia, and in association with autoimmune hemolytic anemia; (2) alloantibodies associated with pregnancy or transfusion; and (3) antibodies associated with the administration of certain drugs such as quinidine, quinine, and sulfonamides.[21]

Table 3. Causes of Thrombocytopenia

Production defect
  Reduced thrombopoiesis (reduced megakaryocytes)
    Marrow injury: Drugs, chemicals, radiation, infection*
    Marrow failure: Acquired*, congenital (Fanconi's syndrome, amegakaryocytic)
    Marrow invasion: Carcinoma, leukemia, lymphoma, fibrosis
    Lack of marrow stimulus: thrombopoietin deficiency
  Defective maturation (normal or increased megakaryocytes)
    B_{12} deficiency, folic acid deficiency*
Sequestration (disordered distribution)
  Splenomegaly
  Hypothermic anesthesia*
Accelerated destruction
  Antibodies
    Autoantibodies
      ITP, systemic lupus erythematosus, hemolytic anemias, lymphoreticular disorders
      Drugs*
    Alloantibodies
      Fetal-maternal incompatibility
      Following transfusions*
  Nonimmunologic
    Injury due to
      Infection
      Prosthetic heart valves*
    Consumption
      Thrombin in disseminated intravascular coagulation
      Thrombotic thrombocytopenic purpura*
    Loss by hemorrhage and massive transfusion

Source: Nossel.[16]
* Some are seen in chemical sensitivity.

Little literature suggests toxic damage to platelets other than by medication, which is legion.[22] Temporary thrombocytopenia after insertion of artificial organs, especially artificial lungs and hearts, however, does occur. Other than trauma, reaction to plastic surfaces has been suggested as a cause of this problem. At the EHC-Dallas, we have now seen six cases of idiopathic thrombocytopenic purpura that were triggered by toxic chemicals (Table 4). Removal of the toxic chemical showed improvement of the platelets, and challenge resulted in depression of the platelet count.

Randolph and other groups have observed cases of thrombocytopenia that appear to be reactive to pesticides.[23,24] A patient treated by Randolph and at the EHC-Dallas is illustrative of an environmentally triggered thrombocytopenia.

Case report — In 1956, this 36-year-old white male presented with a history of malaise, weakness, and a tendency to hemorrhage. On repeated occasions, his initial platelet count was in the range of 10,000. He was worked up and found to be sensitive to pesticides. He also was sensitive to several foods. He was instructed on how to clean up his home environment. He removed the gas heat and carpets and replaced them with electric heat and hard wood floors. He ate a rotary diet with less chemically contaminated foods and drank less polluted water. This patient's platelet count returned to the 200,000 range without medication, and he was able to resume his engineering trade. Over the next 20 years he noticed that if he stayed in a hotel room that had been recently treated with pesticide he would gradually develop fatigue, malaise, and a tendency to bleed. A platelet count taken at these times was always below 20,000. In 1975, this patient was worked up at the EHC-Dallas and again found to be sensitive to pesticides, less than six foods, and natural gas. He continued on his avoidance program and has been well to the present.

It is clear from our modest data that environmental triggers of some cases of thrombocytopenia can be found. Definition and avoidance of these triggers can prevent the use of medication and surgery in some patients (Table 4).

Thrombocytosis

Platelet production can also be stimulated by exogenous sources, resulting in thrombocytosis. We have also seen chemically triggered thrombocytosis as shown in the following case report.

Case report — A 55-year-old white, male dentist was admitted to the hospital with the chief complaint of thrombocytosis including headache, urinary frequency, abdominal bloating, and epigastric pain. He reported that approximately 2 to 3 years previously, he had had a sudden change in his personality. He developed a transient cerebral ischemic episode. He became very anxious and nervous and stated that he had developed a quick temper. In April of that year, his physician found a nodule on his thyroid. He was placed on Synthroid, which made him very nervous. At that time, he discontinued taking it for 2 weeks. Then he was placed on $^1/_2$ dose for 1 month, but still unable to tolerate it, he altogether discontinued using it.

On July 1, 1985, his platelet count revealed a significant elevation at 2,600,000/mm^3. While doing some yard work in outdoor heat of 107°F, he had become very weak and dizzy and collapsed. He reported that he had headaches prior to his awareness of his elevated platelet count. He also experienced severe epigastric pain. Along with the epigastric pain, he had also experienced some bloating and flatulence after eating. His family history was positive for heart disease and emphysema. He was a nonsmoker and nondrinker. He was living

Table 4. Environmentally Triggered Thrombocytopenia after 4 Days Deadaptation in the ECU with Total Load Decreased

Age/Sex/Race	Initial Platelets (mm^3)	After 1 week (mm^3)	Long-Term (1–10 years) (mm^3)	Triggering Agents		
				Biological Inhalants	Food	Toxic Chemicals[a]
60 years/w/f	30,000	200,000	200–400,000	–	–	+
38 years/w/f	15,000	400,000	200–400,000	–	–	+
45 years/w/f	40,000	150,000	150–300,000	–	+	+
40 years/w/f	60,000	170,000	150–300,000	+	+	+
50 years/w/f	50,000	140,000	140–200,000	–	+	+
47 years/w/f	30,000	100,000	100–150,000	–	–	+

[a] Inhaled double-blind: phenol, <0.0024 ppm; formaldehyde, <0.2 ppm; ethanol, <0.50 ppm; chlorine, <0.33 ppm; pesticide 2,4DNP, <0.0034 ppm.

Table 5. Platelet Counts in the ECU for 15
Days with Total Load Reduction

Days		Platelet count (mm³)
Fast days 1–6		2,680,000
6		691,000
7		666,000
8		667,000
9		662,000
10		597,000
11		604,000
12		647,000
14	at 7:08 a.m.	638,000
	at 10:50 a.m.	650,000
	at 1:35 a.m.	621,000
	at 4:30 p.m.	627,000
	at 7:00 p.m.	575,000
	at 9:30 p.m.	624,000
15	at 7:00 a.m.	691,000
	at 10:00 a.m.	657,000
	at 1:00 p.m.	625,000
	at 4:00 p.m.	620,000
Posthospital 2 weeks		508,000

Platelet count through the preceding 3 years in spite of chemotherapy was 2,600,000/ mm³. Admission revealed platelet counts of 2,600,000/mm³ with medication; 2,680,000/ mm³ without medication.

in a 15-year-old home with gas heating and carpeting. His physical exam was normal. He was diagnosed as having thrombocytosis, allergic cephalgia, toxic brain syndrome, irritable bowel syndrome, hyperlipidemia, chemical sensitivity, inhalant sensitivity, and he was admitted to the ECU where he was fed only spring water for 6 days. During his stay in the hospital, his condition improved and symptoms stabilized, including his platelet count (Table 5). His chest X-ray showed mild pleural thickening and scattered calcified granulomas.

Abnormal test results revealed that T_4 was depressed at 30% (32 to 54) and total T_4 at 634/mm³ (670 to 1800). T_4/T_8 ratio was depressed at 0.86 (1 to 2.7). Blood analysis showed elevated triglyceride at 346 mg/dL (30 to 158), elevated uric acid at 10.2 mg/dL (7.5 to 10.1), elevated cholesterol at 322 mg/dL (130 to 290), elevated platelet at 691,000/mm³, elevated EGPT (ery. glut. pyr. transaminase) at 1.6 index (<1.25; indicating low B_6), elevated EGP (ery. glutathione reductase) at 1.3 act. coeff. (0.9 to 1.2) (indicating low B_2), elevated KIN (kynurenic acid) at 3.3 mg/24 hr (indicating low B_6), and depressed vitamin C/WBC at 14.0 (21 to 53 $\mu g/10^8$ cells). His blood toxic chemical analysis revealed toluene at 0.7 ppb, 1,1,1-trichloroethane at 0.5 ppb, and tetrachloroethylene at 0.3 ppb. This patient was orally challenged on a monorotation diet of approximately 3 to 4 foods a day until he had 20 nonreactive, less chemically contaminated foods. He had sensitivity to two foods. As a continuation of his evaluation, the patient underwent intradermal challenge: 7 molds, lake algae, T.O.E., candida, dust, dust mite, 3 danders, 2 smuts, histamine, serotonin, weed mix, tree mix, grass mix, 5 chemicals, terpenes, cotton, fluogen, MRV, and 21 foods were individually tested. The results showed he had sensitivity to two chemicals, five foods, five molds, two smuts, histamine, serotonin, dust mite, weeds, trees, grasses, and three terpenes. Double-blind inhaled chemical challenge was performed in a steel and glass airtight booth. He had no reaction to any of the seven chemicals tested. He was sensitive to the fumes of natural gas. After he was discharged, he received immunotherapy and kept a rotary diversified diet, doing environmental clean up at work and home as best as he could. He felt better than he had in a long time; his platelet was also decreased to 505,000/mm^3, without medication. He removed the gas heat and carpeting from his home, thus creating an oasis. He also took all toxic chemicals from his dental office. He has continued to improve for the past 2 years. He has noted that total excess chemical exposure increases his platelet count. However, within a few days of clearing in a less polluted environment, it returns to normal without medication.

Leukocytes

Acute myelogenous leukemia due to benzene exposure is of particular interest from the standpoint of environmental health. Benzene is a ubiquitous environmental contaminant, being present in gasoline (1 to 5% of total product) and in a wide variety of chemical products and processes. Benzene compounds are found in over 50% of chemically sensitive patients. If other toxic substances with benzene rings such as toluene or xylene are included, this figure increases to 100%. In order to effectively regulate this human leukemogen, we need to know more about the mechanism by which its metabolism leads to bone marrow damage and the relationship of its aplastic actions to its leukemic actions.[25] We show later in this chapter that leukocytes may be damaged by toxic chemicals, especially pesticides, which will result in faulty responses to microbial agents and thus allow an individual to develop recurrent infections.

RETICULOENDOTHELIAL SYSTEM

The reticuloendothelial system appears to be particularly important in chemical sensitivity due to its primary role in disease resistance. Immunologically competent lymphocytes (T and B cells), accessory cells such as antigen-presenting cells of the Langerhan's type, and macrophages recirculate through the blood, lymph, and tissues of the body. They stop in a tightly regulated fashion in certain lymphatic tissues. The reticuloendothelial system composed of cells of the thymus, tonsils, Peyer's patches in the gut, lymph nodes, spleen, liver, and lymphatic can also be effected by pollutant injury. The principle cells in lymph nodes are lymphocytes in the lymphoid follicles and reticuloendothelial cells which line nodule sinuses. Each follicle located in the cortex of the node has a germinal center which contains the rapidly dividing large lymphocytes (B cells) and macrophages. Surrounding the germinal center is a cuff of densely packed small lymphocytes (T cells) which proliferate at a slower rate and which ultimately leave the node. Since the chief function of lymphocytes is to respond to antigen circulating through organs being drained off by lymphocytes, it is evident that they are prone to pollutant injury if a massive or too virulent exposure occurs. The cells either differentiate into plasma cells, B cells, or T cells with their subsets. Even the subsets may be prone to selective injury. Each type cell will be discussed in reference to pollutant injury which may give some insight into chemical sensitivity.

Thymus Gland — Function and Impairment

Basic to understanding the results of the thymus gland being insulted by a variety of factors (e.g., irradiation and chemical, etc.) is a knowledge of the function of this gland.

During fetal and early postnatal life, the thymus gland forms thymic cells that migrate through the body and become the precursors of sensitized lymphocytes. During the first 2 weeks or month postpartum, the thymus secretes a hormone that enhances proliferation of lymphoid tissue throughout the body. This interval is particularly vulnerable to the onset of chemical sensitivity due to the onset of pollutant injury.

A major function of the thymus is to provide the essential microenvironment for the differentiation and expansion of T lymphocyte subpopulations that are necessary for the development of immunoregulatory and cell-mediated effect or functions that are associated with mature T lymphocytes. Two types of distinct thymic function are effector and regulatory. Effector functions include cell-mediated cytotoxicity (killing of foreign antigen-bearing cells — allogeneic, virus-infected, or tumor cells) and lymphokine production (production of mediators of inflammation and nonspecific immunity — delayed-type hypersensitivity). Lymphokine inflammation is frequently seen in the chemically sensitive. Regulatory functions include helper (enhancement of B lymphocyte responses to thymus-dependent antigens), suppressor (diminishing of thymus-dependent B lymphocyte responses and T lymphocyte helper, cytotoxic functions, or delayed-type hypersensitivity), and amplifier (enhancement of T lymphocyte cytotoxic functions). Suppressor CMI cells are often depressed in chemical sensitivity.

Functional T lymphocytes and their precursors develop as a consequence of intrathymic and post-thymic events.[26] This development begins after hemopoietic progenitor cells migrate from the bone marrow to the thymus. The process involves extensive cellular proliferation, primarily in the outer cortex, and intrathymic migration through the cortex to the medulla before exiting the thymus to the peripheral lymphoid compartment, where functional diversification is completed.[27,28] The intrathymic differentiation process requires direct interaction of the lymphoid precursor cells with the thymic epithelial cells, which also produce several hormonal factors that act to promote maturation of precursor cells that have undergone a primary event by direct contact with the epithelium.[29] The molecular events that are operative at this early stage of differentiation are not fully understood. However, the result of these interactions is the selection of a T lymphocyte repertoire that has the capacity to recognize gene products of the individual's own major histocompatibility complex (MHC) antigens, but does not react with them.[30]

Maturing thymocytes express many enzymes of the purine salvage pathway as adenosine deaminase (ADA), purine nucleoside phosphorylase (PNP), and terminal deoxyribosenucleotidyl transferase (TDT). These enzymes mediate several defined functions enabled by subpopulations of T lymphocytes. For example, thymosin is critical to maintenance of immune competency. If thymosin becomes deficient or functionally impaired, severe immunodeficiency occurs and is manifested in altered T cell differentiation. Upon exposure to this hormone, the content of TDT in lymphocytes starts to decline, in keeping with other changes associated with the maturation of T lymphocytes.

When the thymus gland is affected by any agent(s) resulting in a compromised embryogenesis, a migration failure, or improper localization of thymic tissue, the development of primary immunodeficiency diseases may result (Table 6).

The ontogenic differentiation of T cell subpopulation is associated with crucial changes in the distribution of membrane markers. These markers are formed on precursor thymocytes, which are primarily found in the cortical area of the thymus, but are absent from most of the thymocytes that have matured and reached the medullary area. Other markers will appear as the thymocyte reaches further stages of differentiation, finally appearing on a vast portion of mature T lymphoctyes. The thymic gland promotes two categories of effects primarily in T lymphocyte differentiation — the expressions of T cell differentiation antigens and development of the subspecialized T cell functions.[31]

Irradiation, chemicals, or any other substances inimical to the thymus gland will basically impair differentiation and maturation of T lymphocytes, resulting primarily in inhibition of T lymphocyte subpopulation expansion which is necessary for the development of the complex immunoregulatory and cell-mediated effector functions that are attributed to mature T lymphocytes;

Table 6. Immunodeficiency Disorders Associated with an Abnormal Thymus Gland from Compromised Embryogenesis (No Apparent Relationship to Chemical Sensitivity)

Syndrome	Inheritance	Associated Findings
Combined Immunodeficiencies		
Severe combined immundeficiency	AR	Phagocytes absent
Reticular dysgenesis	AR	
Swiss-type	AR/XL	
agammaglobulinemia		
Thymic alymphoplasia	AR	
Adenosine deaminase AR	AR	Cartilage abnormalities
Ataxia telangiectasia	AR	Cerebellar ataxia
		Telangiectasia
		Chromosomal abnormalities
		Endocrine abnormalities
		Dyschondroplasia
Wiskott-Aldrich Syndrome	XL	Thrombocytopenia
Short-limbed dwarfism, type I	AR	Cartilage and hair hypoplasia
		Neutropenia
Cellular Immunodeficiencies		
Thymic hypoplasia	S	HI variable
		Hypoparathyroidism
		Cardiovascular abnormalities
		Megaloblastic anemia
Nezelof syndrome	AR	HI variable
Purine nucleoside phosphorylase deficiency	AR	HI variable
Short-limbed dwarfism, type II	AR	See type I

Source: Griffith.[208]
HI, humoral immunity; *AR*, autosomal recessive; *XL*, X-linked recessive; *S*, sporadic.

consequently, primary immunodeficiency diseases develop. Aldrin, an organo-chlorine pesticide, is known to damage the thymus gland, and it is thought that many other toxic organochlorines and other chemicals also do.[32] The result will be chemical sensitivity often with recurrent infections.

Tonsils and Adenoids

Since tonsils and adenoids are the prime response organs in the pharynx and nodes, they respond to excess food, microbes, biological inhalants, and chemicals.

Peyer's Patches of the Gut and Appendix

The gut-associated lymphatic Peyer's patches are an extensive system of responding lymphatic organs which are now well-known to counteract noxious substances present in the gastrointestinal lumen. These include food and their chemical contaminants, microbials including bacteria, fungus, and virus, and parasites. Damage to these areas may lead to malfunction of the gut and villi.

Macrophages

Monocytes and macrophages circulate and recirculate. Sufficient quantity and function are necessary in order to prevent chemical sensitivity. Monocytes produced in the bone marrow from the differentiated stem cells and precursors as CFU-GM and promoncytes are released into circulation. Under certain circumstances, monocytes-macrophages may be called to the spleen by chemo-tactic factors. In the spleen, monocytes are sequestered in the various filtration beds and there, undergoing transformation, may display a more fundamental function of the interdigitating cell. T cells, which are in close proximity with interdigitating cells, are thought to turn into white pulp for the immunologi-cally significant pathway. A follicular dendritic cell cousin of the interdigitating cell with capacities directed to the B cell and not the T cell constitutes another immunologic cell type. The system of veiled Langernhans interdigitating cells and the systems of monocyte and macrophage giant cells share certain charac-teristics, such as antigen processing, and differ in other respects. The veiled system is not phagocytic. The reticuloendothelial cells, including macroph-ages, may become mildly dysfunctional to being totally overwhelmed and paralyzed by toxic injury.

Pollutant Injury to the Spleen — Reticulocytes and Reticular Fibers

In the spleen, an effective filter of blood contains several types of filtration beds, each of which process the blood that flows through it. Reticular cells are large dendritic fibroblastic cells which form a delicate mesh work laced with argyrophilic reticular fibers that support hematopoietic colonies. Often this cell is included among the null lymphocytes. Hematopoietic cells of different stages of development live in this mesh work. Structurally, point-of-view arterioles open into the reticular mesh work, and veins drain from it. Reticular cells attach to blood vessels, forming an outermost advantitial layer of white pulp. The filters of white pulp are composed of lymphocytes and monocytes-macrophages, while red pulp contains red blood cells. Shunting of the splenic vessels occurs from arterioles to veins with pollutant damage such as by that phenyl hydrazine-affected lymphocytes. The creation of the barrier that leads to shunting depends upon changes in the reticular cells which form the filtration beds and recruitment of circulating stem cells. These cells become activated to intense protein synthesis and secretion. As the cells branch, they line up to form the shunt. The process is reversible and probably occurs often in the chemically sensitive (Figure 5).

The consequences or biological effects of toxic chemicals on the immune system can be measured by increased mortality or morbidity to an infectious agent, rapid onset of neoplasia, altered hematological parameters, changes in lymphoid organ/body weight ratio, changes in lymphoid organ cellularity, and changes in the histology of lymphoid organs. All of these changes have been seen from time to time in many chemically sensitive patients at the EHC-Dallas. Not all changes are present in a given individual.

Genetics may play a part in the development, severity, and clinical manifestations of chemical sensitivity, as discussed in Chapter 3. Clearly, immunological responders and nonresponders to toxic chemical overload exist and are emphasized by organophosphate agents such as the praxon in vitro model system.[33] However, environmental pollutants may overwhelm any individual depending on the conditions and level of exposure, independent of an individual's genetic predisposition.

Pollutant Injury to the Lymph Channels

Pollutants can damage lymph channels, causing swelling of cells and blocking of the channels.

Lymphatics

Lymphatic channels connect throughout the body. They also may be involved in pollutant injury. This injury can occur at any time in life. Once the

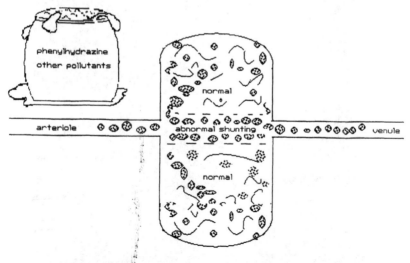

Figure 5.   Pollutant injury causing shunting through the lymph node or spleen from the arteriole directly to the venule.

injury occurs, swelling of the organ or tissue distal to the lymph channel blockage may follow. Substances known to cause lymph channel blockage (other than congenital malformations or absence) are underarm deodorants, toxic chemicals and particulates (coal, asbestos), wrong dose antigen injections, and foods as well as malignancies and parasites. This blockage may occur due to intrinsic endothelial swelling of the lymph channels or intranodal pathways or occasionally due to external compression.

The following is a case of total intrinsic extremity lymphatic blockage.

Case report — A 69-year-old white male presented at the EHC-Dallas with the chief complaint of a 6-month history of swelling of the left leg. Swelling started at the ankle and gradually increased until the whole leg was swollen. Although he experienced no pain, he could barely walk because of the size of his leg. Work-up that included a lymph-angiogram (Figure 6) was done and a total obstruction found. The rest of his work-up for malignancy and other potential causes was negative. A challenge rotary diet was set up, and he was found to be sensitive to wheat. After avoidance of wheat for two weeks, his leg swelling completely subsided, and this patient was able to walk normally. Oral challenge with wheat caused the leg to swell again. After 48 hours of avoidance, intradermal injection tests also confirmed the wheat reaction. The patient remains asymptomatic after 2 years of wheat avoidance.

Lymphocytes

Lymph nodes, spleen, and the reticuloendothelial system in the liver are known to swell in response to pollutant injury. In addition, many toxic

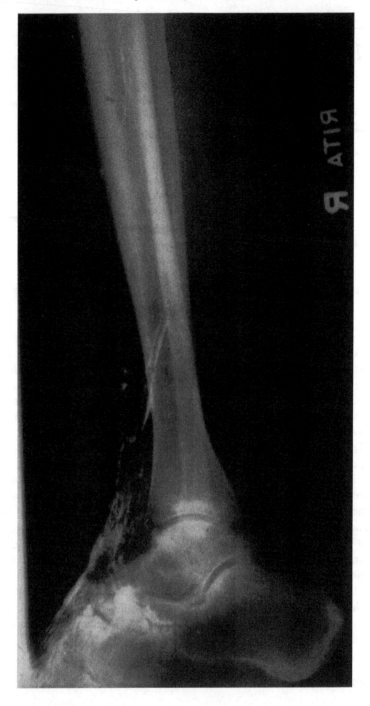

Figure 6. Lymphangiogram showing lymphatic obstruction from wheat.

chemicals are known to effect the cellular components of the immune system such as lymphocytes (Figure 7). Several general categories occur, producing immune suppression mostly via these cells. Included are benzene compounds, halogenated aromatic hydrocarbons (especially PCBs, PBBs, dibenzodioxins), polyaromatic hydrocarbons, urethanes, phorbol diesters, insecticides (organophosphates, organochlorines, and carbamates), ozone, nitrous oxides, and airborne metals (nickel, zinc, magnesium, lead, cadmium, and Hg).[34] (See Chapters 12 aand 13.) Organotins, used primarily as heat stabilizers, catalytic agents, and antifungal antimicrobial compounds, also are immunosuppressors.[35] Drugs like alkylating agents, steroids, antimetabolites, estrogens, and mold toxins also suppress the immune system. Autoimmunity (see Chapter 4) may be induced by metals and organic compounds.[36]

Pollutant Damage to Immune Responses

A number of environmental chemicals and drugs have been investigated for their immunotoxic potential. Growing evidence suggests that various chemicals possess immunotoxic properties which may result in chemical sensitivity. A chemical may be either immunosuppressive or immunostimulatory or both; clearly, they are immunoderegulatory with the same chemical exhibiting both of these properties, depending on the dose regimen, type of chemical,[37] and total toxic load of the individual. All three types of responses are seen upon evaluation of the immune system of selected chemically sensitive patients (Figure 8).

Toxic chemicals can cause malignant or nonmalignant changes. Usually DNA, RNA, or replications are part of malignant change. Both malignant and nonmalignant changes are discussed in this section.

Toxic chemicals can alter the immune system in several ways (Table 7). They can produce damage to an organ(s) or tissue(s). An autoimmune response to the altered tissue follows, or the autoimmune response can occur via cellular or molecular inhibition of the expression of immune responses against antigens. Allergic sensitization to chemicals is well known. Most chemicals, although not antigenic in nature by themselves, can bind with host proteins and then elicit an immune response as haptens. Both cell-mediated and antibody-type responses may be evoked, and based on the way they produce tissue injury, have been characterized as types I through IV mechanisms.[38]

Chemicals can alter protein, causing haptens. They can also change the antigen recognition site on cell membranes. This phenomenon may allow increased sensitivity and/or explain one of the reasons the spreading phenomenon of chemical sensitivity occurs. If antigen recognition is not functioning properly, both related and eventually unrelated substances might cause problems.

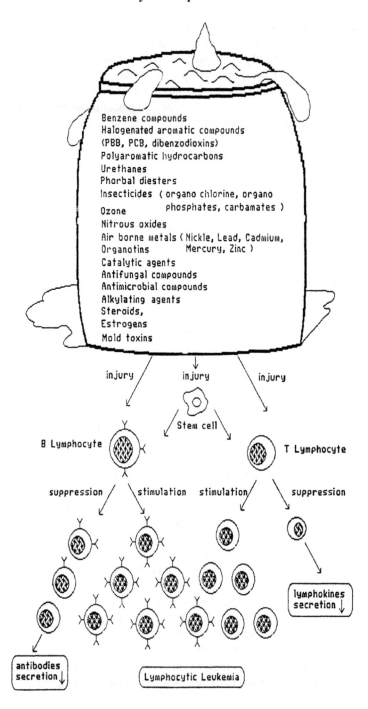

Figure 7.   Pollutants known to affect lymphocytes.

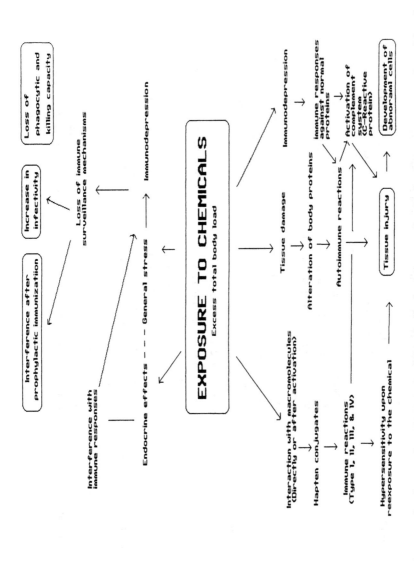

Figure 8.   A general scheme of various types of toxicologic responses in the immune system produced by chemicals.

Table 7. Some Possible Immunotoxicological Mechanisms and Associated Effects and Conditions Possibly Associated with Chemical Sensitivity

Immunologic Nature	Effect or Mechanisms	Chemical Agent or Associated Condition
General	Damage to primary lymphoid organs	Radiation, TCDD, organotins
	Effect on stem-cell-producing site	Chloramphenicol, benzene, radiation
	Deficiency of polymorphonuclear leukocytes	Infections, chronic glomerulomatous disease, myeloperoxidase deficiency
	Inhibition of protein or nucleic acid synthesis	Antimetabolites, alkylating agents (cyclophosphamide)
	General stress mechanisms	Lack of nutrients, steroid hormones
Specific cellular	Interference with interactions between T cell and antigen	Cancer antigens, chronic infections
	Lack of specific lymphocytes (selective destruction)	Thymic aplastia, antilymphocytic serum
	Defective lymphocyte proliferation	Chronic infection
	Defective lymphokine production	Chronic mucocutaneous candidiasis
	Stimulation of T-cells	Levamisole
Nonspecific cellular	Maturation defect	(Hypothetical)
	Lack of lysozymes	Acute viral infection, steroids, phenothiazines
Humoral	Alteration of lymphocyte receptors	Alkylating agents
	Absence of B-lymphocytes	Specific antibodies to B cells; Bruton's sex-linked agammaglobulinemia

Category	Mechanism	Examples
	Failure of differentiation to plasma cells	Defective antigen presentation macrophage depression, lack of IgG and IgA with raised IgM
Complement related	Deficiency of complement (notably C3)	Vinyl chloride, infections, tumors
erythroblastosis	Complement activation	Transfusion reactions, fetalis, hemolytic reactions to drugs or chemicals
Macrophage	Blockade of reticuloendothelial system	Saturating dose of carbon
	Macrophage stimulation	Adjuvant, endotoxin, *Listeria monocytogenes*
Enzyme inhibition	Inhibition of esterases	Organophosphates
	Stabilization of lysosomal membrane	(Hypothetical), metals
Immunotolerance or cellular anergy	Antigen-induced competition with recognition receptors	High dose of monomeric flagellin, tolerance to thyroglobulin
	Immune complex blocking factors	Certain tumor immunity
Autoimmunologic	Antibody-induced immunosuppression	SRBC antibodies
	Production of antibodies or specific cells against normal tissues	Allergic encephalomyelitis, organ damage with specific antibodies

Source: Street, J. C. "Pesticides and the immune system," in *Immunologic Consideration in Toxicology,* Vol. I, R. P. Sharma, Ed. (Boca Raton, FL: CRC Press, 1981) pp. 45–66.

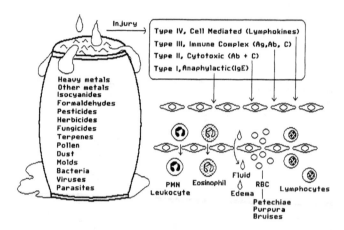

Figure 9.   Pollutant injury to the immune system of the blood vessel wall.

Transport enzymes in the membrane may be altered and cause edema if the mineral pumps, like the sodium pump, are not working. In addition, hormone receptor sites may be altered, explaining why some chemically sensitive patients have severe premenstrual, thyroid, adrenal, and endocrine problems. (See Chapter 24.) Once pollutant injury occurs, immune dysfunctions followed by blood vessel deregulation via the autonomic nervous system will occur.

The immune responses on the blood vessels, cell wall (Figure 9), or other involved organs can be through any of the four types of hypersensitivity or immune mechanisms and frequently can be a combination of two or more.

Type I: Hypersensitivity

Hypersensitivity is usually mediated through the IgE mechanisms on the cell wall. Classic examples are angioedema, urticaria, and anaphylaxis due to sensitivity to pollen, mold, dust, animals, food,[39] or some chemicals such as toluene diisocyanate.[40] Of the immune responses of patients with chemical sensitivity who have been seen at the EHC-Dallas, 10% fall within this category.

Anaphylaxis as a Manifestation of Type I Hypersensitivity. Anaphylaxis is defined by some as an acute reaction to a specific antigen characterized by respiratory distress due to bronchoconstriction followed by vascular collapse. Shock may occur without the antecedent respiratory distress. It results from an explosive generation and release of a variety of potent, biologically active mediators and their concerted effects on various tissues and organs. Cutaneous manifestations exemplified by pruritus and urticaria with and without angioedema are characteristics of such systemic anaphylactic reactions. Anaphylaxis has also been defined as IgE-mediated antigen-induced reactions, but this definition is too narrow. We, as well as Frankland and others, prefer to think of anaphylaxis as a descriptive term for symptoms producing a severe, abrupt event of varied significance.[41,42] Anaphylaxis appears to be a clinical syndrome that has multiplicity of inciting etiological agents with a variety of pathological mechanisms involved. Our observation of patients at the EHC-Dallas makes apparent that usually a series of smaller events leading up to the severe reactions that result in anaphylaxis occur rather than one sudden, unexplained event.

The following case report emphasizes the importance of understanding the multiplicity of triggering agents that lead up to anaphylaxis.

Case report — A 30-year-old nurse entered the EHC-Dallas with the chief complaint of chronically recurring idiopathic anaphylaxis. She presented with a history of having taken antibiotics for vaginitis continuously for 3 years. Prior to admission, she had recurrent, multiple attacks of anaphylactic shock. She presented at the EHC-Dallas wearing a gravity suit and holster containing multiple antianaphylactic medication including epinephrine. She had been told by the local medical school and the National Institutes of Health that there was no known cause or treatment for her anaphylaxis, and the best she could do was wear the gravity suit and be prepared for a short life.

This patient was placed in a less polluted room where she fasted for 4 days and then was worked up. She was given oral, inhaled, and intradermal challenges with biological inhalants, foods, chemicals, and hormones. She was found to be sensitive to biological inhalants (15 molds, lake algae, candida, dust mite, ant bite), foods (17 foods), drugs (penicillin), chemicals (orris root), and hormones (estrogen, progesterone, leutenizing hormone).

After careful history and observation, it became clear that this patient went into shock with each menstrual cycle. She was placed on good home environmental control, a rotary diet of chemically less contaminated foods, and injection therapy for biological inhalants, foods, and hormones. Over a period of 6 months, she gradually improved. She stopped wearing her gravity suit and taking epinephrine. She has been well and working daily for 2 years without any significant episodes of anaphylaxis.

This patient found that she would have to avoid the foods to which she was sensitive and take her foods injections in order to avoid premenstrual anaphylaxis. As she improved, she found that she could be lax on her diet or her hormones or her exposure to mold if each of the other categories were under control. These findings emphasize the multiple causative factors of anaphylaxis and demonstrate the importance of the total load phenomenon.

In our experience, reduction of the total load, as was done with this patient, increases the likelihood that anaphylaxis will not recur. Immunologically, anaphylaxis is designated as those allergic reactions which are initiated by antigen reacting with anaphyllactically sensitized mast or basophil cells in the tissue. Though the IgE antibody is usually involved, IgG anaphylactic antibody may also be seen. Some anaphylactic reactions result from aggregation and mediator release from platelets when induced by antigen antibody complexes. It should be emphasized that in many clinical conditions no antigen or specific antibody can be found, although some pharmacological substances are released which cause the reaction. This type reaction has been referred to as "anaphylactoid." Throughout these subsequent volumes, we show that nonimmune mechanisms for the release of mediators may occur, causing these anaphylactoid reactions.

At the EHC-Dallas, 100 cases of anaphylaxis have been seen. Triggering agents were always multiple, and anaphylaxis occurred when the total load increased. The patients improved with a reduction in their total pollutant load, the use of injection therapy, rotary diets, and a course of management similar to the case reported.

This aggregation and mediator release may be the way some foods and chemicals induce a reaction. It should be pointed out that though they may severely effect other parts of the immune system, some chemicals (e.g., DDT) actually prevent anaphylaxis by degranulating the basophils.

Type II: Hypersensitivity — Cytotoxic Mechanisms

Damages may occur with direct injury to the cell. A clinical example of this is seen in patients exposed to mercury.[43] (See Chapter 11.) In Minimata, Japan, a group developed neurological disease from eating fish exposed to methyl mercury chloride, a toxic substance. The result was in triggering of this mechanism.[44] (See Chapter 26.) At the EHC-Dallas, 20% of immune responses of patients seen fall into this category.

Type III: Hypersensitivity — Immune Complex

In type III hypersensitivity, complement and gamma globulin combine and damage the vessel wall. A clinical example of this mechanism is lupus vasculitis.

More than 19 chemicals including rocaineamide and chlorothiazide[45] are known to trigger the autoantibody reaction of pseudolupus. (See Chapter 23.) Many other toxic chemicals such as beryllium, nickel, and platinum have been shown to trigger the autoimmune response. Other chemicals such as vinyl chloride[46] will produce microaneurysms of small digital arterioles, probably due to this mechanism. Of our chemically sensitive patients, 20% have these type III reactions.

Type IV: Hypersensitivity — Cell-Mediated Immunity

Type IV hypersensitivity occurs with triggering of the T lymphocytes. Numerous chemicals such as phenol, pesticides, organohalides, and some metals will mediate immune responses by activating T lymphocytes, triggering lymphokines, and resulting in type IV reactions.[47] Of the chemically sensitive patients seen at the EHC-Dallas with immune dysfunction, 50% have impaired T cells and manifest environmentally triggered vasculitides. Clinical examples are polyarteritis nodosa, hypersensitivity angiitis, Henoch-Schonlein purpura, and Wegner's granulomatosis.[48] A recent study performed at the EHC-Dallas on 104 proven chemically sensitive individuals (70 vascular, 27 asthmatic, and 7 rheumatoid) compared with 60 normal controls showed that those manifesting a chemical sensitivity through their vascular tree had over 4 S.D. suppression of their suppressor T cells[49] (see later section in this chapter and Chapter 30).

Lymphocytic blastogenesis is impaired in 25% of the chemically sensitive patients who have immune changes seen at the EHC-Dallas. Impaired T cell function as evidenced by delayed skin reactivity is also demonstrated in 25% of our chemically sensitive patients. Though there is significant overlap, many patients with low T cells do not have impaired blastogenesis, while many other patients have normal T cells with impaired blastogenesis. Clearly, the larger portion of our patients' immune responses fall into type III and type IV categories (see Chapter 30). Some 17% of the chemically sensitive patients have positive auto-antibodies. However, 70% of the hospitalized patients have positive autoantibodies. Migratory inhibitory factor to foods was also found to be impaired in some chemically sensitive patients (Table 8).

Proinfectious qualities have been attributed to many pollutants. Some toxic chemicals such as hexachlorobenzene and pentachlorophenol, frequently found in chemically sensitive patients, will suppress the bacteriocidal or viricidal capacity of the neutrophils, thus allowing for recurrent infections. Many toxic chemicals such as DDT have been found in the blood of our chemically sensitive patients[50] and have been shown to damage the killing capacity of polymorphonuclear leukocyte to bacteria and viruses.[51] Toxic chemicals may also decrease the number of responder plasma cells in lymph nodes. Many others such as Milbex and Anthio inhibit phagocytotic activity[52] (Figure 10).

Hexachlorobenzene (HCB) and polychlorinated biphenyl (PCB), found in the blood of 51% of our chemically sensitive patients, now have been shown to reduce resistance to infection by 20 to 30% in studies in mice.[53] This lack of biological killing capacity has been repeatedly observed in a subset of our patients with recurrent respiratory infection and in many patients with recurrent

Table 8. Migratory Inhibitor Factor, Ages
20 to 60 Years, Chemically Sensitive

Patient	Agent	Dilution, Inhibition (%)			
WF	Beef	1:10, 57	1:50, 29		1:500, 68
BF	Corn	1:10, 53	1:50, 13	1:100, 0	1:500, 17
WF	Oranges	1:10, 78	1:50, 43	1:100, 38	1:500, 45
WM	Oranges	1:10, 67	1:50, 36	1:100, 23	1:500, 49
WF	Corn	1:10, 56	1:50, 21	1:100, 0	1:500, 0
WM	Corn	1:10, 34	1:50, 12	1:100, 39	1:500, 27
WM	Corn	1:10, 58	1:50, 0	1:100, 0	1:500, 0
WF	Beef	1:10, 41	1:50, 27	1:100, 66	1:500, 0
BF	Chicken	1:10, 57	1:50, 67	1:100, 50	1:500, 40
WF	Pork	1:10, 60	1:50, 38	1:100, 47	

Blastogenesis PWM, PHA<7.0–28%; 158 patients.
There was a 99.9% correlation with oral food challenge performed under environmentally controlled conditions in the ECU.

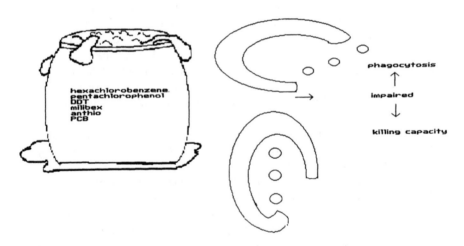

Figure 10.   Effects of toxins on leukocytes.

vaginal infections. As these patients eliminate toxic chemicals from their bodies, recurrent infections stop. Reinfection can almost always be traced to a toxic reexposure. There are overt foods as well as chemicals which consistently trigger acute infection. This was especially noted in women with yeast vaginitis and children with otitis media at the EHC-Dallas.

Mounting evidence supports the involvement of toxic chemicals in either suppressing or stimulating the immune system (Table 9).

**Table 9. Important Chemical and Physical
Agents Shown to Influence the Immune System**

Types	Selected Examples
Dusts and fibers	Cotton, organics, coal dusts, silica, asbestos fibers
Halogenated hydrocarbons	Polychlorinated biphenyls, polybrominated biphenyls, pesticides, dioxins, benzofurans
Organophosphorus carbamate pesticides	Parathion, methylparathion
Plasticizers and plastic monomers	Diisocyanates, organotin compounds, vinyl chloride, styrene, formaldehyde
Metals	Mercury, lead, cadmium, nickel, arsenic, chromium, selenium, zinc
Electromagnetic radiation	X-ray, microwave, ionizing radiation
Food additives and contaminants	Antioxidants, antibiotics, endotoxins, mycotoxins, pesticides, coloring, flavors
Drugs	Alkylating agents, antibiotics, cyclophosphamide, anesthetics, diethylstilbestrol
Biological antigens	Venoms, bacterial toxins, molds, foods
Miscellaneous	Cigarette smoking, marijuana

Hypersensitivity to certain toxic materials, such as endotoxins, has been suggested as a result of an immunosuppression.[54] Chemicals that inhibit the synthesis of protein or nucleic acid (e.g., cyclophosphamide, actinomycin D) can also prevent or modify the synthesis of antibodies, lymphokines, or various enzymes that are important for the expression of immunity.[55] Cyclosporine is made from two strains of ground mold, the fungi imperfecti and the tolypocladiun inflatum. This drug is used in transplantation because it stops the rejection process. It has selective immunosuppression of T cell functions while altering the balance of immune-regulating cells. It reduces T_4 helper cell activity while knocking out the resistance of the Epstein-Barr virus. Also, inhibition of T cytotoxic cells, T helper cells, and some T-independent factors occurs. Differential effects of B cells may be present with either suppression or stimulation occurring.[56] Cyclosporin may be able to produce one type model of chemical sensitivity. In the case of occupational or environmental chemicals, reports have indicated that prolonged depression of immune responses can increase the cancer rate in patients (Table 10).[57]

Numerous other examples of toxic chemicals affecting the immune system and causing malignancy are cited below (Table 10).

Table 10. Substances That Are Carcinogenic in Animals and May Be in Humans or May Be Promoter

2-Acetylaminofluorene
Adriamycin
2-Aminoanthraquinone
1-Amino-2-methylanthraquinone
Amitrole
o-Anisidine and o-a-HCl
Aramite
Benz(a)anthracene
Benzo(b)fluoranthene
Benzotrichloride
Bischloroethyl nitrosourea
Cadmium and certain cadmium compounds
Carbon tetrachloride
1-(2-Chloroethyl)-3-cyclohexyl-1-nitrosourea (CCNU)
Chloroform
4-Chloro-o-phenylenediamine
p-Cresidine
Cupferon
Cycasin
Dacarbazine
DDT
2,4-Diaminoanisole sulfate
2,4-Diaminotoluene
Dibenz(a,h)acridine
Dibenz(a,j)acridine
Dibenz(a,h)antracene
7H-Dibenzo(c,g)carbazole
Dibenzo(a,h)pyrene
Dibenzo(a,i)pyrene
1,2-Dibromo-3-chloropropane
1,2-Dibromoethane (EDB)
3,3′-Dichlorobenzidine
1,2-Dichloroethane
Diepoxybutane
Di(2-ethylhexyl-phthalate
3,3′-Dimethoxybenzidine
4-Dimethylaminoazobenzene
3,3′Dimethylbenzidine
Dimethylcarbamoyl chloride

Indeno(1,2,3-cd)pyrene
Iron dextran complex
Ketone® (Chlorodecone)
Lead acetate and phosphate
Lindane and other isomers
2-Methylaziridine
N,N-dimethyl benzenamine
4,4-Methylenedianile and dHCl
Methyl iodide
Metronidazole
Michler's ketone
Mirex

Nitrilotriacetic acid
5-Nitro-o-anisidine

Nitrofen
2-Nitropropane
N-Nitrosodi-n-butylamine
N-Nitrosodiethanolamine
N-Nitrosodiethylamine
N-Nitrosodimethylamine
p-Nitrosodiphenylamine
N-Nitroso-n-propylamine
N-Nitroso-N-ethylurea
N-Nitroso-N-methylurea
N-Nitrosomethylvinylamine
N-Nitrosomorpholine
N-Nitrosonornicotine
N-Nitrosopiperdine
N-Nitrososarcosine
Norethisterone
Phenazopyridine and HCl
Phenytoin and salt
Polybrominated biphenyls
Polychlorinated biphenyls
Progesterone
1,3-Propane sultone
B-Propiolactone
Propylthiouracil
Reserpine

Table 10 (Continued). Substances That Are Carcinogenic in Animals and May Be in Humans or May Be Promoter

1,1-Dimethylhydrazine
1,4-Dioxane
Direct Black 38
Epichlorohydrin
Estrogens (not conjugated)
Ethylene oxide
Ethylene thiourea
Formaldehyde (gas)
Hexachlorobenzene
Hexamethylphosphoramide
Hydrazine and sulfate
Hydrazobenzene

Saccharin
Safrole
Selenium sulfide
Streptozotocin
Sulfallate
2,3,7,8-Tetrachlorodibenzo-p-dioxin (TCDD)
Thioacetamide
Thiourea
Toluene diisocyanate
Toxaphene
2,4,6-Trichlorophenol
Tris phosphine sulfide
Tris(2,3-dibromopropyl)phosphate
Urethane

Recognized Human Carcinogens

Compound	Site Affected or Tumor Type
4-Aminobiphenyl	Urinary bladder
Analgesic mixtures containing phenacetin	Kidney, urinary bladder
Arsenic and certain arsenic compounds	Skin, lung, liver
Asbestos	Lung, peritoneum or pleura (mesothlioma)
Azathioprine	Lymphoma, skin
Benzene	Leukemia
Benzidine	Urinary bladder
N,N,-Bis(2-chloroethyl)-2-naphthylamine (chlornaphazine)	Urinary bladder
Bis(chloromethyl) ether and technical grade chloromethyl methyl ether	Lung
1,4-Butanediol dimethyl-sulfonate (Myleran)	Breast, leukemia
Certain combined chemotherapy for lymphomas	Leukemia
Chlorambucil	Leukemia
Chromium and certain chromium compounds	Lung

Table 10 (Continued). Substances That Are Carcinogenic in Animals and May Be in Humans or May Be Promoter

Coke oven emissions	Lung, urinary tract
Conjugated estrogens	Liver, endometrium, ovary, breast
Cyclophosphamide	Leukemia, urinary bladder
Diethylstibestrol	Vagina (in utero)
Melphalan	Leukemia
Methoxsalen with ultraviolet A therapy (PUVA)	Skin
Mustard gas	Respiratory tract
2-Naphthylamine	Urinary bladder
Nickel refining	Nasal cavity, lung
Thorium dioxide	Liver
Vinyl chloride monomer	Liver, brain, lung

Probable Human Carcinogens

Acrylonitrile	Nickel and certain nickel compounds
Aflatoxins	Nitrogen mustard
Benzo(a)pyrene	Oxymetholone
Beryllium and beryllium compounds	Phenacetin
Combined oral contraceptives	Procarbazine
Diethyl sulfate	o-Toluidine
Dimethyl sulfate	

ORGANOCHLORINE PESTICIDES

Organochlorine compounds are produced by man and nature in large quantities.[58,59] Organochlorines are persistent chemicals and therefore tend to accumulate in the biosphere. When metabolized, they are mostly bioactivated to give rise to free radicals, which lead to the propagation of further free radicals and/or lipid peroxidation.[60] Hence, organochlorines have the potential of causing damage to the lipid bilayer of biological membranes, as previously shown in Chapter 4.[61] Jensen and Clausen reported that for some organochlorine pesticides there is an age-related increase in their concentration in human adipose tissue.[62] However, in our experience with chemical sensitivity, that may not be entirely true. Our chemically sensitive patients with the highest levels have been in the 20- to 30-year age range and always had extremely high exposures. The Australian patients seen at the EHC-Dallas (Table 11) and by Little et al.[63] seem to have the greatest quantity and highest

Table 11. Chlorinated Hydrocarbons Found
in Some Chemically Sensitive — EHC-Dallas

Hexachlorobenzene
Benzene hexachloride
Chlordane
Aldrin
Dieldrin
DDT
DDE
Endrin
Chloroform
1,1,1-Trichloroethane
Trichloroethylene
Tetrachloroethylene
Dichlorobenzene

levels of organochlorine pesticides. This is true of our patients who worked around the greatest levels of exposure. The presence of organochlorine pesticide suggests that the patients' detoxification systems could not keep up with the overload.

The chlorinated pesticides have a strong tendency to accumulate in animal tissue and to persist in the environment.[64] The impact of these chemicals on immune function has been studied fairly extensively. DDT has been shown to cause suppression of immunological reactivity.[65-69] It also degranulates mast cells, leading to prevention of anaphylactic reactions to immune challenge.[70-72]

Another study by Liang and Rea[73] at the EHC-Dallas, detailed in Chapter 13, shows that the greater the number and the higher the level of organochlorine pesticides, the lower the level of T cells.

Many studies have reported decreased ability to mount a normal antibody response with exposure to DDT;[74-77] this effect appears to be dose and/or phase dependent.[78-81] Other effects reported include decreased leukocyte counts and phagocytic activity,[82,83] reduced levels of IgG with elevated IgM,[84] changes in the spleen and thymus,[85] changes in lymph nodes,[86,87] variations in complement activity,[88] disturbances in fetal and perinatal regulatory mechanisms,[89] and phase variations of specific and nonspecific indices of reactivity.[90]

We have observed a similar phenomena of immune deregulation in some chemically sensitive patients. Of the patients who are sick enough to meet the criteria for hospitalization in the ECU, 50% have no skin whealing when initially tested intradermally, even though symptoms are provoked (Table 12). After a period of 2 to 3 weeks following reduction of the total toxic chemical

Table 12. Skin Sensitivity in 1000 Chemically Sensitive Patients —
EHC-Dallas

Percent Chemically Sensitive Reaction	T-lymphocytes	IgE	Skin Reactivity	Systemic Reaction
10	Normal or low	High	High	Low
30	Normal or low	Low	High	High
10	Normal or low	Low	Delayed from 6–48 hr	High
50	Normal or low	Low	Absent	High

load, however, the skin begins to wheal and intradermal neutralization of symptoms can be accomplished. An occasional very ill patient will not start whealing until 4 to 8 months. Delayed hypersensitivity is also impaired in these patients.

Other organochlorine pesticides, for example, aldrin,[91] lindane,[92,93] hexachlorobenzene,[94-97] mirex,[98] and Arochlor (a PCB)[99-103] are immunosuppressive. Effects include disorders of the adrenal glands and thymus (aldrin),[104] decreased ability to mount a normal antibody response (lindane),[105] impairment of host resistance (Arochlor),[106] and suppression of spleen cell activity (PCB and hexachlorobenzene).[107] All of these toxic pesticides have been found in some chemically sensitive patients. In addition, in our experience, their immune systems improve following the clearance of these toxic substances from their body.

Immunosuppression and deregulation is not limited to the organochlorine pesticides; similar results have been found with the carbamates Sevin,[108-112] and carbaryl[113-115] and with the organophosphates Anthio,[116] malathion,[117] leptophos,[118] chlorofos,[119-122] and parathion.[123124] Parathion has been shown to depress beta cellular and humoral responses.[125] Interestingly, the pyrethroid pesticide resmethrin stimulates cellular immune response.[126]

The polybrominated biphenyls (PBBs) also have been studied extensively, with much of the research stemming from an accidental feeding of a PBB fire retardant to cattle in Michigan in 1973. Tissues from the cattle showed infiltration of lymphocytes into liver, kidney, small intestine, and lungs. Susceptibility to infection and a high rate of immunological activity increased.[127] Farm residents exposed to this chemical showed an increased sensitivity to recall antigens, deviations in lymphocytes, and other defects in immunological cells.[128,129] Laboratory animal studies have shown alterations in B and T cell functions,[130-132] decreases in IgG levels,[133] and other immunosuppressive effects[134,135] after exposure to PBB.

TETRACHLORODIBENZO-p-DIOXIN (TCDD) AND TETRACHLORODIBENZOFURAN (TCDF)

TCDD is a herbicide as well as a commercial chemical. Results of animal studies show this compound to be immunosuppressive,[136] with major effects on cell-mediated immune functions (suppression of T cells).[137-139] Offspring of mice fed TCDD showed increased sensitivity to *Salmonella*.[140]

TCDF, a polychlorinated biphenyl, also suppresses cell-mediated immunity with slight effects on humoral immunity.[141]

Metals

Cutaneous hypersensitivity is the most obvious effect of exposures to metals; however, common effects of Pt, Au, Hg, Cr, Ni, and Be include conjunctivitis, rhinitis, asthma, urticaria, contact dermatitis, nephrotic syndrome, and proteinuria.[142]

In experimental beryllosis, antibody to normal lung tissue has been found; in beryllosis patients, there are antibody reactions to DNA, RNA, heart, spleen, liver, and thyroid tissue.[143] Increased numbers and activity in T cells have been observed in both exposed workers and beryllosis patients in remission.[144] Antiberyllium antibodies are highest in those with beryllosis.[145] Skin sensitivities to Be compounds involve a delayed type IV reaction.[146]

Studies of environmental exposure to cadmium show suppression of antibodies as well as cellular immunity, persisting for several months.[147] Studies suggest that cadmium (as well as lead and mercury) may alter synthesis of cellular DNA, affecting lymphocytes directly and influencing antibody production.[148] The suggestion also has been made that cadmium and lead suppress humoral response through depression of splenic B lymphatic responses.[149] Cadmium suppresses the mononuclear phagocytic system.[150]

Methylmercury has a direct cytotoxic effect, followed by detoxification through macrophages.[151] This metal affects the B cell and plasma-cell synthesis of IgM antibody.[152] Interestingly, methylmercury and selenium act synergistically to increase antibody synthesis.[153] Other metals shown to have effects on the immune system include nickel[154] and copper.[155]

STUDIES AT THE EHC-DALLAS

Studies performed at the EHC-Dallas on patients with chemical sensitivity certainly confirm the deregulation of the immune system by toxic chemicals. The following studies are presented.

T and B lymphocytes and their subsets were evaluated in 70 (ages 21 to 50 years) chemically sensitive patients who had abnormal immune function.

These were compared with 60 (ages 21 to 50 years) nonchemically sensitive control subjects (average individuals — nonsmokers taking no medications and having no allergies). Chemical sensitivity was established in the reactive patients by inhaled challenge testing with different organic and inorganic pollutants such as formaldehyde, <0.2 ppm; phenol, <0.2 ppm; pesticide 2,4-DNP, <0.0034 ppm; chlorine, <0.33 ppm; and petroleum-derived ethanol, <0.5 ppm under rigid environmentally controlled conditions in a glass and steel testing booth. Challenges were double-blind with three placebos acting as control.[156]

Blood was drawn on admission using standard collecting tubes on nonfasting patients. These were all done at the same time of day in both the controls and the chemically sensitive patients. The blood was then immediately processed and measured in our own laboratory using an argon laser-operated EPICS-C Flow Cytometer (Coulter Electronics) using monoclonal antibodies (Coulter Immunology Laboratory). An essential feature of this study was the continuous calibration of the apparatus and the intermittent measurement of healthy, nonsmoking, nonmedicated controls. The chemically sensitive patients were selected without any bias from those patients who were treated in our environmental unit in the last 6 months of 1985. The mean values of the immune parameters measured and their standard deviations were calculated for the different groups of patients as well as for controls.

The results of these measurements were as follows: (1) in the case of chemically sensitive patients with vascular dysfunction (70 patients), both the lymphocyte and T lymphocyte counts showed a slight decrease when compared to the controls (Table 15); (2) the T_4 lymphocyte counts of the chemically sensitive patients were virtually the same as those of the control subjects (Table 13). The T_8 suppressor lymphocyte counts were, however, very significantly lower than those of the control subjects (Table 16); (3) the T_4/T_8 ratio characteristic for the sensitivity or hypersensitivity of the immune system increased from a mean of 1.70 ± 0.06 for control subjects to a mean of 2.20 ± 0.15 for the chemically sensitive patients (Table 14); (4) the T_4/T_8 ratio in the controls (normals) is in very good agreement with the results published recently by others.[157-161] The mean of T_4/T_8 values obtained by these authors in a total of 338 control subjects is very close to 1.7. This gives an assurance for the reliability of the methods we used; (5) the T_4/T_8 ratio obtained for chemically sensitive patients with asthma (27 patients) was 1.55 ± 0.15 (Table 15). It is not significantly different from the controls, and it is in good agreement with the values published recently by Leung et al.[159] ($T_4/T_8 = 1.67 \pm 0.46$; Table 16); (6) the T_4/T_8 ratio obtained for rheumatoid arthritis patients (7 people) was 2.96 ± 0.39, in good agreement with recent values published by Veys et al.,[159] who measured in very active rheumatoid arthritis patients a T_4/T_8 ratio of 3.0 (Table 18).

Various reports have shown that many toxic chemicals do alter the immune response of numerous animals.[164-167] In vitro testing and case reports in

Table 13. Immunological Data of Chemically Sensitive Patients with Vascular Dysfunction and Controls — EHC-Dallas (Average Value)

	Patients (60 Persons)	Controls (60 Persons)	Difference (Patients to Controls)	Signif- icance (p)
WBC($\#mm^3$)	7010.00 ± 290.00	7560.00 ± 220.00	No	>0.05
L($\#/mm^3$)	2420.00 ± 90.00	2770.00 ± 91.00	Smaller	<0.005
L (%)	36.2 ± 1.10	37.30 ± 1.10	No	>0.2
T_{11}($\#/mm^3$)	1780.00 ± 76.00	2080.00 ± 74.00	Smaller	<0.005
T_{11} (%)	72.40 ± 1.00	75.20 ± 0.80	Smaller	<0.05
T_4($\#/mm^3$)	1090.00 ± 48.00	1160.00 ± 43.00	No	>0.1
T_4 (%)	43.70 ± 1.00	42.20 ± 0.70	No	>0.1
T_8($\#/mm^3$)	560.00 ± 30.00	740.00 ± 38.00	Signifi- cantly smaller	<0.0001
T_8 (%)	22.90 ± 0.70	25.80 ± 0.80	Smaller	<0.005
T_4/T_8	2.00 ± 0.15	1.70 ± 0.06	Signifi- cantly larger	<0.0001
B($\#/mm^3$)	220.00 ± 19.00	270.00 ± 23.00	No	>0.05
B (%)	9.40 ± 0.90	9.40 ± 0.60	No	—

affected humans have also been published.[168,169] However, ours is the first large series of patients with proven chemical sensitivity where T and B lymphocytes and subsets have been compared with carefully selected control subjects. Normal individuals in this study were carefully compared with other published control data at a favorable level, $p < 0.001$.[170-175] This correlation substantiates the validity of our laboratory and measuring methods, thereby allowing us to accurately compare our pathologic patients to the general population. The main immunologic defect found in our group of chemically sensitive patients appeared to be in the suppressor cells. This finding was highly significant to over four standard deviations ($p < 0.001$). It should be pointed out that helper cells were generally unaffected in this group. Though a few were changed, it was unclear why the suppressor mechanisms were usually affected in these patients. Perhaps the subpopulation of T cells had been genetically preconditioned or previously damaged by pollutants so that when they were exposed to toxic chemicals they became suppressed.

These T_8 cells might also have had nutritional deficits or have been injured by a previous viral assault, thus rendering them vulnerable to subsequent exposure to toxic chemicals. Regardless of the predisposing factors, the

Table 14. Comparison of Immunological Data of Controls (Mean Values)

	No. of Persons	$T_{11}(\%)$	$T_4(\%)$	$T_8(\%)$	T_4/T_8
EHC-Dallas	60	75.20 ± 0.80	42.20 ± 0.70	25.80 ± 0.80	1.70 ± 0.06
LaVia et al.[148]	193	79.60	44.60	28.30	1.70
Reinherz et al.[149]	42		41.00 ± 2.00	20.00 ± 1.00	2.05 ± 0.14
Leung et al.[150]	17		37.00 ± 3.00	22.00 ± 4.00	1.70 ± 0.30
Tsokos & Balow[151]	8		49.80 ± 2.10	33.00 ± 2.1	1.50 ± 0.4
Veys et al.[152]	78		49.0	38.00	1.30

Table 15. Immunological Data of Asthma Patients and Controls
(Mean Values and Differences)

	Patients (27 Persons)	Controls (60 Persons)	Difference (Patients to Controls)	Significance (p)
WBC(#/mm^3)	7700.00 ± 360.0	7560.00 ± 220.0	No	>0.30
L(#/mm^3)	2610.00 ± 130.0	2770.00 ± 91.00	No	>0.10
L (%)	35.00 ± 2.00	37.30 ± 1.10	No	>0.10
T_{11}(#/mm^3)	1910.00 ± 102.00	2080.00 ± 74.00	No	>0.05
T_{11} (%)	73.10 ± 1.80	75.20 ± 0.80	No	>0.01
T_4(#/mm^3)	1100.00 ± 71.00	1160.00 ± 43.00	No	>0.30
T_4 (%)	38.20 ± 1.80	42.20 ± 0.70	Smaller	<0.05
T_8(#/mm^3)	770.00 ± 58.00	740.00 ± 38.00	No	>0.30
T_8 (%)	29.10 ± 2.40	25.80 ± 0.80	No	>0.05
T_4/T_8	1.55 ± 0.15	1.70 ± 0.06	No	>0.10
B(#/mm^3)	320.00 ± 48.00	270.00 ± 23.00	No	>0.10
B(%)	12.30 ± 1.80	9.40 ± 0.60	No	>0.05

suppressor population responded adversely in this subset of patients. Why chemically sensitive asthmatics did not show a change vs. the control is also puzzling. We have no explanation for this because they had about the same degree of severity of chemical sensitivity as the two other groups.

Whether or not the immune system is the site of primary defects in chemical sensitivity is still unknown, but in view of previous studies from our center, we think it is highly unlikely. Rather, the primary defect seems more likely to involve changes in either protective surfaces or defects in biochemical enzyme detoxification systems. Clearly, there exists a subset of chemically sensitive patients who have a convincingly significant suppression of T cells vs. controls. These results will allow future comparison with named disease processes, i.e., arthritis, asthma, vasculitis, collagen diseases, which have few known causes or triggering agents. Data strongly supports these diseases as environmentally triggered.

We have performed further studies evaluating T and B lymphocytes in 19 chemically sensitive patients who had measurable levels of toxic chemicals (TVOCs) in their blood. These patients (10 male, 9 female) were selected from the patients admitted to the ECU. Their mean age was 44 years with a range from 10 to 63 years. The findings in these patients were compared with those of 70 (14 males, 56 females) proven chemically sensitive patients previously studied in whom TVOCs had not been measured. The mean age of this group was 42 years with a range from 21 to 50 years.[176] The chemical sensitivity of these patients was proven by double-blind challenges. All were nonsmokers.

Table 16. Immunological Data of Rheumatoid Arthritis Patients and
Controls — EHC-Dallas (Mean Values and Differences)

	Patients (7 Persons)	Controls (60 Persons)	Difference (Patients to Controls)	Sig- nif- icance (p)
WBC(#/mm^3)	7090.00 ± 810.00	7560.00 ± 220.00	No	>0.20
L(#/mm^3)	2250.00 ± 280.00	2770.00 ± 91.00	Smaller	<0.05
L (%)	32.10 ± 2.30	37.30 ± 1.10	Smaller	<0.05
T_{11}(#/mm^3)	1600.00 ± 250.00	2080.00 ± 74.00	Smaller	<0.05
T_{11} (%)	69.60 ± 2.90	75.20 ± 0.80	Smaller	<0.05
T_4(#/mm^3)	1120.00 ± 210.00	1160.00 ± 4.30	No	>0.40
T_4 (%)	47.90 ± 4.00	42.20 ± 0.70	No	>0.05
T_8(#/mm^3)	370.00 ± 37.00	740.00 ± 38.00	Signifi- cantly smaller	<0.0001
T_8 (%)	17.10 ± 1.30	25.80 ± 0.80	Signifi- cantly smaller	<0.0001
T_4/T_8	2.96 ± 0.39	1.70 ± 0.06	Signifi- cantly larger	<0.001
B(#/mm^3)	170.00 ± 51.00	270.00 ± 23.00	Smaller	<0.05
B(%)	7.40 ± 1.40	9.40 ± 0.60	No	>0.05

Sixty nonsmoking, healthy individuals (26 males, 34 females) without a history of chemical sensitivity were selected as a control group for this study. Their mean age was 35 years, with a range from 21 to 50 years.[177]

This study was performed from May 1983 to December 1986. Twelve volatile organic chemicals (benzene, toluene, ethylbenzene, xylenes, styrene, trimethylbenzenes, chloroform, dichloromethane, 1,1,1-trichloroethane, trichlorethylene, tetrachloroethylene, dichlorobenzene) were measured in whole blood using purging traps combined with gas chromatography and mass spectrometry.[178,179] The T and B lymphocytes and their subsets were measured by an argon laser-operated EPICS-C Flow Cytometer (Coulter Electronics) using monoclonal antibodies (Coulter Immunology Laboratory).[180,181]

The mean values of the immune parameters were measured, and their standard deviations were calculated for this group of patients and were compared with the corresponding values obtained in the controls and in a larger group of chemically sensitive patients studied previously[182] in whom TVOCs had not been measured.

Table 17. Frequency of Distribution of TVOCs of 19 Selected Chemically Sensitive Patients — EHC-Dallas

Chemical	Patients with Detectable Blood Levels of TVOCs (%)
Benzene	11
Toluene	42
Ethylbenzene	16
Xylenes	37
Styrene	5
Trimethylbenzenes	5
Chloroform	11
Dichloromethane	16
1,1,1-trichloroethane	32
Trichloroethylene	11
Tetrachloroethylene	53
Dichlorobenzene	11

One or more TVOCs were found in the blood of the 19 selected chemically sensitive patients. The frequency distribution of the TVOCs is shown in Table 17.

Chemically sensitive patients with TVOCs in their blood represent a subset of the chemically sensitive individuals. For this reason, our first step in analyzing our data was to compare the immunological parameters of these two groups, using data obtained from an earlier study[182] on 70 chemically sensitive patients in whom TVOCs had not been measured. The main results of this comparison are shown in Table 20.

All the immunological data found in the 19 subjects was in good agreement with the corresponding data found earlier in a larger group of 70 chemically sensitive patients with immune dysfunction, except for B cell counts; these were greater in the study group than in the larger group of 70 chemically sensitive patients (Table 18).

The 19 chemically sensitive patients were also compared with the control group of 60 normal individuals. The main results of this comparison are shown in Table 19.

The total lymphocyte counts (L) and T_{11} lymphocyte counts were somewhat lower in the chemically sensitive patients with TVOCs in their blood than in the control subjects, in agreement with the general pattern found in the previous study, when the general group of chemically sensitive patients was compared to the controls.[184]

Table 18. Immunological Data of Chemically Sensitive Patients with TVOCs in Their Blood and Chemically Sensitive Patients not Tested for TVOCs (Average Values) — EHC-Dallas

	Patients with TVOCs (19 Persons)	Patients Not Tested (70 Persons)	Difference[a]	Significance (p)
WBC(#/mm^3)	7053.00 ± 600.00	7010.00 ± 290.00	None	>0.40
L(#/mm^3)	2372.00 ± 196.00	2420.00 ± 90.00	None	>0.40
T_{11}(#/mm^3)	1744.00 ± 143.00	1780.00 ± 76.00	None	>0.40
T_4(#/mm^3)	1125.00 ± 95.00	1090.00 ± 48.00	None	>0.30
T_8(#/mm^3)	563.00 ± 56.00	560.00 ± 30.00	None	>0.40
T_4/T_8	2.15 ± 1.70	2.20 ± 0.15	None	>0.40
B(#/mm^3)	306.00 ± 38.00	220.00 ± 19.00	Greater	<0.05

[a] Patients with TVOCs to patients not tested.

Table 19. Immunological Data of Chemically Sensitive Patients with TVOCs in Their Blood and Controls (Average Values) — EHC-Dallas

	Patients (19 Persons)	Controls (60 Persons)	Difference[a]	Significance (p)
WBC(#/mm^3)	7053.00 ± 600.00	7560.00 ± 220.00	None	>0.20
L(#/mm^3)	2372.00 ± 196.00	2770.00 ± 91.00	Fewer	<0.05
T_{11}(#/mm^3)	1744.00 ± 143.00	2080.00 ± 74.00	Fewer	<0.05
T_4(#/mm^3)	1125.00 ± 95.00	1160.00 ± 43.00	None	>0.30
T_8(#/mm^3)	563.00 ± 56.00	740.00 ± 38.00	Fewer	<0.05
T_4/T_8	2.15 ± 17.00	1.70 ± 0.06	Greater	<0.01
B(#/mm^3)	306.00 ± 38.00	270.00 ± 23.00	None	<0.20

[a] Patients to controls.

Agreement similar to our previous findings was shown by the lower value of T_8 lymphocyte counts and the correspondingly higher T_4/T_8 ratio found in the 19 subjects when compared to the controls. Of particular importance was the T_4/T_8 ratio, which increased from 1.70 ± 0.06 for control subjects to a mean of 2.15 ± 0.17 for the 19 chemically sensitive subjects.

No other immunological parameters measured showed any significant difference between the chemically sensitive patients and the controls.

The control group had two or less toxic chemicals in their blood.

The result of this study was that the chemically sensitive patients with TVOCs in their blood showed the same major immunological changes in T

Table 20. Multiple Regression for Cell Mediated Immunity and Immune Cell Parameters in 68 Cases of Chemical Sensitivity — EHC-Dallas

Ind. Var.	B Coeff.	Std. Err. (B)	Prob.
WBC	0.711099	0.000022	<0.0001
Lym (%)	−0.008222	0.003506	0.0227
LymC	0.076276	0.000032	<0.0001
T_{11} (%)	−4.165069	0.003181	<0.0001
$T_{11}C$	0.763768	0.000036	<0.0001
$T_4\%$	−0.001671	0.002809	0.5545
T_4C	−79.23036	0.000046	<0.0001
$T_8\%$	27.82866	0.003612	<0.0001
T_8C	0.005937	0.000087	<0.0001
T_4/T_8	9.035378	0.022853	<0.0001
BLym (%)	−1.756029	0.002289	<0.0001
BLymC	0.000618	0.000239	0.0125

Dependent variable: CMI-N   Multiple correlation summary:

			Multiple R	R-Square
Multiple R =	0.99			
Std. Err. Est. =	0.3023	Unadjusted	0.99	0.9801
Constant =	28.8061	Adjusted	0.9878	0.9758

Std. error of estimate = 0.3023
Sample size = 68

cells that have been found in the general group of chemically sensitive patients in whom TVOCs had not been measured,[185] i.e., there was a reduced number of T_8 lymphocytes and an increased T_4/T_8 ratio from the mean of 1.70 ± 0.06 obtained for the control subjects to a mean of 2.15 ± 0.15 for the general group of chemically sensitive patients in whom TVOCs had not been measured. This agreement shows that the previous group behaved very much like a similar subset of the general group, at least in respect to the T_4/T_8 ratio.

The major difference between the group of 19 in whom TVOCs had been measured and the general group of 70 chemically sensitive in whom TVOCs had not been measured was the elevated B count in the 19 subjects. The general group of 70 probably included a certain number of individuals with elevated TVOC levels in their blood. The above-mentioned differences in the B count between the two groups may be explained either as an extreme statistical fluctuation combined with some systematical errors in the measurement or as a real difference in the immune response of the two groups. We think that the first explanation is unlikely, but plan to test it by using a larger number of patients and a detailed study of the correlations between B cell counts and the TVOC levels in the blood.

The second hypothesis of a real difference in the immune response of the two groups opens a number of interesting possibilities with respect to the immunological response produced by elevated levels of TVOCs in the blood. For instance, TVOCs in the blood may account for low T cells,[186] low B cells,[187] inhibited antibody production and the mitogenic response of lymphocytes,[188] reduced T lymphocytes,[189] inhibited RNA synthesis,[190] altered killer cells,[191] altered immune responses,[192-195] delayed hypersensitivity,[196] depressed T cell function, associated with an increase in suppressor T lymphocyte expression and loss of T lymphocyte cytotoxicity for tumor target cells,[197] depressed cell-mediated immunity,[198] and reduced T suppressor cells in atopic disease.[199-201]

Apparently, some people who have become chemically sensitive with an accumulation of various organic hydrocarbons develop alterations in the immune system,[202] but these alterations differ depending upon specific chemical molecule, total load, and individual response.

As shown in Chapter 4, the major pathway for the metabolism of drugs and chemicals is that of mixed-function oxidation; this pathway is primarily cytochrome P-450 dependent. Different toxic chemicals can be degraded to form the same reactive intermediates and may give rise to the same antigens. This generation of antigens might explain the cross-reactivity observed with many drugs and chemicals by which exposure to one chemical appears to induce an immune response to a number of related compounds. For example, chemicals that contain an aldehyde group, such as glucose, or a carbonyl moiety adjacent to the hydroxy group, such as cortisol or 16-alpha-hydroxyestrone, can react nonenzymatically with proteins to form intermediates which are antigenic.[203] Bucala et al.[204] have proposed, for example, that covalent adducts between cortisol or 16-alpha-hydroxyestrone and tissue or plasma proteins may be antigenic and result in immunological injury, as seen in systemic lupus erythematosis. An alternative mechanism of interaction is that of reactive intermediates made by modulation of the cytochrome P-450 system forming a superoxide anion or hydrogen peroxide. These free radicals may, without direct injury to the immune system, produce cellular damage and destruction of the endoplasmic reticulum. Hence, toxic chemicals may give rise to reactive intermediates, and these may produce covalent antigens which could stimulate an immune reaction and produce prostaglandins autacoids and active oxygen; or a toxic chemical may produce reactive intermediates and thereby generate hydrogen peroxide and active oxygen.[205] Both reactions can produce chemical sensitivity.

Studies by Liang and co-workers[206] at the EHC-Dallas show the effect of pollutants on immune cell-mediated immunity (CMI; Table 20). Sixty-eight proven chemically sensitive patients were measured for immune cell count, including WBC, lymphocytes, T_{11} cell, T_4 (helper) cell, T_8 (suppressor) cell, and B cells. At the same time, the CMI was also measured by Multitest CMI skin test kits. These data were analyzed with Straight Multiple Regression, Anova, and Standard Regression. The results showed that CMI, including both the number of antigens and the sum of averages of the diameters of skin reaction wheals, had a strong correlation with these measured immune cells (adjusted multiple R = 0.99, p <0.0001). The number of WBCs had a positive correlation with both the number of antigens producing CMI reactions and the sum of averages of CMI reaction (beta coefficient = 2.16, p <0.0001). The percentage of T_8 cells had a positive correlation with CMI (beta coefficient = 0.52, p <0.0001). The absolute number of T_8 cells also had a positive correlation with CMI (beta coefficient = 0.16, p <0.001). The ratio of T_4/T_8 was positively correlated with CMI, too (beta coefficient = 0.26, p <0.001). However, the number of T_4 cells was negatively correlated with CMI (beta coefficient = –0.57, p <0.0001). The percentage of B cells had a negative correlation with CMI (beta coefficient = –0.55, p <0.001). These results indicate that WBC, percent of T_8 cells, and absolute number of T_8 cells and T_4/T_8 were very significant and had an enhanced effect on the CMI, while the absolute number of T_{11} cells and T_4, as well as percent of B cells, had a significantly weakened effect on the CMI.

Another study was performed by Liang[207] on the correlation between whole blood intracellular minerals and CMI in 68 chemically sensitive patients (Table 21). The multiple regression analysis showed a very significant correlation of 14 kinds of whole blood intracellular minerals with the number of skin positive antigens of the CMI (adjusted multiple R = 0.9873, R square = 0.9749, p <0.0001). Intracellular calcium, chromium, phosphorus, and selenium had a positive correlation with the number of positive antigens of the CMI. In other words, these minerals could be correlated with the increased number of skin positive antigens of CMI (p <0.0001). Intracellular potassium, copper, zinc, manganese, silicon, barium, and sodium had a negative correlation with the number of skin positive antigens of CMI, i.e., these minerals could be correlated with the decrease in the number of skin positive antigens of CMI (p <0.0001). There was a significant correlation of 14 kinds of whole blood intracellular minerals with the sum of average skin positive diameters of CMI (adjusted R = 0.7976, R square = 0.6362, p <0.0001). Calcium, iron, and chromium had a positive correlation with the sum of average skin positive antigens of CMI (p <0.0001 to 0.03), and copper, zinc, manganese, silicon, and sodium had a negative correlation with the sum of average skin positive diameters of CMI (p <0.0001 to 0.04). These results indicate that whole blood cell minerals have a strong correlation with CMI in a group of chemically sensitive individuals; these results can be correlated with increased CMI reactions (Ca, Fe, and Cr), and some of them can decrease CMI reaction (Cu, Zn, Mn, Si, and Na). It is not known whether the mineral deregulation seen in chemically sensitive patients is a result or the cause of their chemical sensitivity.

It is clear from these studies and those presented in the literature that toxic chemical exposure has many effects of the immune system, some of which result in or propagate chemical sensitivity. The problem of chemical sensitivity is partly due to dysfunction of the immune system as well as the nonimmune detoxification systems and the presence or absence of nutrient fuels.

Table 21. Multiple Regression for Minerals and CMI Whole Blood Intracellular in 68 Cases of Chemical Sensitivity — EHC-Dallas

Ind. Var.	B Coeff.	Std. Err. (B)	Prob.
Ca	0.711099	0.01926	<0.0001
Mg	−0.008222	0.008912	0.3604
K	−0.076276	0.009239	<0.0001
Cu	−4.165069	0.481355	<0.0001
Zn	−0.763768	0.035203	<0.0001
Fe	−0.001671	0.001089	0.1308
Mn	−79.23036	2.146435	<0.0001
Cr	27.82866	1.254754	<0.0001
P	0.005937	0.000989	<0.0001
Se	9.035378	0.795307	<0.0001
Si	−1.756029	0.076615	<0.0001
Su	0.000618	0.000613	0.318
Ba	−5.771563	1.038106	<0.0001
Na	−0.95802	0.024682	<0.0001

Dependent variable: CMI-N
Multiple R = 0.99
Std. err. est. = 0.308
Constant = 28.8061

Multiple correlation summary:

	Multiple R	R-Square
Unadjusted	0.99	0.9801
Adjusted	0.9873	0.9749

Std error of estimate = 0.308
Sample size = 68

REFERENCES

1. Wintrobe, M. M. "Anemia of bone marrow failure," in *Principles of Internal Medicine*. 5th ed., T. R. Harrison, R. D. Adams, I. L. Bennett, W. H. Resnik, G. W. Thorn, and M. M. Wintrobe, Eds. (New York: McGraw-Hill, 1966), pp. 644–645.
2. Rees, R. B., J. H. Bennett, E. M. Hamlin, and H. I. Maibach. "Aminopterin for psoriasis," *Arch. Dermatol.* 90:544–552 (1964).
3. Webster, L. T., Jr. "Drugs used in the chemotherapy of protozoal infections," in *Goodman and Gilman's The Pharmacological Basis of Therapeutics,* 8th ed., A. G. Gilman, T. W. Rall, A. S. Nies, and P. Taylor, Eds. (New York: Pergamon Press, 1990), p. 986.
4. Gormsen, J. "Agranulocytosis and thrombocytopenia in case of tetanus treated with curate and chlorproniazine," *Dan. Med. Bull.* 2:87–89 (1955).
5. Lassen, H. C. A., E. Henriksen, F. Neukirch, and H. S. Kristensen. 1950. "Treatment of tetanus severe bone-marrow depression after prolonged nitrous-oxide anaethesia," *Lancet* 1:527–530 (1950).

6. Kieler, J. "Cytologic effect of nitrous oxide at different oxygen tensions," A*cta. Pharmacol. Toxicol.* 13:301–308 (1957).

7. Lassen, H. C., and S. H. Kristensen. "Remission in chronic myeloid leuckaemia following prolonged nitrous oxide inhalation," *Dan. Med. Bull.* 6:252–255 (1959).

8. Okamoto, S. "Effects of prolonged administration of nitrous oxide upon hemopoietic organs in rats," Personal communication from K. Iwatsuki, Tohoku University School of Medicine, Sendai, Japan (1985).

9. Eastwood, D. W., C. D. Green, M. A. Lambodin, and R. Gardner. "Effect of nitrous oxide on the white-cell count in leukemia," *N. Engl. J. Med.* 268(6):297–299 (1963).

10. Brabec, M. J., and I. A. Bernstein. "Cellular the common subcellular common and molecular targets of foreign compounds," in *Toxicology Principles and Practice*, Vol. 1, A. L. Reeves, Ed. (New York: John Wiley & Sons, 1981), p. 39.

11. Silver, F. "Carbon monoxide," in *Clinical Ecology*, L. D. Dickey, Ed. (Springfield, IL: Charles C Thomas, 1976.

12. Smith, R. P. 1986. Toxic responses of the blood, in *Casarrett and Doull's Toxicology: The Basic Science of Poisons*, 3rd ed., C. D. Klaassen, M. O. Amdur, and J. Doull, Eds. (New York: Macmillan, 1986).

13. National Academy of Sciences. "The Health Effects of Nitrate, Nitrite, and N-Nitroso Compounds" (Washington, DC: National Academy Press, 1981).

14. Leighton, F. A., O. B. Peakall, and R. G. Butler. "Heinz body hemolytic anemia from the ingestion of crude oil, a primary toxic effect in marine birds,"*Science* 220:871–873 (1983).

15. Leighton, F. A., O. B. Peakall, and R. G. Butler. "Heinz body hemolytic anemia from the ingestion of crude oil, a primary toxic effect in marine birds," *Science* 220:871–873 (1983).

16. Nossel, H. L. "Platelet disorders," in *Harrison's Principles of Internal Medicine,* 8th ed., G. W. Thorn, R. D. Adams, E. Braunwald, K. J. Isselbacher, and R. G. Petersdorf, Eds. (New York: McGraw Hill, 1977), p. 1714.

17. Nossel, H. L. "Platelet disorders," in *Harrison's Principles of Internal Medicine,* 8th ed., G. W. Thorn, R. D. Adams, E. Braunwald, K. J. Isselbacher, and R. G. Petersdorf, Eds. (New York: McGraw Hill, 1977), p. 1714.

18. Nossel, H. L. "Platelet disorders," in *Harrison's Principles of Internal Medicine,* 8th ed., G. W. Thorn, R. D. Adams, E. Braunwald, K. J. Isselbacher, and R. G. Petersdorf, Eds. (New York: McGraw Hill, 1977), p. 1714.

19. Nossel, H. L. "Platelet disorders," in *Harrison's Principles of Internal Medicine,* 8th ed., G. W. Thorn, R. D. Adams, E. Braunwald, K. J. Isselbacher, and R. G. Petersdorf, Eds. (New York: McGraw Hill, 1977), p. 1714.

20. Nossel, H. L. "Platelet disorders," in *Harrison's Principles of Internal Medicine*, 8th ed., G. W. Thorn, R. D. Adams, E. Braunwald, K. J. Isselbacher, and R. G. Petersdorf, Eds. (New York: McGraw Hill, 1977), p. 1714.

21. Nossel, H. L. "Platelet disorders," in *Harrison's Principles of Internal Medicine*, 8th ed., G. W. Thorn, R. D. Adams, E. Braunwald, K. J. Isselbacher, and R. G. Petersdorf, Eds. (New York: McGraw Hill, 1977), p. 1714.

22. Wintrobe, M. M. *Clinical Hematology*. 5th ed. (Philadelphia: Lea & Febiger, 1961).

23. Randolph, T. G., and G. F. Kroker. "Allergic thrombocyteopenia," 12th Advanced Seminar in Clinical Ecology, 103 (Abstr.) (1978).

24. Karpiuski, R. E. "Purpura following exposure to DDT," *J. Pediatr.* 37:373–379 (1950).

25. Clayson, D. B. "Carcinogenic hazards due to drugs," in *Drug Induced Disease*, Vol. 4, L. Meyler and R. Peck, Eds. (Amsterdam: Excerpta Medica Press, 1972), pp. 91–109.

26. Stutman, O. "Intrathymic and extrathymic T-cell maturation," *Immunol. Rev.* 42:138 (1978).

27. McPhee, D., J. Pye, and K. Shortman. "The differentiation of T-lymphocytes. V. Evidence for intrathymic death of most thymocytes," *Thymus* 1:151 (1979).

28. Scollay, R. G., E. C. Butcher, and I. L. Weissman. "Thymus cell migration: Quantitative aspects of cellular traffic from the thymus to the periphery in mice," *Eur. J. Immunol.* 10:210 (1980).

29. Bach, J. F., and M. Popiernik. "Cellular and molecular signals in T-cell differentiation," in *Microenvironments in Haemopoietic and Lymphoid Differentiation* (Ciba Foundation Symposium 84), R. Porter and J. Whelan, Eds. (London: Pitman Medical Publishing Co., Ltd., 1981), p. 368.

30. Zinkernagel, R. M., A. Althage, E. Waterfield, B. Kindred, R. M. Welsh, G. Callahan, and P. Pincetl, "Restriction specificities, alloreactivity, and allotolerance expressed by T-cells from nude mice reconstituted with H-2-compatible or -incompatible thymus grafts,"*J. Exp. Med.* 151(2):376–399 (1980).

31. Bach, J. F., and M. Popiernik. "Cellular and molecular signals in T-cell differentiation," in *Microenvironments in Haemopoietic and Lymphoid Differentiation* (Ciba Foundation Symposium 84), R. Porter and J. Whelan, Eds. (London: Pitman Medical Publishing Co., Ltd., 1981), p. 368.

32. Smith, R. P. "Toxic responses of the blood," in *Casarett and Doull's Toxicology. The Basic Science of Poisons*, 3rd ed., C. D. Klassen, M. O. Amdur, and J. Doull, Eds. (New York: Macmillan, 1986), p. 226.

33. La Du, B. N., and H. W. Eckerson. "Could the human paraoxonase polymorphism account for different responses to certain environmental chemicals?," in *Banbury Report 16. Genetic Variability in Responses to Chemical Exposure*, G. S. Omenn and H. Gelboin, Eds. (Cold Spring Harbor, NY: Cold Spring Harbor Laboratory, 1984), p. 167.

34. Luster, M. L., D. R. Germolee, and G. T. Rosenthal. "Immunotoxicology: Review of current status," *Ann. Allergy* 64:427–432 (1990).

35. Dean, J. H., M. J. Murray, and E. C. Ward. "Toxic responses of the immune system," in *Casarett and Doull's Toxicology. The Basic Science of Poisons*. 3rd ed., C. D. Klassen, M. O. Amdur, and J. Doull, Eds. (New York: Macmillan Publishing Co., Inc., 1986), p. 245.

36. Dean, J. H., M. J. Murray, and E. C. Ward. "Toxic responses of the immune system," in *Casarett and Doull's Toxicology. The Basic Science of Poisons*. 3rd ed., C. D. Klassen, M. O. Amdur, and J. Doull, Eds. (New York: Macmillan Publishing Co., Inc., 1986), p. 256.

37. Descotes, J., Ed. *Immunotoxicology of Drugs and Chemicals* (Amsterdam: Elsevier, 1986).

38. Burrell, R. "Immunology of occupational lung diseases," in *Occupational Lung Diseases*, 2nd ed., W. K. C. Morgan and A. Seaton, Eds. (Philadelphia: W. B. Saunders, 1984), pp. 196–211.

39. Theorell, H., M. Blambock, and C. Kockum. "Demonstration of reactivity to airborne and food antigen in cutaneous vasculitis by variation in tibrino peptide and others, blood coagulation, fibrinolysis, and complement parameters," *Thrombo. Haemo. Sts. (Stattz)* 36:593 (1976).

40. Butcher, B. T., R. N. Jones, and C. E. O'Neill. "Longitudinal study of workers employed in the manufacture of toluene — diisocyanate," *Am. Rev. Respir. Dis.* 116:411 (1977).

41. Frankland, A. W. "Anaphylaxis in relation to food allergy," in *Food Allergy and Intolerance*, J Brostoff and S. J. Challacombe, Eds. (London: Bailliere Tindall, 1987), pp.456–466.

42. Prausnitz, C., and H. Kustner. "Studien über uberemfindlichkeit," Central bl Bateriol 1921, Abt. Orig. 86 160–9. Transl. from the German by Carl Prausnitz, in *Clinical Aspects of Immunology*, P. G. H. Gell and R. R. A. Coombs, Eds. (Oxford: Blackwell Scientific Publications, 1962), 808–816.

43. Gaworski, C. L., and R. P. Sharma. "The effects of heavy metals on (3H)thymidine uptake in lymphocytes," *Toxicol. Appl. Pharmacol.* 46:305–313 (1978).

44. Takenchi, T., N. Morikawa, H. Matsumoto, and Y. Shiraishi. "A pathological study of minamata disease in Japan,"*Acta Neuropathol.* 2:40 (1962).

45. Mannik, M., and B. C. Gilland. "Systemic lupuserythematosus," in *Harrison's Principles of Internal Medicine*, 8th ed., G. W. Thom, P. D. Adams, E. Braunwald, K. J. Isselbacher, and R. G. Petersdorf, Eds. (New York: McGraw-Hill, 1977), p. 429.

46. Lelbach, W. K., and H. J. Marsteller. "Vinyl chloride associated disease," *Ergeb. Inn. Med. Kinderheilkd.* 47:1–110 (1981).

47. Winslow, S. G. *The Effects of Environmental Chemicals on the Immune System: A Selected Bibliography with Abstracts* (Oak Ridge, TN: Toxicology Information Response Center, Oak Ridge National Laboratory, 1981).

48. Rea, W. J. "Environmentally triggered small vessel vasculitis," *Ann. Allergy* 38:245–251 (1977).

49. Rea, W. J., A. R. Johnson, S. Youdim, I. J. Fenyves, and N. Samadi. "T and B lymphocyte parameters measured in chemically sensitive patients and controls," *Clin. Ecol.* 4(1):11–14 (1986).

50. Laseter, J. L., H. R. Deleon, W. J. Rea, and J. R. Butler. "Chlorinated hydrocarbon pesticides in environmentally sensitive patients," *Clin. Ecol.* 2(1):3–12 (1983).

51. Aripdzhanov, T. M. "Effects of the pesticides anthio and milbex on the immunological reactivity and certain autoimmunological reactivity and certain autoimmune processes of the body," *Gig. Sanit.* 7:37–42e (1973).

52. Aripdzhanov, T. M. "Effects of the pesticides anthio and milbex on the immunological reactivity and certain autoimmunological reactivity and certain autoimmune processes of the body," *Gig. Sanit.* 7:37–42e (1973).

53. Aripdzhanov, T. M. "Effects of the pesticides anthio and milbex on the immunological reactivity and certain autoimmunological reactivity and certain autoimmune processes of the body," *Gig. Sanit.* 7:37–42e (1973).

54. Vos, J. G. "Immune suppression as released to toxicology," *CRC Crit. Rev. Toxicol.* 5(1):67–101 (1977).

55. Salmon, S. E. "Drugs and the immune system," in *Review of Medical Pharmacology,* F. J. Meyers, E. Jawetz, and A. Goldfien, Eds. (New York: Lange, 1970), p. 446.

56. Penn, I. "Tumors occurring in organ transplant recipients," in *Advances in Cancer Research*, Vol. 28, G. Klein and S. Weinhouse, Eds. (New York: Academic Press, 1978), p. 446.

57. Polak, L. "Antigenic competition in the induction of contact sensitivity in the guinea pig," *Int. Arch. Allergy Appl. Immunol.* 76(3):275–281 (1985).

58. Bryson, P. D. *Comprehensive Review in Toxicology,* 2nd ed., (Rockville, MD: Aspen Publishing, Inc., 1989), p. 527.

59. Hayes, W. J. *Pesticides Studied in Man* (Baltimore: Williams & Wilkins, 1982), p. 173.

60. Hayes, W. J. *Pesticides Studied in Man* (Baltimore: Williams & Wilkins, 1982), p. 173.

61. Stevens, J. T., K. M. Oberholser, S. R. Wagner, and T. E. Greene. "Content and activities of microsomal electron transport components during the development of dieldrin-induced hypertrophic hypoactive endoplasmic transport," *Toxicol. Appl. Pharmacol.* 39:411–421 (1977).

62. Jensen, G., and J. Clausen. "Organochlorine compounds in adipose tissue of greenlanders and southern danes," *J. Toxicol. Environ. Health* 5:617–629 (1979).
63. Little, C. H. Personal communication (1985).
64. Van Miller, R., R. Marlar, and J. Allen. "Tissue distribution and excretion of tritiated tetrachlorodibenzo-p-doxin in non-human primates and rats," *Food Cosmet. Toxicol.* 14:31–34 (1976).
65. Banerjee, B. "Effects of sub-chronic DDT exposure on humoral and cell-mediated immune responses in albino rats," *Bull. Environ. Contam. Toxicol.* 29:827–834 (1987).
66. Banerjee, B., M. Kamachandran, and Q. Hussain. "Sub-chronic effect of DDT in mice," *Bull. Environ. Contam. Toxicol.* 37:433–440 (1986).
67. Street, J. C., and R. P. Sharma. "Quantitative aspects of immunosuppression by selected pesticides," *Toxicol. Appl. Pharmacol.* 29(1):135–136 (1974).
68. Atabaev, S. T., I. B. Boiko, and V. A. Ilima. "Effect of pesticides on the immunological state of the body," *Gig. Sanit.* 8:7–10 (1978).
69. Perelygin, V. M., M. B. Shpirt, O. A. Aripov, and V. I. Ershova. "Effect of some pesticides on immunological response reactivity," *Gig. Sanit.* 12:29–33 (1971).
70. Aaskari, Z. M., and J. Gabliks. "DDT and immunological responses. I. Altered histamine levels and anaphylactic shock in guinea pigs," *Arch. Environ. Health* 26(6):309–331 (1976).
71. Gabliks, J., T. Al-Zuhaidy, and E. Askari. "DDT and immunological responses. A reduced anaphylaxis and mast cell population in rats fed DDT," *Arch.Environ.Health* 30(2):81–84 (1975).
72. Gabliks, J., E. M. Askari, and N. Yolen. "DDT and immunological responses. I. Serum antibodies and anaphylactic shock in guinea pigs," *Arch. Environ. Health* 26(6):303–308 (1973).
73. Liang, H.-C., and W. J. Rea. "Pesticide effects on cell-mediated immunity in chemically sensitive patients," (1991), in press.
74. Klotz V. I., R. A. Brysin, V. G. Brysin, and A. A. Safarova. "Effect of pesticides of the immunological reactivity of the body of animals and man," *Gig. Sanit.* 9:35–36 (1978).
75. Aaskari, Z. M., and J. Gabliks. "DDT and immunological responses. I. Altered histamine levels and anaphylactic shock in guinea pigs," *Arch. Environ. Health* 26(6):309–331 (1976).
76. Perelygin, V. M., M. B. Shpirt, O. A. Aripov, and V. I. Ershova. "Effect of some pesticides on immunological response reactivity," *Gig. Sanit.* 12:29–33 (1971).
77. Street, J. C., and R. P. Sharma. "Quantitative aspects of immunosuppression by selected pesticides," *Toxicol. Appl. Pharmacol.* 29(1):135–136 (1974).

78. Banerjee, B., M. Kamachandran, and Q. Hussain. "Sub-chronic effect of DDT in mice," *Bull. Environ. Contam. Toxicol.* 37:433–440 (1986).
79. Latimer, J. W., and H. S. Siegel. "Immune response in broilers fed technical grade DDT (Antibodies)," *Poultry Sci.* 53(3):1078–1083 (1974).
80. Aaskari, Z. M., and J. Gabliks. "DDT and immunological responses. I. Altered histamine levels and anaphylactic shock in guinea pigs," *Arch. Environ. Health* 26(6):309–331 (1976).
81. Wiltrout, R. W., C. D. Ercegovich, and W. S. Ceglowski. "Humoral immunity in mice following oral administration of selected pesticides," *Bull. Environ. Contam. Toxicol.* 20(3):423–443 (1978).
82. Evdokimov, E. S. "Effect of organochlorine pesticides on animals," *Veterinarya* 12:94–95 (1974).
83. Aaskari, Z. M., and J. Gabliks. "DDT and immunological responses. I. Altered histamine levels and anaphylactic shock in guinea pigs," *Arch. Environ. Health* 26(6):309–331 (1976).
84. Roa, D. S. V. S., and B. Glick. "Pesticide effects on the immune response and metabolic activity of chicken lymphocytes," *Proc. Soc. Exp. Biol. Med.* 154(1):27–29 (1977).
85. Street, J. C., and R. P. Sharma. "Alteration of induced cellular and humoral immune responses by pesticides and chemicals of environmental concern: Quantitative studies of immunosuppression by DDT, arochlor 1254, carbaryl, carbofuran, methylparathion," *Toxicol. Appl. Pharmacol.* 32(3):587–602 (1975).
86. Street, J. C., and R. P. Sharma. "Quantitative aspects of immunosuppression by selected pesticides," *Toxicol. Appl. Pharmacol.* 29(1):135–136 (1974).
87. Aaskari, Z. M., and J. Gabliks. "DDT and immunological responses. I. Altered histamine levels and anaphylactic shock in guinea pigs," *Arch. Environ. Health* 26(6):309–331 (1976).
88. Olefir, A. I. "Effect of chemical substances on the formation of acquired immunity," U. S. NTIS Report, AD-A008261 (1974).
89. Sonntag, A. C. "Xenobiotics and molecular teratology," *Clin. Obstet. Gynecol.* 18(4):199–207 (1975).
90. Banerjee, B., M. Kamachandran, and Q. Hussain. "Sub-chronic effect of DDT in mice," *Bull. Environ. Contam. Toxicol.* 37:433–440 (1986).
91. Giurgea, R., C. Witterberger, and G. Frecus, S. Manciulea, M. Borsa, D. Coprean, and S. Ilyes. "Effects of some organochlorine pesticides on the immunological reactivity of white rats," *Arch. Exp. Veterinaermed.* 32:769–774 (1978).
92. Giurgea, R., C. Witterberger, and G. Frecus, S. Manciulea, M. Borsa, D. Coprean, and S. Ilyes. "Effects of some organochlorine pesticides on the immunological reactivity of white rats," *Arch. Exp. Veterinaermed.* 32:769–774 (1978).

93. Desi, I., L. Varga, and I. Farkas. "Studies on the immunosuppressive effect of organochlorine and organophosphoric pesticides in subacute experiments," *J. Hyg. Epidemiol., Microbiol., Immunol.* 22(1):115–122 (1978).
94. Loose, L. D., K. A. Pittman, K. F. Benitz, and J. B. Silkworth. "Polychlorinated biphenyl and hexachlorobenzene-induced humoral immunosuppression," *RES J. Reticuloendothel. Soc.* 22(3):255 (1977).
95. Loose, L. D., K. A. Pittman, K. F. Benitz, J. B. Silkworth, W. Mueller, and F. Coulston. "Environmental chemical-induced immune dysfunction," *Ecotoxicol. Environ. Safety* 2(2):173–198 (1978).
96. Loose, L. D., J. B. Silkworth, K. A. Pittman, K. F. Benitz, and W. S. Mueller. "Impaired host resistance to endotoxin and malaria in polychlorinated biphenyl and hexachlorobenzene-treated mice," *Infect. Immun.* 20(1):30–35 (1978).
97. Silkworth, J. B., and L. D. Loose. "Environmental chemical-induced modification of cell-mediated immune responses," *Adv. Exp. Med. Biol.* 12/A:499–522 (1980).
98. Glick, B. "Antibody-mediated immunity in the presence of mirex and DDT," *Poultry Sci.* 53(4):1476–1485 (1974).
99. Atabaev, S. T., I. B. Boiko, and V. A. Ilima. "Effect of pesticides on the immunological state of the body," *Gig. Sanit.* 8:7–10 (1978).
100. Loose, L. D., K. A. Pittman, K. F. Benitz, J. B. Silkworth, W. Mueller, and F. Coulston. "Environmental chemical-induced immune dysfunction," *Ecotoxicol. Environ. Safety* 2(2):173–198 (1978).
101. Loose, L. D., J. B. Silkworth, K. A. Pittman, K. F. Benitz, and W. S. Mueller. "Impaired host resistance to endotoxin and malaria in polychlorinated biphenyl and hexachlorobenzene-treated mice," *Infect. Immun.* 20(1):30–35 (1978).
102. Thomas, R. T., and R. D. Hinsdsill. "The effect of perinatal exposure to tetrachlorodibenzo-*p*-dioxin on the immune response of young mice," *Drug Chem. Toxicol.* 2($^1/_2$):77–78 (1979).
103. Glick, B. "Antibody-mediated immunity in the presence of mirex and DDT," *Poultry Sci.* 53(4):1476–1485 (1974).
104. Desi, I., L. Varga, and I. Farkas. "Studies on the immunosuppressive effect of organochlorine and organophosphoric pesticides in subacute experiments," *J. Hyg. Epidemiol., Microbiol., Immunol.* 22(1):115–122 (1978).
105. Loose, L. D., K. A. Pittman, K. F. Benitz, and J. B. Silkworth. "Polychlorinated biphenyl and hexachlorobenzene-induced humoral immunosuppression," *RES J. Reticuloendothel. Soc.* 22(3):255 (1977).
106. Loose, L. D., K. A. Pittman, K. F. Benitz, J. B. Silkworth, W. Mueller, and F. Coulston. "Environmental chemical-induced immune dysfunction," *Ecotoxicol. Environ. Safety* 2(2):173–198 (1978).

107. Loose, L. D., K. A. Pittman, K. F. Benitz, J. B. Silkworth, W. Mueller, and F. Coulston. "Environmental chemical-induced immune dysfunction," *Ecotoxicol. Environ. Safety* 2(2):173–198 (1978).

108. Banerjee, B., M. Kamachandran, and Q. Hussain. "Sub-chronic effect of DDT in mice," *Bull. Environ. Contam. Toxicol.* 37:433–440 (1986).

109. Aaskari, Z. M., and J. Gabliks. "DDT and immunological responses. I. Altered histamine levels and anaphylactic shock in guinea pigs," *Arch. Environ. Health* 26(6):309–331 (1976).

110. Dinoeva, S. K. "Dynamics of changes in the immune structure of lymphatic follicles of the spleen during pesticide poisoning," *Gig. Sanit.* 3:85–87 (1976).

111. Sonntag, A. C. "Xenobiotics and molecular teratology," *Clin. Obstet. Gynecol.* 18(4):199–207 (1975).

112. Olefir, A. I., O. P. Mintser, and R. E. Sova. "Interrelation of indexes of natural body resistance during chronic poisoning with chlorophos, polychloropinene, and sevin," *Gig. Sanit.* 4:25–28 (1977).

113. Atabaev, S. T., I. B. Boiko, and V. A. Ilima. "Effect of pesticides on the immunological state of the body," *Gig. Sanit.* 8:7–10 (1978).

114. Street, J. C., and R. P. Sharma. "Quantitative aspects of immunosuppression by selected pesticides," *Toxicol. Appl. Pharmacol.* 29(1):135–136 (1974).

115. Fan, A., J. C. Street, and R. M. Nelson. "Immune suppression mice administered methyl parathion and carbofuran by diet," *Toxicol. Appl. Pharmacol.* 45(1):235 (1978).

116. Aripdzhanov, T. M. "Effects of the pesticides anthio and milbex on the immunological reactivity and certain autoimmunological reactivity and certain autoimmune processes of the body," *Gig. Sanit.* 7:37–42e (1973).

117. Loose, L. D., K. A. Pittman, K. F. Benitz, and J. B. Silkworth. "Polychlorinated biphenyl and hexachlorobenzene-induced humoral immunosuppression," *RES J. Reticuloendothel. Soc.* 22(3):255 (1977).

118. Koller, L. D., J. H. Exon, and J. G. Roan. "Immunological surveillance and toxicity in mice exposed to the organophosphate pesticide, leptophos," *Environ. Res.* 12(3):238–242 (1976).

119. Banerjee, B., M. Kamachandran, and Q. Hussain. "Sub-chronic effect of DDT in mice," *Bull. Environ. Contam. Toxicol.* 37:433–440 (1986).

120. Sonntag, A. C. "Xenobiotics and molecular teratology," *Clin. Obstet. Gynecol.* 18(4):199–207 (1975).

121. Atabaev, S. T., I. B. Boiko, and V. A. Ilima. "Effect of pesticides on the immunological state of the body," *Gig. Sanit.* 8:7–10 (1978).

122. Shubik, V. M., M. A. Nevstrueva, S. A. Kalnitskii, R. E. Levshits, G. N. Merkushev, E. M. Pilschik, and T. V. Ponomareva. "Effect of chronic enteral administration of radioactive and chemical substances on the immune response," *Gig. Otsenka Faktorov Radiats Neradiats Prir Ikh Komb* 87–91 (1976).

123. Atabaev, S. T., I. B. Boiko, and V. A. Ilima. "Effect of pesticides on the immunological state of the body," *Gig. Sanit.* 8:7–10 (1978).

124. Evdokimov, E. S. "Effect of organochlorine pesticides on animals," *Veterinarya* 12:94–95 (1974).

125. Dankliker, W. B., A. N. Hicks, S. A. Levinson, K. Stewart, and R. T. Brawn. "Effects of pesticides on the immune response. U.S. NTIS report PB80-1309834," *Environ. Sci. Technol.* 16(2):204–210 (1979).

126. Street, J. C. "Pesticides and the immune system," in *Immunologic Consideration in Toxicology,* Vol. 1, R. P. Sharma, Ed. (CRC Press, Inc.: Boca Raton, FL, 1981), pp. 45–66.

127. Cook, H., D. R. Helland, B. H. Vanderweele, and R. J. Dejong. "Histotoxic effects of polybrominated biphenyls in Michigan dairy cattle," *Environ. Res.* 15(4):82–89 (1978).

128. Bekesi, J. G., J. Roboz, H. A. Anderson, J. P. Robez, A. S. Bischbein, I. J. Selikoff, and J.F. Holland. "Impaired immune function and identification of polybrominated biphenyls (PBB) in blood compartments of exposed Michigan dairy farmers and chemical workers," *Drug Chem. Toxicol.* 2($^{1}/_{2}$):179–191 (1979).

129. Brody, J. E. "Immunological effects found in people in Michigan who ate food contaminated by PBB," *New York Times* (August 2, 1977).

130. Carter, J. W. "The effects of polychlorinated biphenyls on T-cell mediated immunity in mice," *Anat. Rec.* 193(3):501 (1979).

131. Allen, J. R., and L. Lambrect. "Responses of rhesus monkeys to polybrominated biphenyls," *Toxicol. Appl. Pharmacol.* 45(1):340–341 (1978).

132. Farber, T., L. Kasza, and A. Giovetti. "Effects of polybrominated biphenyls (Firemaster BP-6) on the immunologic system of the beagle dog," *Toxicol. Appl. Pharmacol.* 45(1):343 (1978).

133. Moore, J. A., M. I. Luster, and B. N. Gupta. "Toxicological and immunological effects of a commercial polybrominated biphenyl mixture (Firemaster FF-1)," *Toxicol. Appl. Pharmacol.* 45(1): 295–296 (1978).

134. De Carlo, F. J., J. Seifter, and V. J. DeCarolo. "Assessment of the hazards of polybrominate biphenyls," *Environ. Health Pers.* 23:351 (1978).

135. Luster, M. I., R. E. Faith, and J. A. Moore. "Effects of polybrominated biphenyls (PBB) on immune response in rodents," *Environ. Health Pers.* 23:227–232 (1978).

136. Allen, J. R., and J. P. VanMiller. "Health implications of 2,3,7,8-tetrachloridibenzo-p-dioxin (TCDD) exposure in primates," in *Pentachlorophenol,* K. R. Kao, Ed. (New York: Plenum Press, 1977).

137. Faith, R. E., and J. A. Moore. "Impairment of thymus dependent immune functions by exposure of the developing immune system to 2,3,7,8-tetrachlordibenzo-p-dioxin (TCDD)," *J. Toxicol. Environ. Health* 3(3):451–564 (1977).

138. De Carlo, F. J., J. Seifter, and V. J. DeCarolo. "Assessment of the hazards of polybrominate biphenyls," *Environ. Health Pers.* 23:351 (1978).

139. Thigpen, J. E., R. E. Faith, E. E. McConnell, and J. A. More. "Increased susceptibility to bacterial infection as a sequela of exposure to 2,3,7, 8-tetrachlorodibenzo-p-dioxin," *Infect. Immun.* 12(6):1314–1324 (1975).

140. Glick, B. "Antibody-mediated immunity in the presence of mirex and DDT," *Poultry Sci.* 53(4):1476–1485 (1974).

141 Luster, M. I., R. E. Faith, and M. I. Luster. "Effects of 2,3,7,8-tetrachlorodibenzofuran (TCDF) on the immune system in guinea pigs," *Drug Chem. Toxicol.* 2($^1/_2$):49–60 (1979).

142. Kazntzis, G. "The role of hypersensitivity and the immune response in influencing susceptibility to metal toxicity," *Environ. Health Pers.* 25:111–118 (1978).

143. Alekseeva, O. G., E. V. Vasileva, and A. A. Orlova. "Abolition of natural tolerance and the influence of the chemical allergen beryllium on auto-immune processes," *Bull. World Health Organization* 51(1):51–58 (1974).

144. Ermakeva, N. G., and E. V. Vasileva. "Determination of the T and B-lymphocytes in workers exposed to the effect of the chemical allergen beryllium," *Gig. Tr. Prof. Zabol.* 4:32–35 (1978).

145. Vasileva, E. V., N. G. Ermakova, and H. A. Orlova. "Specific humoral and cellular responses in berylliosis," *Gig. Tr. Prof. Zabol.* 7:8–12 (1977).

146. Krivanek, N., and A. Reeves. "The effect of chemical forms of beryllium on the production of the immunologic response," *Am. Hyg. Assoc. J.* 33(1):245–252 (1972).

147. Koller, L. D., J. H. Exon, and J. G. Roan. "Immunological surveillance and toxicity in mice exposed to the organophosphate pesticide, leptophos," *Environ. Res.* 12(3):238–242 (1976).

148. Gaworski, C. L., and R. P. Sharma. "The effects of heavy metals on (3H)thymidine uptake in lymphocytes," *Toxicol. Appl. Pharmacol.* 46:305–313 (1978).

149. Koller, L. D. "Effect of Environmental Contaminants on Cell-Mediated Immunity," U.S. NTIS Report PB 292934:37 (1978).

150. Barnes, D. W., and A. E. Munson. "Cadmium-induced suppression of cellular immuity in mice," *Toxicol. Appl. Pharmacol.* 45(1):350 (1978).

151. Hirokawa, K., and Y. Hayaski. "Acute methylmercury intoxication in mice-effect on the immune system," *Acta Pathol. Japon.* 30(1):23–32 (1980).

152. Koller, L. D., J. H. Exon, and B. Arabogast. "Methylmercury: Effect on serum enzymes and humoral antibody," *J. Toxicol. Environ. Health* 2(5):115–123 (1977).

153. Koller, L. D., N. Issacson-Keskvliet, and J. H. Exon. "Synergism of methylmercury and selenium producing enhanced antibody formation in mice," *Arch. Environ. Health* 34(4):248–252 (1979).

154. Graham, J. A., D. E. Gardner, F. T. Miller, M. J. Daniels, and D. L. Coffin. "Effect of nickel chloride on primary antibody production in the spleen," *Environ. Health Pers.* 12:109–113 (1975).

155. Rosales, R. R., and A. Perlmutter. "The effects of sublethal doses of methylmercury and copper, applied singly and jointly, on the immune response of the blue gourami *(Trichogaster trichoptenes)* to viral and bacterial antigens," *Arch. Environ. Contam. Toxicol.* 5(3):325–331 (1977).

156. Rea, W. J., A. R. Johnson, S. Youdim, I. J. Fenyves, and N. Samadi. "T and B lymphocyte parameters measured in chemically sensitive patients and controls," *Clin. Ecol.* 4(1):11–14 (1986).

157. LaVia, M. F., P. E. Hartabise, and J. W. Parker. "T-lymphocyte subset phenotypes: A multisite evaluation of normal subjects and patients with AIDS," *Diagn. Immunol.* 3:75–82 (1985).

158. Reinherz, E. L., H. L. Weiner, and S. L. Hauser. "Loss of suppressor T cells in active multiple sclerosis—analysis with monoclonal antibodies," *N. Engl. J. Med.* 303:125 (1980).

159. Leung, D. Y. M., A. R. Rhods, and R. S. Guha. "Enumeration of T cell subsets in atopic dermatitis using monoclonal antibodies," *J. Allergy Clin. Immunol.* 67:450 (1981).

160. Tsokos, G. C., and J. E. Balow. "Phenotypes of T lymphocytes in systemic lupus erythematosus: Decrease cytotoxic/suppressor subpopulation is associated with deficient allogeneic cytotoxic responses rather than with concanavalin A-induced suppressor cells," *Clin. Immunol. Immunopathol.* 26:267–276 (1983).

161. Veys, E. M., P. Hermans, G. Verbruggan, and H. Mielants. "Immunoregulatory changes in autoimmune disease," *Diag. Immunol.* 1:224–232 (1983).

162. Tsokos, G. C., and J. E. Balow. "Phenotypes of T lymphocytes in systemic lupus erythematosus: Decrease cytotoxic/suppressor subpopulation is associated with deficient allogeneic cytotoxic responses rather than with concanavalin A-induced suppressor cells," *Clin. Immunol. Immunopathol.* 26:267–276 (1983).

163. Tsokos, G. C., and J. E. Balow. "Phenotypes of T lymphocytes in systemic lupus erythematosus: Decrease cytotoxic/suppressor subpopulation is associated with deficient allogeneic cytotoxic responses rather than with concanavalin A-induced suppressor cells," *Clin. Immunol. Immunopathol.* 26:267–276 (1983).

164. Giurgca, R., Witterberger, C. G. Frecus, S. Manciulea, M. Borsa, D. Coprean, and S. Ilyes. "Effects of some organochlorine pesticides on the immunological reactivity of white rats," *Arch. Exp. Veterinaermed.* 32(5):769–776 (1978).

165. Aripdzhanov, T. M. "Effects of the pesticides anthio and milbex on the immunological reactivity and certain autoimmunological reactivity and certain autoimmune processes of the body," *Gig. Sanit.* 7:37–42e (1973).

166. Frash, V. N., and B. G. Karaulov. "Leukosisi promoting effects of benzene (state of stem and immunocompetent cells under the effects of small doses of benzene)," in *Rd Stvolovyk Kletok Leikozo-Kantsero-Geneze*, R. E. Kaketskii, Ed. (Kiev: Akademiya Nauk Ukrains'koi, RSR, 1977), pp. 79–80.

167. Loose, L. D., J. B. Silkwoorth, K. H. Pittman, K. F. Benitz, and W. S. Mueller. "Impaired host resistance to endotoxin and malaria in polychlorinated biphenyl- and hexachlorobenzene treated mice," *Infect. Immun.* 20(1):30–35 (1985).

168. Hruskoner, M. S., J. A. Rasanen, H. Markonen, and S. Ass. "Asbestos-exposed workers as a cause of immunological stimulation," *Scand. J. Respir. Dis.* 59:326–332 (1978).

169. Moszcynski, P. "Evaluation of the total immunity of workers exposed to organic solvents containing benzene and its homologues," Med. *Prag.* 3D:225–229 (1979).

170. Dueva, L. A. "Immunological manifestations of delayed-type hypersensitivity and problems of specific immunodiagnosis in occupational allergy of chemical etiology," *Gig. Tr. Prof. Zabol.* 1:19–23 (1979).

171. Rea, W. J., A. R. Johnson, S. Youdim, I. J. Fenyves, and N. Samadi. "T and B lymphocyte parameters measured in chemically sensitive patients and controls," *Clin. Ecol.* 4(1):11–14 (1986).

172. Reinherz, E. L., H. L. Weiner, and S. L. Hauser. "Loss of suppressor T cells in active multiple sclerosis—analysis with monoclonal antibodies," *N. Engl. J. Med.* 303:125 (1980).

173. Leung, D. Y. M., A. R. Rhods, and R. S. Guha. "Enumeration of T cell subsets in atopic dermatitis using monoclonal antibodies," *J. Allergy Clin. Immunol.* 67:450 (1981).

174. Tsokos, G. C., and J. E. Balow. "Phenotypes of T lymphocytes in systemic lupus erythematosus: Decrease cytotoxic/suppressor subpopulation is associated with deficient allogeneic cytotoxic responses rather than with concanavalin A-induced suppressor cells," *Clin. Immunol. Immunopathol.* 26:267–276 (1983).

175. Veys, E. M., P. Hermans, G. Verbruggan, and H. Mielants. "Immunoregulatory changes in autoimmune disease," *Diag. Immunol.* 1:224–232 (1983).

176. Rea, W. J., A. R. Johnson, S. Youdim, I. J. Fenyves, and N. Samadi. "T and B lymphocyte parameters measured in chemically sensitive patients and controls," *Clin. Ecol.* 4(1):11–14 (1986).

177. Rea, W. J., A. R. Johnson, S. Youdim, I. J. Fenyves, and N. Samadi. "T and B lymphocyte parameters measured in chemically sensitive patients and controls," *Clin. Ecol.* 4(1):11–14 (1986).

178. Dowty, B. J., D. R. Carlisle, J. L. Laseter, and J. S. Storer. "Halogenated hydrocarbons in New Orleans drinking water and blood plasma," *Science* 187:75 (1975).

179. Laseter, J. L., and B. J. Dowty. "Association of biorefractories in drinking water and body burden in people," *Ann. N.Y. Acad. Sci.* 298:546–547 (1977).

180. Leung, D. Y. M., A. R. Rhods, and R. S. Guha. "Enumeration of T cell subsets in atopic dermatitis using monoclonal antibodies," J. Allergy *Clin. Immunol.* 67:450 (1981).

181. Fletcher, M. A., G. C. Baron, M. R. Ashman, M. A. Fischl, and N. G. Klimas. "Use of whole blood methods in assessment of immune parameters in immunodeficiency states," *Diagn. Clin. Immunol.* 1(5):69–81 (1987).

182. Rea, W. J., Y. Pan, A. R. Johnson, and E. J. Fenykes. T and B lymphocytes in chemically sensitive patients with toxic volatile organic hydrocarbons in their blood. *Clinical Ecology* 5(4):171–175 (1987/88).

183. Rea, W. J., A. R. Johnson, S. Youdim, I. J. Fenyves, and N. Samadi. "T and B lymphocyte parameters measured in chemically sensitive patients and controls," *Clin. Ecol.* 4(1):11–14 (1986).

184. Rea, W. J., Y. Pan, A. R. Johnson, and E. J. Fenykes. T and B lymphocytes in chemically sensitive patients with toxic volatile organic hydrocarbons in their blood. *Clinical Ecology* 5(4):171–175 (1987/88).

185. Rea, W. J., Y. Pan, A. R. Johnson, and E. J. Fenykes. T and B lymphocytes in chemically sensitive patients with toxic volatile organic hydrocarbons in their blood. *Clinical Ecology* 5(4):171–175 (1987/88).

186. Schwartze, G., and K. D. Wozniak. "Behavior of human lymphocytes under the influence of benzene in vitro and vivo. 2," *Ges. Hyg. Ihre Grenzgeb.* 31(3):171–173 (1985).

187. Rozen, M. G., and C. A. Snyder. "Protracted exposure of C5TBL/G mice to 300 ppm benzene depresses B and T lymphocyte numbers and mitogen responses. Evidence for thymic and bone marrow proliferation in response to the exposures," *Toxicology* 37(1–2):13–26 (1985).

188. Wierda, D., R. D. Irons, and W. F. Greenlee. "Immunotoxicity in C57BL/G or 6 mice exposed to benzene and aroclor 1254," *Toxicology Appl. Pharmacol.* 64:410–417 (1981).

189. Dueva, L. A. "Immunological manifestations of delayed-type hypersensitivity and problems of specific immunodiagnosis in occupational allergy of chemical etiology," *Gig. Tr. Prof. Zabol.* 1:19–23 (1979).

190. Post, G. B., R. Snyder, and G. F. Kalf. "Inhibition of RNA synthesis and interleukin-2 production in lymphyocytes in vitro by benzene and its metabolites, hydroquinone and p-benzoquinone," *Toxicol. Lett.* 29:161–167 (1985).

191. Grayson, M. H., and S. S. Gill. "Effect of in vitro exposure to styrene, styrene oxide, and other structurally-related compounds on murine cell-mediated immunity," *Immunopharmacology* 11:165–173 (1986).

192. Loose, L. D., J. B. Silkwoorth, K. H. Pittman, K. F. Benitz, and W. S. Mueller. "Impaired host resistance to endotoxin and malaria in polychlorinated biphenyl- and hexachlorobenzene treated mice," *Infect. Immun.* 20:30–35 (1985).

193. Loose, L. D., J. B. Silkworth, K. A. Pittman, K. F. Benitz, and W. S. Mueller. "Impaired host resistance to endotoxin and malaria in polychlorinated biphenyl and hexachlorobenzene-treated mice," *Infect. Immun.* 20(1):30–35 (1978).

194. Moszcynski, P. "Evaluation of the total immunity of workers exposed to organic solvents containing benzene and its homologues," *Med. Prag.* *3D*:225–229 (1979).

195. Rea, W. J., A. R. Johnson, S. Youdim, I. J. Fenyves, and N. Samadi. "T and B lymphocyte parameters measured in chemically sensitive patients and controls," *Clin. Ecol.* 4(1):11–14 (1986).

196. Chang, K. J., J. S. Chen, P. C. Huang, and T. C. Tung. "Study of patients with polychlorinated biphenyl poisoning. I. Blood analysis of patients," *Taiwan I Hsuen Hui Tsa Chih* 79(3):304–313 (1980).

197. Clark, D. A., G. Sweeney, S. Safe, E. Hancock, D. G. Kilburn, and J. Gauldie. "Cellular and generic basis for suppression of cytotoxic T-cell generation by haloaromatic hydrocarbons," *Immunopharmacology* 6(2):143–153 (1983).

198. Hoffman, R. E., P. A. Stehr-Green, K. B. Webb, R. G. Evans, A. P. Knutsen, W. F. Schramm, J. L. Staake, B. B. Gibson, and K. K. Steinberg. "Health effects of long-term exposure to 2,3,7,8-tetrachlorodibenzo-p-dioxin," *JAMA* 255(15):2031–2038 (1986).

199. Chandra, R. K., and M. Baker. "Numerical and functional deficiency of suppressor T cells precedes development of atopic eczema," *Lancet* 2:1393–1394 (1983).

200. Valverde, E., J. M. Vich, J. Huguet, J. V. García-Calderón, and P. A. García-Calderón. "An in-vitro study of lymphoctues in patients with atopic dermatitis," *Clin. Allergy* 13(1):81–88 (1983).

201. Strannegard, O., and I. Strannegard. "T lymphocyte abnormalities in atopic disease," *Immunol. Allergy Pract.* 5:13–19 (1983).

202. Leung, D. Y. M., A. R. Rhods, and R. S. Guha. "Enumeration of T cell subsets in atopic dermatitis using monoclonal antibodies," *J. Allergy Clin. Immunol.* 67:450 (1981).

203. Bucala, R., J. Fishman, and A. Cerami. "Formation of covalent adducts between cortisol and 16 alphahydroxyestrone and protein: Possible role in the pathogenesis of cortisol toxicity and systemic lupus erythematosus," *Proc. Natl. Acad. Sci. U.S.A.* 79:3320–3324 (1982).

204. Bucala, R., J. Fishman, and A. Cerami. "Formation of covalent adducts between cortisol and 16 alphahydroxyestrone and protein: Possible role in the pathogenesis of cortisol toxicity and systemic lupus erythematosus," *Proc. Natl. Acad. Sci. U.S.A.* 79:3320–3324 (1982).

205. Gibson, G. G., R. Hubbard, and D. V. Park, Eds. *Immunotoxicology* (New York: Academic Press, 1983).

206. Liang, H.-C., and W. J. Rea. "Pesticide effects on cell-mediated immunity in chemically sensitive patients," (1991), in press.

207. Liang, H.-C., and W. J. Rea. "Pesticide effects on cell-mediated immunity in chemically sensitive patients," (1991), in press.

208. Griffith, R. C. "Thymus gland," in *Anderson's Pathology.* Vol. II, 8th ed., J. M. Kissane, Ed. (St. Louis, MO: C. V. Mosby Co., 1985), p. 1359.

6 NUTRITIONAL STATUS AND POLLUTANT OVERLOAD

INTRODUCTION

In the thousands of observations and scientific studies performed in the Environmental Control Unit (ECU) on chemically sensitive patients, pollutant-nutrient interactions were found to be basic to understanding chemical sensitivity. Several principles involving these interactions have evolved, and their use will aid the clinician in evaluating the effects of pollutant overload upon the body's nutrient pool. These principles emphasize that (1) both an individual and his specific end-organs are biochemically and nutritionally unique and that there are many reasons for pollutant deposition and injury; (2) an energy loss can be seen in the chemically sensitive and is now understandable due to pollutant interactions upon the energy regulators (adensosine triphosphate [ATP], phosphorous, calorie and proteins metabolism) of the body; (3) cellular and organelle excesses as well as deficiencies of nutrients occur due to pollutant injury (e.g., solvents), resulting in damage to the membranes and regulators of cells, nucleus, mitochondria, etc.; (4) pools of nutrients such as vitamins, minerals, amino acids, and enzymes exist which are available for a finite amount of detoxification, and when they are depleted by poor nutrient intake or excess demand (due to pollutant overload), the system dysfunctions and illness starts. Only at this point will nutrient deficiencies appear, but even then, they may not be obvious or measurable due to end-organ specificity; (5) pollutants may adversely affect food quality, eating habits, intestinal flora, transport across membranes in the gut and bloodstream as well as transport across extracellular fluid and within the cells; (6) pollutants may disturb or destroy nutrients directly; (7) pollutants compete with nutrients for absorption and utilization; (8) pollutants damage the reabsorption mechanism in the

221

kidney and possibly the gastrointestinal tract; and (9) they can also damage the repair mechanisms. Knowledge of these principles allows better understanding of the nature of chemical sensitivity and affords the clinician a methodology by which to evaluate and treat systematically the chemically sensitive patient.

POLLUTANT DAMAGE TO BIOCHEMICAL NUTRIENT INDIVIDUALITY

Due to their genetic makeup, their state of nutrition and pollutant exposure in utero, and their state of nutrition and total load at the time of exposure, individuals may react differently to identical pollutant loads (see Chapter 3). Further, individual responses may be even more selective and varied for specific end-organs. The clinician, therefore, must consider individual differences of end-organ involvement when evaluating the chemically sensitive.

The level of an individual's pollutant load may vary due to his total body biochemical individuality and the specific biochemical individuality of each target organ. The detoxification ability of target organs may vary due to quantitative and qualitative cellular differences for vitamins, minerals, amino acids, lipids, carbohydrates, and enzyme content both in the blood and in each end-organ. Due to the higher fat content in one organ, which attracts lipophilic xenobiotics and holds onto them, or its being nearer to the portal of injury of a pollutant or its landing on an already damaged organ or just random pollutant deposition, one organ can have a higher pollutant load than another and thus different problems in different individuals may manifest. Due to normal biochemical uniqueness, individual responses to different environmental pollutants which lodge in specific end-organs may also vary considerably and also cause increased problems. These responses may be accentuated or even altered in the chemically sensitive (Table 1). For example, a specific xenobiotic such as chlordane may deposit equally on end-organs A and B, yet it might end up that one of these organs will ultimately have amounts of chlordane ten times higher than the other. By sheer volume and toxic nature, detoxification of either organ would be difficult, but since organ A also normally has less nutrients, including enzymes, to drive the detoxification process, it has difficulty fending off the pollutant load. Thus, resistance may break down and chemical sensitivity occur or existing chemical sensitivity may be exacerbated. Once this breakdown of resistance occurs, a downward spiral of ill health follows with clinical manifestations. Organ B may be better able to cope with this same pollutant overload because it contains more of the specific types of nutrient-derived enzyme detoxification substances, e.g., cytochrome P-450, than organ A. If the same amount of pollutant were delivered to organ B, no adverse clinical manifestations would occur. Since each end-organ has its own nutrient-to-pollutant ratios available for detoxification, injuries that result in different symptoms and signs in different individuals occur in different parts of the

Table 1. End-Organ Nutrient Specificity and Individuality Resulting in the Chemically Sensitive Individual

	Organ A	Organ B
Xenobiotic characteristic organ specificity and, therefore, level	$10X$	1
Biochemical individuality		
(A) Nutrients	$5X$	$10X$
(B) Enzymes	1	$5X$

body; thus the varied clinical responses seen in some chemically sensitive individuals. Localized muscle cramps in a chemically sensitive individual, for example, may develop with magnesium deficiency while the occurrence of problems with bronchospasm may be nonexistent because of the local differences in end-organ magnesium levels available for function or repair. Another example is that taurine, which is higher in the liver, leukocytes, brain, and heart, may be deficient due to localized pollutant injury in only one organ, and this deficiency may affect the function of this end-organ before it effects other organs. Then symptoms and dysfunction may occur in that end-organ only without signs of peripheral deficiency. If an organ such as tissues of the nervous system or the bone marrow is more prone to acquiring pollutants and also has less normal nutrient content to use in detoxification, an individual may be even more prone to pollutant injury followed by the onset of chemical sensitivity.

Pollutant overload may induce secondary food intolerance due to disturbance of the metabolism, and it can cause deficiencies or loss of nutrient reserve pools of vitamins, minerals, lipids, carbohydrates, proteins, and enzymes. These nutrient deficiencies characterize the chemically sensitive.

POLLUTANT INJURY TO ENERGY METABOLISM

One of the biggest complaints of chemically sensitive patients is loss of energy. Many patients complain of extreme fatigue after pollutant exposure. Many are chronically weak or experience sudden episodes of weakness. However, weakness may be just a switching from aerobic to anaerobic metabolism, which renders inefficient function of energy metabolism decreasing overall energy at least 36 times. This switch may occur in many ways, resulting in poor function with the symptoms of fatigue, weakness, and poor memory. In total physical combustion of food stuffs including carbohydrates, fats, and protein, energy is distributed in many ways. Fifty percent of energy is used for fat released in catabolism or futile cycles (resynthesis in cycles resulting in two rather than three ATPs). This process may be a greater percentage in the chemically sensitive, producing less ATP and resulting in weakness and fatigue. Futile cycles are hydrolysis and resynthesis of triglycerides and glycogen, metabolic shunts that fuel electrons into the coenzyme site of electron transport (via $FADH_2$) rather than into the NADH site, resulting in a maximum of two rather than three ATPs per electron pair, and a deliberate heat production via brown fat and/or direct uncoupling of oxidative phosphorylation in other tissues (Figure 1). Between 1 to 9% of energy is lost in the feces because of nondigestion or utilization by bacteria. In the chemically sensitive, imbalance of gastrointestinal flora and poor digestion due to pollutant injury will adversely modify the percentage of these energy patterns. Another 6 to 10% of energy is required for digestion, absorption, distribution, modification, and storage of digestible nutrients. More energy may be utilized in the chemically sensitive for these processes; thus, symptoms of energy loss result. The final 25 to 40% of energy is energy for basal metabolism and physical and mental energy. At times, some chemically sensitive patients can barely muster enough energy for physical survival, and they have extreme difficulty solving mental problems probably due to the total need and lack of sufficient energy for survival.

The actual energy captured in the form of high-energy phosphate bonds is less than 40% of the general population with the cost from protein breakdown being highest. Frequently the chemically sensitive are unable to tolerate, absorb, and incorporate adequate amounts of protein; they then may experience a loss of energy. What remains of the inherent metabolizable energy of the ingested nutrients (approximately 50%) is released as heat, which is largely used to maintain body heat. If more energy is utilized for basic function in the chemically sensitive than is released as heat, energy expenditure may lessen for heat production. This reduction of energy expenditure may account for the usually cold body temperature (around 95 to 96°F) found in the chemically sensitive. The heat, being the product of built-in inefficiency of the metabolic pathways for catabolism of carbohydrates, fats, and other fuels, plus mechanisms that result in futile cycles,[1] may not be present in the chemically sensitive because there is not enough excess energy to allow for wasted energy. For other tissues, thyroid hormones may also be important in reducing the efficiency of ATP vs. heat production. The coldness of the chemically sensitive may be partially due to inappropriate thyroid function, but more often due to other factors such as peripheral vasospasm. Perhaps in part, this action is caused by increasing the shunting of glycolysis electrons and/or increasing membrane ATPase activities. These factors illustrate some flexibility and/or variability in the production of warmth vs. high-energy phosphate bonds, which then influence the energy needs of individuals and perhaps allow some adaptation to circumstances such as pollutant overload. These bonds are always inadequate in the chemically sensitive (Figure 2).

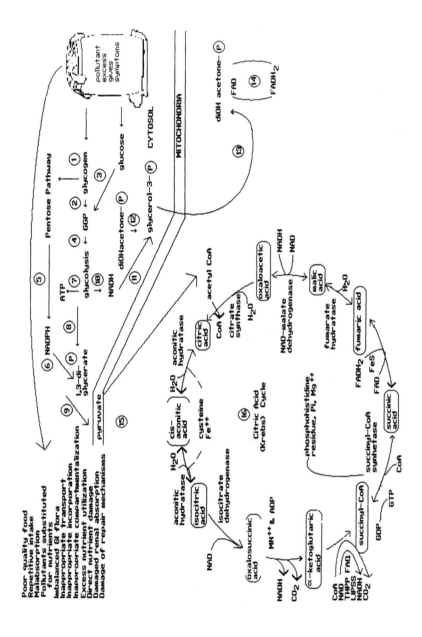

Figure 1.   Pollutant injury to metabolic sites for loss of energy.

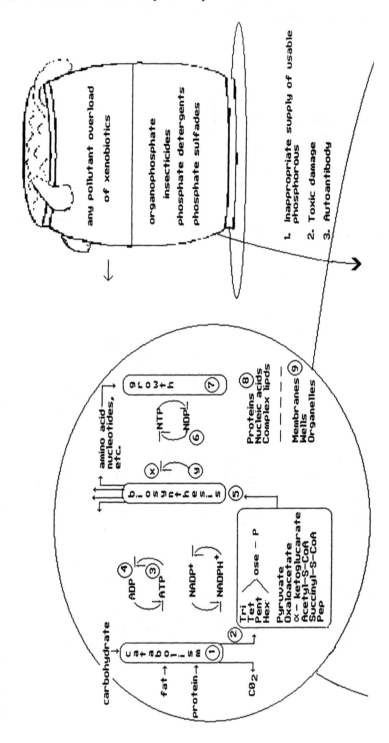

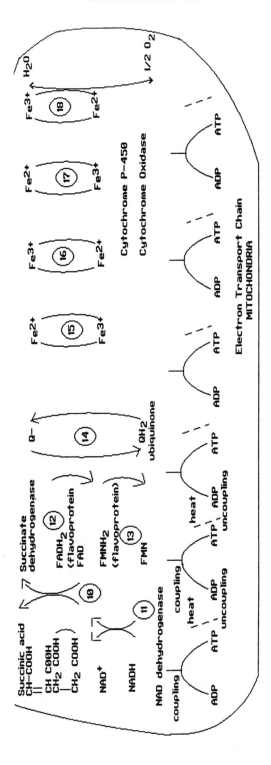

Figure 2.   Potential areas of pollutant injury in the chemically sensitive patient.

Pollutant injury in the chemically sensitive may occur at many steps of energy storage, development, or utilization. An increased load of toxic chemicals may only strain the efficient operation of the metabolic systems without prolonged pathology occurring. In a chemically sensitive individual who develops cold sensitivity with resultant subnormal temperature, as is seen in a large number of the chemically sensitive studied at the EHC-Dallas, less heat is generated. This hypermetabolism that uses up nutrient fuels but does not generate heat and is often seen with pollutant injury is not as efficient as regular metabolism; thus, overuse of nutrients may occur. Poor intake due to overuse of nutrients, poor food quality, and malabsorption may lower energy output. When the body has increased utilization of nutrients in order to combat pollutants, a vicious cycle develops and nutrient depletion increases. The whole cycle continues and metabolism becomes increasingly less efficient until energy is extremely poor. In addition to cold sensitivity and energy loss, the patient at this point develops weight loss and eventual protein caloric malnutrition. Further, increases in total body pollutant load require that more acetylation enzymes and acetyl molecules be used for detoxication and detoxification; thus, a shift away from efficient food metabolism occurs. More metabolic shunts are used, which causes further inefficient use of energy. In addition to these malfunctions, direct toxic damage, which may halt or severely alter the process or damage nutrients, may occur. For example, when organophosphate pesticides get into the phosphorous pool, which results in less production of proper high-energy phosphorous,[2] direct interference and damage occurs. Phosphorous binds to tissue proteins decreasing function. In addition to incorporation of toxic chemicals into pools, antibodies build to cytochrome (P-450) and other systems, resulting in loss of energy and precise function.[3] These antibodies may be generated after excess load of phosphorous or sulfur pesticides. As shown in Chapter 4, damage to mitochondria also occurs with pollutant overload, causing energy depletion. Antibodies to enzymes can be the cause of autoimmune diseases and are seen in the autoimmune liver disease.

Caloric deficiency may also occur with chemical exposure. Chemically sensitive patients often first become sensitive to the ingestion of fat due to the lipophilic nature of most toxic chemicals; they then develop sensitivity to proteins and next carbohydrates. For example, phenol marasmus, which is marked by caloric deficiency, was well known in English hospitals after Lister introduced antiseptic techniques using carbolic acid (phenol).[4] (See Chapter 12.) This syndrome has been seen in many of our patients. Further, an individual who develops caloric deficiency as a result of chemical exposure becomes malnourished and may initially experience symptoms of malaise, energy loss, subtle errors in judgment, poor memory, or unexplained anxiety. Frequently, there is also a gradual, then increasing, inability to tolerate particular foods. This growing intolerance is usually followed by nausea, vomiting and bloating, often accompanied by sudden weight loss. Because of the patient's malnutrition, treatment at this point with food avoidance will generally exacerbate problems. Eventually, severe deficiency can result in neurological, cardiac, genitourinary, and gastrointestinal end-organ failure. This downward spiral of nutritional depletion continues unless the chemical load is decreased and nutrients replaced. *Severe malnutrition at times may be the end-stage of chemical sensitivity.* The EHC-Dallas has seen over 100 patients who were so severely malnourished they had to have intravenous hyperalimentation or tube feeding for extended periods of time in order to reestablish their equilibrium (see Chapter 37). These patients are similar to those identified in early descriptions of phenol marasmus.

Lack of conversion of β-alanine to alanine, a B_6-dependent step, may disturb metabolism causing loss of energy. Conversion of alanine in the muscle to pyruvate to transport to the liver and convert back to alanine is an extremely essential step in gluconeogenesis in the liver. Disordered metabolism in this process, along with altered glycolysis and gluconeogenesis, will result in energy loss in the chemically sensitive.

MEMBRANE DAMAGE RESULTING IN EXCESS CELLULAR NUTRIENTS AND DAMAGE TO MEMBRANE POTENTIALS OF ENERGY PRODUCTION

Any lipophilic toxic chemical can invade or attack the lipid membrane of a cell or its organelle, causing either membrane inhibition or proliferation. Though any toxic substance may cause some damage, various substances such as solvents are more prone to do so than others. The most damaging are chlorinated ethanes and ethylenes or other aliphatic and aromatic hydrocarbons, which may penetrate the membranes causing damage to the membrane selectivity as well as to its mineral pumps and energy generators. Because membrane damage is one of the most common results of pollutant injury in the chemically sensitive, the clinician must be on the alert for clinical changes such as edema, inability to detoxify, and sodium-calcium-magnesium channel dysfunction. Damage to the mitochondrial and endoplasmic membrane as well as the cell membrane can readily create areas of imbalance disturbing the ordering of water and, thus, markedly disturbing functions. Clinically, a failure of nutrient absorption from the gastrointestinal tract, incorporation for nutrient transport, or failure of nutrient reabsorption in the kidney may be seen. Both the ligand-gated channels and voltage-gated ionic channels may be damaged by pollutant injury. Ligand-gated channels that may be damaged by pollutant injury are the acetylcholine, the glycine, and the gamma-aminobutyric acid (GABA) receptors. Since the ligand usually contains minerals, its active attachment may become damaged, resulting in memory malfunction, loss of judgment, and paraethesis. Pollutant injury may also cause antibody formation, which then causes malfunction of the receptors. The voltage-gated channels that can be injured include those of sodium, potassium, and calcium transport. Organophosphate insecticides can damage acetyl cholinesterase, thereby damaging the acetylcholine channel, while alkaloids like strychnine, benzo diazapines, barbiturates, and picrotoxins are known to harm the glycine to butyric channels.[5] Calcium channel blockers like 1,4-dihydropyridine may inhibit calcium flow. A myriad of solvents and other hydrocarbons may damage sodium and potassium channels, resulting clinically in edema and potassium maldistribution with weakness and fatigue resulting.

Another glaring reason for understanding membrane damage in the chemically sensitive is that of ATP generation, which requires proper membrane integrity for its redox potentials to work. The bulk of ATP is generated by the oxidation of reduced substrates coupled to the phosphorylation of adenosine diphosphate (ADP) to ATP, with oxygen serving as the terminal acceptor which may be altered in the chemically sensitive. The chief function of respiration in ATP production catalyzed by ATP synthetase in the respiratory chain consists of a complex of redox carriers including flavoproteins, quinones, and cytochromes (Figure 2). ATP synthetase catalyzes the oxidation of NADH, succinate, and a few reduced substrates by passing electrons sequentially from one carrier to the next. The reduced substrates that serve as electron donors may be present in the external environment, but are more commonly produced by metabolic reaction within the cells, particularly by the citric acid cycle. The driving force for ATP generation during oxidative phosphorylation is the difference in redox potential between two half reactions linked by the electron transport cascade across membranes. The respiratory chain is an elaborate device to translocate protons across the membrane in which it is housed, and the electrochemical potential of protons across membranes is the driving force of ATP synthesis. The complete oxidation of glucose to CO_2 yields 36 moles of ATP per mole of glucose, while glucose to lactic acid per day yields 2 moles. ATP production is 300,000 times daily, but the pool of ATP production is 1,000 times per day. The mean lifetime of an ATP molecule is one third of a second. The rate of energy production of a human being per unit weight is 10^4 times greater than the sun. In the chemically sensitive, interference of membrane function of xenobiotics can alter this generation.

The biological role of NAD is analogous to ATP in that both are carriers of activated electrons that cycle from oxidized to reduced forms. Generation of NADH and NADPH constitutes a form of energy conservation that is no different in principle from free production of ATP and is frequently altered in the chemically sensitive. The adenine and nicotinamide nucleotides function both as carriers of metabolic energy and as regulators of energy flux.

Damage to the membranes containing these functions as well as any of the enzymes or other components either by nutrient depletion or direct pollutant injury will decrease production of ATP and thus exacerbate or induce the weakness and fatigue of chemical sensitivity.

Results of studies at the EHC-Dallas suggest that any nutrient can be imbalanced in the cell or organelle after pollutant damage. Therefore, not only is there an excess in xenobiotics, but in some cases, an excess of minerals occurs intracellularly and perhaps intraorganellely with pollutant injury as well. Hard evidence from the EHC-Dallas indicates that exits for minerals such as calcium, manganese, barium, aluminum, copper, potassium, selenium, and zinc may be deficient or excessive in some cells of the chemically sensitive after pollutant injury occurs. Some of these mineral alterations in themselves can then further cause target organ damage with exacerbation of the chemical sensitivity. Also, amino acids such as β-alanine have been found in excess in the cells of some chemically sensitive patients, causing a derangement in orderly valine metabolism, resulting in a loss of taurine with resultant gastro-intestinal upset and an exacerbation of the chemical sensitivity. Our studies of nutrients in the chemically sensitive clearly demonstrate that a patient could simultaneously have excess and deficient nutrients due to pollutant injury (see sections Minerals and Amino Acids in this chapter).

NUTRITIONAL DEFICIENCIES AND POOL DEPLETION

Either deficiencies or pool depletions (ready reserves) may cause or exacerbate chemical sensitivity, upon pollutant exposure. Outright measurable deficiencies may occur in chemical sensitivity but so can pool depletion that is difficult to measure. Both can be measured, the former by static or dynamic analysis of blood levels, the latter by stressed pollutant challenge, where the patients are exposed to a pollutant and their reactions are cleared by supplementation of nutrients. Understanding the nature and limited availability of nutrient pools will enable the clinician to manipulate pollutant unloading as well as perform nutrient analysis and therapy in the chemically sensitive. Therefore, both aspects will be discussed together.

Many environmental pollutants have been implicated in the depletion of different vitamins,[6] minerals, amino acids, lipids, and enzymes. Numerous toxic chemicals such as benzene,[7] carbon monoxide,[8] and pesticides[9] apparently increase requirements for vitamins A, C, E, and all the B vitamins in order to detoxify. Folate, riboflavin, pyridoxine, and niacin also can ameliorate the effects of some pollutants and/or be depleted by chemical exposure.[10]

Deficiency and ready reserve pool depletions may be caloric and non-caloric and may compound the original pollutant injury. Loss of nutrient reserve pools increases vulnerability to disease and results in suboptimum function (Table 2). Each aspect of nutrient availability (i.e., vitamins, minerals, amino acids, lipids, and enzymes) will be discussed separately in order to aid clinicians in finding an orderly way to evaluate the nutrient status of the chemically sensitive.

Table 2. Vitamins and Their Role in Enzyme Function Which May Malfunction Due to Pollutant Damage in the Chemically Sensitive

Vitamin	Coenzyme Form (or Active Form)	Type of Reaction or Process Promoted
Water-soluble		
Thiamine (B_{13})	Thiamine pyrophosphate	Decarboxylation of α-keto acids
Riboflavin (B_2)	Flavin mononucleotide, flavin adenine dinucleotide	Oxidation-reduction reactions
Nicotinic acid (B_3)	Nicotinamide adenine dinucleotide, nicotinamide adenine dinucleotide phosphate	Oxidation-reduction reactions
Pantothenic acid (B_5)	Coenzyme A	Acyl-group transfer Acetyl-group transfer
Pyridoxine (B_6)	Pyridoxal phosphate	Amino-group transfer
Biotin	Biocytin	CO_2 transfer
Vitamin B_{12}	Deoxyadenosyl cobalamin	1,2 Hydrogen shifts
Ascorbic acid	Not known	Cofactor in hydroxylation reactions, antioxidant
Fat soluble		
Vitamin A	Retinal	Visual cycle, skin, antioxidant
Vitamin D	1,25-Dihydroxychole-calciferol	Regulation of Ca^{2+} metabolism
Vitamin E	Not known	Protection of membrane lipids, free radical scavenger
Vitamin K	Not known	Cofactor in carboxylation reactions

Eating patterns that deprive the body of nutrients and bring about depletion of nutrient pools such as failure to vary the diet or excessive and monotonous intake of foods low in nutritional value (i.e., repetitious eating of beef, sweets, milk, eggs, etc.) will stress any one type of metabolic system. Excluding certain nutrients by eating processed foods or foods high in sugar and fat results in selective nutrient overload and depletion and leaves an individual vulnerable to illness. At the EHC-Dallas, we have observed that these eating patterns have a direct effect on the onset and severity of chemical sensitivity. We suspect also that as the American diet becomes increasingly nutritionally unbalanced due to the prevalence of these poor eating habits, significant portions of the population may experience increased susceptibility for chemical sensitivity. Many individuals may be nutritionally marginal, and when pollutant stress occurs, overload may follow, resulting in symptoms of overt nutrient deficiency.

Generally, selective nutrient intake is common in the U.S. population; thus, selectivity appears accentuated in the chemically sensitive. This lack of adequate nutrient intake may produce or exacerbate some of the initial problems of the chemically sensitive. According to the major nutritional surveys over the last 25 years, there are significant shortages in the nutrients identified in Table 3 among various population groups[11-15] which may predispose them to the onset of chemical sensitivity.

The population groups at highest risk for malnutrition and probably chemical sensitivity include women, the elderly, adolescents, low-income groups, ethnic minorities, and infants and children. Early chemical sensitivity is seen in women at a 7:1 ratio to men presenting at the EHC-Dallas, and its onset would be compatible with the high incidence of women with nutrient deficiency seen in these surveys. The nutrients most often lacking among these groups include women: vitamins A, C, and B_6, calcium, iron, magnesium; teenagers (male): iron, magnesium, and vitamin C; teenagers (females): adolescents who ate less than 70% of their energy needs met the RDA for only 1 of 12 nutrients studied (protein); elderly: vitamin B_6 and C, calcium, thiamine, riboflavin, and niacin; and infants/children: iron, vitamins C, A, niacin; low-income groups: vitamin C, thiamine, riboflavin, niacin, iron; ethnic minorities: vitamin A (Hispanic); vitamins A and C, thiamine, riboflavin, niacin, and iron (blacks).[16] These specific nutrients are often essential for pollutant detoxification, and their deficiencies may explain the increasing incidence of chemical sensitivity in these groups.

The 1977–1978 Nationwide Food Consumption Survey showed that significant percentages of the population were consuming less than 70% of the RDA for the nutrients listed which would make them prone to malnutrition and then susceptible to xenobiotic overload (Table 3).

Regardless of age, income, race, urbanization, or region, nutrients consumed by Americans in both 1985 and 1986 were below the RDA for seven nutrients including vitamin B_6, calcium, magnesium, iron, vitamin E, folacin, and zinc. Nutrient intake by children was lower in 1986 than in 1985; however, the intake of only two nutrients — iron (86% RDA) and zinc (82% RDA) — failed to meet the RDA in both years. Varying degrees of inadequate intake are seen in poor and black children. These findings correlate with our direct measurements in the chemically sensitive, who were found to have deficiencies of chromium, zinc, magnesium, manganese, and vitamins B_1, B_2, B_3, B_6, and C.[17] If intake in the chemically sensitive, who may originally be part of these groups, worsens or is heaped upon the deficiencies seen in these groups, severe nutritional deficiencies or diminished nutrient pools are fairly certain to result. Since nutrient deregulation occurs in the chemically sensitive, we wonder if the

Table 3. American Population with Nutrient Intake below the RDA with Resultant Nutrient Deficiency — A Predisposition for Chemical Sensitivity?

National Survey	Nutrient Shortage
USDA Household Food Consumption Survey 1965–1966	Vitamins A and C; calcium
Ten State Nutrition Survey 1968–1969 (lower income groups)	Vitamins A and C; riboflavin, calcium, iron
HANES I 1971–1972	Vitamins A and C; calcium, iron, magnesium
USDA Nationwide Food Consumption Survey 1977–1978	Vitamins B_6, A, C; calcium, iron, magnesium
HANES II 1977–1978	Vitamin C; thiamin; riboflavin; iron
1985–1986	B_6, calcium, magnesium, iron, zinc, folacin

Nutrient	% (Individuals with Intakes <70% RDA)
Vitamin B_6	51
Calcium	42
Magnesium	39
Iron	32
Vitamin A	31
Vitamin C	26
Thiamin	17
Vitamin B_{12}	15
Riboflavin	12

groups reported to be deficient by these national surveys are more prone to develop chemical sensitivity. We expect that this is the case.

Physiologically, these nutrient deficiencies in the chemically sensitive are probably the result of a combination of factors including intake of food contaminated by commercial growing and processing as well as monotonous eating patterns. Pollutants may initially effect the food chain, yielding nutrient-deficient foods. Damage (i.e., vitamin and mineral and protein deficiencies) to food that occurs during commercial food growing and processing is well established.

Foods grown using toxic chemicals may look good but be deficient in certain vitamins and minerals (see Chapter 8). For the chemically sensitive, food ingestion may be complicated by intake of toxic substances in the food.

Nutritionally deficient and chemically contaminated food may cause an imbalance in the gastrointestinal tract and flora, exacerbating the chemically sensitive. Even high-quality foods can be destroyed by additives used as preservatives, colorings, and flavorings. For example, EDTA is a substance that, when added to some vegetables such as peas, will deplete minerals.[18] Many chemically sensitive patients exhibit gastrointestinal upset and malabsorption, which is another reason for nutrient depletion in the chemically sensitive. Clearly the vitamin and mineral swings seen in the studies on our chemically sensitive patients suggest a selective type of malabsorption and membrane instability that results in severe dysfunction (e.g., an organochlorine pesticide such as chlordane will damage a cell membrane; the calcium pump then becomes damaged and excess calcium ends up in the cell). Most of our chemically sensitive patients are unable to tolerate many foods; therefore, they ingest a limited variety. This limited variety may increase their chemical load and may disturb gut flora, resulting in inadequate presentation of nutrients to the bloodstream for absorption. Vitamin depletion has been attributed to chemical overexposure of either an individual or the food he eats, and it results in overuse of the detoxification systems or direct damage to a nutrient.[19] It appears from our observations at the EHC-Dallas that environmental chemicals are the main contributing factor to vitamin deficiencies or pool depletion in the chemically sensitive.[20-22] Our observations are strongly supported by the fact that as blood levels of toxic chemicals decrease, nutrient values normalize in the chemically sensitive, and the chemically sensitive improve clinically. It has become clear from our observations that environmental overload has become a fact of life for many people.

In the chemically sensitive, pollutants can have a direct destructive effect upon a nutrient,[23] or they can directly interfere with transport mechanisms, resulting in inadequate ambient absorption from the intestine or incorporation within a cell. Competitive absorption may occur, whereby a toxic chemical may be preferentially absorbed in larger quantity than, or in place of, a needed nutrient. For example, environmental chemicals including hydrazine,[24] isoniazid,[25] hydralazine,[26] and phenelzine[27] are suspected of being antimetabolites which contribute to apparent B_6 deficiency. Attempting to fuel the hypermetabolism required for detoxification, the immune and enzyme systems constantly utilize nutrients. The result of this activity is a gradual depletion of nutrients over time.

The relationship of nutritional status of vitamins to the effects of pollutant toxicity has been extensively reviewed by Calabrese.[28] Although the contribution of genetic factors to the observed nutrient deficiencies in this population is not completely known, genetic variation in susceptibility to environmental agents is evident.[29] (See Chapter 3.)

Three of the four major processes for the detoxication of harmful chemicals are oxidations, reductions, and degradations. They are also the major pathways for food breakdown (see Chapter 4). These systems can become overloaded, producing inappropriate breakdown of food stuffs,[30] which in turn can yield secondary toxic and allergic effects,[31] as are often seen in the chemically sensitive (see Chapter 4). In addition, selective depletion of amino acids, fats, or carbohydrates can also occur from overloading these systems, resulting in nutrient depletion and exacerbation of chemical sensitivity. Depletion of nutrients and toxic injury can also form leaks in the kidney and gastrointestinal tract due to direct pollutant injury to these organs, thus preventing essential reabsorption of vitamins, minerals, and amino acids.

Many enzyme systems use cofactors that are derived from specific vitamins (both food and water soluble), and from minerals such as Zn^{2+}, Fe^{2+}, Cu^{2+}, Mn^{2+}, and others (see Enzymes of Biotransformation in this chapter), resulting in increased sensitivity. For example, some patients who have food and chemical sensitivity and who are depleted of nutrients necessary for sulfoxidations are poor sulfonators. (See Chapter 4.) All enzyme systems need amino acids for their formulation, and these may be damaged or overutilized by pollutants, resulting in insufficient enzyme formation and thus inadequate detoxication and detoxification.

We have just discussed the many ways nutrient deficiency or pool depletion that results in increased susceptibility to pollutant exposures can occur. Additional evidence for this phenomenon is available in the literature. Marginal deficiencies of protein, vitamins A, C, E, and B_6, and folacin have been shown to result in greater vulnerability to infection from viral and bacterial processes.[32] This kind of vulnerability is frequently seen in the chemically sensitive. These vitamins help to maintain immune function, and inadequacies can impair the body's ability to resist disease. Similarly, vitamins C, A, K, and B-complex and zinc, iron, and copper are critical for proper wound healing,[33] which may be altered in the chemically sensitive. *The role of nutritional factors as modifying influences on pollutant toxicity resulting in chemical sensitivity is a critical issue in environmental health.* Despite the need to know more about nutrient-pollutant interactions, an amazing wealth of knowledge already exists that has important implications.[34] Clearly, because complexities of nutrient interactions occur in so many diverse areas of the body, symptoms are varied. Of necessity, these symptoms include weakness, malaise, depression, exhaustion, anorexia, paresthesia, etc. The seemingly vague nature of the aforementioned symptoms once cleared in the ECU takes on new meaning with the realization that these symptoms can be signs of nutritional dysfunction when compared with a normal, asymptomatic baseline. The aforementioned symptoms, which occur after pollutant challenge, are no longer vague. They are clearly defined because the patient has been symptom free until a pollutant challenge overstressed his nutrient fuels and reproduced symptoms. In addition, the deficiency or pool depletion which might have been overlooked because blood levels were "normal," can now be proven by individual nutrient challenge where the clinician will give one essential nutrient, e.g., vitamin C or B_6, etc., and signs and symptoms disappear. This testing procedure allows the clinician to reassess his perception of what is considered "normal," thereby

Essential nutrient pools shrink with pollutant stress in the chemically sensitive; therefore, localized and transient nutrient deficiency may develop with lowered resistance occurring with exacerbation of symptoms.

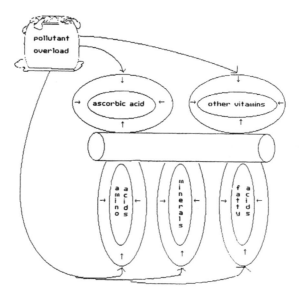

Figure 3.   Pollutant injury to nutrient pools.

producing a new concept of optimum health. This reassessment integrates the concepts of subclinical and pathognomonic, etc.

Our observations in the ECU suggest that the effects of pollutants on nutrients in the chemically sensitive may be subtle and insidious or blatant, and a slow downward progression of bodily function may result from deficiencies, pool depletions, or imbalances. Currently, numerous studies involving animals, as well as humans, confirm these observations and show the influence of vitamins A, B-complex, C, D, and E on pollutant toxicity.[35]

It should be emphasized that the total body nutrient pools (e.g., sulfur pool, phosphorous, ascorbic acid, amino acid, etc.) which are utilized for quick response to the various needs of vital organs can be rapidly depleted by constant pollutant overload (Figure 3). These pools appear to be at different levels in the chemically sensitive and the normal individual, with the chemically sensitive becoming rapidly depleted when stressed by pollutants. Thus, the clinician will sometimes have difficulty in completely assessing a pollutant-derived nutrient deficiency because blood levels may be normal while end-organs or merely their pools may be deficient. Stressing the chemically sensitive patient by pollutant challenge may reproduce symptoms and reveal these subtle deficiencies or define the limits of nutrients available to respond to the pollutant stress.

VITAMIN AND MINERAL LEVELS IN THE CHEMICALLY SENSITIVE

Studies at the EHC-Dallas have shown that the chemically sensitive patient is deficient in many vitamins and minerals in spite of self-supplementation in 50% of the cases presenting. Many others are deficient in amino acids, lipids, and antipollutant enzymes. Our studies have been spread over several years and have repeatedly confirmed the deficiencies in the chemically sensitive similar to but accentuated as defined by the National Nutritional Surveys. One group of 120 patients showed deficiency in 10 to 60% of various vitamins. This depletion is astounding since nutrient pools are difficult to evaluate and were not included. Therefore, these figures represent the absolute minimum deficiency seen in these chemically sensitive patients.

The preliminary data at the EHC-Dallas on serum vitamin concentrations and functional assay were obtained prior to 1983 from Kirkpatrick Medical Laboratory[36] or the New Jersey Medical School,[37] and Monroe Medical Laboratory.[38] Methodologies used by these labs are summarized in Chapter 30.

The criteria for deficiency were taken in most cases from the standards recommended by the RDA, Monroe, and Pangborn.[39] Levels measured were by enzyme stimulation indices, urinary metabolites, and direct blood and serum concentrations.[40-42]

In our preliminary nutritional assessment study of 120 chemically sensitive patients using serum vitamin concentrations, a significant number exhibited vitamin deficiencies. To verify these deficiencies, a second group of 333 environmentally sensitive patients was assessed by a panel of 10 assays offered by Monroe Medical Laboratory. These included functional assays and serum vitamin concentrations as described earlier. The results of this second study are summarized in Figure 4. Of these patients, 64% had a deficiency of B_6; 30% had a deficiency of B_2; 29% had a deficiency of B_1; 28% had a deficiency of C; 27% had a deficiency of folate; 24% had a deficiency of D; 19% had a deficiency of B_3, and 4% had a deficiency of B_{12}. The functional vitamin assays were performed on patients prior to treatment, which included nutrition replacement therapy. The results of these studies suggest that the chemically sensitive have a particular susceptibility to B_6 deficiency.

The combined group of 453 patients presented with a myriad of symptoms related to the neurological, cardiovascular, respiratory, gastrointestinal, genitourinary, or integumentary systems. Their common feature was a marked and often extreme sensitivity to toxic environmental chemicals, foods, food additives, inhalants, and molds.

These results support the view that vitamin deficiencies are widespread in environmentally sensitive patients, and B_6 deficiency is most pronounced. This deficiency might have been predicted because B_6 is so essential in the myriad of transamination reactions used for metabolism as well as detoxification. Only

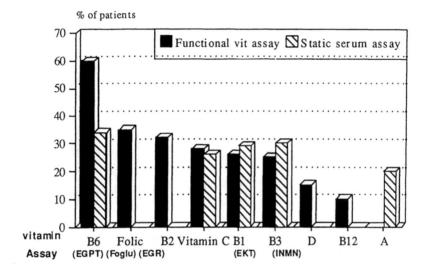

% of patients

Figure 4.   Functional vitamin assay vs. static serum levels (120 chemically sensitive patients).

3 of the 333 patients were deficient in vitamin A and/or β-carotene (Table 4).
This small number plus 12 of the preliminary 120, however, does not negate
the possibility of end-organ deficiency or pool depletion of vitamin A, especially in the liver, which is the main area for detoxification. Many of our
patients respond to β-carotene even when their serum levels are normal.[43]

Individual vitamin, mineral, amino acid, and fatty acid deficiencies and
excesses and enzyme abnormalities are discussed within their respective sections in this chapter.

Because of the importance of nutrient deficiencies and pool depletion and
the benefit and complexity of supplementation, an analysis of each vitamin,
mineral, amino acid, fatty acid, and enzyme is given in the following pages.

In the chemically sensitive, impaired cholesterol metabolism results from
chemical injury to the liver and direct pollutant injury to nutrients. Also,
inappropriate amounts of B_3, B_5, C, biotin, magnesium, and omega-3 fatty
acids appear to disturb their fatty acid metabolism, resulting in high cholesterol
and triglycerides. A more detailed discussion of impaired cholesterol is presented later in this chapter in the section on lipids.

INDIVIDUAL NUTRIENTS

As emphasized in the previous pages, individual and complex interactions
of xenobiotics with the body's metabolic processes fueled by nutrients are
paramount in understanding chemical sensitivity. Most interactions require

Table 4. Vitamin Assay in 333 Environmentally Sensitive Patients

Vitamin (range)	71 men					262 women					333 Combined				
	No. Men Tested	No. Men Low	% Men Low	Mean	S.D.	No. Women Tested	No. Women Low	% Women Low	Mean	S.D.	Total No. Tested	Total Low	Total % Low	Mean	S.D.
B_1 (0–17.3%)	69	13	18.8	13.03	17.00	253	71	28.1	11.59	13.8	322	84	26.1	11.87	14.53
B_2 coefficient (0.9–1.20)	67	26	38.8	1.17	0.19	260	88	33.8	1.18	0.46	327	114	34.9	1.18	0.42
B_3 (3–17 mg/24 hr)	68	11	16.2	12.64	12.92	250	52	20.8	11.87	26.83	318	63	21.4	12.04	21.52
B_6 index (<1.25)	69	26	37.7	1.44	0.47	253	160	63.2	1.44	0.55	322	186	57.8	1.44	0.53
B_{12} (0–3 mg/24 hr)	67	11	16.4	4.35	9.48	242	35	14.5	4.43	11.12	309	46	14.9	4.49	10.76
Folate (0–3 mg/24 hr)	68	24	35.3	3.41	5.47	252	79	31.3	4.45	11.23	320	103	32.2	4.25	10.27
Vit. A (25–75 μg/dl)	68	3	4.4	55.35	31.99	252	15	5.95	48.54	22.93	320	18	5.63	50.60	25.26
Vit. C (21–53 μg/10^8 cells)	71	22	31	35.40	22.88	254	68	26.77	31.67	16.88	325	90	27.7	33.4	18.46
Vit. D (12–41 μg/ml)	68	23	33.8	26.82	17.55	237	92	38.8	25.91	18.64	305	115	37.7	28.00	18.40
β-Carotene (50–250 μg/ml)	69	3	4.35	143.12	86.69	256	12	4.7	149.64	95.61	325	15	4.62	149.56	93.70

Source: Ross, Rea, Johnson, Maynard, and Carlisle.[43]

many nutrients and are complex, while a few require only one nutrient and are simple. Each nutrient will be discussed separately in order to allow the clinician an opportunity to have an orderly knowledge of the possible response of each nutrient to pollutant exposure.

Vitamins

Vitamins along with minerals are usually the cofactors that allow the detoxification systems to counteract xenobiotics. Each will be considered separately.

Vitamin A

Although vitamin A is necessary to the prevention and treatment of chemical sensitivity in that it may severely effect metabolism and some detoxification reactions, it is often depleted by pollutant injury (Figure 5). In fact, 1 to 15% of chemically sensitive patients will have a vitamin A deficiency. Therefore, understanding its physiology will aid understanding chemical sensitivity.

Vitamin A is a fat-soluble, potent antioxidant that affects free radicals adversely[44] by disarming singlet oxygen (see Chapter 4). In the chemically sensitive, it supports growth and health and is necessary for vision, reproduction, mucous secretion, and the maintenance of differentiated epithelia.[45] It also fights bacteria[46] and infections,[47] both of which are often seen in a subset of the chemically sensitive. Furthermore, it is essential for a normal estrogen cycle, and low vitamin A may adversely affect the premenstrual tension syndrome experienced by many chemically sensitive women.

Compatible with our findings are the results of other studies which have shown that vitamin A supplementation reduced death in patients with measles[48] and, conversely, that mild vitamin A deficiency in children has been linked to increased mortality from all causes.[49]

β-Carotene is the precursor of vitamin A, which has a protective function in humans as well as plants. It protects plants and the chemically sensitive against singlet oxygen generated by oxidation (in plants photooxidation or photosynthesis).[50] The use of pesticides, which cause the formation of free radicals or the disturbance of other mechanisms, may cause loss of β-carotene in plants[51] and subsequently diminish vitamin A available from plants for humans. Several pesticides such as tetrachlorofenvinphos, lindane, and fenitrotion decrease the β-carotene content of wheat, carrots, radishes, and flax.[52-55] It is believed that herbicides block the pathway of β-carotene formation.[56,57] Some herbicides may vary on stimulating or decreasing β-carotene.[58,59] Because pesticide usage may alter the β-carotene content of crops significantly, values given in many nutrient content books may be inaccurate.

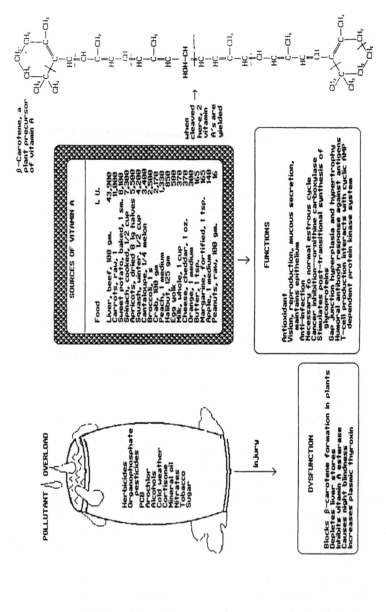

Figure 5.   Pollutant injury with excess total load.

A supply good for several months is usually stored in the liver of a well-nourished person. This storage may not be the case in the chemically sensitive who have a high herbicide and pesticide content in their blood and tissues.

Absorption. The dietary vitamin A esters are hydrolyzed in the lumen of the small intestine to form retinol. Hydrolysis can be disturbed by pollutant overload in the intestine, and malabsorption is seen in the chemically sensitive. Retinol passes across the mucosal cell wall where it is again esterified and carried as retinyl ester to the liver where it is stored. Carotenoids are particularly absorbed as such from the intestine and contribute to the yellow color of the blood serum. Most of the carotene is converted in the intestinal mucosa to retinal (vitamin A aldehyde).

Although vitamin A is useful in preventing and treating chemical sensitivity, many chemically sensitive (especially formaldehyde sensitive) patients cannot tolerate it when it is administered orally. One factor contributing to vitamin A intolerance may be that both the liver and intestinal mucosa have enzymes which catalyze reduction of the aldehyde to alcohol.[60] These enzymes are used both for converting the vitamin A in a food and its contaminants to aldehydes. Therefore, food contaminated by toxic chemicals could contribute to this intolerance in that if the pathways used for detoxifying aldehydes are already overloaded, they may be unable to convert more vitamin A aldehydes or toxic additive aldehydes to alcohols. This inability to function may result in increased chemical sensitivity.

β-Carotene can be enzymatically split apart by β-carotene dioxygenase to form two molecules of vitamin A in the small intestine. An oxidized form of retinol is present in rhodopsin, a retinolopsin (a protein) complex, which is present in stacked intracellular membranes of rod cells in the retina. Rhodopsin helps in the firing in the impulse of the optic nerve which sends its message to the brain. Sometimes dioxygenase may be damaged or insufficient in the chemically sensitive, resulting in insufficient formation of vitamin A with resultant deficiency in the aforementioned.

Each of the three major retinoids, retinol, retinal, and retenoic acid, appears to have its own unique biologic function. Retinol in the lowest oxidation state probably serves as a hormone, a sterol for reproductive activity, which may be disturbed by pollutant overload. Retenoic acid and its metabolites affect differentiation of epithelia and may serve as carriers for oligosaccharides in the synthesis of glycoproteins. Pollutant overload may selectively damage this synthesis of glycoproteins. Thus, when retinol enters its target cell, it is promptly bound to a cellular retinol binding protein (CRBP) distinct from the retinol binding protein present in serum. Then the CRBP transports the retinol within the cell where the latter appears to bind to nuclear proteins, and this complex may be disturbed by pollutant injury. When this process happens, immune dysfunction results.

Vitamin A and Thyroid Status. Many people with chemical sensitivity have chemically induced thyroiditis and subsequent hypothyroidism. This damage may be due in part to reduction of vitamin A levels in the thyroid.

Interrelations between retinoids and thyroid function appear in two forms: (1) those relating to the conversion of carotenes to retinoids (e.g., dioxygenase activity) and (2) those relating to the plasma circulation, both of which are bound noncompetitively to prealbumin. Stimulation (or regulation) of dioxygenase activity appears to be one of the roles of thyroid hormone, in that hypercarotonemia tends to occur with hypothyroid states and also symptoms of night blindness are associated with a lack of thyroid function.[61] Vitamin A deficiency can increase plasma thyroxine levels, but not result in hyperthyroidism.[62] Also, supplemental intake of retinoids appears to decrease plasma levels of thyroid hormone in hyperthyroid individuals.[63] Vitamin A lowers serum-bound protein iodine (PBI) in rats. The exact relationships here remain to be elucidated, but they may involve a derangement of thyroid gland feedback regulation by thyroid-stimulating hormone (TSH).[64,65] Pollutant injury may disturb this interrelationship selectively or generally, damaging the dioxygenase activity. Gaitan[66] has shown that resorcinol and other toxic chemicals trigger thyroiditis, which results in hypothyroidism (see Chapter 38).

Deficiency. Pollutant damage may deplete the liver store of vitamin A by various mechanisms.[67] Deficiency in the liver may only become apparent with toxicant stress and may account for the few deficiencies we have observed at the EHC-Dallas. Numerous reports have suggested that marginal vitamin A deficiency occurs with pollutant exposure.

Seshadry and Ganguly[68] reported that organophosphorus pesticides inhibit vitamin A esterase in vitro. Later studies have noted that intramuscular injections of 1.5×10^6 international units of vitamin A in a quick-absorbing formula were a useful adjunct treatment of yearling cattle poisoned by the organophosphate insecticide Ruelene.[69] Ember et al. observed a disruption of vitamin A metabolism in suicidal patients suffering from acute parathion poisoning and found that following exposure to elevated levels of pesticides, serum cholinesterase and vitamin A levels decreased markedly, although no clinical symptoms of poisoning were recorded.[70,71] Removal of the worker from the pollutant exposure brought about a resumption of normal serum cholinesterase activity and vitamin A levels. The authors suggested that latent intoxication may be detected by determining plasma cholinesterase activity and vitamin A levels. DDT also affects vitamin A metabolism because it inhibits aliesterase, which detoxifies the O-P pesticides.[72]

Villeneuve et al.[73] demonstrated an interaction between polychlorobiphenyls (PCBs) and vitamin A showing that the concentration of liver vitamin A was lower in animals receiving Aroclor 1254 than in control animals. This report was followed by one from Cecil et al., who showed that rats exposed to Aroclor 1242 had vitamin A storage in the liver that was diminished by as much as

50%, even though no incidence of toxicity or deficiency symptoms was noted.[74] These investigators suggested that if only marginal quantities of vitamin A were present in the diets of some susceptible rats, hypovitaminosis could result.

Many other substances have been identified as antivitamin A including alcohol,[75] coffee,[76] cold weather,[77] cortisone,[78] diabetes,[79] excessive iron,[80] infections,[81] laxatives,[82] liver disease,[83] mineral oil,[84] nitrates,[85] sugar,[86] tobacco,[87] vitamin D deficiency,[88] and zinc deficiency.[89]

Dietary Sources. The dietary sources of preformed vitamin A are chiefly liver (especially fish liver), butter and fortified margarine, egg yolk, whole milk, cream, and fortified skim milk. The carotene forms are found in dark green, leafy and yellow vegetables (collards, turnip greens, carrots, sweet potatoes, squash) and yellow fruit (apricots, peaches, cantaloupe). The deeper the green or yellow of a vegetable, the more carotene (provitamin A) it contains. Cod and halibut fish oils are usually sources for the therapeutic doses of fat-soluble vitamin A. Grains, meat, raisins, potatoes, radishes, unfortified pasteurized dairy products, and mushrooms are poor sources of vitamin A.[90]

Possible Clinical Mechanisms for Vitamin A in Prevention and Treatment of Chemical Sensitivity and Cancer. Many chemically sensitive patients develop cancer as the end stage of their chemical sensitivity. It is important, therefore, to understand positive effects of vitamin A in relation to chemical sensitivity and cancer. The possible mechanisms of action of vitamin A and other retinoids in cancer prevention and therapy are numerous. Each will be discussed individually.

Malignant transformation involves the loss of cellular differentiation, and retinoids play a role in both the induction and enhancement of such differentiation. Retinoids do not appear to influence the initiation phase of carcinogenesis, but they may inhibit the later stages of tumor promotion. Several mechanisms for retinoid action in cancer will now be discussed.

Retinoid inhibits the enzyme ornithine decarboxylase, which controls the generation of polyamines involved in the growth of normal and neoplastic tissues. The steroid hormone-like activity of vitamin A may result in its binding with intracellular proteins which translocate to the nucleus to affect control of carcinogenesis and differentiation. Retinoids directly stimulate the post-transitional synthesis of glycoproteins which mediate cellular adhesion and growth. They also stimulate gap junction hyperplasia and hypertrophy. The gap junction connects adjacent epithelial cells and plays an important role in regulating tissue organization and growth control. Most gap junctions disappear during carcinogenesis. Retinoids may exert their control on neoplastic alterations via their stimulation of the immune system. They stimulate humoral antibody responses against antigens and stimulate T-cell production. The interaction of retinoids with cyclic adenosine monophosphate (AMP)-dependent protein kinase system may be a possible link between retinoids and malignant transformation and stimulation of cell differentiation. Reversibility of breast cancer cells is seen with vitamin A supplementation. Retinoids effectively inhibit the production of superoxide anions in polymorphonuclear leukocytes (PMNs) stimulated by mitogens, which may result in alteration of cell membrane fluidity. β-Carotene (provitamin A) may have a direct protection effect against carcinogenesis by deactivating excited molecules, particularly excited or singlet oxygen.

Mandani and Elmongy[91] suggest that the most likely mechanism for possible modification of gene expression, which controls the synthesis of dozens of proteins, involves the retinoids. The process by which the retinoids modify the expression is still, however, a mystery.

Because of the importance of vitamin A to the optimal function of major systems and processes as discussed throughout this section, vitamin A supplementation is often advantageous in the chemically sensitive. Because many chemically sensitive patients cannot tolerate vitamin A, however, treatment is often difficult. Therefore, lowering of the total body load to reduce the overall stress placed on the major systems becomes essential to these patients' treatment.

B-Complex Vitamins

B-complex vitamins are water soluble. They are needed for the oxidation-reduction, degradation, and conjugation systems to function properly. However, many reports from the last 15 years have suggested that pollutant overload leads to depletion of these vitamins and a variety of adverse effects, not the least of which is a breakdown of the conjugation systems.

Innami et al.[92] have clearly demonstrated that diets deficient in thiamine and vitamin B_6 will enhance the toxicity of PCBs.

In addition, we have observed at the EHC-Dallas a large proportion of chemically sensitive patients who are intolerant of B vitamin administration. We have usually attributed this to the fact that the B vitamins are made from yeasts. Indeed, some patients can tolerate the micelles drops made from bacteria. Also, many patients have been found able to tolerate injectable preservative-free vitamin Bs. However, there may be metabolic reasons only beginning to be understood that account for some intolerance. For example, chemically sensitive patients may have insufficient levels of thiamine pyrophosphate (TPP) due to formaldehyde overload. These patients may need thiamine for TPP. Flavin adenine dinucleotide (FAD) from B_2 is needed for aldehyde dehydrogenase. Pyridoxal phosphate from B_6 is needed for aldehyde detoxification. Also, this reaction needs taurine. It may be that B intolerance occurs because of excess pollutants or inadequate cofactors which then makes the vitamin a pollutant rather than nutrient.

Though they may often be needed together, each B vitamin will be discussed separately.

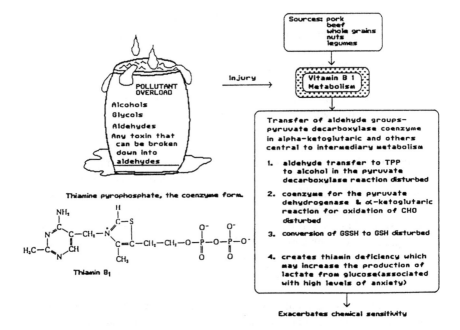

Figure 6. **Pollutant overload causes dysfunction of vitamin B$_1$ metabolism system.**

Vitamin B$_1$ (Thiamin)

Thiamin function is extremely important to understanding some changes that occur in chemical sensitivity.

In 2 separate studies at the EHC-Dallas involving over 400 chemically sensitive patients, 20% of the patients were B$_1$ deficient vs. normal controls. Supplementation with B$_1$ intravenously often resulted in acute improvement of these chemically sensitive patients.

Pollutant injury to thiamin occurs from excess intake of formaldehyde or overuse of the aldehyde detoxifying mechanism. In addition, products such as alcohols, glycols, and other aldehydes may overload the system (Figure 6).

An understanding of the biochemistry of thiamin metabolism helps make clear the pathophysiology of pollutant overload. For example, the coenzyme form of thiamin is TPP, which functions in several enzyme reactions in which aldehyde groups are transferred from a donor to an acceptor molecule. Magnesium is also used in this series of reactions. When TPP serves as a transient intermediate carrier of the aldehyde group, e.g., in the pyruvate decarboxylase reaction (fermentation of sugar by yeast to yield alcohol), the carboxyl group of pyruvate is lost as CO_2, and the rest of the molecule sometimes referred to as active acetaldehyde is simultaneously transferred to the thiamin-thiazole ring. Also B$_1$ serves as coenzyme (TPP) in the more complex pyruvate dehydrogenase and α-ketoglutarate dehydrogenase reactions taking place in the main pathway of oxidation of carbohydrate in cells. Both pathways may be disturbed in some chemically sensitive patients by pollutant overload.

Deficiency. Lack of thiamie may decrease the conversion from glucose to pyruvate, resulting in an increase in lactate (associated with high levels of anxiety often seen in a subset of chemically sensitive patients) by decreasing the activity of the pyruvate dehydroycolase.[93] B_1 deficiency in its extreme form causes beriberi, which is rarely seen in the 20th century. Many chemically sensitive patients with premenstrual syndrome (PMS), however, have been observed to have their central nervous system symptoms improved by the administration of B_1 and B_2 in addition to B_{12}. *Candida* species (often seen in some chemically sensitive) and some raw fish elaborate thiaminases that destroy vitamin B_1 in the gut. This condition gives a tendency for the depletion of the B_1 pool as seen in the chemically sensitive. B_1 is also important along with glutathione synthetase and ATP in its role of restoring oxidized gluta- thione (GSSG) to reduced glutathione (GSH), which when absent or depleted results in decreased tolerance to some chemicals in the chemically sensitive.

Sources. Pork has the greatest supply of thiamine. Good sources of thiamine are beef, organ meats, whole wheat or whole grain cereals, nuts, and legumes (especially peas and beans). Moderate thiamine sources include milk, avoca- dos, cauliflower, spinach, and dried fruits. Poor sources are unenriched white flour and pastas, polished rice, molasses, blueberries, corn, and cheese.[94]

Vitamin B₂

Most detoxication reactions of conjugation of xenobiotics are energy depen- dent. If the nutrient fuels to the mitochondria are disturbed by pollutants, weakness occurs with resultant symptoms of chemical sensitivity. Over 20% of chemically sensitive patients are riboflavin (B_2) deficient and may exhibit this weakness (Figure 7).

Riboflavin is important in several reactions within the body. The coenzyme activity of riboflavin in oxidation-reduction reactions involves electron trans- port (oxidative phosphorylations). Succinate + E + FAD → fumarate + E- $FADH_2$ in which E-FAD designates the succinate dehydrogenase in its bound FAD. Riboflavin is a component of two closely related enzymes, flavin mono- nucleotide (FMN) and FAD, which function as a tightly bound prosthetic group of a class of dehydrogenase known as flavoproteins or flavin dehydrogenase. In the reactions catalyzed by these enzymes, the isoalloxazine ring of the flavin nucleotides serves as a transient carrier of a pair of hydrogen atoms removed from substrate molecules. Succinate dehydrogenase, a member of the group of enzymes catalyzing the citric acid cycle, which is the final metabolic pathway for the oxidative degradation of carbohydrates and fats in the mitochondria, is

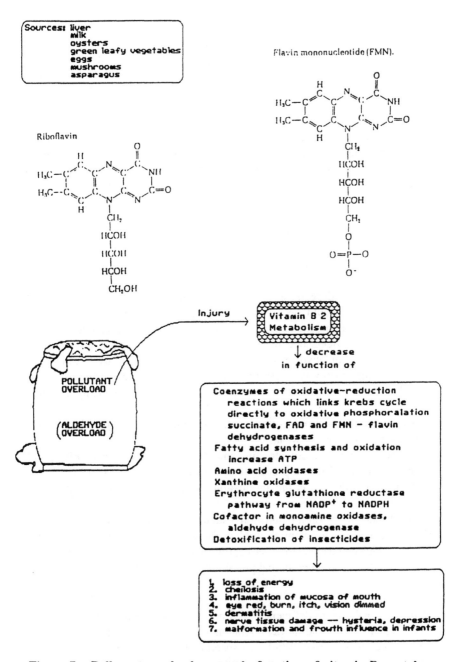

Figure 7.   Pollutant overload causes dysfunction of vitamin B_2 metabolism system.

a member of the flavin dehydrogenase. It contains a covalently bound prosthetic group of FAD and catalyzes the reaction. Some flavin dehydrogenase also contain iron or some other metal as part of their active sites.[95]

Riboflavin has antioxidant power as an indirect function of its activity and therefore influences the reactions of the chemically sensitive. Riboflavin is used in the erythrocyte glutathione reductase pathway to stimulate the conversion of NADH and NADP, which works along with superoxide dismutase (SOD) to trap superoxide anion radicals and prevent pollutant damage to the cell by free radical pathology. Glutathione reductase regenerates glutathione disulfide to reduced glutathione, its active (detoxifying) form for reacting with peroxides (See Chapter 4). GSH is essential in preventing pollutant damage to the chemically sensitive.

FAD is a cofactor in monoamine oxidase reactions which deaminate (oxidize) catecholamines and serotonin to stop their actions and thus reduce their levels[96] (see Chapter 4). If these are not functioning well, mood swings, affective disorders, behavioral abnormalities, and periods of depression occur as seen in a subset of the chemically sensitive.

Aldehyde dehydrogenase metabolizes aldehydes. One possible effect of subnormal FAD would be a lowered aldehyde dehydrogenase which is an FAD-linked liver enzyme activated by iron, sulfur, manganese, and molybdenum of FAD. Patients become hypersensitive to resins, glues, plywood, or other sources of aldehyde vapors (see Chapter 14). Sensitivity to aldehydes usually means sensitivity to ethanol since the same systems are involved. At the EHC-Dallas, we usually see this phenomenon in the chemically sensitive patient.

Detoxification of certain sulfur-containing insecticides may be impaired if FAD monooxygenase is weak.[96,97]

Riboflavin is also a cofactor fundamental to all areas of metabolism and intimately involved in the processes by which the oxidation of glucose and fatty acid is utilized for the production of ATP and support of anabolic processes.

A 10-state survey on the nutritional status of Americans found insufficient riboflavin intake is a potential problem for young people and certain ethnic groups, which could make them susceptible to developing chemical sensitivity. If milk products and other animal protein sources are curtailed, a deficiency is possible.

Deficiency. Riboflavin deficiency does not occur in isolation, but is found as a component of multiple-nutrient deficiency states, as shown in our studies of the chemically sensitive. While no specific disease has been attributed to riboflavin deficiency,[98] several symptoms have been associated with an inadequate intake. These include cheilosis (cracks at the corners of the mouth) and inflammation of the mucous membranes in the mouth accompanied by a smooth, purple-tinged glossitis, reddening of the eyes (due to increased vascularization), and eyes that tire easily, burn, itch, and are sensitive to light. Vision may also be dimmed. Further, vascularization of the cornea, an unusual dermatitis characterized by simultaneous dryness and greasy scaling, nerve tissue damage that may manifest as depression and hysteria, and malformations and retarded growth in infants and children may all be seen in some chemically sensitive patients.

Patients report symptoms of deficiency when their daily intake falls below 0.6 mg. The symptoms occur predominantly in alcohol abusers[99] and those individuals who have excess intake of petroleum alcohols.

Sources. Excellent sources of riboflavin are liver, milk, and milk products, all of which the chemically sensitive are often sensitive to. Moderate sources are oysters, meat, dark green leafy vegetables, eggs, mushrooms, asparagus, broccoli, avocado, brussels sprouts, and fish (such as tuna and salmon). Poor sources are apples, grapefruit, unenriched pastas, cereals, grains, cabbage, and cucumbers.

Because riboflavin is destroyed by light, milk and enriched pastas should be stored in opaque containers. Fresh vegetables should be stored in a dark, cool environment and cooked in a covered pot. Some riboflavin is lost when cooking water and drippings are discarded.

Vitamin B₃ (Niacin or Nicotinamide)

Vitamin B_3 (Niacin or Nicotinamide)

Niacin deficiency has been documented in over 20% of chemically sensitive patients (Figure 8). Although their nutrient pool deficiency cannot be directly measured, we suspect it may be even greater. Lack of B_3 results in inadequate detoxification, thus exacerbating chemical sensitivity.

Vitamin B_3 is produced from tryptophan. Of a diet of tryptophan, 1/60 produces sufficient niacin. Lack of B_3 results in inadequate detoxification, thus exacerbating chemical sensitivity. Because of its importance to the detoxification process, the physiology of vitamin B_3 is now presented.

Nicotinamide is a component of the two related coenzymes nicotinamide adenine dinucleotide (NAD) and nicotinamide adenine dinucleotide phosphate (NADP). Oxidized forms are NAD^+ and $NADP^+$ and reduced forms are NADH and NADPH. The nicotinamide portions of these coenzymes serve as the transient intermediate carrier of a hydride ion (H^+) that is enzymatically removed from a substrate molecule by the action of certain dehydrogenase. These are needed in most of the Krebs cycle. An enzymatic reaction catalyzed by malate dehydrogenase (zinc dependent), which dehydrogenates malate to yield oxaloacetate, a step in the oxidation of carbohydrates and fatty acids, is an example of how nicotinamide works. This enzyme catalyzes the reversible transfer of a hydride ion from malate to NAD^+ to form NADH; the other hydrogen atom leaves the hydroxyl group of malate to appear as a free H^+ ion:

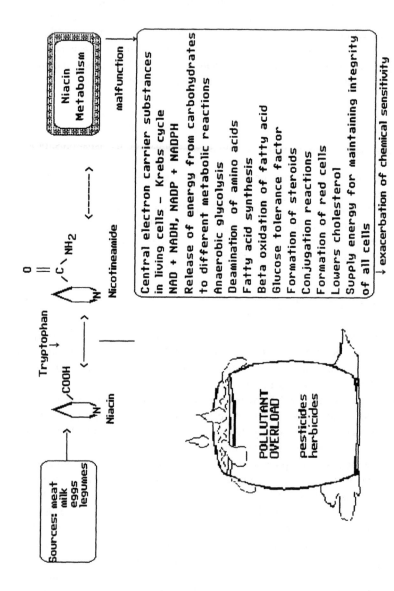

Figure 8. Pollutant overload causes dysfunction of niacin metabolism system.

L-malate + NAD$^+$ ←—→ oxaloacetate + NADH + H$^+$. Disturbance of this orderly process has been seen to occur in the chemically sensitive.

Niacin is involved in the functions of over 50 metabolic reactions that are important in the release of energy from carbohydrates and, thus, in disturbance of niacin metabolism, which results in energy loss as seen in the chemically sensitive. Niacin is also important in the deamination of amino acids, fatty acid synthesis, and β-oxidation of fatty acids. It is essential for the formation of steroids, the metabolism of several drugs and toxicants, and in the formation of red blood cells. Because of its diverse and critical role in so many metabolic pathways, niacin is vital in supplying energy to and maintaining the integrity of all body cells.[100] Niacin also mobilizes chemicals through lipolysis. When it is first administered, inhibition of fatty acids followed by their release occurs (see Chapter 35). Since niacin is widely used throughout the body, disturbance of its function by pollutant overload exacerbates many types of symptoms in the chemically sensitive.

Sources. The best dietary sources are protein foods such as organ meats, peanuts, muscle meats, poultry, legumes, milk, and eggs. Milk and eggs are especially good sources because of their high protein and tryptophan content. Moderate sources are whole grains and breads.[101]

Deficiency. Niacin deficiency, known as pellagra, affects every cell, but is most critical in tissues with rapid cell turnover such as the skin, the gastrointestinal tract, and the nerves. The initial symptoms are weakness, lassitude, anorexia, and indigestion. The classic symptoms of pellagra are "the 3Ds," dermatitis, diarrhea, and dementia. The fourth "D" is death. Pellagra can be used as a model only for early niacin pool deficiency since its full course is rarely seen in 20th century society.

The dermatitis is a scaly, dark pigmentation that develops on areas of the skin exposed to sunlight, heat, or mild trauma (such as the face, arms and elbows, back of the hands, feet, or parts of the body exposed to body secretions or mild irritations). Similar but less severe symptoms are often seen in some chemically sensitive individuals. Other parts of the body are pale in color. All parts of the digestive tract are affected. The tongue is swollen, corroded, and brilliant red. Diarrhea, if it develops, may be accompanied by vomiting and severe inflammation of the mouth. In addition, diarrhea results in faulty fat and fat-soluble vitamin absorption, resulting in further exacerbation of pollutant injury in the chemically sensitive. Achlorhydria, which is also seen in 50% of our chemically sensitive patients, contributes to intestinal infection and lesions.

Nervous system disorders due to niacin deficiency include irritability, headache, insomnia, pain in the extremities, loss of memory, and emotional instability, all of which are seen in a subset of the chemically sensitive. In advanced stages, convulsions may develop. Shortly before death, convulsions and coma may occur. Because of its effect on the nervous system, niacin therapy has been used in the treatment of some schizophrenia. Although pellagra is not seen in First World society, many of its early symptoms are frequent in the chemically sensitive. This phenomenon may be due to stress on the energy generation steps in which niacin is involved.

Deficiency symptoms are seen in diets containing less than 7.5 mg/day. A niacin deficiency seldom occurs alone, and treatment of pellagra with this B vitamin will not cure all symptoms. Often, deficiencies of riboflavin, thiamine and other B vitamins, protein, and iron simultaneously compound the condition, and supplementation of all must be included in the therapy.[102]

Vitamin B₅ (Pantothenic Acid)

Pantothenic acid deficiency has not been observed in the chemically sensitive, although supplementation of pantothenic acid produces good responses in some chemically sensitive patients, perhaps due to a pool depletion that is relatively unmeasurable. Therefore, its physiology may be important in chemical sensitivity.

Pantothenic acid is an amide of pantoic acid and β-alanine (Figure 9). Pantothenic acid is bound to coenzyme A, a transient carrier of acetyl groups. The coenzyme A molecule has a reactive thiol (–SH) (cysteine) group to which acetyl groups become covalently linked to form thioesters during acetyl group transfer reactions. In the oxidative decarboxylation of pyruvate by the pyruvate dehydrogenase complex, acetyl-CoA is formed.

Coenzyme A is important in acetylation of choline and specific aromatic amines (often toxic to the chemically sensitive), oxidation of fatty acids, pyruvate, α-ketoglutarate, and acetaldehyde, and synthesis of fatty acids, sphingosine, citrate, acetoacetate, phospholipid, and cholesterol (and all substances made from it, such as bile, vitamin D, and steroid hormones). Pantothenic acid participates in a variety of pathways involved in the metabolism of carbohydrates, fats, and protein. In addition, coenzyme A functions in the synthesis of porphyrin, a heme component of red blood cells, and the neurotransmitter acetylcholine. It is also essential in acetyl conjugations of phenol amines, sulfuramino, aliphatic amines (e.g., ethylamine), and hydrazines. Acetylation is disturbed in some chemically sensitive patients.

Sources. The word "pantos" means "everywhere" and reflects the ubiquitous role of the vitamin in the body as well as its presence in the diet and may explain why deficiency has not been observed in the chemically sensitive. This B vitamin is found in a wide variety of foods representing three food groups. Good sources include liver and organ meats, fish, chicken, eggs, cheese, whole grain cereals and breads, avocados, cauliflower, green peas, dried beans, nuts, dates, and sweet potatoes. Other foods contain pantothenic acid in smaller but contributory amounts. Fruits are not a good source. Because refined grains are not enriched with pantothenic acid, the significant losses in milling make processed grains a poor source.[103]

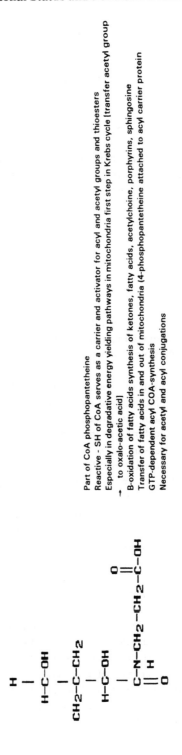

Sources:
beef, pork, chicken,
fish, whole grain
cereals, avocado,
cauliflower, legumes

Pantoic Acid
+
Beta Alanine

Part of CoA phosphopantetheine
Reactive - SH of CoA serves as a carrier and activator for acyl and acetyl groups and thioesters
Especially in degradative energy yielding pathways in mitochondria first step in Krebs cycle [transfer acetyl group
  to oxalo-acetic acid]
B-oxidation of fatty acids synthesis of ketones, fatty acids, acetylchoine, porphyrins, sphingosine
Transfer of fatty acids in and out of mitochondria (4-phosphopantetheine attached to acyl carrier protein
GTP-dependent acyl COA-synthesis
Necessary for acetyl and acyl conjugations

Figure 9.   Vitamin B$_5$ (pantothenic acid).

Deficiency. Pantothenic acid deficiency has not been reported in humans. A laboratory-induced deficiency can be created if subjects are fed a synthetic diet complete in all nutrients except pantothenic acid and are simultaneously given a pantothenic acid antagonist that further depletes the body. The resultant deficiency produces fatigue, cardiovascular and gastrointestinal problems, upper respiratory infections, depression, and numbness and tingling in the extremities. Gastrointestinal disturbances are further complicated by a reduction in bile synthesis. All of these symptoms can be found in some chemically sensitive patients. We speculate that these symptoms may be the result of low vitamin B_5 since supplementation gives relief in some chemically sensitive patients.

In animals, deficiency symptoms include dermatitis, graying or hair loss, hemorrhage, neurological lesions, inflammation of the nasal mucosa, adrenal cortex atrophy, corneal vascularization, and sexual dysfunction.[104] All of these can be found in some chemically sensitive patients.

Vitamin B_6

Sixty percent of chemically sensitive patients are B_6 deficient, whether they are taking oral supplementation or not (Figure 10). Therefore, understanding some of its physiology is necessary to understanding the nature of chemical sensitivity.

The active form of vitamin B_6 is pyridoxal phosphate. It also occurs as pyridoxamine phosphate in its amine form. Pyridoxal phosphate catalyzes a number of transamination reactions in which an amino group of an amino acid is reversibly transferred to the carbon atom of a keto acid. In such reactions catalyzed by transaminases or aminotransferases, the pyridoxal phosphate serves as a transient intermediate carrier of the amino group from its donor, the amino acid, to the amino group acceptor, the α-keto acid. Transaminases typically catalyze double-replacement reactions. There is a great tendency for the aldehyde group on the pyridoxal phosphate to form a Schiff's base with the amino acid, which in effect neutralizes its toxicity. If this reaction is overloaded with excess exogenous chemicals, the chemically sensitive may react.

When amino group transfer steps are retarded (usually by pollutant injury), the degree and extent of the resulting amino acidemias or acidurias are dependent upon dietary input of protein. The resulting symptoms also can be strongly influenced by diet, and allergic-like reactions to foods can develop as seen in the chemically sensitive. Retarded amino group transfer can be caused by the following: deficiency in intake of B_6, reduced ability to transform the vitamin into the coenzyme (for phosphorylation, pyridoxal 5-phosphate rather than B_6 may have to be supplemented), reduced binding of the coenzyme to the apoenzyme (the transaminase), or subnormal α-ketoglutaric acid (the amino group receptor). Pollutant injury can cause these with a resultant exacerbation of chemical sensitivity.

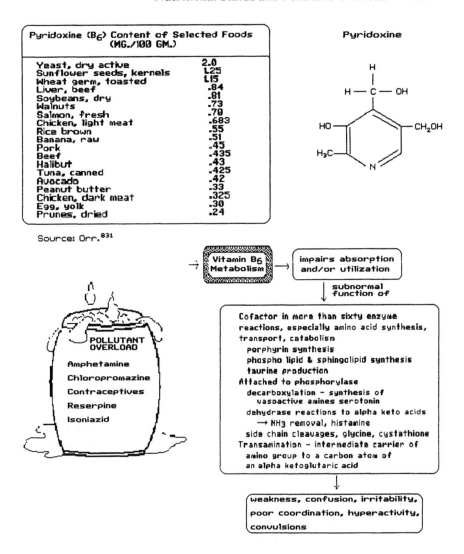

Figure 10.   Pollutant overload causes dysfunction of vitamin B₆ metabolism system.

Subnormal coenzyme activity of pyridoxal 5-phosphate may lead to subnormal taurine. When taurine is acutely low, extreme sensitivities to environmental chemicals, such as chlorine, chlorite (bleach), aldehydes, alcohols, petroleum solvents, and ammonia can develop. Taurine mediates the vigorous chemical oxidation sequence that is initiated by the respiratory burst occurring with neutrophilic phagocytosis of microbes or foreign substances. When taurine levels are low, the body's oxidant defense chemistry is unregulated with respect to scavenging excess levels of the oxidant OCl (hypochlorite), and endogenous formation of aldehydes can become excessive. If low taurine occurs, impaired xenobiotics occur as seen in the chemically sensitive. If poor methylation occurs due to impaired methionine metabolism, which is a result of the low B_6 and, hence, low taurine, poor ability to conjugate epinephrine and serotonin occurs and chemical sensitivity is then exacerbated.

Chronically impaired methionine metabolism is associated with the development of clinical disorders (atherosclerosis and osteoporosis) over a period of time. Subtle clinical symptoms of this are headache, eyestrain, muscle weakness, fatigue, mild myopathy, and myopia.

B_6 catalyzes over 60 reactions in intermediary metabolism (usually transamination) and has been shown to be of value in relieving the symptoms of PMS, carpal tunnel syndrome, depression, exhaustion, hyperemesis gravidism, and edema in both chemically and nonchemically sensitive patients.

Pyridoxal kinase is activated by zinc and magnesium. Suboptimal activity of this enzyme, or low levels of zinc or magnesium, or impaired membrane transport can make supplements of B_6 (pyridoxal phosphate) desirable.

Bioavailable forms of B_6 are apparently more reactive than other B vitamins and are consequently readily depleted during food processing. Vitamin B_6 depletion during food processing can be quite dramatic in that not only are the vitamins inactivated or destroyed, but some food additives are, in fact, antimetabolites of B_6[105] (e.g., EDTA is added to frozen vegetables such as peas and broccoli to make them a brilliant green; it depletes B_6). Phosphopyroxyllysine in cells produces pyrodoxyllysine during thermal food processing, which is itself an antimetabolite that can accentuate vitamin B_6 deficiency symptoms.[106]

Although the number of known B_6-dependent enzymatic reactions has stabilized, recent data also implicate B_6 in the production of PG1 prostaglandin-like substances, which are thought to be anti-inflammatory.[107] Inflammation is present in chemical sensitivity, and this may be one of the reasons increased requirement of B_6 is recognized. Certain segments of the population, including breast-fed infants,[108] women after prolonged use of oral contraceptives,[109] the elderly, those with carpal tunnel syndrome,[110] and alcoholics[111] have B_6 deficiency, which will make these groups susceptible to chemical overload. Whether or not the extent of vitamin B_6 deficiency observed in our study may, in part, reflect a widespread deficiency state in the general population is unclear. The deficiency may be specific to the chemically sensitive individual, since 60% of the chemically sensitive were deficient. The fact that over 60 enzyme reactions are dependent on B_6 activity strongly suggests that depletion could result in a variety of different metabolic abnormalities that exacerbate the chemically sensitive.

Antivitamins include alcohol, coffee, estrogens, oral contraceptives, penicillamine, postmenopausal drugs, radiation exposure, tobacco.

A deficiency in B_6 can produce a high value in kyneuric and xanthurenic acid and ornithine. That urine analysis can be used to identify these amino acids and, hence, a B_6 deficiency makes this a useful diagnostic tool in treating the chemically sensitive.

Sources. Protein foods that are good sources of vitamin B_6 include meats and organ meats, poultry, fish, egg yolk, soybeans and dried beans, peanuts, and walnuts. Other good sources are bananas, avocados, cabbage, cauliflower, potatoes, whole grain cereals and bread, and prunes. Poor sources include egg whites, fruits, refined grains, lettuce, milk, and beer. Significant amounts can be lost during cooking and improper storage of foods.[112]

Deficiency. A vitamin B_6 deficiency produces profound effects upon amino acid metabolism, resulting in a wide spectrum of possible effects ranging from a reduced synthesis of niacin to impaired production of neurotransmitters and hemoglobin.

The symptoms of a vitamin B_6 deficiency are generalized and diffuse. No particular disease has been associated with a vitamin B_6 deficiency, although many symptoms have. These include weakness, mental confusion, irritability and nervousness, insomnia, poor coordination in walking, hyperactivity, convulsions, abnormal electroencephalogram, declining blood lymphocytes and white blood cells, anemia, and skin lesions. Symptoms similar to those of riboflavin and niacin deficiencies (seborrheic dermatitis, glossitis, cheilosis, and stomatitis) have been reported. All of these have been observed in some chemically sensitive patients.

Some drugs can impair vitamin B_6 absorption and utilization. Amphetamine, chlorpromazine, reserpine, and oral contraceptives affect either the concentration of the vitamin in tissue or its enzyme form. Tryptophan metabolism is impaired in women using the birth control pill, possibly due to reduced circulating vitamin B_6.[113]

Some genetic diseases are related to abnormalities in vitamin B_6 metabolism. These include infant convulsive seizures caused by a reduced glutamic decarboxylase activity and resulting in decreased synthesis of GABA, vitamin B_6-responsive anemia due to decreased formation of aminolevulinic acid in heme synthesis, cystathionuria from reduced activity of cystathionase, xanthurenic aciduria resulting from decreased conversion of kynurenine to anthranilic acid because of reduced kynureninase activity, and homocysteinuria caused by reduced conversion of homocysteine to cystathionine.

Vitamin B_{12} (Cobalamin)

Because 15% of the chemically sensitive are folic acid deficient, knowledge of the physiology of folic acid can aid understanding of their problems related to this deficiency (Figure 11).

The coenzyme form of B_{12} is called "S'deoxyadenosylcobalamin." The S'deosyadenosyl group replaces the cyano group of the usual cyanocobalamin. Enzymes requiring the coenzyme of B_{12} commonly have the ability to carry out the shift of a hydrogen atom from one carbon atom to an adjacent one in exchange for an alkyl, carboxyl, hydroxyl, or amino group. Methylcobalamin,

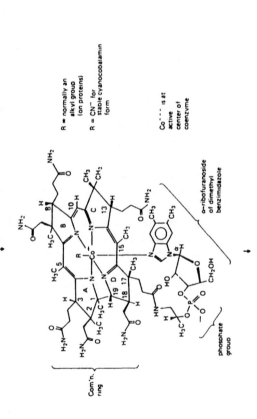

bacterial synthesis of organ meats, milk, eggs, fermented soy, clams in the gut

R = normally an
alkyl group
(on proteins)

R = CN⁻ for
stable cyanocobalamin
form

Co⁻⁻⁻ is at
active
center of
coenzyme

α-ribofuranoside
of dimethyl
benzimidazole

phosphate
group

Corr'n.
ring

exchange of hydrogen for akyl, carboxyl, hydroxyl or amino group
reforms tetrahydrafolate
allows catabolism of medium chain length fatty acids, carbohydrates, proteins
production of glutathione, methionine, and choline
activation of amino acids dairy protein formation
proper DNA replication
anaerobic degradation of lysine

POLLUTANT
OVERLOAD

DAMAGE

Figure 11. Sources of B₁₂.

another coenzyme form of vitamin B_{12}, participates in some enzymatic reactions involving the transfer of methyl groups.

In amino acid metabolism, transfer of methyl group may require assistance of tetrahydrofolic acid and cobalamin (B_{12}). Transfer of methyl groups is done in forming choline or creatinine.

Methylated tetrahydrofolic acid $\xrightarrow{\hspace{1cm} B_{12} \hspace{1cm}}$ Tetrahydrofolic acid
    + homocysteine acid   + methionine

One-carbon transfer reactions require B_{12}, folic acid, and a source of a one-carbon group such as methionine and serine. A high value for formiminoglutamic acid can indicate a deficiency in the one-carbon transfer reactions. (Methylhistidine is often high when there is a deficiency in one-carbon transfer reactions.)

$$\text{Histidine} \xrightarrow[H_4 \text{ folic}]{C_1} \text{Glutamic acid}$$
$$|$$
$$\text{Formiminoglutamic}$$

Vitamin B_{12} plays a role in the activation of amino acids during protein formation and in the anaerobic degradation of the amino acid lysine and is thus important in anabolism of the chemically sensitive. The coenzyme of vitamin B_{12} is a carrier of methyl groups and hydrogen and is necessary for carbohydrate, protein, and fat metabolism, and it may be disturbed in some cases of chemical sensitivity. Because of its methyl transfer role, vitamin B_{12} is active in the synthesis of the amino acid methionine from its precursor, homocysteine. The coenzyme-dependent synthesis of methionine occurs by first removing a methyl group from methyl folate, a derivative of the biologically active form of folic acid. Then this methyl group is transferred to homocysteine and methionine is formed. Because methionine is needed for choline synthesis, B_{12} plays a secondary role in this lipid pathway. A choline deficiency that causes fatty liver can be prevented by vitamin B_{12} or the other methyl donors (betaine, methionine, folic acid). Disturbed B_{12} metabolism in this cycle again can result in impaired methylation of xenobiotics with exacerbation of chemical sensitivity. Clinically, sensitivity to chlorines and other ambient chemicals may arise in the chemically sensitive.

Sources. The only source of vitamin B_{12} in nature is microbial synthesis. The vitamin is not found in plants, but is produced by bacteria in the digestive tract of animals or by microbial fermentation of foods. Since the chemically sensitive have disturbed microflora, inadequate production of B_{12} may occur.

Sources containing more than 10 μg/100 g serving are organ meats (liver, kidney, heart), clams, and oysters. Good sources (3 to 10 μg/100 g serving) are nonfat dry milk, crab, salmon, sardines, and egg yolk. Moderate amounts (1 to 3 µg/100 g serving) are found in meat, lobster, scallops, flounder, swordfish, tuna, and fermented cheese. Other sources include fermented soybean products, poultry, and fluid milk products.

Deficiency. Deficiency is more often caused by improper absorption than by dietary lack. This malabsorption is seen in the chemically sensitive. Because vitamin B_{12} is affected at temperatures above 100°C, some of the vitamin is lost when meat is cooked on a hot grill.[114]

Impaired fatty acid synthesis, observed in vitamin B_{12} deficiency states, can result in impairment of brain and nerve tissue functions, which occurs in some chemically sensitive. The myelin sheath is malformed in a vitamin B_{12} deficiency, and this contributes to faulty nerve transmission. Ultimately, neurological disturbances result from prolonged vitamin B_{12} deficiency. These are characterized by numbness, tingling, weakness, etc.

Proper DNA replication is dependent on the function of coenzyme vitamin B_{12} as a methyl group carrier. Megaloblastic anemia (characterized by large, immature red blood cells) and changes in bone marrow associated with a vitamin B_{12} deficiency are due to the role of the vitamin in DNA synthesis. Improper cell replication and inadequate DNA translation cause the large cells observed in this disorder. The result is anemia, leukopenia, thrombopenia, and fewer but larger and less mature blood cells. Poor cell division in the gastrointestinal tract and other epithelial tissues produces glossitis and megaloblastosis. General growth and repair are curtailed as well.

Pernicious, or megaloblastic, anemia is the characteristic symptom of a vitamin B_{12} deficiency. This condition is caused by either inadequate intake or reduced gastric secretion of a mucoprotein intrinsic factor that is necessary for proper vitamin B_{12} absorption. Intrinsic factor is produced by the parietal cells of the stomach. It binds onto the vitamin and is transported into the small intestine. In the presence of calcium, this transport complex attaches to the intestinal wall, facilitating absorption of the vitamin.

Pernicious anemia may also develop from several other conditions, including gastrectomy, surgical removal of the portion of the lower intestine responsible for vitamin B_{12} absorption, the development of antibodies to intrinsic factor, hereditary and environmental malabsorption, and a diet devoid of animal products (strict vegetarianism).

In addition to anemia, deficiency symptoms include glossitis, degeneration of the spinal cord, loss of appetite, gastrointestinal disturbances, fatigue, pallor, dizziness, disorientation, numbness, tingling, ataxia, moodiness, confusion, agitation, dimmed vision, delusions, hallucinations, and, eventually, "megaloblastic madness" (psychosis).

Vitamin B_{12} deficiency symptoms are generally found in mid to late life, and are often a result of the reduced secretion of intrinsic factor. This condition can be corrected by vitamin B_{12} injection. A large proportion of patients with chemical sensitivity have hypochlorhydria. Whether this is part of the problem or concomitant is unknown. Certainly, disturbed gastrointestinal flora is a problem, and these patients often respond to B_{12} injections.

B_{12} is stored in appreciable amounts (1.0 to 10.0 mg), primarily in the liver. One third of the body's stores are in the muscles, skin, bone, lung, kidney, and spleen. Because daily needs are small, and little vitamin B_{12} is excreted except through the bile, a deficiency takes years to manifest.

Vitamin B_{12} absorption can be inhibited by many gastrointestinal disorders, such as gluten-induced enteropathy, tropical sprue, regional ileitis, malignancies and granulomatous lesions in the small intestine, tapeworm, bacteria associated with blind loop syndrome, and other disorders that impair normal intestinal function. The need for vitamin B_{12} is increased by hyperthyroidism, parasitism, and pregnancy. We have observed that some patients with chemical sensitivity have improved markedly with B_{12} injections, experiencing increased energy and decreased neurological symptoms.

Folic Acid (Folacin, Pteroylglutamic Acid)

Folic acid appears to have a role in some cases of chemical sensitivity, but it is not clearly defined (Figure 12). About 35% of the chemically sensitive are low in folate. A better understanding of the physiology of folic acid may help in our overall understanding of chemical sensitivity.

Folic acid was first isolated from spinach leaves, but has a very broad biological distribution including yeast, wheat, and lettuce. It has three major components: glutamic acid, para-aminobenzoic acid (PABA), and a derivation of pteridine. Deficiency of folic acid (pterolglutamic acid) causes a type of anemia in which red blood cells do not mature properly. Folic acid has no coenzyme activity itself, but it is enzymatically reduced in the tissues to tetrahydrofolic acid (FH_4), its active coenzyme form. Tetrahydrofolate functions as an intermediate carrier of one-carbon groups in a number of complex enzymatic reactions in which methyl ($-CH_3$), methylene ($-CH_2^-$), methenyl ($-CH=$), formyl ($-CHO$), or formimino ($-CH=NH$) groups are transferred from one molecule to another. Tetrahydrofolate is an essential coenzyme in the biosynthesis of thymidylic acid, a nucleotide building block of DNA. Thus, its function if disturbed by pollutants could adversely effect some chemically sensitive.

Some bacteria do not require preformed folic acid as a growth factor because they can produce it if PABA, one of folic acids components, is available. Thus, PABA is a vitamin for these bacteria. This was a valuable discovery because it led to our understanding at the EHC-Dallas of the mode of action of sulfanilamide, an important drug that inhibits growth of pathogenic bacteria requiring PABA.

Sources	Folacin uG./GM.	
	Free	Total
Yeast		
Brewer's	1.75	38.50
Active, dry	1.40	40.90
Wheat germ	2.57	3.28
Wheat bran	1.34	2.58
Egg yolk	2.30	2.73
Romaine lettuce	.60	1.87
Cabbage	.43	.96
Whole wheat bread	.37	.62
Whole wheat flour	.59	.67

Source: Butterfield and Calloway.[832]

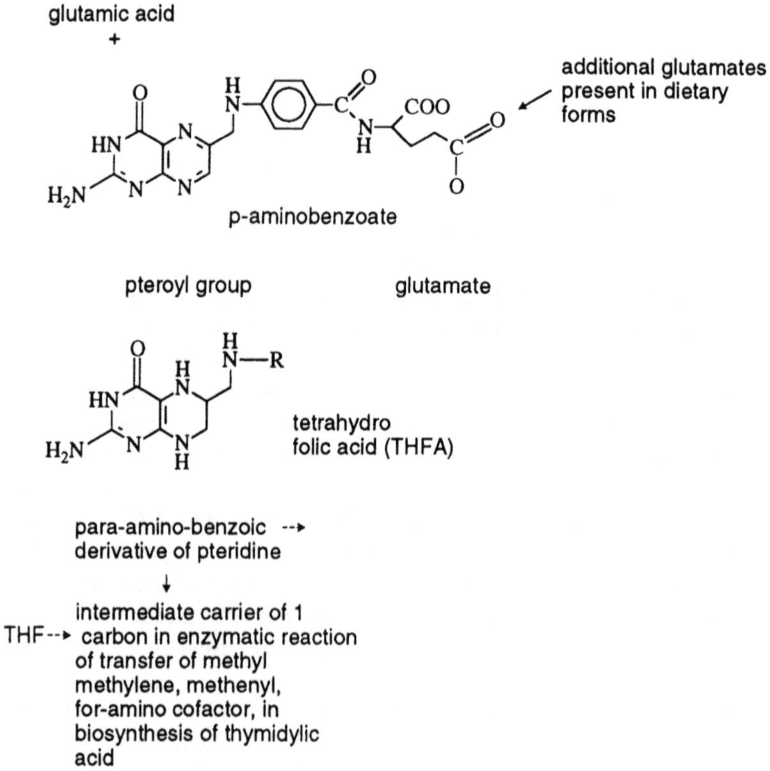

glutamic acid
+

additional glutamates present in dietary forms

p-aminobenzoate

pteroyl group   glutamate

tetrahydro
folic acid (THFA)

para-amino-benzoic -->
derivative of pteridine

↓

intermediate carrier of 1
THF--> carbon in enzymatic reaction
of transfer of methyl
methylene, methenyl,
for-amino cofactor, in
biosynthesis of thymidylic
acid

Figure 12.   Potential pollutant damage to folic acid metabolism.

Source. The best sources of folic acid are liver, kidney beans, lima beans, fresh, dark green leafy vegetables, especially spinach, asparagus, and broccoli. Good sources are lean beef, potatoes, whole grain wheat bread, and dried beans. Poor sources include most meats, milk, eggs, most fruits, and root vegetables.[115]

Deficiency. Folic acid deficiency is one of the most common vitamin deficiencies. The symptoms are similar to those of a vitamin B_{12} deficiency: megaloblastic anemia, irritability, weakness, weight loss, apathy, anorexia, dyspnea, sore tongue, headache, palpitation, forgetfulness, hostility, paranoid behavior, glossitis, gastrointestinal tract disturbances, and diarrhea. However, inadequate folacin intake does not result in the irreversible nerve damage seen in vitamin B_{12} deficiency. Because of the vitamin's vulnerability to destruction, as much as 100% may be lost if foods are improperly stored, if cooking water is discarded, or if foods are reheated or overcooked.

Deficiency can result from poor dietary intake, defective absorption, or abnormal metabolism. These can in turn result from sprue, intestinal dysfunction, or gastric resection. Many medications, including aspirin[116] and anticonvulsants,[117] may also interfere with folacin absorption and metabolism.

One of the symptoms of folic acid deficiency, megaloblastic anemia, is not uncommon during pregnancy, especially during the last trimester. Elevated blood levels in the fetus at birth suggest an increased drain on maternal stores. Hormonal changes during pregnancy may also play a role in folacin status, since polyglutamate absorption is reduced by as much as 50% in women taking the birth control pill, a medication that mimics the hormonal status of pregnancy.

Dietary restriction of folacin will result in depressed serum levels in less than a month. Erythrocyte and liver stores are depleted within 3 months and formiminoglutamate, urocanate, formate, and aminoimidazole carboxamide urinary excretion increases. Care is needed in treating the chemically sensitive since they often do have folic pool depletion.

The symptoms of a folic acid deficiency quickly respond to therapy. If necessary, doses of up to 1 mg may be administered. Maintenance therapy is 100 µg for 1 to 4 months, including the amounts obtained from folacin-rich foods in the diet. In most cases, folacin deficiency may be prevented by the ingestion of one to two folacin-rich fruits or vegetables daily.[118]

PABA

PABA is thought to have some influence on chemical sensitivity, but its role is unclear (Figure 13). The following review may be of some help in understanding its role.

Levine et al.[119] evaluated and showed the capacity of PABA to mitigate the toxicity of ozone in the rat model, as well as with human in vitro

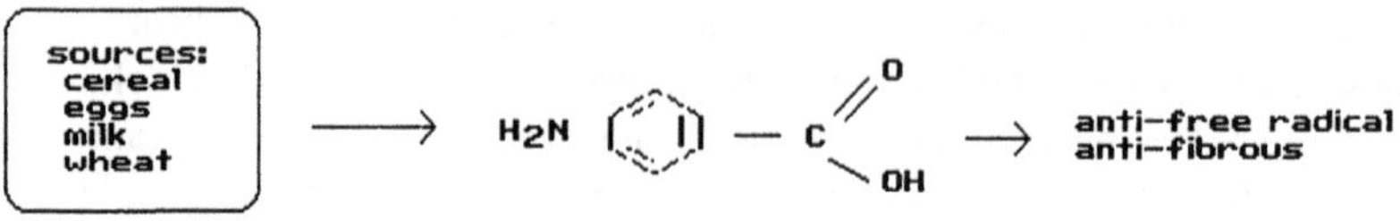

Figure 13.   Potential pollutant injury to PABA metabolism.

studies. This study was based on the premise that certain aromatic amines (e.g., PABA) may be able to prevent free radical formation,[120] which is so devastating in the chemically sensitivity. Protection from ozone-induced mortality was found in intraperitoneal injection of rats with 2 ml of 20 mM PABA before ozone exposure (15 ppm) as compared to controls. The PABA-treated rats died in an average time of 426 min whereas the untreated rats died in 261 min. Deficiencies in thiamine, B_1, riboflavin, B_2, niacin, B_5, and B_6 enhance the toxicity of PABA. Therefore, care should be taken to supplement PABA early in the treatment of the chemically sensitive.

PABA is a member of the vitamin B-complex group of vitamins. PABA is a component of several biologically important systems, and it participates in a number of fundamental biological processes. It has been suggested that the antifibrosis action of potassium p-aminobenzoate is due to its mediation of increased oxygen uptake at the tissue level. Fibrosis is believed to occur from either too much serotonin or too little monoamine oxidase activity over a period of time. Monoamine oxidase requires an adequate supply of oxygen to function properly. By increasing oxygen supply at the tissue level, potassium p-aminobenzoate may enhance monoamine oxidase activity and prevent or bring about regression of fibrosis.[121]

Source. Small amounts of PABA are found in cereal, eggs, milk, and meats. Detectable amounts are normally present in human blood, spinal fluid, urine, and sweat.[121]

Supplementation. Sieve[122] reported that 48 vitiligo patients were treated with PABA for 10 months. They received PABA (100 mg, three to four times daily) in addition to vitamin B complex. Due to the slowness of the effect, parenteral monoethanolamine PABA (100 mg) was added twice daily. Results were slow but striking by 6 to 7 months of treatment. Hughes[123] reported that oral PABA can reduce vitiligo.

Zarafonetis[124] reported that 200 mg of PABA was given four to five times daily to his patients, who then showed striking clinical improvement in gluten enteropathy. Even patients who were not controlling their intake of gluten showed improvement. Doses of 15 to 20 g/day were easily taken and well tolerated.[125] The dose in treatment of chemical sensitivity is unknown and empirical.

Some clinical evidence suggests that PABA supplementation might be important in treatment of the chemically sensitive. For example, Zarafonetis[126] reported that 104 patients of scleroderma were treated with PABA. Grace[127] presented two case reports of scleroderma and dermatomyositis. Treatment involved 10 to 20 g of PABA daily. These patients improved significantly. In another study, 21 patients with Peyronie's disease were placed on potassium p-aminobenzoate therapy for periods ranging from 3 months to 2 years. Pain disappeared from 16 of 16 cases in which it had been present. Objective improvement in penile deformity occurred in 10 of 17 patients and decreased in plaque size in 16 of 21.[128]

Potential side effects of PABA also include anorexia, nausea, fever, and skin rash.[129] Kantor[130] reported a patient who developed a hepatitis most likely due to PABA therapy for lichen sclerosis et atrophicus.

Deficiency. PABA deficiency is associated with constipation, depression, digestive disorders, fatigue, graying hair, headaches, and irritability.[131] Supplementation of B_5 and copper also have been reported to restore the natural color of gray hair.

Biotin

In biotin-dependent enzymes, the biotin molecule is covalently attached to the enzyme protein through an amide linkage with the E-amino group of a specific lysine residue at the enzyme active site. The name biocytin is given to this biotinyllysine residue. Biotin is a transient carrier of a carboxy (–COO–) group in a number of enzymatic carboxylation reactions requiring ATP. An example of a biotin-dependent carboxylation reaction is that catalyzed by pyruvate carboxylase, which carboxylates pyruvate to yield oxaloacetate. Chemical overload can cause spontaneous bruising responsive to biotin supplementation in the chemically sensitive (Figure 14). Presumably this pathway is disturbed in the blood vessel wall.

Biotin enzyme system includes acetyl CoA carboxylase, β-methyl crotonyl CoA carboxylase, propionyl CoA carboxylase, pyruvate carboxylase, and methylmalonyloxalacetic transcarboxylase.

Biotin affects the metabolism of protein, fats, and carbohydrates. It plays a role in several functions, including incorporating amino acids into protein, the synthesis of fatty acids, deriving energy from glucose, the synthesis of pyrimidine (nucleic acid), the conversion of folic acid to its biologically active form, reducing the symptoms of a zinc deficiency. Pollutant damage to these functions could have adverse effects in the chemically sensitive, resulting in an exacerbation of symptoms. Adequate amounts of biotin along with B_3, B_5, C, Mg, and omega-3 fatty acids are necessary in preventing hypercholesterolemia.[132]

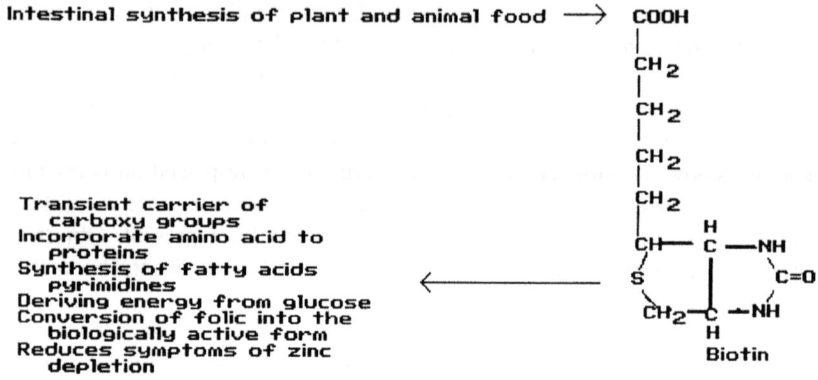

Intestinal synthesis of plant and animal food ⟶

Transient carrier of
   carboxy groups
Incorporate amino acid to
   proteins
Synthesis of fatty acids
   pyrimidines
Deriving energy from glucose
Conversion of folic into the
   biologically active form
Reduces symptoms of zinc
   depletion

$$\begin{array}{l} COOH \\ | \\ CH_2 \\ | \\ CH_2 \\ | \\ CH_2 \\ | \\ CH_2 \end{array}$$

Biotin

Figure 14.   **Potential pollutant injury to biotin.**

Sources. Intestinal synthesis is a significant source of biotin unless antibiotics or other agents that interfere with microbial action are present. Often overuse of antibiotics leading to biotin deficiency may account for the onset of some chemical sensitivity. Both plant and animal foods contain biotin. Good dietary sources include liver and other organ meats, molasses, and milk.[132]

Deficiency. Deficiency symptoms are uncommon, even if a biotin-deficient diet is consumed. However, the ingestion of large quantities of raw egg whites will produce deficiency symptoms by inhibiting absorption in the intestinal tract. (Avidin, a protein-carbohydrate compound in raw egg white, binds with biotin to inhibit absorption.) As a result, a nonpruritic dermatitis, hypercholesterolemia, electrocardiograph changes, anemia, anorexia, nausea, lassitude, and muscle pain can develop. Avidin is deactivated by cooking the egg white, thus eliminating its biotin-binding capabilities.

Infants may develop a deficiency from poor absorption of biotin or from improper binding to mucosal cell receptors. In these cases, infants develop hepatomegaly, lactic acidosis, and skin rash. This deficiency may then make them prone to chemical sensitivity. Biotin supplementation produces prompt cessation of symptoms.

The average American diet contains 150 to 300 meq of biotin per day. This amount, plus what is absorbed from microbial synthesis in the intestines, alleviates clinical deficiency symptoms,[132] but data on supplementation are not clear in the chemically sensitive.

Vitamin C (Ascorbic Acid) — (Water Soluble)

Vitamin C deficiency as measured by static blood levels occurs in 20% of the chemically sensitive (Figure 15). However, oral and parental challenges will often acutely diminish some cases of acute sensitivity, suggesting a need

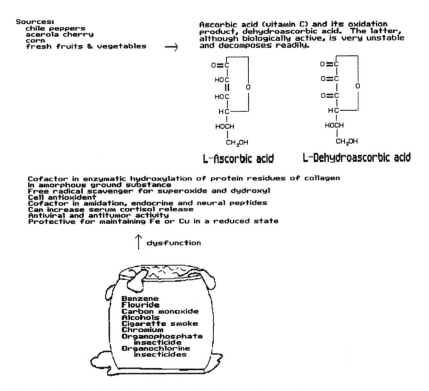

Figure 15.   Potential pollutant effects on vitamin C.

in a much higher proportion of the chemically sensitive patients. Vitamin C appears to be needed in virtually every pollutant-exposed patient because of its free radical and antioxidant function. Since vitamin C appears to have such a paramount role in chemical sensitivity, we review a large part of its physiology to facilitate understanding of its involvement.

Physiology. Ascorbic acid appears to act as a cofactor in the enzymatic hydroxylation of protein residues of collagen of connective tissue of vertebrates to form 4-hydroxyproline. Hydroxyproline residues are found only in collagen and no other animal proteins. Ascorbic acid appears to function in the formation and maintenance of a major component of connective tissue of higher animals. It is needed for tyrosine metabolism and the conversion of tryptophan to niacin both of which have been noted to be disturbed in some chemically sensitive.

Amorphous ground substance of the vessel wall is somewhat dependent on vitamin C.[133] Vitamin C supplement can be used not only to strengthen the blood vessel wall, but also as a free radical scavenger and antioxidant in the chemically sensitive. It is one of the best tools in the physician's armamentarium to treat antioxidant stress in the chemically sensitive.

Vitamin C functions as a cellular antioxidant and is therefore important for people with environmental sensitivities, because the vitamin C may act as a chelating agent to pull chemicals such as pesticides and other toxins out of the body.[134] *Whereas antipollution enzymes usually act intracellularly, vitamin C as an antioxidant acts mostly on external cellular membranes.* In addition to being a scavenger of the free radical superoxide, ascorbate is an effective scavenger of the highly reactive hydroxyl radical and has a host of metabolic effects. It also is important in many detoxification reactions. Many factors including acute emotional or environmental stress, the use of oral contraceptives, and smoking affect the need for and utilization of vitamin C. Heavy cigarette smoking has been observed to lower plasma levels of vitamin C by as much as 40%. The reasons for vitamin C reduction in heavy smokers are unknown. Conceivably, these reduced body levels could be the result of increased utilization of L-ascorbic acid for the direct detoxification of aldehyde toxicants in vivo, possibly by way of a condensation reaction such as with formaldehyde or increased synthesis and release of tissue catecholamines induced by ethanol and the aldehyde toxicants requiring vitamin C to counteract them. Rat studies by Sprince indicate that acetaldehyde protectants (reducing agents such as L-ascorbic acid) can serve in varying degrees as protectants against other aldehydes, notably acrolein and formaldehyde (toxicants in the gas phase of cigarette smoke).[135]

Less ascorbic acid is available in smokers for utilization and storage, indicating the lower amount absorbed. Smokers appear to oxidize more ascorbic acid to dehydroascorbic acid, which isomerizes in the gastrointestinal tract due to the secretion of an oxidative enzyme, ceruloplasmin, involved in the oxidation of serotonin, which is known to be released by nicotine.[136]

Pathophysiology of Ascorbic Acid. There is now overwhelming scientific evidence that ascorbic acid is essential for human function and disease prevention. The integrity of vitamin C metabolism partly influences endocrine function. A part of the problem of the chemically sensitive patient appears to be in the pituitary-hypophysial/neuro axis. Evidence shows that the uptake of dehydroascorbic and ascorbic acid is higher in the hypophysis (a prime target for pollutant injury in the chemically sensitive) than the liver, kidney, spleen, pancreas, and other organs.[137] In addition, transfer of the ascorbate of most organs including leukocytes, liver, spleen, etc. is dependent upon the concentration of ascorbic acid in the extracellular pool, pH, temperature, and activation of the sodium pump (or the transfer is sodium dependent).[138-140] However, incorporation into the cell is also accomplished by diffusion.[141] In addition, chromatin granules will shuttle ascorbate from the extravesicular to the intravesicular space.[142] Cytosolic ascorbate is the source of the intragranular reducing equivalents required during norepinephrine synthesis.[142] Some organs such as the brain and eye are now being vigorously studied. See Chapter 27 for information on vitamin C and the eye.

Brain. The uptake of ascorbic acid in the hypophysis is usually sodium dependent, but it can occur with diffusion. The uptake in the brain is higher than in most tissues in the body,[143] and this fact may explain why pollutant damage in the chemically sensitive manifests itself in autonomic nervous system and endocrine dysfunction. In addition, sodium and ascorbic acid must be transported into the intermediate pituitary lobe to participate in the peptidylalpha-amidation.[144] Ascorbate is obviously a cofactor in the synthesis of α-amidation, endocrine, and neural peptides,[145] and its presence explains some of the regulating effects of ascorbate supplementation on the pollutant-injured chemically sensitive patient. Ascorbyl radicals are not stable in aqueous solution.[146] Although ascorbic acid is regenerated in the rat brain, it is not in humans. Therefore, an adequate supplementation is necessary in the chemically sensitive. $SO_3 + SO_5$ oxidizes ascorbate.

Ascorbic acid metabolism may malfunction in the chemically sensitive for many reasons. First, semidehydrascorbate reductase appears to be the major enzyme catalyzed for the regeneration of reduced ascorbic acid in the central nervous system.[147] It may be damaged by pollutant injury in the chemically sensitive. In hypophysectomized rats, ascorbic acid concentrations are lower in the adrenals, liver, blood, and urine, emphasizing the potential for endocrine deregulation in the chemically sensitive. In the liver, ascorbic acid release may also be caused by ACTH, but other unknown pituitary factors may also be involved,[148] and these may be disturbed by pollutant overload. Cerebral capillaries are the main route of movement of ascorbic acid into the rat brain[149] and probably the human. Therefore, the blood vessel deregulation seen in the chemically sensitive may cause uneven distribution of vitamin C. Pollutant injury is highlighted when one looks at experimental data as in the study where multiple amphetamine injections reduced the release of ascorbic acid in the neostriatum of the rat brain.[150] Crus cerebri lesions abolish amphetamine-induced ascorbate release.[151] Lack of ascorbate homeostasis reflects irreversible loss of neurologic function after spinal cord injury.[152] Similar release occurs with stimulation in the nucleus accumbens and olfactory tubercle.[153] Since most chemically sensitive are supersensitive to odors, overstimulation of their olfactory tubercule may result in depletion of vitamin C in this area from overrelease. Then a chain reaction of dysfunction to the thalamus, hypothalamus thypophysis, and pituitary could clear. Ascorbic acid concentration in the fetal rat brain varies from 374 to 710 mg/g with an 18% drop occurring after birth.[154] High-pressure-exposed rats showed a decrease of ascorbic acid of 25 to 75% of controls. This decrease supports the view that ascorbic acid participates in the defense against reactive O_2 intermediates, lipid peroxidations, and neurotransmission.[155] There is also a decrease in plasma levels of ascorbate in marine and human-like animals. This plasma level decrease is caused by increased utilization as well as increased excretion. Schizophrenic patients have been shown to have significantly lower fasting plasma vitamin C levels and 6-h urinary vitamin C excretion after ascorbic load test than controls.[156]

This observation agrees with other studies that schizophrenia is associated with impaired ascorbic acid metabolism. A small minority of chemically sensitive patients exhibit schizophrenia.

Kidney. Ascorbic acid is reabsorbed in the kidney by a sodium-dependent active transport mechanism that operates by concentrating ascorbic acid in the cellular fluid.[157] This appears to be the renal brush border membrane vesicles.[158] There is evidence that overload will decrease in ascorbic acid presence, such as that seen in helminth infestation. Ascorbic renal loss is present in some chemically sensitive. Patients who form chronic kidney stones showed decreased (urinary) ascorbate and citrate with increased calcium and oxalate.[159] In addition, urine sodium, mucopolysaccharide, and protein were significantly increased. Also, concurrent administration of ascorbic acid inhibited ascorbic absorption and increased oxalate excretion. These findings suggest that hyperoxaluria in stone patients can be explained on the basis of a mechanism involving malabsorption of citrate ascorbate and possibly hydroxycarboxylic acids. Results of oral and intravenous administration of ascorbate indicate an enhanced production of oxalate from ascorbate in those who form renal stones as compared with normal populations. This phenomenon must be rare since we have not observed stone formation in 20,000 chemically sensitive patients supplemented with ascorbate. Most ascorbate is generated from the gut.[160] Alcohol enhances vitamin C excretion in the urine,[161] and thus those chemically sensitive patients who have high alcohol (pollutant from any source) in their blood will lose more.

It has been shown that activation of the mitochondrial NADH:ascorbate radical oxidoreductase in the extravesicular medium causes a decrease in intravesicular ascorbate radical in chromaffin granule ghosts, but not in liposomes. This data provides direct experimental evidence for the hypothesis that the adrenal medullary mitochondrial NADH:ascorbate radical oxidoreductase could drive the re-reduction of ascorbate free radical generated inside the chromaffin granule by the turnover of dopamine β-hydroxylase, without the ascorbate radical ever having to leave the granule.[162] This process may be disturbed in the chemically sensitive due to direct pollutant damage and also adrenal stress demands.

There is evidence of an ascorbate shuttle for the transfer of reducing equivalents across chromaffin granule membranes. Incubation of intact chromaffin granules with tyramine results in a time-dependent decrease in reduced intragranular ascorbate and production of octapamine. The stoichiometry of this relationship of octapamine synthesized and ascorbate oxidized closely approximates unity. The addition of ascorbate extragranularly 30 min after addition of tyramine reverses the oxidation of intragranular ascorbate and is another reason supplementation helps the chemically sensitive when oxidant stress occurs. Extravesicular NADH had no significant effect on matrix ascorbate levels during β-hydroxylation and emphasizes the need for specific supplementation in the chemically sensitive. These data provide new in vitro evidence that chromaffin granules shuttle reducing equivalents inwardly from an extravesicular to an intravesicular ascorbate pool and that cytosolic ascorbate is the source of the intragranular reducing equivalents required during norepinephrine biosynthesis.[163] Both under and oversynthesis is a possibility in some chemically sensitive as indicated by their under- and over-reactions to certain pollution.

In one study, ascorbic acid and catecholamine release from digitonin-treated chromaffin cells occurred. Catecholamine release from permeabilized chromaffin cells was dependent on the free calcium concentration and the temperature of the incubation mixture of ascorbate with digitonin, a steroid glycoside. By contrast, ^{14}C-labeled ascorbic acid, preloaded into the cells, was released by digitonin treatment in a manner independent of the concentration of free calcium and with only moderate regard to the incubation temperature. The sensitivity of ascorbic acid release to digitonin treatment was identical to that of calcium-dependent catecholamine release. These results thus suggest that ascorbic acid preloaded into the cells may directly efflux from the cell cytoplasm as a result of the permeabilization of the plasma membrane. This conclusion supports the clinical observation in the chemically sensitive that full pools of vitamin C will more easily counteract pollutant injury than depleted pools, thus the need for constant supplementation at high levels, which will keep tissue levels adequate and acting in the preventative mode.

Dimethylepinephrine, a permanently positively charged catecholamine analog that is known to be excluded from vesicular fractions, was also released by digitonin treatment in a manner independent of calcium. The time course of dimethylepinephrine release was very similar to that of ascorbic acid release. Thus newly accumulated ascorbic acid in chromaffin cells may be localized to a free pool in the cell cytoplasm rather than in a vesicular compartment.[164]

It has been shown that ascorbic acid and not ACTH can increase serum cortisol concentration during etomidate infusion. Ascorbic acid even restores the ACTH:cortisol ratio to preoperative values.[165] This restoration may also happen in exposure to other pollutants in the chemically sensitive.

Heart. In the metabolism and function of ascorbic acid and its metabolites, an interesting study reported the polarographic measurement of ascorbate washout in isolated perfused rabbit hearts. In the isolated hearts, ascorbate infused into the aorta was detected in a right ventricular by the electrode as well as by the use of ^{14}C-labeled ascorbate. Both recorded time courses were similar except for a scaling factor dependent on flow velocity. During continuous infusion, the arteriovenous difference of ascorbate was $2 \pm 2\%$ (SD), indicating a relatively low consumption of ascorbate by the isolated heart.[166] The ascorbate effect on the heart in the chemically sensitive may be due to its cholesterol lowering as well as its free radical scavaging ability.

Liver. In one study, plasma ascorbic acid levels were higher in the phenytoin group than in the controls, but tissue levels and the rate of ascorbic acid synthesis were similar in the two groups. This observation suggests a shift response in the group exposed to pollutants. Also, copper concentration in liver and kidney was significantly higher in phenytoin-treated rats than in controls. Iron, zinc, and manganese levels were unchanged in comparison to control values in liver, kidney, heart, and brain.[167] The level of vitamin C in the mouse liver, kidney, spleen, and serum decreases after infection with T-crassiceps. Diet affects the content of hepatic lipid, plasma minerals, and tissue ascorbic acid in hens and estrogenized chicks. Plasma and hepatic ascorbic acid were significantly increased by the various ascorbic-rich diets, but no significant differences in hepatic ascorbic acid were observed when calculated per unit of fat-free dry matter.[168] If decreased liver levels of ascorbate occur in the chemically sensitive as they do in these animals, a need for excess ascorbate acid to counteract pollutant overload would be expected.

Spleen. It was demonstrated that helminth infestation affects the ascorbic acid content in various tissues of experimental mice. Maximum decrease in blood serum was observed on the day of the maximum intensity of infection 15 and 30 cysti-cerci per mouse. There was a correlation between the intensity of infection and decrease in the ascorbic acid level in the spleen.[169]

Lung. Effects of protein deficiency and food restriction on lung ascorbic acid and glutathione were studied in rats exposed to ozone. In the liver from weanling rats, glutathione concentrations were also reduced in response to protein deficiency. Exposure to ozone produced no additional response.[170] If pollutant exposure continued, however, resulting in more nutrient depletion, effects would be expected. At the EHC-Dallas, we have noted that the chemically sensitive respond to intravenous ascorbate supplementation after pollutant exposure in cases of environmentally triggered asthma.

Prostate. The effect of castration on the metabolism of L-ascorbic acid in rat prostate has been measured. An appreciable decrease in the levels of prostatic ascorbic acid and dehydroascorbic acid, along with an increase in diketogulonic acid, was seen in rats 10 days after castration. Castration caused a decrease in the activities of such biosynthetic enzymes as L-gulono-gamma-lactone oxidase and D-glucuronolactone-delta-hydrolase with no significant alteration in the activity of L-gulono-gamma-lactone hydrolase in the rat prostate.[171] Evidence of change in the chemically sensitive is unknown at this time.

Trachea. Autoradiographic localization of ascorbic acid-dependent binding sites for $[I_{125}]$-iodocyanopindolol in guinea pig trachea has been shown. Light microscopic autoradiography showed that the supposedly β-adrenoceptor-selective radioligand $[I_{125}]$-iodocyanopindolol (I-CYP) bound to sites in both the guinea pig tracheal epithelium and smooth muscle were sensitive to propranolol and isoprenaline. Low levels of binding were associated with subepithelial mucosal cells. Ascorbic acid caused a concentration-related increase in total I-CYP binding, which was predominantly associated with the subepithelial mucosa, was not inhibited by propranolol, and was thus not associated with β-adrenoceptors.[172] Our experience at the EHC-Dallas, however, has been that ascorbate supplementation increases the recovery of tracheitis in the chemically sensitive.

Estrogen. Effects of estrogen and progestogen on the ascorbic acid status of female guinea pigs have been assessed. In vitro studies revealed a markedly higher rate of oxidation of ascorbic acid in the presence of either estinyl or progestogen than in untreated controls. An in vivo dose-related effect of ascorbic acid indicated that the steroid-mediated lowering effect of the vitamin level could be counteracted by increasing the dose of ascorbic acid from 1 to 10 mg/day for 2 weeks. These results suggest that the interactions between oral contraceptive hormones and ascorbic acid may be of clinical importance only in the case of borderline intake of the vitamin,[173] which is seen in many chemically sensitive females.

Vitamins and Minerals. Hydroxylation of pyridoxine in the presence of food components, especially ascorbic acid, causes loss of vitamin B_6 in plant foods during food processing, storage, and cooking.[174] These results elucidate the in vitro radical-scavenging functions attributed to vitamin E and vitamin C, as well as their synergism in lipid antioxidation.[175] This study also supports observations at the EHC-Dallas that ascorbate supplementation helps prevent oxidation in the pollutant exposed, chemically sensitive patient.

Site-specific modification of albumin by free radicals occurs in the reaction with copper (II) and ascorbate. The rate of utilization of molecular oxygen and ascorbate as a function of Cu(II) concentration is nonlinear at copper:albumin ratios of greater than one. It appears that Cu(II) bound to the tightest albumin-binding site is less available to the ascorbate than the more loosely bound cation. The absence of Cu(II) or the presence of metal-chelating agents causes inhibition. There was no evidence of intermolecular crosslinking or of the formation of insoluble, albumin-derived material. This reaction generates OH radicals, which rapidly interreact with protein and modify it in a "site-specific" manner.[176] Possibly the chemically sensitive, when exposed to pollutants, may develop such a reaction, allowing for hapten formation.

Specific cleavages of DNA by ascorbate in the presence of copper ion or copper chelates have been noted. The DNA cleavage specificity of ascorbate in the presence of copper ion is analyzed with end-labeled pBR322 DNA fragments. The nonenzymatic reaction of Cu(II)/ascorbate and DNA shows certain degrees of cleavage preference toward purine-containing short segments in the labeled DNA under mild conditions (at 0°C, 10 min). The segments of pyrimidine clusters are least susceptible to cleavage. This scission activity in relation to the antiviral and antitumor activities of vitamin C reported in the literature deserves careful consideration,[177] at least with the chemically sensitive population, since one segment of these experience recurrent infections.

A mechanism of the inhibition of catalase by ascorbate involves the presence of active oxygen species, copper, and demidehydroascorbate. Ascorbate reversibly inhibits catalase, and this inhibition is enhanced and rendered irreversible by the prior addition of copper(II)-gishistidine. In the absence of copper, the inhibition was prevented and reversed by ethanol, but not by superoxide dismutase, benzoate, mannitol, thiourea, desferrioxamine, or DETAPAC. In the presence of the copper complexes, mannitol, benzoate, and superoxide dismutase still had no effect, but thiourea, desferrioxamine, DETAPAC, or additional histidine decreased the extent of inactivation to that seen in the absence of copper. In the presence of copper, ethanol protected at [ascorbate] less than 1 mM, but was ineffective at [ascorbate] greater than 2 mM, even in the absence of oxygen. Although in the absence of copper, complete removal of oxygen provided full protection against inactivation by ascorbate, this protection was not seen if the catalase was briefly preincubated with H_2O_2 prior to flushing with nitrogen, or if copper was present. In fact, if copper was present, inactivation was enhanced by the removal of oxygen. Increasing the concentration of oxygen from ambient to 100% slowed the inactivation, whether or not copper was present. It is concluded that the initial reversible inactivation involves reaction with H_2O_2 to form compound I, followed by one-electron reduction of compound I to compound II. In the presence of added copper, the initial (reversible) inactivation allows H_2O_2 to accumulate sufficiently to permit irreversible inactivation.[178] This mode of interaction suggests that ascorbate reduces the enzyme-bound iron through an "inner-sphere" mechanism.[179]

Malignancy. The dehydroascorbic acid (DHA) uptake properties of lymphocytes of B cell origin serve to distinguish this lineage from T cell CLL or normal lymphocytes.[180] The cytotoxicity of the mixture of ascorbic acid with 2,6-dimethoxy-p-benzoquinone is concluded to result from a loss of NAD(P)(H) reducing power in the cells in vitro.[181] In the study of ascorbate status in murine and human leukemia, it is known that because mice can synthesize ascorbic acid but man cannot, the ascorbate status in murine and human leukemia can be compared. The decline in plasma ascorbate concentration in both cases indicates that vitamin C deficiency occurs in malignancy.[182] Vitamin C was assayed in sera samples from normal subjects and patients with cancer of the uterine cervix or ovary and leukemia and lymphoma patients. Among the murine group the tumors included sarcoma 180 in solid and ascitic form, benzo[a]pyrene-induced fibrosarcoma, Dalton's ascitic lymphoma, and Schwartz lymphoblastic leukemia. The serum level of vitamin C was found to be lower than that of the normal controls in all cases studied.[183]

Yeast. Ascorbic acid is specifically utilized by some yeast. Some 180 strains of yeast belonging to 17 genera and 53 species were screened for their ability to grow on ascorbic acid and iso-ascorbic acid as the sole carbon source. Most of the tested strains (157) were unable to grow on either compound. Strains of seven species of the genus *Cryptococcus,* of two *Candida* species, of *Filobasidiella neoformans, Trichosporon cutaneum, Lipomyces starkeyi, Hansenula capsulata,* and one strain of *Aureobasidium pullulans* were able to grow on ascorbic as well as on iso-ascorbic acid. In addition, four strains of *Aureobasidium pullulans, Candida blankii,* and *Cryptococcus dimennae* could use only ascorbic acid for growth.[184] This information may be highly significant in the chemically sensitive who are also sensitive to mold and candida since supplementation, although usually helpful, can add to their problems.

Peroxidation. Hemolysis of sheep erythrocytes is seen with the cell membrane of liquid paraffin-induced guinea pig macrophages. In addition, L-ascorbate was found to be replaced by NADPH, a substrate of the membrane-bound NADPH oxidase, showing that L-ascorbate, when combined with the phospholipids isolated from the membrane fraction by extraction with chloroform-methanol and thin-layer chromatography, acts as a donor of active oxygen.[185] The ascorbate-dependent lipid peroxidation in the microsomal fraction of at-term placentas and fetal membranes were studied. In preliminary experiments, the optimum conditions for the measurement of this reaction were determined. Lipid peroxidation was significantly higher in the chorion and amnion compared to the placenta. In both fetal membranes, but not in the placenta, the reaction was enhanced by arachidonic acid and inhibited by indomethacin. The results indicate that the ascorbate-dependent lipid peroxidation in the fetal membranes is specific for prostaglandin synthetase activity and that these tissues are a major site of prostaglandin synthesis.[186] Integrity of this process is important in the chemically sensitive.

Changes in ascorbate-induced lipid peroxidation of hepatic rough and smooth microsomes are seen during postnatal development and ageing of rats. Smooth microsomes are also more sensitive to inhibitors of lipid peroxidation. Microsomal content of phospholipid increases during postnatal development and decreases during ageing, whereas that of ascorbic acid and α-tocopherol does not show any particular trend.[187] There was an appreciable decrease in the rate of lipid peroxidation under the scorbutic condition. In the tissue fraction of scorbutic guinea pigs, the activities of biosynthesizing enzymes, measured in vitro, under optimum conditions were found to be higher with no significant alterations in the catabolizing enzymes.[188] DHA is taken into the first cellular boundary of the placenta between maternal and fetal circulations by the sodium-independent monosaccharide transporter. In contrast to DHA, L-ascorbic acid, the reversibly reduced form of vitamin C, was taken into these vesicles much more slowly. This uptake was not affected by cytochalasin B nor by a sodium concentration gradient; it appeared to occur by simple diffusion.[189] It is quite clear that if this occurs in the chemically sensitive, ascorbate will be needed.

Forsman and Frykholm showed that "exposure to benzene produces an increased requirement of vitamin C and that an extra supply of vitamin C increases resistance to the effects of benzene vapors."[190] Of the patients with chemical sensitivity at the EHC-Dallas, 10% have benzene in their blood. If these patients are added to the group of patients whose blood contains ethyl benzene, we have another 31% of patients containing benzene. If those patients with toluene, xylene, and other benzene compounds in their blood are added to this group, virtually 100% of the chemically sensitive patients are contaminated by these substances and therefore have extra vitamin C requirements. Given the extent of widespread chemical contamination these patients experience and the additional demands this load places on vitamin C reserves, it is hardly surprising to find that more than 25% of chemically sensitive patients are vitamin C deficient by measurement. Overall, however, the evidence for vitamin C deficiencies in the chemically sensitive suggests that all will need supplementation in spite of measurements.

That vitamin C is useful in treating pollutant injury is evident from the results of a number of studies. Gabovich and Maistruk,[191] for instance, investigated the influence of ascorbic acid supplementation (100 to 500 mg/day) on fluoride excretion levels in industrially exposed workers. They noted that a low-level ascorbic acid supplementation of 50 to 150 mg/day (in addition to basal level of 25 mg/day) resulted in an enhanced excretion of fluorine in workers exposed to 0.0021 mg/L fluorine gases. Workers receiving a supplementary 100 mg/day of vitamin C showed higher rates of fluorine excretion than those receiving 50 mg/day.

Several investigators have examined the usefulness of ascorbic acid as a potential treatment in cases of carbon monoxide asphyxiation.[192-194] In their papers, Zaffini and his associates report on a series of case histories in which a number of individuals suffering from acute asphyxiation were successfully treated with massive intravenous doses of ascorbic acid (which ranged from 4 g to greater than 50 g) along with other traditional carbon monoxide treatment procedures such as administration of oxygen and intravenous glucose solution. This has been our experience in carbon monoxide exposed, chemically sensitive patients.

Yunice and Linderman[195] have also reported that ascorbic acid offered significant protection against the lethal effects of elevated levels of ethanol; thus, only 13 of 40 mice injected with a fixed quantity of ethyl alcohol on two consecutive days survived, whereas 100% of the ascorbic acid-treated (25 mg) mice survived.

Pelletier[196] confirmed previously published reports[196-198] that were based on a limited number of volunteers when he reported that cigarette smokers have lower serum vitamin C levels than nonsmokers. Varghese et al.[199] have shown that supplementation of 4 g ascorbic acid each day can reduce up to 70% of the carcinogenic nitroso compounds normally found in feces.

In 1965, Samita et al.[200] reported that a significant advance in the protection of chromium workers could be made via the use of filters impregnated with ascorbic acid. They found the effectiveness of the respiratory drive was improved by 35 to 50% depending on the concentration of ascorbic acid employed, with an effective time of up to 3 h. Filters impregnated with 20% ascorbic acid can retain this effectiveness for at least 1 month.

Chakraborty et al.[201] evaluated the influence of two organophosphate insecticides (parathion and malathion) on a number of health indicators in rats, and how ascorbic acid administration may affect the toxicity of these insecticides. They noted that both insecticides diminished growth rate by approximately 25%. However, the rats given both the insecticide and ascorbic acid treatment exhibited only 10% decrease in growth rates. This suggested that ascorbic acid reduces this adverse effect caused by malathion and parathion.

While there have been no longitudinal or cross-sectional research efforts on effects of pesticides in humans, Rea et al.[202] provided data on the incidence of chlorinated pesticides in the blood of patients and indicated certain immune and behavioral correlates. Forty hospitalized patients treated at the EHC-Dallas were tested for the presence of 19 organochlorine pesticides in the blood stream. Of the 40 patients, 39 were found to have measurable levels of pesticides in their blood. From the 19 different pesticides, 12 were evident in varying numbers and combinations in the blood samples. Pesticides found were BHC (alpha, beta, gamma), chlordane, DDT, DDE, endosulfan I, endosulfan II, endosulfan sulfate, endrin, heptachlor, heptachlor epoxide, and hexachlorobenzene, with the most common being DDT and hexachlorobenzene. It should be noted that DDT levels have been estimated to be fatal at an oral dose of 500 mg/kg of body weight of the solid material. When placed on rigid environmental control and initial intravenous vitamin C, up to 15 g two times a day, these patients decreased their pesticide levels and increased their brain function.[203]

Deficiency. The earliest signs of ascorbic deficiency may begin during the first month of deprivation, depending on the rate of catabolism. Deficiency appears after the serum level has fallen below 0.2 mg/100 ml. Severe deficiency of ascorbic acid causes scurvy, but a milder decrease causes ascorbic pool depletion. Pool depletion is often seen in the chemically sensitive, resulting in an inability to rapidly respond to pollutant injury and has been discussed thoroughly earlier in this chapter.

Deficiency is characterized by decreased urinary excretion, plasma concentration, and tissue and leukocyte concentration. Other symptoms include weakness, poor appetite and growth, anemia, tenderness to touch, swollen and inflamed gums, loosened teeth, swollen wrist and ankle joints, shortness of breath, petechial hemorrhages from the venules, bleeding or fracture of ribs at costochondral junctions, fracture of epiphysis, and multiple subcutaneous and subperiosteal hemorrhages with pain on motion of the body. Secondary infections develop easily in the bleeding areas. All these characteristics can be attributed primarily to collagen defects.

Neurotic disturbances consisting of hypochondriasis, hysteria, and depression followed by decreased psychomotor performance have been reported in ascorbic acid deficiency. Although scurvy is rare today, dietary surveys indicate that many Americans receive insufficient amounts of this vitamin for optimum health,[204] and this fact is certainly true in the chemically sensitive.

Source. Vitamin C is found in fresh fruits and vegetables, especially citrus, berries, chili peppers, cabbage family, and corn.

Toxicity. Very little has been reported on vitamin C toxicity. Diarrhea occurs if an excess is taken by mouth. One case of iron overload has been reported.[205] Anticoagulant assays may be off.[206] Vitamin C may interfere with copper metabolism[207] (Figure 6.19). Some 15,000 patients have been given an average of 6 g and in a few up to 100,000 g of vitamin C at the EHC-Dallas without any severe complications and no toxicity. Unfortunately, some patients are sensitive to the corn or palm residue that the vitamin C is made from and the intravenous has to be stopped. In these patients, intradermal injection neutralization therapy may be needed in order to continue supplementation.

Vitamin D (Fat Soluble)

Vitamin D is needed to help regulate calcium and phosphorous metabolism both in the general population and in the chemically sensitive (Figure 16). In the chemically sensitive, however, vitamin D poses a unique set of problems since approximately 30% of the chemically sensitive have excessive (15%) or deficient (15%) vitamin D. Those who live in northern climates have more difficulty generating vitamin D due to less exposure time to the sun. It has been shown that those persons living where the oxidant pollutant levels are high may have a concomitant decrease in vitamin D accumulations by as much as 15% over a 25-year period,[208] and if they lived in these areas it would certainly stress the chemical sensitivity. Pasteurization also eliminates vitamin D.

Vitamin D_3, or cholecalciferol, is normally made in the skin of people and animals from an inactive precursor, 7-dehydrocholesterol, by reactions that are promoted by exposure to the ultraviolet component of sunlight. A sun-sensitive subset of the chemically sensitive exists who are particularly prone to vitamin D deficiency.

Vitamin D_3 itself is not biologically active, but it is the precursor of 1,25-dihydroxycholecalciferol, which is hydroxylated in two stages, first in the liver and then in the kidney. It is transmitted particularly to the small intestine and bones, where it regulates calcium and phosphate metabolism. Pollutant damage

Formation and function of the active form of vitamin D3, 1,25-dihydroxycholecalciferol.

1,25-Dihydroxy-cholecalciferol

→

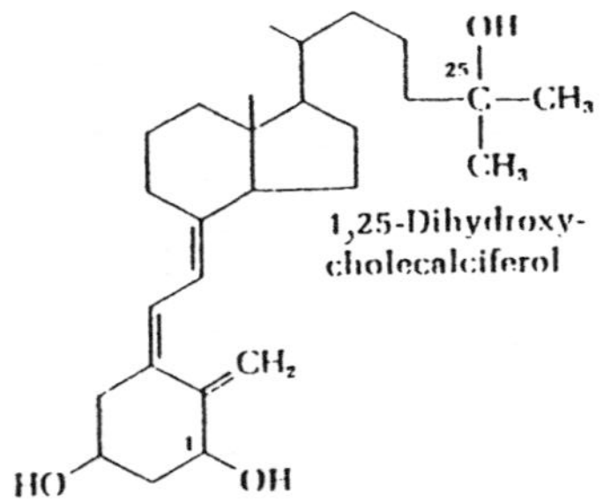

Regulates calcium and PO4

Sources:
sunlight stimulates skin   ──────────→

unpasteurized milk (cream)

Vitamin D Content of Unfortified Foods
(international units/100 G. Unless otherwise stated)

Butter	35
Cheese	12–15
Cream	50
Egg yolk	25 I.U./average yolk
Halibut	44
Herring	
Fresh, raw	315
Canned	330
Liver	
Beef, raw	9–42
Calves, raw	0–15
Lamb, raw	17–28
Pork, raw	44–45
Chicken, raw	50–67
Mackerel	
Fresh raw	1100
Milk	
Cow's	0.3–4 I.U./100 ml.
Human	0–10 I.U./100 ml.
Oyster	5 I.U./3–4 medium-sized
Salmon	
Fresh, raw	154–550
Canned	220–440
Sardines	
Canned	1150–1570
Shrimp	150

Source: Auiola.[833]

Figure 16.   Potential pollutant damage to vitamin D.

to the liver may account for the abnormal vitamin D levels seen in many chemically sensitive patients.

Absorption. Dietary vitamin D is absorbed with the fats from the intestine with the aid of bile. Often, however, bile metabolism in the chemically sensitive is disturbed by inadequate taurine. In this instance, natural and xenobiotic peptide conjugation will not occur either. Under these circumstances, malabsorption occurs, resulting in lessened absorption of vitamin D and requiring not only vitamin D supplementation in the chemically sensitive, but also taurine.

Vitamin D is absorbed into the bloodstream through the skin. Regardless of its initial route of entry, via the intestines or the skin, vitamin D is subsequently carried into the bloodstream, to the liver, and transformed into its active form.[209]

Deficiency. A vitamin D deficiency is responsible for rickets in children and osteomalacia in adults. Both conditions are a result of defective ossification leading to reduced rigidity in bones and ultimately causing bones to become soft and pliable and to bend readily. All types are caused by metabolic abnormalities in absorption or metabolism of vitamin D or end-organ responsiveness.

Osteomalacia is found in women with closely spaced, multiple pregnancies and in confined individuals not exposed to sunlight. In osteomalacia, the calcium to phosphorus ratio changes, and calcium losses outweigh those of phosphorus. Serum calcium levels drop, sometimes resulting in tetany, which is frequently seen in the chemically sensitive, and a need for calcium and continuous supplementation of vitamin D in selected patients arises.

Celiac disease (gluten-sensitive enteropathy) is indirectly related to vitamin D deficiency. The impaired mineralization that results in structural deformities is due to steatorrhea. Because vitamin D absorption depends on normal bile secretion and fat absorption, a deficiency results from unabsorbed fats, calcium soaps, and vitamin D that are flushed out in the steatorrheic stool. In our experience at the EHC-Dallas, many c eliacs can be controlled by diet and injection therapy and environmental control without the need for vitamin D supplementation.

If there is not enough D_3 to put calcium in the bone, calcium can be found in areas of inflammation such as blood vessels from the brain, coronary, iliacs, chest wall, or lung muscles. Some patients with chemical sensitivity and low calcium and magnesium also have low vitamin D. Elevated D_3 has been observed in 15% of chemically sensitive patients. The cause and treatment for this is unknown at the present.

All vitamin D deficiency diseases respond to vitamin D therapy. However, many chemically sensitive patients are intolerant of vitamin D supplementation. Thus, they remain deficient in spite of attempts at supplementation. While some damage cannot be rectified and supplementation may not work and of necessity have to be stopped, further damage can be prevented by reducing the total pollutant load, thereby reducing stress on the system.

Sources. Vitamin D can be acquired either as preformed vitamin D by ingestion or by exposure to sunlight. It is found in only small amounts in butter, cream, egg yolk, and liver. The best food sources are the fish liver oils (cod, salmon, herring). In recent years, approximately 98% of all pasteurized milk has been fortified with vitamin D, usually 400 IU per quart. Plants are a poor source, with mushrooms and dark green leafy vegetables containing minute amounts. It is abundant in fish oils. The other common form is vitamin D_2, or ergocalciferol, a commercial product made by ultraviolet irradiation of ergosterol from yeast.

Strict vegetarians have few dietary choices in meeting vitamin D requirements. Vitamin D can be obtained by irradiating some of the aforementioned foods containing the precursors and by exposing the body to ultraviolet light.[210]

Toxicity. Excess vitamin D may lead to arteriosclerosis, cardiomyopathy, and/or renal failure.

Vitamin E (Fat Soluble)

The role of vitamin E in the chemically sensitive is extremely important because of its antioxidant effects (Figure 17). The following facts about its function and physiology will allow the clinician to better understand this role.

Vitamin E has been used in some vascular and PMS patients, as well as any chemically sensitive patients with pollutant injuries. It is particularly important in maintenance of lipid membrane integrity and the prevention of tissue polyunsaturated fatty acid peroxidation, which occur with pollutant exposure and injury in many chemically sensitive patients (see Chapter 4). *Despite a low molar concentration in membranes, vitamin E serves as the major lipidsoluble, chain-breaking antioxidant, preventing lipid peroxidation and modulating the metabolism of the arachidonic acid cascade initiated by lipoxygenase and/or cyclooxyenase.*

Essentially, the tocopherol molecule interrupts lipid peroxidation by converting lipid peroxyl radicals to lipid hydroperoxides, which are further acted upon by glutathione peroxidase to convert them to stable alcohols. Though in the normal person the presence of these alcohols is innocuous, their presence in the chemically sensitive can be quite harmful due to the already damaged tissue or an increase in total body load.

Vitamin E appears to be highly efficient as an antioxidant and is thus necessary in counteracting chemical sensitivity. Further, chemical and biochemical studies suggest that vitamin E can be reduced after it has been oxidized and prior to its decomposition. Ascorbic acid and glutathione are

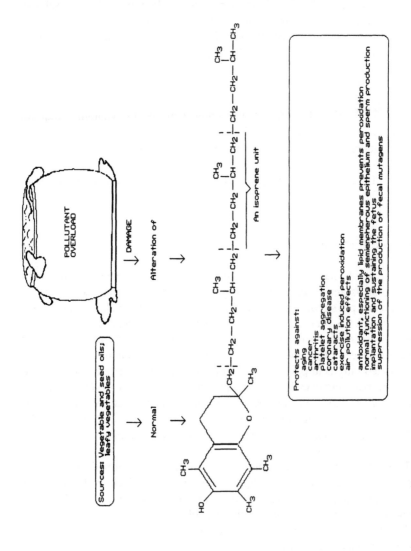

Figure 17.   Potential pollutant damage to vitamin E.

major water-soluble intracellular antioxidants (reductants) in the cytosol that generate reduced vitamin E. This reaction is dependent on the concentration of these substances and/or the enzymes that remain in their reduced form.[211] If the precise interaction of these vitamins and enzymes in the chemically sensitive is disturbed, adequate reduction of vitamin E is interfered with and vitamin E becomes ineffective in combatting new pollutant exposures.

The role of vitamin E in the prevention of free radical damage is clear both in the individual and chemically sensitive patient. Using breath pentane output as a measure of lipid peroxidation, studies have demonstrated that intake of 1000 IU of vitamin E for 10 days significantly decreased breath pentane excretion in healthy adults consuming a normal diet. Study results may be significant in view of research evidence showing a role for free radical-related damage in normal body processes and in certain diseases and protective effects of the vitamin in controlling peroxidation in body tissues.[211] Studies have now shown that vitamin E is protective against aging,[212-214] cancer,[215] arthritis,[216,217] platelet aggregation[218-220] and coronary disease,[221] cataracts,[222,223] exercise-induced peroxidation,[224,225] and air pollution effects.[226-228] It clearly decreases pollutant effects in the chemically sensitive in our studies at the EHC-Dallas.

Vitamin E protects lipid-rich regions of cells from free radical peroxidation. This is the area where lipophilic xenobiotics would be depositing in the chemically sensitive. Vitamin E achieves this action by breaking the chain reaction of downward antioxidant pollutant injury. It works along with the enzymes glutathione peroxidase and glutathione reductase in the glutathione regeneration cycle to help counteract free radicals and other oxidants before they initiate free radical pathology. It is required for superoxide dismutase to function to trap superoxide anion radical and prevent damage to the cell by free radical pathology.[229] Clearly, vitamin E is not only a free radical scavenger, but also a stabilizer of the cellular membranes (Figure 17), which may be a target of pollutants in the chemically sensitive.

A recent study showed vitamin E to be protective against the damaging effects of high concentrations of oxidant pollutants from city air.[230] Since ozone causes lipid peroxidation in a similar fashion to NO_2, a deficiency of the lipid antioxidant vitamin E enhances the toxicity of ozone. Goldstein et al.[231] exposed rats that had been reared on diets either normal or deficient in vitamins, to either 10.4 or 14 ppm ozone. The rats reared on the vitamin E-deficient diets exhibited significant earlier death via pulmonary edema. Vitamin E will prevent oxygen-induced hemolysis in artificial organs by strengthening the red cell membranes.

Levander et al.[232] reported that selenium deficiency by itself did not enhance splenomegaly, anemia, and mechanical fragility of the red cells in lead-exposed rats, nor did it offer a significant sparing effect in lead-exposed rats maintained on a vitamin E-deficient diet. In contrast, excess levels of selenium in the vitamin E-deficient diet partially prevented the splenomegaly and anemia of lead-poisoned rats (Figure 17).

Deficiency. Vitamin E deficiency is difficult to diagnose because of its presence in diverse ways. Although the major influence of the deficiency is on the reproductive system, nervous system, muscle tissue, and blood erythrocytes, not all species manifest a deficiency in one or all of these areas. Some symptoms are amplified by dietary polyunsaturated fatty acids. Other symptoms may be prevented by nonspecific antioxidants (such as selenium or the sulfur-containing amino acids)[233] (Figure 18).

Vitamin E deficiency can result in a variety of free radical-produced disorders, all of which are associated with heightened lipid peroxidation. Deficiency results in more inflammation in animals, while oral vitamin E has been shown to stop lipid peroxidation.[234] Vitamin E deficiency is observed in patients with malabsorption syndromes (choleastatic liver disease),[235-237] abetalipoproteinemia,[238] short bowel syndrome,[239] cystic fibrosis,[240] and patients with peripheral neuropathy (ataxia and areflexia).[241] Also, familial vitamin E deficiency has been reported. These lipid malabsorption syndromes cause alterations in levels of plasma lipoproteins; therefore, plasma levels of tocopherol may reflect the abnormal lipoprotein levels. Two studies showed vitamin E-deficient rats and mice were killed by smaller doses of paraquat than the controls.[242-244]

Clinical evidence of peripheral-nerve dysfunction in the vitamin E-deficient patients was demonstrated by various combinations of hyporeflexia, decreased vibratory and position sensations, and paresthesia.[245] In addition, tests of nerve-conduction velocity showed mild abnormalities in sensory-nerve function in two patients. Histologic examination of the sural nerves obtained from the vitamin E-deficient patients revealed a mild loss of larger caliber, myelinated fibers in the two patients with homozygous hypobetalipoproteinemia. The sural nerves in the three patients with familial isolated vitamin E deficiency were normal. A wide spectrum of morphologic abnormalities was seen in the nerves obtained from the control subjects (who were selected because of symptoms of peripheral neuropathy), including demyelination, inflammation, fibrosis, and axonal degeneration. In general, the sural nerves from the vitamin E-deficient patients showed less histologic damage than those of the control subjects.[246,247]

Vitamin E deficiency is often associated with symptoms of a peripheral neuropathy. One study evaluated whether vitamin E deficiency affects the vitamin E content of the peripheral nervous system. The α-tocopherol content has been measured in biopsy specimens of sural nerve and adipose tissue from 5 patients with symptomatic vitamin E deficiency (2 with homozygous hypobetalipoproteinemia, and 3 with familial isolated vitamin E deficiency) and 34 control patients with neurologic diseases without vitamin E deficiency.[248-250] A significant reduction in tissue tocopherol content was present in the vitamin E-deficient patients, as compared with the controls, both in sural nerves (1.8 ± 1.2 vs. 20 ± 16 ng/µg of cholesterol [$p < 0.001$], or 7.7 ± 5.4 vs. 64 ± 44 ng/mg of wet weight [$p < 0.01$]) and in adipose tissue (46 ± 43 vs.

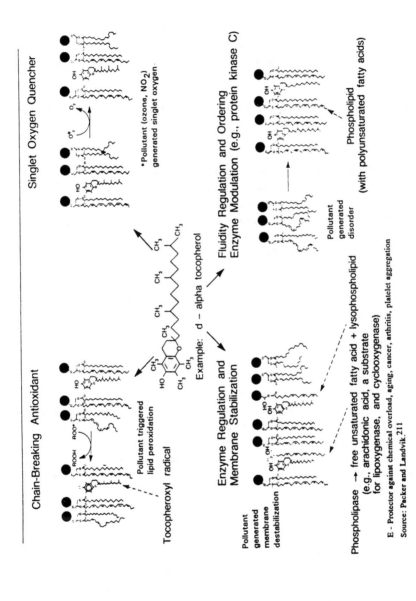

Figure 18.   Possible ways vitamin E acts against pollutant exposure in the chemically sensitive.

222 ± 111 ng/mg of triglyceride [p <0.001]). Levels of tocopherol in adipose tissue were significantly correlated (p <0.001) with levels in peripheral nerves.[251] Many chemically sensitive patients develop neuropathy with solvent exposure. At this time, we do not know whether the local lesion involves a vitamin E deficiency, but we expect so since some respond to supplementation.

The cholesterol contents in the biopsy specimens of sural nerve were not significantly different in the two groups.[252,253]

The vitamin E-deficient subjects had significantly less tocopherol in their adipose tissue than did the control subjects, whether levels were expressed in relation to triglyceride, cholesterol, or weight. These facts emphasize the importance of having high tissue levels in the chemically sensitive in order to readily combat pollutants.

The patients with the lowest tocopherol levels in adipose tissue had the lowest levels in the sural nerve.[254]

Vitamin E deficiency in humans results in a striking reduction in tocopherol in the peripheral nerves, which precedes from the axonal degeneration of the nerves reported in histologic studies.[255-257] This study suggests that damage to the peripheral nerves during vitamin E deficiency results from an inadequate amount of the antioxidant tocopherol,[258] and that the destruction of the nerves is a result of free radical damage.

Vitamin E supplementation has been shown to be an antipollutant in prevention of lung pathology. Fariss[259] evaluated the ability of vitamin E to reverse the effect of chemically induced toxicity on liver cells.[260] In animal studies, he found that when the animal was treated with a drug or chemical such as carbon tetrachloride (seen in many chemically sensitive patients), which is known to enhance free radical oxidation of the liver, prevention occurred by giving vitamin E as an antioxidant. These results show that the addition of vitamin E succinate to the incubation medium was capable of protecting isolated hepatocytes against chemically-induced cell death and that vitamin E may be a very important chemical antioxidant in the liver. The basic implication is that vitamin E, working in the fat-rich region of the body, serves as both a specific antioxidant and, potentially, as a modulator of the arachidonic acid cascade and leukotriene formation. Furthermore, administration of vitamin E to patients with a deficiency of this vitamin raises the level of tocopherol in adipose tissue[261] and halts the progression of neurologic disease.[262-265] Similar results have been seen with supplementation in the chemically sensitive at the EHC-Dallas.

Reports have documented that supplementation with vitamin E in pharmacologic amounts (400 IU to 10 g/day) in patients with abetalipoproteinemia,[266] cholestasis,[267] or familial isolated vitamin E deficiency[268,269] results in improved neurologic function, or at least cessation of further neurologic deterioration.[270]

Dietary Sources. The vegetable and seed oils (wheat germ) are the richest sources of this vitamin. Different tocopherols are not uniformly distributed in foods: the vitamin E content of safflower oil is 90% α-tocopherol; the content of corn oil is only 10%. Vitamin E content is often related to the linoleic acid content of the oil. Animal products are medium to poor sources of vitamin E and are largely dependent on the fat composition of the diet of the animals. Cooking and processing foods can substantially reduce their vitamin E content. For example, tocopherols are removed during the milling of white flour, and in making white bread, all the vitamin E can be lost if chloride dioxide is used in the bleaching process.

Good sources of vitamin E include whole grains, green leafy vegetables, margarine and shortening made from vegetable oil, milkfat, butter, egg yolks, liver, and nuts.

In the customary U.S. diet, about 64% of vitamin E intake is supplied by salad oils, margarine, and shortening. Another 11% is supplied by fruits and vegetables, and 7% comes from grains and grain products. Diets for the chemically sensitive may be altered, thus allowing different amounts of vitamin E vs. the general population.

Absorption. Vitamin E is absorbed in the same way as the other fat-soluble vitamins in the presence of bile salts and fat, and this process may be damaged by pollutant injury as seen in the chemically sensitive. Vitamin E is stored primarily in the fatty tissues and not in the liver, unlike the other fat-soluble vitamins.[271] Since the chemically sensitive have problems with malabsorption, they may have a deficiency in the vitamin E itself, but more likely they will have a deficiency with the total body pool or the area of local chemical injury. Solvent damage to peripheral nerves may well cause local vitamin E deficiency.

Vitamin K

Although the role of vitamin K in chemical sensitivity is not known, vitamin K physiology should be understood generally (Figure 19).

Vitamin K deficiency is uncommon in humans because of the its wide distribution in plants and animals and the microbial synthesis in the intestinal lumen. Vitamin K is found in alfalfa, green beans, turnip greens, broccoli, cabbage, lettuce, spinach, peas and beans, liver of beef, and chicken. Vitamin K is absorbed from 10 to 70% depending on the amount of accompanying fat in the diet and the help of bile acids. It is distributed on chylomicrons and then on ULDL and LDL. The liver is the main area of storage. Vitamin K is lost through urine and feces.

Animals deficient in vitamin K tend to bleed profusely, and small bruises can escalate into major hemorrhages. Blood clotting is slowed due to a lack of prothrombin and other factors important in blood clotting (Table 5).

Ingestion of antibiotics or other agents that interfere with microbial activity will curtail intestinal synthesis, as seen in many chemically sensitive patients.

Sources:
microbial synthesis in GI
alfalfa
green beans
turnip greens
broccoli
cabbage
spinach
peas
beef and chicken liver

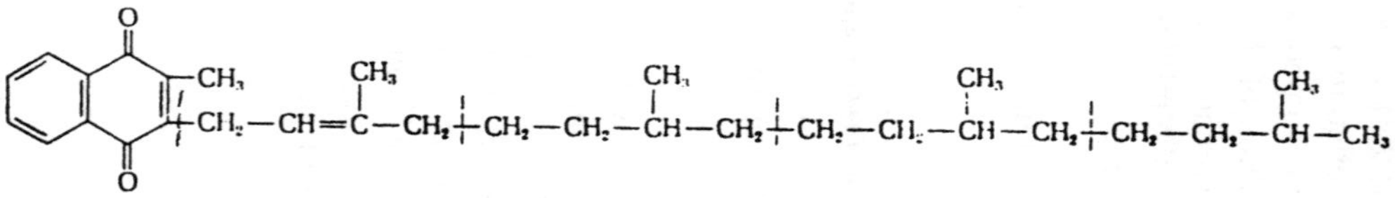

Clotting mechanism - r-carboxylation of glutamine
Bone metabolism, (hydroxylapetite dissolution) reduces
Kidney function - reabsorption of CA^{2+}

Figure 19.   Potential pollutant injury to vitamin K.

**Table 5. Causes of Deficiencies of Vitamin K-Dependent Coagulation
Factors Able to Exacerbate Some Forms of Chemical Sensitivity**

Hemorrhagic disease of the newborn
Dietary inadequacy (low-fat diets, protein-calorie malnutrition)
Total parenteral nutrition deficient in the vitamin
Biliary obstruction (gallstones, stricture, fistulas)
Malabsorption syndromes (cystic fibrosis, sprue, celiac disease, ulcerative
colitis, regional ileitis, short-bowel syndrome)
Liver disease
Drug therapy
Cumarin anticoagulants and related drugs (warfarin indanediones,
phenprocumon, hydantions, salicylates)
Antibiotics including cephalosporins
Megadoses of vitamin E

Source: Olson.[276]

However, unless adults are given bowel sterilizing agents and fed a vitamin K-
deficient diet for several weeks, vitamin K deficiency is not likely.[272] Dicoumeral
is a direct antagonist of vitamin K (Table 5).

Choline

Dysfunction of choline metabolism may have adverse effects on the chemically sensitive (Table 6). Organophosphate pesticides disturb cholinesterases,
thus exacerbating orderly nerve function and chemical sensitivity. Other substances such as solvents and chlorinated compounds may also disturb nerve
function and exacerbate chemical sensitivity, resulting in hyper- and
hypoesthesthia, total loss of sensation and motor function, burning, numbness,
and tingling.

Deficiency. Chronic ingestion of a diet deficient in choline has major consequences that include hepatic, renal, memory, and growth disorders, all seen in
some chemically sensitive patients. In the rat, hamster, pig, dog, and chicken,
choline deficiency results in fatty infiltration of the liver.[273,274] This process is
probably due to a disturbance in the synthesis of phosphatidylcholine, which
is needed to export triglycerides as part of lipoproteins. Fatty infiltration of the
liver begins in the central area of the lobule and spreads peripherally. This
process is different than what occurs in kwashiorkor or essential amino acid
deficiency, where fatty infiltration usually begins in the portal area of the
lobule.[275]
Choline is produced from serum via phosphotydal ethanolamine and with
the help of methyl groups from methionine. Thus, disturbance of methionine

Table 6. Potential Pollutant Damage to Choline

Food Sources of Choline	Choline Chloride (mg/100 g food)	Phosphatidyl-choline
Peanuts	—	1113
Pecans	—	333
Peanut butter	—	966
Wheat germ	—	2820
Wheat	—	613
White flour	—	346
Polished rice	—	586
Cornmeal	—	280
Spinach	10	10
Cauliflower	78	2
Kale	89	2
Potatoes	40	1
Lettuce	18	0.2

Food		
Brussels sprouts	43	2
Calf liver	650	850
Lamb chops	—	753
Beef round	—	453
Ham	—	800
Trout	—	580
Cheese	—	50–100
Eggs	0.4	394

Lecithin → or Phosphotidal ethanolamine / Methionine → Choline $(CH_3)_3-N^+CH_2CH_2O$

A. A major substituent of phospholipids in cell membrane and serum lipoprotein (emulsion)

B. Donor of fatty acid to cholesterol in disposal of LDL cholesterol (L–CAT)

C. Source of methyl groups for synthesis of methionine

D. Substrates for the formation of neurotransmitter acetylcholine

or B_6 metabolism, as frequently seen in the chemically sensitive, may inhibit the efficient production of choline, thereby exacerbating the chemical sensitivity.

Olson[276] reported that choline deficiency in adult rats causes a decrease in plasma choline, cholesterol, and phospholipid, a reduction in high-density lipoproteins, and a marked decrease in low-density β-lipoprotein. Toxic chemical overload seen in chemically sensitive patients may give a similar scenario.

Rogers and Newberne[277] reported that induction of hepatocarcinoma by chemical carcinogens such as vinyl chloride is enhanced in animals fed choline-deficient diets. Renal function is also compromised with abnormal concentrating ability, free water reabsorption, sodium excretion, glomerular filtration rate, renal plasma flow, and gross renal hemorrhage.[278,279] Kratzing and Perry[280] reported that infertility, growth impairment, bone abnormalities, decreased hematopoiesis, and hypertension are associated with low-choline diets. Certainly chemically sensitive patients have a need for choline.

A large number of clinical studies using choline or phosphatidylcholine therapy for neurologic diseases have already been undertaken. Choline and phosphatidylcholine therapy benefits patients with tardive dyskinesia, a syndrome thought to involve deficient cholinergic neurotransmission.[281,282] This disorder is characterized by involuntary choreic movements of the tongue, lips, and jaw. Similar findings have been seen in a few chemically sensitive patients who seem to respond to choline supplementation.

Cholinergic neurons are vital for normal memory in rats and humans.[283] Bartus et al.[284] reported that choline deficiency in mice is associated with memory impairment. Choline administration improves the memory of humans who are poor initial learners in serial-learning and selective-reminding tasks.[285] This type of brain function occurs in a subset of the chemically sensitive, but only partial success has been seen with choline supplementation in the group. Perhaps this is due to permanent damage or to pollutants that remain contained in the body preventing total clearing. A number of reports have described memory function in Alzheimer's disease patients treated with choline or phosphatidylcholine by double-blind studies.[286,287] Many patients with chemical sensitivity have been seen to have this brain dysfunction, which is due to pollutant overload and presumably causes a local choline deficiency in their brains (see Chapter 26).

In a few cases, large doses of pure lecithin, 3 g (about 6 parts choline) with or without anticholine drugs, have improved memory, presumably because of increase in brain acetylcholine concentrations.[288] In rats, a high choline intake leads to increases in brain acetylcholine concentrations.[289,290] It should be noted that federally regulated food standards allow the lecithin sold in commercial market to be a mixture of various phospholipids and to contain less than actual lecithin. Efforts are currently underway to make pure lecithin more available. Improvements have been seen in only a few patients using lecithin at the EHC-Dallas, perhaps due to inadequate delivery or as yet undefined, unknown causes.

When combined with acetyl, choline forms acetylcholine a neurohumoral transmitter of intestinal peristalsis. It, of course, has many other parasympathetic nervous system functions that are extremely prone to pollutant injury in the chemically sensitive (see Chapters 26 and 27).

Oral administration of choline has a slight hypotensive effect in humans.[291] This effect is prevented, however, by pretreatment with hemicholinium-3. Choline must be transported via a carrier into some cells, possibly a neuron, to lower blood pressure.[292]

Sources. Free choline is found in some foods such as liver, oatmeal, soybeans, cauliflower, kale and cabbage, and phosphotidylcholine-containing foods — eggs, liver, soybeans, and peanuts.[293]

Pseudovitamins

Flavonoids (Bioflavonoids, Vitamin P)

The flavonoids are a large group of generally insoluble phenolic compounds widely distributed in plants. The basic flavone structure consists of a 1,4-benzopyrene with a phenyl substitution at the two position. Flavonoids may help some patients, as observed at the EHC-Dallas with chemical sensitivities, but may make others ill, apparently due to their chemical configuration since they are benzopyrenes. At present, empirical trials are the only way of telling which will help.

Inositol

Physiological Function. Free inositol, phosphates of inositol, and inositol phospholipids are the principal forms of inositol in nature. Of the nine possible isomers of hexahydrocyclohexane, myoinositol is the sole isomer of nutritional and metabolic consequence. Its structure and stereochemical relationship to D-glucose is shown in Figure 20.[294]

Free myoinositol arises in animal cells both from dietary sources and from biosynthesis from D-glucose. The overall reactions, stemming particularly from work in Eisenberg's laboratory, involve an internal cyclization of glucose-6-phosphate yielding inositol-1-phosphate with subsequent hydrolysis to inositol.[295] The absolute requirement for NAD^+ but with no net gain of NADH suggests a tightly coupled oxidation reduction mechanism; 5-ketoglucose-6-phosphate and inosose 2,1-phosphate have been suggested as reaction intermediates[296] (Figure 20).

Of prime importance is the conversion of inositol to its phosphatides, a process occurring in all animal organs studied to date. The major route is the cytidine diphosphodiacylglycerol (CDP) pathway, wherein inositol reacts with

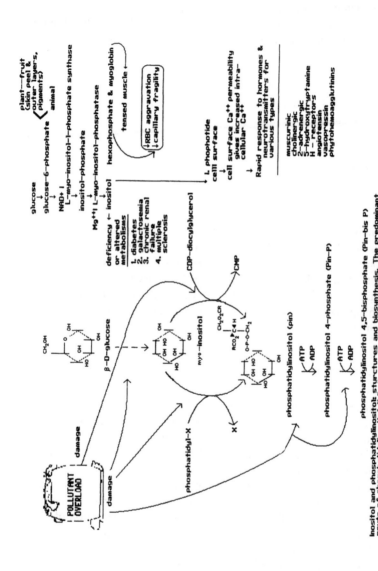

Figure 20. Pollutant injury to inositol in the chemically sensitive.

Inositol and phosphatidylinositol structures and biosynthesis. The predominant route of phosphatidylinositol (1,2-diacyl-sn-glycero-3-phosphoryl-inositol, Pin) biosynthesis is via a transferase reaction of myo-inositol with the liponucleotide, cytidine diphosphodiacylglycerol (CDP-diacylglycerol), giving Pin and cytidine monophosphate (CMP). Some free inositol may also react in an exchange reaction with the base moiety (expressed in the general form, X) of endogenous microsomal phospholipid to give Pin and X.

CDP-diacylglycerol in the presence of a transferase enzyme to yield phosphatidylinositol.[297]

Inositol hexaphosphate can act as an allosteric effector in interacting with human adult myoglobin to shift the quaternary equilibrium from the relaxed (R) state toward the tensed (T) state.[298]

The predominant physiologically active form of inositol is as the phosphatide, which functions primarily at the membrane level. Membrane phosphatidylinositol has a special function in the response of various cells to external stimuli such as certain hormones and neurotransmitters. Stimuli in which major effects are to produce rapid physiologic responses include muscarinic, cholinergic, 2-adrenergic, 5-hydroxytryptamine, histamine (H_1) receptors, angiotensin, vasopressin, or those that bring about longer term stimulation of cell proliferation, e.g., phytohaemagglutinins and other mitogens in high serum concentrations. Such events appear to involve degradation of membrane phosphatidylinositol[299] and may control cell surface Ca^{++} permeability, giving rise to an elevation in intracellular Ca^{++} concentration.[300] Some evidence was the discovery that Li^+ is a potent inhibitor of inositol phosphatase, which permitted a demonstration of the accumulation of inositol-1-phosphate released concomitantly by appropriate receptor activity from phosphatidylinositol in the presence of Li^+.[301] Evidence that pollutants can injure this function is strong and certainly borne out by our studies at the EHC-Dallas where we have shown excess intracellular calcium in 24% of our chemically sensitive patients.

Deficiency. There is growing evidence of altered metabolism of inositol in patients with diabetes mellitus, chronic renal failure, galactosemia, and multiple sclerosis.[302] All of these conditions are found in some chemically sensitive patients. There is some evidence for an impaired synthesis of inositol phosphatides and of inositol transport in nerve tissue from diabetic rats. The hyperglycemia associated with untreated human diabetes may impair inositol transport,[303] presumably by an unfavorable competition with glucose for a common transport mechanism, thus resulting in a gross intracellular inositol deficiency.[304] It was found that an improvement in neurophysiologic measurements occurred in diabetic patients given inositol orally (500 mg twice daily for 2 weeks).[305]

The catabolism of inositol via the kidney inositol oxygenase system is markedly decreased in experimental diabetes, as evidenced by elevated levels of inositol in the kidney and by an inositoluria, and there is a dramatic elevation of serum inositol in human subjects with chronic renal failure.[306]

Sources. Inositol is present mainly as phytic acid in plant and animal tissues. A mixed North American diet may consume approximately 1 g of inositol per day.[307]

Most flavonoids are concentrated in the skin peel and outer layers of fruits and vegetables and areas most accessible to light. Beverages such as tea, coffee, wine, and beer also contain significant amounts. As pigments that produce the colors of many flowers, fruits, and vegetables, flavonoid compounds range from the pale yellow and colorless flavanones in citrus fruit to the red and blue anthocyanins in berries.[308,309] The chemically sensitive often have problems tolerating these types of substances and may well be prone to bioflavonoid deficiency.

Because an average daily intake of about 1000 mg of flavonoids occurs in a typical American mixed diet, there is no bioflavonoid deficiency condition as long as proper absorption occurs. However, in the pollutant-damaged, chemically sensitive there is always malabsorption. About one half of the amount ingested is absorbed in a pharmacologically active form; the rest is degraded by intestinal bacteria. By pharmacologic action, bioflavonoids may reduce red blood cell aggregation[310] or decrease bleeding associated with capillary fragility.[311] Some of the observed effects may be explained by the ability of bioflavonoids to chelate metals or by their ascorbate-protecting antioxidant effects.[312] Stillbirths[313,314] can be prevented with bioflavonoid supplementation.

Typical flavonoids are colorful antioxidants. Certain bioflavonoids inhibit aldosereductase, which converts glucose and galactose to their polyols. (These polyols have been implicated in the neuropathy of diabetes and in cataract formation that accompanies diabetes and galactosemia.) Flavonoids like quercitin also inhibit phosphodiesterases (which break down cyclic nucleotides) and thus effect smooth muscle relaxation.[315]

Pills containing bioflavonoids may be unsafe, not only for the general public, but especially for the chemically sensitive, because quercin, a common bioflavonoid, is a mutagen[316] and may exacerbate chemical sensitivity.

Minerals

In the chemically sensitive patient, some of the elements in the periodic table are needed in relatively large amounts, while others are needed in smaller or trace amounts because of their unique biochemical susceptibility. This varied need is not only true for the chemically sensitive; it is also true for plants and animals. Normally plants obtain their share of elements from the soil. Plants produced in a soil deficient in elements may lack nutrients even though they appear unharmed. This deficiency can have an impact on human and animal health[317] and especially the chemically sensitive due to their already present nutrient disturbances, thus the need for high-quality organic food.

The chemically sensitive are often mineral deficient or have intracellular excess, which may reflect not only their damage by toxic chemicals on the body (especially transport and membrane changes), but also poor nutrition due to poor dietary intake of low-quality foods. In the series at the EHC-Dallas in over 200 random chemically sensitive patients, intracellular deficiencies were

Table 7. Mineral Study[a] (Percent of ECU Patients with Abnormal Whole Blood Cell Trace Mineral Assays)

Percent of Chemically Sensitive ECU Outpatients	2 S.D. Below Mean	3 S.D. Below Mean	2 S.D. Above Mean	3 S.D. Above Mean
Magnesium*	3.3	0.0	30.0	25.0
Zinc*	5.0	0.8	20.0	5.0
Chromium	44.0	44.0	0.8	0.8
Selenium	10.0	3.3	11.6	12.5
Sulfur	34.0	2.5	0.8	0.0
Calcium	6.6	0.0	26.6	2.5
Phosphorus	0.8	0.0	9.1	0.0
Sodium	5.0	0.0	2.5	0.0
Potassium	4.1	0.8	34.1	6.6
Copper	5.0	1.6	14.1	5.8
Iron	0.8	0.8	3.3	0.8
Manganese	0.0	0.0	5.8	0.0
Silicone	10.0	0.0	0.0	0.0

[a] This study consisted of 120 patients who had been hospitalized in the Environmental Care Unit (ECU) and 80 outpatients of the Environmental Health Center-Dallas (EHC-Dallas). Overall, both inpatients and outpatients had similar mineral assay results.

* 40% deficiency by I.V. challenge.

present: 89% in chromium, 33% in sulfur, 30% in silicon, 14.5% in selenium, 7.5% in zinc, and 3% in magnesium, as evidenced by intracellular red blood cells studied. In another series of 120 patients, mineral deregulation was present (Table 7). Some intracellular minerals were low, but others were too high, suggesting pollutant damage to the cell membrane.

Since zinc and magnesium changes are not found in large quantities by serum and cellular measurements, other studies have to be performed to further evaluate these in the chemically sensitive. In another study at the EHC-Dallas, 19% of the chemically sensitive patients with muscle cramps and nerve irritability were magnesium deficient by a combination of serum and cellular measurements. Additional deficiencies (40% of chemically sensitive patients) were found by oral and intravenous challenge of zinc and magnesium. Oral, intramuscular, or intravenous challenge is often necessary since organ-specific deficiency is impossible to measure at this time. Therefore, the only alternative will be to see how a clinical challenge works. Further studies are shown in the Laboratory chapter.

Table 8. General Role of Ions in Metabolic Processes Potentially Prone to Pollutant Injury

Ion	Na, K	Mg, Ca, Mn	Zn, Cd, Co	Cu, Fe, Mo, Mn
Bond strength Biological function	Weak Charge transfer nerves	Moderate Trigger reactions hydrolysis, phosphate transfer	Strong Hydrolysis pH control	Strong Oxidation Reduction Reaction
Ligand	0	0	N and S	N and S

The animal body contains millions of ligands to which the metals are attached. The site of biological activity often is the metal itself. The trace metals are part of the metalloproteins (metalloenzymes and metallo-activating enzymes). These ultraefficient catalysts are capable of regenerating themselves in living systems. They may become damaged by pollutant injury exacerbating chemical sensitivity. The general role of metal ions in biological processes are shown in Table 8.

The essential nature of metals is well established. The metal concentration is controlled within limits in the body for optimal well-being by proteins and hormones. If the controls fail, such as happens in pollutant injury, disorders or disease are likely to arise. The disorders may be due to a deficiency or an excess of the metal as well as to damage to interactions between cations and anions, as seen in the chemically sensitive. The excesses cause toxicity or imbalances. The ratio between requirement and toxicity may be narrow or wide depending on the mineral involved. *In a study done at the EHC-Dallas, one of the most outstanding features in the chemically sensitive vs. controls was that some minerals were too high (aluminum, barium, manganese, copper, potassium, selenium, zinc) and others too low (often the same minerals).* (See Chapter 30.) These mineral abnormalities tend to explain the fragility seen in the chemically sensitive patient, since pollutants often disturb the cell membranes. The problems with minerals occur routinely in the chemically sensitive, but again, biochemical individuality makes expression as either deficiency or excess random and unpredictable. The key to diagnosis and treatment is to know this is a common problem and be on the lookout for it.

The animal body contains a host of minerals. Many of the metal enzymes are important for the life processes. Other metals may be associated with chelates as in the skeletal system or with metal-binding proteins in transport systems. Dietary intake of phytates, fiber, oxalates, and clays reduces mineral availability to the body. However, there are also enhancers for mineral absorption such as histidine, cysteine, ascorbic acid, citric acid, fructose, lactose, and certain proteins.[318] Though minerals may be needed in combination, each will be discussed separately.

Sodium and Potassium

The majority of aspects of sodium and potassium flux are now well understood in medicine and are discussed in many texts. Pollutant injury to some pumps and membranes will disturb function of these elements. Sodium and potassium channels are voltage sensitive. Some channels open in response to polarization, and pollutant injury may change the charge, allowing for inappropriate flow. Some potassium channels are known to respond to transmitter binding and second messengers. Pollutant injury in the chemically sensitive may change the kinetic properties, resulting in inappropriate distribution of the ion. Clearly, pollutant injury to the sodium pump results in the periorbital, hand and pedal edema seen in the chemically sensitive patient, while damage to the potassium channels results in one of the main symptoms, inappropriate distribution with weakness. Disturbance of sodium and potassium functions will exacerbate chemical sensitivity. An adequate ratio of sodium, potassium, calcium, magnesium, and lithium is necessary for membrane stability and can be disturbed by pollutant injury. Further discussion is reserved for medical texts.

Calcium

Calcium is the most abundant mineral in the human body, and about 99% of it resides in the bones and teeth. The ratio of calcium to phosphorus is important for healthy bone structure, and vitamins A, C, and D are needed to help control the metabolism of calcium. The calcium that is not fixed in structural tissue is extremely important in the chemically sensitive, and it has many physiological involvements. Some of these involvements include enzyme activation,[319] electrolytic nerve function,[320] blood pH mediation[321] and clotting mechanism chemistry,[322] muscle contraction,[323] assistance in regulating heart beat, and cell membrane permeability. Calcium plays vital roles in directing cell function.[324] Any of these functions may be disturbed by pollutant overload, thus exacerbating chemical sensitivity (Figure 21). Knowledge of the metabolic influences of calcium is extremely important in managing the chemically sensitive.

Calcium serves as an almost universal ionic messenger, conveying signals received from the cell surface to the inside of the cell, and can be disturbed by pollutant injury. The calcium ion is involved in such diverse processes as the

Sources: Calcium and phosphorus content of foods			
	Average Serving		
Food	Approximate Measure	Milligrams of Calcium	Milligrams of Phosphorous
Peanuts, roasted with skins	2/3 cup	69	391
Turkey, roasted fresh only	3 oz.	7	213
Fish (halibut, broiled with butter or margarine	4.5 oz.	20	310
Pork loin, broiled, med. fat	2.0 oz.	7	181
Milk, nonfat (skim), fluid	8 oz.	296	233
Milk, whole, fluid	8 oz.	288	227
Chicken, roasted	3 1/3 oz.	12	242
Loin lamb chop, broiled	3 1/3 oz.	9	163
Beef, hamburger, cooked (reg. ground)	3 oz.	10	196
Oysters, raw	6 oysters	81	123
Cheese, cheddar	1 oz.	213	136
Peas, cooked	2/3 cup	23	105
Egg, poached	1 large	51	121

Source: Czajka-Narins. 934

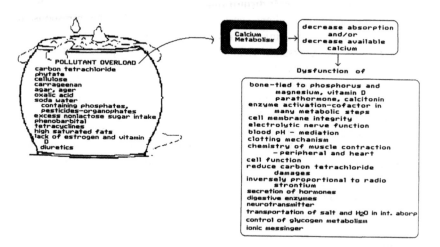

Figure 21.   Pollutant overload causes dysfunction of calcium metabolism.

regulation of muscle contraction,[325] the secretion of hormones, digestive enzymes,[326] neurotransmitters,[327] the transportation of salt and water across
intestinal linings,[328] and the control of glycogen metabolism in the liver,[329] all
of which have been observed to malfunction in selected chemically sensitive
patients. This messenger function involves minute flows of calcium across
membranes of cells. The cells have a set of mechanisms by which they regulate
intracellular calcium. The mechanisms work mainly by controlling the movement of calcium across the plasma, the mitochondria, and the sarcoplasmic
membranes (in muscles or calcisomes in nonmuscle). The amount of influx of
calcium flow varies greatly, often depending on the available calcium, the
amount of pollutant damage to the membrane, and its nutritional state. Such
cycling across the plasma membrane is part of a complex chain of events
by which cells generate sustained responses to their environment and are
thus extremely important in the chemically sensitive. The sensitivity of a cell
to very small changes in calcium ion concentration reflects the very low

concentrations of ions in the cell and the 10,000 times greater amount around the cell. Maintenance of these concentrations of ions depends upon the low permeability of the plasma membrane to calcium and membrane bound pumps that drive calcium out of the cell (see Chapter 4). Often membrane damage occurs due to toxic chemicals and alters the pumps or the calcium permeability in the chemically sensitive. This alteration appears to occur in the chemically sensitive as evidenced by our studies where 25% of our chemically sensitive patients were found to have excess calcium (above 2 S.D.) in their cells. When a cell is stimulated by an external signal, channels in the membrane open and allow calcium to enter at 2 to 4 times the normal rate. Some channels open when neurotransmitters change voltage, others when hormones interact at the cell surface.[330] Calcium abnormalities may be responsible for hormone dysfunction experienced by some chemically sensitive patients, neurotransmitter responses experienced by others, and a combination of effects experienced by still others.

Calmodulin

Calmodulin is a calcium-binding protein that regulates intracellular calcium. Calmodulin itself is not active. Its active form is calmodulin-Ca complex. Intracellular physiologically active calcium ion is calmodulin bound. Calmodulin serves as an intracellular calcium receptor and regulator of intracellular cyclic AMP through adenylate cyclase, in glycogen metabolism, intracellular motility of microtubules and microfilaments, and in Ca-dependent protein kinases.[330] These facts again emphasize why the pollutant-damaged, chemically sensitive patient may have weakness, fatigue, and edema (Figure 22).

Calmodulin is synthesized on both free and bound ribosomes, then is discharged into the cytosol, and then a large part of it becomes attached to intracellular membranes. It is present at multiple sites in the cell. Calmodulin is localized into smooth and rough endoplasmic reticulum (ER), as well as the nuclear envelope, mitochondrial membranes, and membrane vesicles. Due to this location, it or its output is particularly prone to pollutant damage since pollutants frequently attack these areas in the cell (see Chapter 4). Calmodulin also regulates the red blood cell calcium pump, the ER calcium pump, and the sarcoplasmic calcium pump.[330] Pollutant injury to these pumps can cause diffuse dysfunction in the chemically sensitive.

Calmodulin seems to be transported along the neuronal process to the postsynaptic membrane and may mediate the Ca effects on synaptic transmission and thus have a role in the release of neurotransmitters.[330] Inappropriate neurotransmitter release may be seen in some chemically sensitive patients.

Calmodulin regulates synthesis and degradation of cyclic AMP by controlling adenylate cyclase and phosphodiesterase by influx of calcium ions in

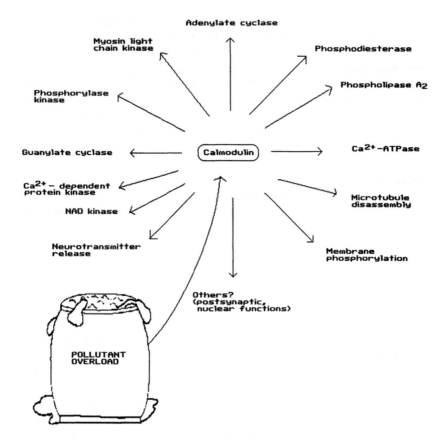

Figure 22.   Pollutant injury to calmodulin may affect any of these functions.

response to stimuli. Weakness and fatigue, being the prime symptoms of chemical sensitivity, may in part be triggered by pollutant injury to this mechanism. It also regulates synthesis and degradation of the enzyme phospholipase A_2, and therefore synthesizes prostaglandins. Phospholipase A_2 controls metabolism and endoperoxides and thromboxane A_2.

Calmodulin also controls the calcium pump to extrude calcium from cytoplasmic membrane. Increased cytosolic Ca activates Ca ATPase system, and it increases Ca efflux. Calmodulin regulates the activity of actomyosin ATPase in nonmuscle cells as well as activity of myosin light-chain kinases in smooth, skeletal, and cardiac muscles. Calmodulin also regulates activity of phosphorylase kinase and glycogen syntase, which controls the degradation and synthesis of glycogen. Phosphorylation of phosphorylase causes an increase in the activity of glycogen synthesis, while that of glycogen synthase causes a decrease.[330]

Pollutants can cause injury to calmodulin-Ca complex or may cause injury to ribosomes and their production or may injure any of the calcium pumps (see Chapter 4). As a result, abnormal neurotransmitter response or abnormality in cyclic AMP regulation may be seen, which in turn can cause a chain reaction of abnormalities in different enzymatic reactions.[330] These abnormal responses and regulation are apparently what happens in some chemically sensitive patients.

Deficiency/Excess Symptoms and Disorders. A calcium deficiency in the chemically sensitive can result in afflictions of the nervous system: muscle spasms, cramps, numbness, tingling in the arms and legs, and convulsions. Nervousness, insomnia, irritability, and headaches have been relieved with calcium, magnesium, and vitamin D intake[332] as well as reduction of total pollutant load in some chemically sensitive patients. Also, osteoporosis,[333] arthritis,[334] rheumatism,[335] and premenstrual tension[336] may be related to low calcium.

Hypercalcemia can result in calcification of tissue and hardened deposits in the kidney, resulting in impaired renal function.[337] Usually this condition relates to other mineral imbalances as well.[338]

Calcium is clearly one mineral that is necessary for membrane stability and thus vascular wall tone. Obviously it is essential for bone. It also is a cofactor in many metabolic steps. Its intake and environmental influence should be carefully evaluated.

The relationship of calcium absorption to diet is important. During World War II, the English observed that Ca was poorly used with the consumption of bread with 92% extraction of wheat flour in comparison to bread with 69% extraction in which more of the outer fraction of the wheat berry was removed.[339] Sodium phytate added to the 69% extraction flour depressed the absorption of Ca, indicating that phytate in the bran might be responsible for the poor Ca utilization. Several conflicting reports[340-342] suggest that calphytates both antagonize and have no effect on calcium absorption.

A typical vegetarian diet may contain sufficient uronic acids to bind about 360 mg of dietary Ca per day,[343] but over 80% of the acids may be fermented in the intestine.

Carrageenan (an additive in many sweets and candy) and agar-agar reduced the absorption of Ca in short-term studies with rats.[344] Possibly, a part of the adaptation process may involve a reduction in the urinary excretion of Ca, as was shown in adult male chickens fed a diet very low in Ca.[345] Pectin, which also contains a high level of uronic acids, had no effect on Ca balance when consumed at 36 g/day by young men for a 6-week period.[346] Again, pectin is metabolized in the lower digestive tract, which frees the Ca for absorption. Malabsorption is already a problem for the chemically sensitive, and it may be accentuated by additives.

Abnormal levels of oxalic acid in food (beets, tea, spinach, rhubarb)[347] have been a nutritional concern, since calcium oxalate is extremely insoluble.

Spinach is a notoriously rich source of oxalic acid. It is questionable whether oxalic acid from one foodstuff will affect the availability of Ca from another source. Frequently, the chemically sensitive have problems with these foods, and their abstinence may be an advantage to them.

Lactose is known to enhance Ca utilization.[348-351] It increases the diffusional component of the transport system rather than the active transport component, which is dependent upon vitamin D.

Anticalcium. Calcium can be depleted by diuretics,[352] excessive nonlactose sugar intake (junk food, phosphate sodas, processed foods, organophosphate pesticides on foods),[353] lack of exercise,[354] phenobarbital,[355] high phosphorus intake,[356] stress (excessive),[357] tetracycline,[358] oxalic,[359] phytic acids,[360] high saturated fatty acids,[361] vitamin D deficiency,[362] and lack of estrogen in the female.[363] All of these substances will exacerbate the chemically sensitive over a period of time.

Sources. Excellent sources of calcium (providing approximately 300 mg per serving) include the following: milk and yogurt (1 cup), hard cheeses ($1^1/_2$ ounces), cottage cheese ($^1/_4$ cup), dark green leafy vegetables (1 to 2 cups cooked), and broccoli (2 cups cooked). Citrus fruits, canned fish with edible bones, and dried peas and beans are good calcium sources also. Meats and nuts that are excellent sources of many other nutrients are poor sources of this mineral. Whole grains are not a good source, unless they comprise a large portion of the diet, in which case they can be a major contributor. Hard water provides some calcium; commercial mineral waters are a poor source.[364]

Diagnosis. The gradual loss of calcium in the chemically sensitive is difficult to evaluate because of the large pool in bones. Frequently this loss is not manifested for years until the bones show demineralization.

Serum levels should be measured, but they will not reflect a long-term slow leak. They will be of value, however, for the increased levels of calcium seen in hyperparathyroidism. Intracellular levels will aid in finding more deficiencies or excesses which are seen in the chemically sensitive. However, they also will not show a slow loss; 24-h urine with balance studies will help evaluate the loss.

Treatment. (See Chapter 37.) One to three grams of calcium have been given daily to chemically sensitive patients at the EHC-Dallas with vascular disease and PMS and have been found in many cases to stabilize their symptoms, presumably by stabilizing their cell membranes. Calcium has also been given to patients with osteoporosis. It has reduced the damage caused by carbon tetrachloride in animals,[365] menstrual tension, and peripheral muscular cramps.[366-368] We have given calcium as an intravenous challenge to chemically sensitive patients with the aforementioned symptoms. Some of these

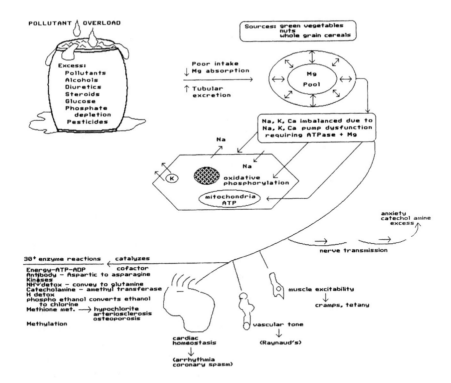

Figure 23.   Pollutant injury to magnesium metabolism.

patients respond well with the elimination of symptoms. Those who do not, often respond to intravenous magnesium. Calcium has been found to be inversely proportional to radiostrontium; thus, it is useful in protecting a patient against this pollutant.

Magnesium

Magnesium deficiency has been found in up to 40% of the chemically sensitive at the EHC-Dallas. Because of its prevalence in this population, we believe understanding magnesium deficiency and deregulation of metabolism involving magnesium are extremely important in diagnosing and treating chemical sensitivity (Figure 23).

Magnesium is the fourth most abundant cation in the human body. It is needed for good bone formation. It helps catalyze over 30 intermediary metabolic reactions and metalloenzymes along with coenzymes and is thus extremely important in patients with chemical sensitivity since this element is necessary in many detoxification reactions. Magnesium is essential for cardiac homeostasis, neurochemical transmission, skeletal muscle excitability, and the maintenance of normal intracellular calcium, sodium, and potassium levels and, therefore, is extremely important in treating chemical sensitivity. Because of space limits, we discuss only the better understood reactions involving Mg here.

The body stores only 1% of the magnesium in the extracellular fluid. Some 50% is stored in bone, and the rest is stored in intracellular fluid. Intra- and extracellular fluid content vary independently.

Magnesium is essential for many metabolic steps and metallo-enzymes.Magnesium is an essential cofactor in chronically impaired methionine metabolism, which is associated with the development of clinical disorders over a period of time (arteriosclerosis and osteoporosis). Symptoms of deficiency include headache, eyestrain, muscle weakness, fatigue, mild myopathy, and myopia. Protein-bound magnesium is the mineral activator of adenosyltransferase enzyme.[368] Acute impairment will exacerbate sensitivity to chemicals in the chemically sensitive patient. If this reaction is deficient, aspartic acid would be expected to be high and asparagine low. Asparagine is a frequent component of antibody proteins; low asparagine can affect synthesis of antibodies and other proteins.[368]

Mg forms a complex with ATP and ADP and, therefore, is a mandatory cofactor for all kinases and other enzymes with nucleotides as a substrate. Kinases are Mg-dependent enzymes. If this kinase reaction is deficient, phosphoethanolamine would be expected to be low and ethanolamine high. Phosphoethanolamine reacts to produce phosphatidylcholine, which in turn produces choline. Choline is the precursor of acetylcholine,[368] which may function inappropriately in some chemically sensitive patients.

In the intestine, acetylcholine acts as a neurohumoral transmitter that maintains peristalsis. A decrease in acetylcholine can lead to bowel irregularity.

Ammonia detoxification can be impaired due to impaired formation of glutamine. Magnesium is needed as a cofactor for ammonia detoxification. A toxic effect of ammonia is that it interferes with oxidation in the citric acid cycle. It does this by depleting the supply of α-ketoglutarate. Common symptoms in subacute hyperammonenia are headache, tiredness, lethargy, irritability, allergy-like food reactions (especially to high-protein foods), and perhaps occasional diarrhea or nausea.[368] High ammonia is seen in a subset of chemically sensitive patients.

Mg is important for metabolic processes of oxidative phosphorylation, enzymatic reactions involving ATP, deoxyribonucleic and ribonucleic acid metabolism, and fat and protein synthesis. In particular, Mg is required for action of membrane sodium-potassium ATPase. Na:K ATPase operates at the Na-K-pump and is necessary for maintaining normal intracellular (with proper potassium concentration) ionic balance. These imbalances are frequently seen in the chemically sensitive, resulting in edema, hormone receptor problems, and antigen recognition site impairments.

Catechol O-methyltransferase requires magnesium for activity, although other divalent cations (e.g., Co^{++}, Mn^{++}, Cd^{++}, Fe^{++}, Ni^{++}) can be substituted.

Magnesium is also inhibited by excess Mg^{++} and by Ca^{++}. This enzyme is widely distributed in mammalian tissue and plays a primary role in the extraneuronal inactivation of endogenous catecholamines (dopamine, norepinephrine, epinephrine).

Deficiency. Special tests are revealing unexpectedly high rates of magnesium deficiency in patients with hypertension,[369] heart problems,[370] tetany,[371] cramps,[372] and anxiety[373] as well as in chemical sensitivity. Standard serum tests have revealed that 11% of the county hospital admissions and 65% of the ICU admissions are magnesium deficient.[374] In our experience, standard serum testing is inadequate. Since magnesium, like potassium, is primarily an intracellular cation, it is not surprising that magnesium deficiency can cause atrial and ventricular arrhythmias, potassium deficiency, electrocardiographic alterations, and neuromuscular abnormalities, which are dominant in some cases of chemical sensitivity (Table 9).

Approximately 30 to 40% of ingested magnesium is absorbed, probably in the small intestine. The rest is excreted. Malabsorption is a big problem in the chemically sensitive. Therefore, it is not surprising that these patients require increased oral or parenteral supplementation. In addition, 95% of the stored magnesium filtered load is reabsorbed in the kidney. Frequently, leaks occur due to toxic chemical damage, resulting in magnesium loss. Any disorder that interferes with absorption in the small intestine or with reabsorption in the kidney will result in Mg deficiency. Some of these include pesticide overload, diuretics, digitalis, alcohol consumption, increased glucose, phosphate depletion, and gluco- and mineral corticoids (Table 9).

Mg deficiency is the most underdiagnosed electrolyte deficiency. Whang[375] found Mg deficiency in several types of patients as shown in Table 11.

Rea et al.[376] reported 65% of vasculitis patients seen at the EHC-Dallas who were chemically sensitive were Mg depleted. Whang[377] also found an increased occurrence of magnesium deficiency with other electrolyte abnormalities (Table 12). These included patients presenting with hypokalemia, hypophosphatemia, hyponatremia, and hypocalcemia.

Since serum levels may not accurately reflect total body balance, thus having clinical implications for heart patients, investigators examined over 100 consecutive admissions to the coronary care unit. They compared serum magnesium levels to cellular magnesium content. Observation of intracellular content clearly showed that more individuals were Mg depleted than were revealed by other tests. Accordingly, serum levels indicated that only 5% of patients were hypomagnesemic, while a check of lymphocyte magnesium levels boosted the incidence to 53%. This finding is in agreement with studies at the EHC-Dallas, where we have found that only 5% were deficient if serum Mg were measured. Measurement of intracellular blood, however, showed 14% more were Mg deficient, while intravenous challenge indicated deficiency in 49%.

Table 9. Potential Causes of Magnesium Deficiency in the Chemically Sensitive

---

I. Decreased intake
    A.   Starvation
    B.   Protein-calorie malnutrition
    C.   Chronic alcoholism
    D.   Prolonged intravenous therapy is inadequate Mg^{++}
    E.   Intake of magnesium-deficient junk foods
II. Decreased intestinal absorption
    A.   Surgical resection
    B.   Severe diarrhea
    C.   Intestinal bypass
    D. H_2-receptor antagonist therapy
    E.   Tropical and nontropical sprue
    F.   Celiac disease
    G.   Invasive and infiltrative process
    H.   Prolonged gastrointestinal suction
III. Excessive urinary losses
    A.   Diuretics
    B.   Diabetic ketoacidosis
    C.   Acute tubular necrosis — diuretic phase
    D.   Chrone alcoholism
    E.   Hypercalcemia
    F.   1° Aldesteronism
    G.   Inappropriate ADH
    H.   Aminoglycocide toxicity
    I.   Cisplatin therapy
    J.   Idiopathic renal magnesium wasting
    K.   Pesticide renal magnesium wasting

Magnesium depletion predisposes to hypertension and cardiac arrhythmias and may predispose to coronary vasospasm and perhaps myocardial infarction (Table 10). This type of vascular response is often seen in the chemically sensitive.

In 1957, Kobayashi first noted that the nature of drinking water might influence death rates from cardiovascular disease;[378] the incidence of strokes is high in areas with soft water. Since then, many studies have supported this thesis.[379] A 1983 South African study[380] found a 10% reduction in coronary heart disease mortality for every 6 mg/L increase in water magnesium levels. A similar Dartmouth study[381] in 1960 reported the same effect per 8 mg/L increase.

Moreover, in a study connected at Glostrup Hospital,[381] MI patients were shown to have decreased levels of magnesium as judged by a

Table 10. Clinical Conditions Associated with
Magnesium Depletion

Gastrointestinal disorders
Malabsorption
  Inflammatory bowel disease
  Gluten enteropathy; sprue
  Intestinal fistulas or bypass
  Ileal dysfunction with steatorrhea — regional enteritis
  Immune diseases with villous atrophy
  Short-bowel syndrome
  Radiation enteritis, irritable bowel; ulcerative colitis
Renal tubular dysfunction
  Metabolic
  Hormonal
  Drugs
Endocrine disorders
  Hyperaldosteronism
  Hyperparathyroidism with hypercalcemia
  Postparathyroidectomy period
  Hyperthyroidism
Genetic and familial disorders
  Primary idiopathic hypomagnesemia
  Renal wasting syndromes
  Bartter's syndrome
  Infants born of diabetic or hyperparathyroid mothers
Inadequate intake or provision of magnesium
  Alcoholism
  Protein-calorie malnutrition (usually with infection)
  Prolonged infusion of magnesium-low nutrient solutions
Hypercatabolic states (burns, trauma)
Excessive lactation
Cardiovascular dysfunction

parenteral magnesium retention test. This phenomenon was also observed in intravenous challenge studies of chemically sensitive patients at the EHC-Dallas.

Low magnesium can cause vasospastic phenomenon. For example, Rude[382] used a magnesium tolerance test by which small doses of magnesium are administered over several hours, and magnesium excretion was measured in the urine. Magnesium retention in controls was 22% while the infarct group showed a magnesium retention of 42%. He also showed normal people will excrete up to 90% or more of the magnesium which is infused, that patients with hypomagnesemia retain 80 to 90% of infused magnesium,[382] and that patients measured for both lymphocyte magnesium concentrations and magnesium tolerance presented results that correlated.[382]

Vasomotor action of magnesium is most likely due to interaction with the transport of calcium across the smooth muscle cell membrane and with the release of calcium from the inner surface of the sarcolemma and/or from intracellular depots. The vasomotor reactivity of coronary arteries appears to be more sensitive than is the aorta in magnesium-deficient conditions.[383]

Magnesium may also be lost by vomiting, diarrhea, long-term use of diuretics or ammonia chloride, alcoholism, and protein malnutrition. Significant magnesium may be lost during diabetic acidosis.

Sources. Agriculture, monoculture, and herbicide use, as well as the use of water softeners, have reduced the concentration of magnesium in soil and water, respectively.[384] Therefore, commercial food does not always have as high a content as desired.

Good sources of magnesium are nuts, legumes, whole-grain cereals and breads, soybeans, seafood, and seaweed. Magnesium is important in photosynthesis, so dark green vegetables high in chlorophyll are good sources. Milk, although not a good source of magnesium, provides approximately 22% of daily needs for the average American. Again, the amount of Mg obtainable from milk depends on the quality of the milk supply, which is directly dependent on the quality of grass and feed fed the cows. Small amounts are supplied by pork, meat, poultry, and eggs.[385]

The average daily normal adult diet contains approximately 20 meq of magnesium obtained mostly from green vegetables. In order to maintain a positive magnesium balance, a daily intake of 0.5 meq/kg is critical. The density of magnesium in some foods is so low that contemporary diets of high fat and carbohydrate intake may be insufficient for adequate maintenance and thus further damage the chemically sensitive. Supplementation is often necessary. See Chapter 37.

Magnesium is one of the most important minerals in metabolism of foods and detoxication of xenobiotics in the chemically sensitive; therefore, attention should be paid to its status in evaluating the chemically sensitive.

Diagnosis. Many clinical conditions are associated with magnesium depletion (Table 11). A knowledge of these will increase the index of suspicion for the diagnosis of magnesium deficiency. They are often associated with chemical sensitivity.

Magnesium deficiency will often be evident if an adequate health history is taken. Since 95% of the Mg that is filtered through the kidney is reabsorbed (70% absorbed in the thick ascendary limb of the loop of Henle), any factor that blocks the reabsorption of sodium chloride in this part of the nephron will also promote the urinary excretion of magnesium. Therefore, individuals on diuretics or with renal lesions are suspect for low magnesium. We have seen many

**Table 11. Metabolic, Hormonal, and Drug Influences on
Magnesium Excretion**

Increased excretion
Hypermagnesemia
Hypercalciuria
Hyperaldosteronism
Hyperparathyroidism
Familia renal wasting syndromes
Primary, Bartter's, and related syndromes
Potassium depletion
Alcoholism
Increased extracellular fluid volume
Phosphate depletion
Diuresis
   Osmotic (diabetes, glucose, mannitol)
   Postrenal obstruction
   Postrenal transplantation
Nephrotoxic drugs
   Amphotericin, cisplatin
   Aminoglycosides, cyclosporin
Acidosis
   Fasting, ketoacidosis, NH_4Cl
Mineralocorticosids
Hyperthyroidism
Decreased excretion
Hypomagnesemia
Parathyroid hormone
Hypocalcemia
Alkalosis
Hypothyroidism
Contracted extracellular fluid volume
Antidiuretic hormone
Calcitonin
Glucagon
K^+, Mg^{2+} sparing diuretics

chemically sensitive patients with renal magnesium leaks after exposure to pesticides and solvents.

Inadequate magnesium most severely affects cardiovascular, neuromuscular, and renal tissues. Therefore, anyone with spasms of coronary arteries, cerebral arteries, peripheral vessels, umbilical vessels, or who has myocardial infarction, hypertension, and preeclampsia will be suspect. Once a history

Table 12. Symptoms of Magnesium Deficiency
Seen in the Chemically Sensitive

Weakness	Tetany
Confusion	Anorexia
Personality changes	Nausea
Anxiety attacks	Lack of coordination[a]
Muscle tremor	Gastrointestinal disorders
Cramps	

[a] Uncoordinated muscle movements; twitches (especially in the face and eye muscles); alopecia; swollen gums; skin lesions; lesions of the small arteries; myocardial necrosis; spasm of coronary arteries, cerebral arteries, and umbilical vessels; arrhythmia with sudden death.

reveals a potential deficiency, laboratory work should be performed. Dietary evaluation of proper intake is also necessary. Intake of a diet low in magnesium for 3 months will lower serum magnesium, calcium, and potassium. These deficiencies normalize with magnesium therapy.

The symptoms of magnesium deficiency are shown in Table 12. Alcoholic hallucinations may be caused or aggravated by a magnesium deficiency, and high intakes of calcium can increase the severity of deficiency symptoms.

Serum deficiency is revealed in serum levels below 1.4 meq/L and will yield only about 5 to 10% of the deficiencies in chemically sensitive patients. Total blood intracellular magnesium will reveal another 10 to 15% of the deficient patients. Renal balance studies before and after oral or parenteral challenge will give the highest yield for this diagnosis.

Intravenous challenge is occasionally necessary to correct a total body deficiency, especially in chemically sensitive patients with muscle cramps and nervous irritability. Magnesium is an integral factor for vascular membrane function. Since the chemically sensitive have many problems with the vascular membrane and thus edema, magnesium deficiency has been studied at the EHC-Dallas over the last several years (Table 13).

In one study, magnesium deficiency in patients with chemical sensitivity was shown. Magnesium levels of 51 consecutive patients at the EHC-Dallas were evaluated. This evaluation was done by serum analysis, RBC analysis, and intravenous challenge. A total of 47 patients were allowed to eat their consistent diet, while 4 fasted. After informed consent, serum and RBC magnesium were drawn; 24-h urine for creatinine and magnesium was then measured before and after intravenous challenge; $^2/_{10}$ meq/kg of either magnesium chloride or magnesium sulfate was given in 500 cc of normal saline over a 4-h period. Signs and symptom scores were also measured before and after challenge. Challenge magnesium (urine) excretion values were compared with prechallenge 24-h values in addition to serum and RBC levels below 2 S.D. Of

**Table 13. Results of Magnesium Evaluation in 51
Chemically Sensitive Patients EHC-Dallas**

No. of Patients	Magnesium Levels	Control
1 Patient	Serum 2 S.D. below control	2 ± 0.6 meq.
5/33 (15%)	R.B.C. 2 S.D. below control	44 ± 5.0 ppm
37/47 (79%)	Mg excretion below 20%	
4/51 (7.8%)	No Mg excretion level measured[a]	
20/47 (42%)	Mg excretion between 60–80%	
12/47 (26%)	Mg excretion between 40–60%	
5/47 (11%)	Mg excretion below 40%	

[a] Four patients were unable to tolerate the magnesium challenge due to immediate severe reactions. The challenge was terminated and no excretion measurement was obtained.

the total nonfasting patients, 79% (37/47) were below 60% excretion; 25% (12/47) were below 80% excretion; 11% (5/47) were below 40% excretion; and 38% (18/47) showed immediate improvement in clinical response (Table 13). The most common response noted was improvement in neck/shoulder pain, muscle spasm, energy, sleep patterns, ability to clear sensitivity reactions, and decrease in tension and anxiety. Overall improvement with magnesium supplement seemed to be a significant improvement acutely and chronically and should be considered in selected patients. Those patients with chronic kidney leaks, for instance, sometimes need to have constant or intermittent intravenous supplementation.

Zinc

Proper zinc (Zn) balance is a necessary element in the treatment of the chemically sensitive, since it is essential in so many enzyme reactions and is abnormal in 7–54% of the chemically sensitive. Understanding its physiology is, therefore, paramount to understanding chemical sensitivity (Figure 24).

Physiology. Zinc is needed for wound healing,[386] protein,[387] CHO,[388] lipid,[389] and nucleic acid synthesis[390] and is part of over 90 metallic enzymes.[391] Zinc is needed for prostatic function and is often low in prostatism. Research has shown zinc supplements are capable of reducing lipid peroxidation,[392] probably through enhanced synthesis of superoxide dismutase, and is thus extremely important in reducing chemical sensitivity. Because zinc supplements can reduce lipid peroxidation, zinc loading has been used to stabilize cell membranes, a much needed process in the chemically sensitive since one of

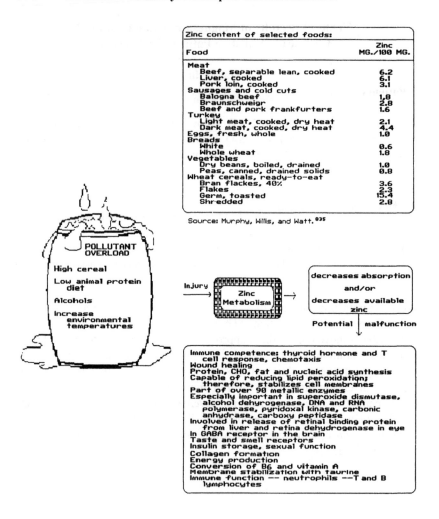

Figure 24. Pollutant injury to zinc metabolism.

their main problems appears to be membrane instability. Zinc has reduced the damage induced by carbon tetrachloride in animals.[393] Chapril et al.[394,395] have shown that the oral administration of zinc to intact rats markedly diminished injected carbon tetrachloride-induced liver damage. This decrease in damage was reflected in part by the reduced level of lipid peroxides formed in liver microsomes as well as enhanced endogenous peroxidation resulting from normal metabolism without in exogenous source of oxidants. Many of our chemically sensitive patients have been seen to have blood levels of chlorinated hydrocarbons, e.g., tetrachloroethylene, trichloroethylene, and trichloroethane (all similar to carbon tetrachloride), often coming from the fumes of their dry-cleaned clothes, in which zinc supplementation has helped.

Zinc, along with copper, Mn, and Fe, is in superoxide dismutase, which is an enzyme designed to trap carcinogens or prooxidants, and is thus necessary in the chemically sensitive. This enzyme occurs in two forms. The extramitochondrial form contains zinc and copper. The mitochondrial form contains manganese and iron (see Enzymes in this chapter). Zinc also catalyzes many metabolic reactions in the body and is an essential macronutrient.

Zinc is present in many NAD- and NADP-linked dehydrogenases, e.g., alcohol dehydrogenase in the liver catalyzes dehydrogenation of ethanol to yield acetaldehyde (contains two atoms of Zn^{2+}, which appear to bind the NAD^+ coenzyme to the active site of the enzyme). Often the chemically sensitive are intolerant to alcohol and will respond to Zn supplementation. Also, zinc is an essential component of DNA and RNA polymerase. It is apparent that zinc deficiency in any of these three enzymes could conceivably alter their function so that a jumping gene phenomenon occurs. McClintock[396] showed that this phenomenon of jumping genes occurs due to the response to the organism's attempt to adapt to its environment. This phenomenon could be the explanation for the sudden, overnight change in the onset of new chemical sensitivity.

Zinc is involved in the retinal binding protein in the liver and its release into the plasma. Zinc is in pyridoxal kinase, which is used to convert B_6. It is in retinal dehydrogenase in the retinal involved in the metabolism of vitamin A continuing visual pigments. A zinc-taurine conjugate acts as a membrane stabilizer. Again, membranes are prime targets for pollutant injury and almost universally damaged in the chemically sensitive. Zinc is necessary for GABA receptor function in the brain. The GABA receptor is where valium works. It may be that some of the anxiety and emotional lability seen in some of the chemically sensitive patients may be due to distortion of this receptor with zinc deficiency.

Zinc is present in carbonic anhydrase, which catalyzes the hydration of CO_2 and H_2CO_3, and is in the proteolytic enzyme carboxypeptidase, which is secreted into the small intestine. Zinc deficiency in some chemically sensitive people may be one explanation for the lack of sufficient buffering through this mechanism. Insulin is stored as a zinc complex. Zinc participates in the functioning of the taste and smell receptors of the tongue and nasal passages and other zinc-dependent digestive enzymes, including aminopeptidase. Zinc is a cofactor in the carnosine reaction.

Deficiency. A zinc deficiency has extensive metabolic effects in the chemically sensitive because of its wide variety of functions. Although the exact occurrence of zinc deficiency in the chemically sensitive is uncertain, deficiency ranges between 7 and 54%. Protein synthesis is impaired as is energy production. Collagen formation and alcohol tolerance are impeded. Alcohol intolerance and low energy are common in many of the chemically sensitive, and zinc deficiency may be one of the reasons for these.

These restricted functions alone result in diverse manifestations often seen in the chemically sensitive, including changes in hair and nails, dwarfism, sterility, skin inflammation, lethargy, anemia, poor wound healing, and a loss of taste and smell.

Zinc levels are suppressed during acute and chronic infections,[397] pernicious anemia,[398] alcoholism,[399] cirrhosis (zinc levels are 50% of normal),[400] renal disease,[401] cardiovascular disease,[402] some malignancies,[403] protein-calorie malnutrition,[404] and parenteral feeding.[405]

A prenatal deficiency increases the risk of spontaneous abortion and restricts fetal growth. Congenital malformations include changes in skeletal, brain, heart, gastrointestinal tract, eye, and lung tissue. The brain of a zinc-deficient infant contains less DNA than that of a healthy infant.

If a deficiency occurs during a period of rapid growth, the clinical manifestations (such as failure to grow or develop sexually) are more severe. Prostate gland, seminal vesicle and sperm degeneration from a zinc deficiency are reversible; testicular degeneration is not.

Zinc deficiencies seen in children in the U.S. suggest inadequate intake in additional segments of the population, which may increase the possibility of chemical sensitivity developing if the right set of environmental pollutants occurs. Zinc deficiencies may exist in preschool children, hospital patients, and low-income or elderly populations due to low dietary zinc. Athletes and strict vegetarians may also have depressed zinc levels. Contributing factors to a low trace mineral diet are low meat consumption combined with refined grains, convenience foods, and a high-fat, high-sugar intake. This legacy of the modern American diet makes Americans prone to chemical sensitivity.

A diet high in cereal and low in animal protein has produced zinc deficiency symptoms in Middle Eastern populations. The cause may be the high phytate diet, in which phytate binds with available zinc and reduces absorption. Geophagia and intestinal parasites common in these regions may also contribute to poor zinc absorption. Elevated environmental temperatures compound the problem, causing increased zinc loss through sweat.[406]

Many defects of the immune system have been documented in subjects with zinc deficiency,[407] including T cell response and chemotaxis of neutrophils. Thymic hormone activity is very low.[408,409] These characteristics are often found in the chemically sensitive.

Rogers[410] has shown that 54% of a selected group of chemically sensitive patients had red blood cell zinc deficiencies. Her studies reinforce ours in that zinc is an extremely important element in treating the chemically sensitive.

Sources. Animal foods are a good source of zinc. Excellent sources include oysters, herring, milk, meat, and egg yolks. The zinc in whole grains, even though it is not well absorbed, supplies a substantial contribution to the diet, especially for the vegetarian with a reduced protein intake. Fruits and vegetables are poor sources of the mineral.

Breast milk contains a zinc-binding protein that increases absorption in the infant's intestinal tract. The zinc in infant formula is not absorbed as well as the zinc in breast milk.[411]

Copper

The copper-to-zinc ratio is extremely important in the chemically sensitive for many reasons.

Copper (Cu) is a cofactor for superoxide dismutase and some mitochondrial respiratory chain components. Copper plays an important role in the catalytic activity of cytochrome oxidase. Thus, it is extremely important in detoxification reactions with pollutant overload in the chemically sensitive. The copper atoms of cytochrome oxidase undergo cyclic Cu (II) to Cu (I) valence transitions as they participate in carrying electrons to oxygen. One method of xenobiotic detoxification appears to depend upon the direct oxidation of sulfides in the presence of transition metals such as iron or copper. Metabolically available iron-containing compounds such as heme, ferritin, or several metalloprotein complexes stimulate H_2O oxidation. However, metal-catalyzed oxidation is too slow to protect against massive acute exposure. This copper balance is essential and often precarious in the chemically sensitive (Figure 25).

Most copper found in the blood is tightly bound to ceruloplasmin (protein). A small portion is loosely bound to amino acids (histidine, threonine, and glutamine) and to albumin. A key step in iron assimilation is the conversion of ferrous to ferric ions for incorporation into transferrin in plasma is attributed to the catalytic activity of ceruloplasmin. In serum the cuproprotein, ferroxidase II, also catalyzes the oxidation of ferrous ions. Copper is excreted through the bile, so increased Cu could mean biliary obstruction. Hypercupremia may be observed in some infectious states in some chemically sensitive patients. Redistribution of copper from liver to blood is believed to occur as a result of leukocytic endogenous mediator (LEM), which stimulates synthesis of ceruloplasmin in the liver.

Copper is also present in the active group of lysyl oxidase, an enzyme that makes the crosslinkage between polypeptide chains in collagen and elastin, and is also needed for proper utilization of iron in the body. Copper is an important activator of certain enzymes such as glutathione reductase, tyrosinase acid dioxygenase (pyruvic, tryptophan), and β-hydroxylase. Disturbance in those processes will exacerbate some cases of chemical sensitivity.

The copper ion has been shown to be a necessary factor in angiogenesis. For example, a diet depleted of copper has been shown to halt brain tumor growth in rabbits.[412]

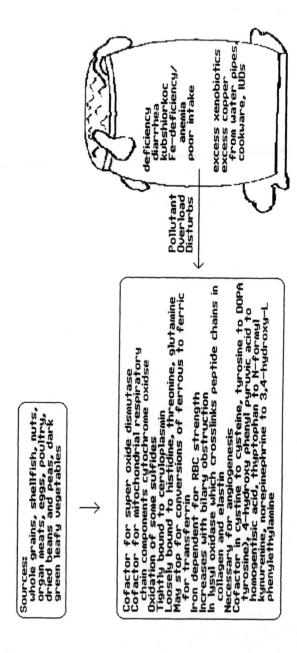

Figure 25.   Pollutant injury to copper.

Deficiency. Clinical copper deficiency is rare, but has been reported in children with kwashiorkor, chronic diarrhea, or iron-deficiency anemia. Inadequate copper intake is common, as are subclinical deficiencies (especially in hospital and parenteral feedings). Because copper is important to the normal development of nerve, bone, blood, and connective tissue, deficiency can result in a decline in red blood cell formation and subsequent anemia.

A copper deficiency may result in the following findings: a low white blood cell count associated with reduced resistance in infection; faulty collagen formation; fragile connective tissue that is easily damaged, resulting in damage to blood vessels, epithelial linings, and numerous other tissues; bone demineralization; central nervous system impairment because of reduced energy metabolism, disintegration of nerve tissue, or alterations in neurotransmitter concentrations; diminished skin pigmentation because of the role of copper in synthesizing melanin from tyrosin; copper deficiency anemia (seen in infants fed a cow milk diet exclusively after the first 3 months);[413] Menke's syndrome is a sex-linked recessive defect of copper absorption. The infants have retarded growth, defective keratinization and pigmentation of the hair, hypothermia, degenerative changes in aortic elastin, abnormalities of the metaphyses of long bone, and progressive mental deterioration. Brain tissue is practically devoid of cytochrome C oxidase, a cuproenzyme, and this syndrome shows a marked accumulation of copper in the intestinal mucosa. Parenteral administration of copper results in transient improvement.[414]

Copper deficiency has been seen in chemically sensitive patients. Probably it effects the cytochrome P-450 systems, exacerbating sensitivity. One case refractive to supplementation at the EHC-Dallas was noted to be associated with high levels of pentachlorophenol. Only after the pentachlorophenol decreased did the copper loss stabilize.

Copper Excess. Copper excess has been found with copper water pipes, copper coated, cookware, copper intrauterine birth control devices, and in Wilson's disease due to ceruloplasmin abnormalities. Copper excess disturbs copper-zinc balance and leads to zinc deficiency syndrome.

This disturbed balance of copper with zinc has been seen in the chemically sensitive often making it difficult to treat them because at times it is difficult to balance the ratio between the two.

Sources. Good dietary sources of copper are whole-grain cereals and breads, shellfish, nuts, organ meats, eggs, poultry, dried beans and peas, and dark green, leafy vegetables. Fresh and dried fruits and vegetables are moderate to poor sources. Milk and milk products are poor sources.[415]

Diagnosis. Serum and intracellular copper levels are available, as are ceruloplasmins. Intracellular levels surpass serum determination.

Phosphorus

Large quantities of phosphorus are found in bones. It is also mainly involved in energy metabolism as a part of ATP, which is the energy currency of the body as well as part of other nucleotides, nucleic acids, and various phosphorylated compounds. The chemically sensitive have energy problems, which indicates disturbed phosphorus metabolism and its relationship to ATP, which can be the prime target of pollutant injury. Phosphorus participates in activation reactions of all areas of metabolism. Intracellular concentrations are higher than extracellular, being present in 5 to 20 mM within various cell compartments. It is the main intracellular buffer. As part of phospholipids, phosphorus is responsible for the charge that makes them detergents. Intracellular concentrations are much higher than extracellular, since phosphorylated compounds do not easily cross cell membranes except with pollutant injury (Figure 26).

Phosphate detergents, organophosphate insecticides, and excess in phosphate drinks all have a potential to overload the phosphate pools within the body. Organophosphate insecticides in the phosphate pool can cause dysfunction of the cytochrome P-450 system and the generation of ATP and acts as a negative phosphate rather than a positive force in the body (see Chapter 4). We have observed at the EHC-Dallas that the chemically sensitive usually have severe problems handling these exogenous sources of phosphates. Avoidance of organophosphates, phosphates in sodas, and exposure to phosphate detergents helps stabilize the chemically sensitive.

Sources. Phosphate is abundant in processed foods, colas and other soft drinks, and organophosphate pesticides. Also, it is found in high-protein foods, like meat. Free organic phosphate is readily absorbed. Homeostasis is achieved by regulating excretion via the urine or gastrointestinal tract. The renal reabsorption of phosphate is increased by vitamin D, glucocorticoids, and growth hormone. It is decreased by estrogen, long-term glucocorticoids, thyroid and parathyroid hormones, as well as elevated plasma Ca^{2+}.[416] (See Table 1 in this chapter for more information on energy phosphate.)

Chromium

The role of chromium in the chemically sensitive is unknown, but since it is the most common deficiency in chemical sensitivity, its physiology should be studied.

Chromium (Cr) is essential in animal and human nutrition. In animals, Cr deficiency can cause a diabetic-like state,[417-421] impaired growth, elevated blood lipids,[417-419,421,423,424] increased aortic plaque formation, and decreased fertility and longevity. Cr functions biologically as an organic complex with nicotinic acid and amino acids named glucose tolerance factor (GTF). In

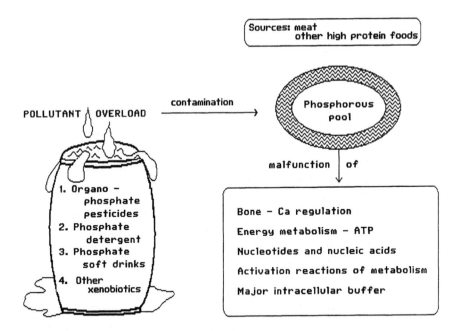

Sources: meat
        other high protein foods

POLLUTANT Λ OVERLOAD   —contamination→   Phosphorous pool

malfunction | of

Bone – Ca regulation

Energy metabolism – ATP

Nucleotides and nucleic acids

Activation reactions of metabolism

Major intracellular buffer

1. Organo –
   phosphate
   pesticides
2. Phosphate
   detergent
3. Phosphate
   soft drinks
4. Other
   xenobiotics

Figure 26.   Pollutant injury to phosphorous metabolism.

humans, any of these changes may or may not occur.[417-421,423-429] Both food Cr and Cr administered as trivalent Cr salts are absorbed at a rate no greater than 1%.[430] Use of stainless steel equipment in food processing and in cooking and eating utensils may add Cr to the diet, but it may not be utilized.[431]

In the human, Cr deficiency has been demonstrated unequivocally in only one clinical situation, patients on total parenteral nutrition without added Cr. In such patients, impaired glucose tolerance, hyperglycemia, relative insulin resistance, peripheral neuropathy, and a metabolic encephalopathy have been noted with reversal of the clinical phenomena by Cr repletion. Many studies have been performed to determine whether Cr deficiency may be important in other clinical conditions, namely, diabetes mellitus, pregnant and parous women, and the aged population. Available data indicate that Cr supplementation can improve glucose metabolism in glucose-intolerant individuals and decrease the total HDL cholesterol ratio regardless of the status of glucose tolerance. This improvement appears to be true in the chemically sensitive who have glucose intolerance. However, whether Cr supplementation has long-term health benefits is unknown. It is still unclear whether Cr deficiency, latent or overt, is common in any human situation other than generalized malnutrition and total parenteral nutrition without added chromium.[430]

Kumpulainen et al.[432] used newer instrumentation, improved background correction, and the method of standard additions to find that the normal serum Cr ranged from 0.1 to 0.2 ng/ml, normal urine Cr ranged from 0.06 to 0.20 ng/

ml, and 24-h urine Cr averaged 170 ng/day. These values are very similar to those which have been reported by others.[433-443] Bunker et al.[444] have reported valid Cr balance studies using similar methods. In a group of 22 elderly people, they found the self-selected Cr intake to average 24.5 µg/day, considerably less than the usually recommended 50 to 200 µg/day. However, the group as a whole was in Cr balance with an average net intestinal absorption of 600 ng/day (2.4% of intake) and an average urinary excretion of 400 ng/day. The apparent positive balance of 200 ng/day is within the error of the methods. Only one subject was in severe negative Cr balance. These data indicate that low Cr intakes can occur among the elderly, but they absorb twice as much Cr as generally thought and therefore do not incur negative Cr balance in the absence of other stresses.[445]

Lack of chromium has been the primary mineral deficiency in the chemically sensitive patient in a series of consecutive mineral analyses performed at the EHC-Dallas. Of the chemically sensitive, 89% have chromium deficiency. Chromium supplementation appears to aid in these patients.

Sources. Good sources of chromium are whole grains, brewers' yeast, pork kidney, meats, and cheeses. Little information exists on the chromium content of vegetables. Hard water can supply from 1 to 70% of the daily intake (Table 14).

Selenium

Proper selenium (Se) balance is necessary in the chemically sensitive because selenium has immune-stimulating properties that enhance the capacity of PHA blastogenic transformation of lymphocytes,[446] as evidenced by phytohemoagglutin (PHA) stimulation. Selenium is an important antagonistic nutritional element to many toxic chemicals such as mercury, especially methyl mercury, and it helps prevent against its toxicity. Glutathione peroxidase, which is often suppressed in the chemically sensitive, requires selenium and this enzyme. This enzyme converts glutathione disulfide and is, along with superoxide dismutase and catalase, in the first line of defense against pollutant injury. Frequently deficiencies, which exacerbate their problems are present in the chemically sensitive (Table 15).

Actions between vitamin E and selenium overlap. A study of the influence of selenium on ozone toxicity was reported by Chow.[447] He found that exposed rats reared on either selenium or vitamin E-deficient diets exhibited greater biochemical changes than animals given normal or supplemented diets. It is thought that calcium might serve as a "biologic carrier" for Se.[448] Toxicity of selenium occurs at too high an intake.

Studies in chemically sensitive patients often show a need for selenium from suspect diet and pollutant overload, both of which cause damage.

Table 14. Chromium

Sources:
Pork, kidney, brewer's yeast, meats, cheese

Specific Foods and Their Chromium Content:

Food	µg/g	µg/100 kcal
Vegetable oils		
Margarine, corn oil	0.23	2.56
Corn oil	0.12	0.33
Cottonseed oil	0.05	1.00
Safflower oil	0.07	0.80
Butter, unsalted	0.21	2.30
Grains		
Buckwheat	0.38	11.0
Wheat	0.03	0.8
Cereal Products		
All bran	0.25	8.1
Puffed rice	0.71	18.2
Wheat germ	0.07	2.0
Molasses	0.22	10.0

↓

Chromium Function:

Aids in glucose metabolism (TF)
Decreases HDL/cholesterol

Source: Schroeder, Nason, and Tipton.[836]

Some chemically sensitive patients have been seen to improve with supplementation.

Deficiency. Poor soil deficiency symptoms similar to those of vitamin E deficiency have been reported in animals raised on selenium. Laboratory-induced muscular dystrophy in lambs has been treated with selenium and vitamin E. And hepatic necrosis in rats, produced by a diet low in cystine and vitamin E, has been found to be only partially responsive to the two nutrients, although very responsive to selenium.

In humans, a selenium deficiency in the soil and water has resulted in cardiomyopathy and myocardial deaths. Keshan cardiomyopathy in China is

Table 15. Selenium: Estimated Human Daily Intake of Selenium from Dietary Sources (μg/day)[a]

Sources: Country	Plant		Animal		Total	Excess → intoxication Lack → heart disease
	Vegetables, Fruits, Sugars	Cereals	Dairy Products	Meat, Fish		
Canada	6.9	74.0	23.4	46.0	150.7	
Canada, Halifax	7.4	105.0	21.8	90.0	224.2	
Canada, Toronto-1	5.1	62.0	6.5	24.7	98.3	
Canada, Toronto-2	1.3	111.8	5.0	30.4	148.5	
Canada, Winnipeg	9.1	79.8	27.6	64.3	180.8	
Egypt	2.0	17.0	1.0	1.0	21.0	
Japan	6.5	23.9	2.3	55.6	88.3	
Japan	34.8	72.7	5.7	94.6	207.8	
New Zealand	1.0	4.0	11.0	12.0	28.0	
New Zealand	5.8	4.3	8.4	37.7	56.2	
United Kingdom	4.0	30.0	4.0	22.0	60.0	
U.S., Maryland	5.4	44.5	13.5	68.6	132.0	
U.S., "Market Basket"	2.1	102.9	4.7	58.8	168.5	
U.S., S. Dakota	10.4	57.0	47.6	101.4	216.4	
Venezuela	14.6	88.2	70.4	152.6	325.8	

Source: Diplock and Chaudhry.[837]

[a] Increases blastogenesis; antagonistic to Hg; essential for glutathione peroxidase; function overlaps with vitamin E.

Table 16. Selenium

| Place — China | Daily Selenium Intake | | |
	Minimum	Maximum	Average (μg)
High-selenium area with a history of intoxication reported as chronic selenosis	3200	6690	4990
High-selenium area reported as without selenosis	240	1510	750
Moderate-selenium area (Beijing)	42	232	116
Low-selenium area with Keshan disease	3	22	11

Source: Yang, Wang, Zhou, and Sun.[838]

Selenium and Cancer: Prospective Studies

	U.S., Willett et al. 1983[a]		Finland, Salonen et al. 1985[b]	
Plasma Se (ng/ml)	>154	<115	>47	<47
Relative risk of cancer	1	2.4	1	5.8
Total cancer (/10^5)	406		362	

Source: Casey.[839]
[a] See Willett.[840]
[b] See Salonen.[841]

prevented with selenium supplementation. Low-selenium soil has also been associated with an increased risk of cancer (Table 16).

Sources. Excellent sources of selenium are liver, kidney, meats, and seafood. Grains and vegetables will vary in their selenium content, depending on the soil in which they were grown.[449]

Manganese

While limited information is available on the effects of manganese (Mn^{2+}) in chemical sensitivity, the fact that it is essential in some enzyme reactions which are necessary to the prevention of chemical sensitivity makes understanding its basic physiology important (Table 17).

Table 17. Manganese

Sources
   Liver, kidney, spinach, muscle meats
   Whole grain cereals, dried beans and peas, nuts

Functions
   Enhances intestinal absorption when combined with L-histadine or citrate
   Component of tisssue-superoxide dismutase
   Associated with some of the enzymes of mucopolysaccharide,
   glycoprotein and lipopolysaccharide production, glycosyl transferases
   Component of arginase (arginine-urea hydrolyase) for conversion of
   arginine to urea
   Cofactor in some phosphate transferring enzymes (pentose phosphate shunt)
   Cofactor in glycine to serine; isocitric to a-ketoglutaric
   Pyruvic to oxaloacetic acid
   Pancreatic and brain function

The enzyme arginase, which hydrolyzes arginine to form urea, an end-product of human amino group metabolism, contains Mn^{2+} which is essential for its activity. Mn^{2+} is also a component of tissue superoxide dismutase. Mn^{2+} also serves as a cofactor of some phosphate-transferring enzymes and in the enzymatic reaction by which oxygen is produced during plant photosynthesis in chloroplasts. Blood distribution is via transferrins. Major organs of accumulation are liver, kidney, bone, pancreas, and the pineal, pituitary, and lactating mammary glands.

Mn^{2+} is implicated in melatonin and dopamine production, in fatty acid synthesis, and in formation of membrane phosphatidal inositol. Chelation with bile constituents in the intestinal tract probably aids absorption.[450]

Davis et al.[451] studied the effect of EDTA on growth and the severity of leg abnormalities in chicks fed diets containing isolated soybean protein and supplemented with various levels of Mn. They found definite evidence that 0.07% EDTA improved growth and reduced the severity of perosis when added to a diet marginal in Mn.

There are differences between human milk, cow milk, and infant formula in the ligands which bind Mn.[452-454] Some of the ligands are proteins with molecular weights in the range of 80,000 to 407,000, but some low-molecular-weight Mn-binding ligands are also present in cow milk and infant formula. The intestinal absorption of Mn in the rat in the presence of either L-histidine or citrate was absorbed more rapidly from the ileum or jejunum than when no ligands were included in the Mn-containing perfusion solution.[455] The kinetics of the absorption process were calculated and were compatible with a high-affinity, low-capacity, active transport mechanism for Mn absorption in the rat intestine. These data suggested a limited role for small-molecular-weight ligands associated with both diffusion and active transport.[456]

In rats, a manganese-deficient diet produces sterility and testicular degeneration. Manganese-deprived pregnant rats produce weak offspring with poor survival rates. The surviving offspring show growth retardation and abnormal otoliths of the inner ear, resulting in poor balance, convulsions, and epilepsy-like seizures. Lactation is also impaired. Guinea pigs deprived of manganese in utero develop a small pancreas and reduced glucose tolerance. In chicks, shortened legs and vertebral columns are the result of a manganese deficiency.[457]

Deficiency. Manganese deficiency has substantial effects on the production of hyaluronic acid, chondratin sulfate, heparin, and other forms of mucopolysaccharides that are important in growth and maintenance of connective tissue, cartilage, and bone.[458] Other effects of Mn deficiency are an impairment of glucose tolerance and B cell granulation of the pancreas. Manganese-deficient rats are unable to lactate. Manganese is in glutamine synthetase, an essential enzyme for brain function.

Magnesium can substitute for many of the functions of Mn.

Sources. The richest sources of manganese are liver, kidney, lettuce, spinach, muscle meats, tea, whole-grain cereals and breads, dried peas and beans, and nuts. Moderate amounts are found in leafy green vegetables, dried fruits, and the stalk, root, and tuber parts of vegetables. Small amounts are provided in meats, fish, and other animal products.[459] Mn is less abundant in the body than Mg, Fe, Zn, or Cu. The average American consumes 3 to 4 mg/day.

Diagnosis. Intracellular manganese analysis seems to be the best method of diagnosis at the present time. Some chemically sensitive patients with manganese deficiency have responded with supplementation. Those with excess manganese are a problem to treat.

Iron

Iron (Fe) is extremely important in preventing an exacerbation of chemical sensitivity due mainly to its metabolic activity (Figure 27). The most important is in the cytochrome P-450 system where it is an integral part. Low ferritin levels are seen in many chemically sensitive patients at the EHC-Dallas. Supplementation decreases their sensitivity.

Iron is needed to catalyze several reactions in the body in addition to being used in the Hb molecule.

$$\beta\text{-carotene} \xrightarrow[\text{Nicotinamide}]{\text{Fe, NADH, O}_2} \text{Vitamin A}$$

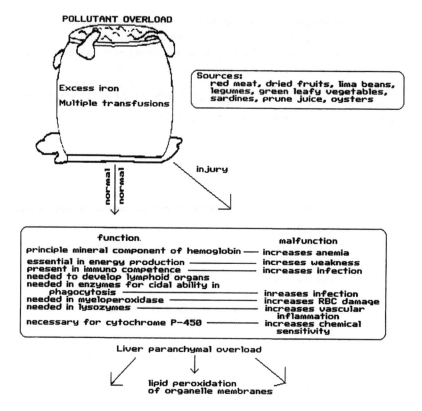

Figure 27. Iron.

If β-carotene is excessively high and vitamin A is at its lower end, the above reaction would be suspected to be deficient, resulting in vitamin A deficiency as is seen in 15% of our chemically sensitive patients.

$$\text{Acetaldehyde} \xrightarrow[\substack{\text{Fe, Mo}}]{\substack{\text{NAD}\quad\quad\quad\text{NADH} + \text{H}^+ \\ \text{Aldehyde dehydrogenase, FAD} \\ \text{H}_2\text{O}}} \text{Acetic acid}$$

$$\text{Tyrosine} \xrightarrow[\substack{\text{Fe}^{+++}}]{\substack{\text{O}_2\quad\quad\quad\text{H}_2\text{O} \\ \text{Tyrosine hydroxylase}}} 3,4\text{-Dihydroxy-L-phenylalanine (DOPA)}$$

$$\xrightarrow[\substack{\text{NADH}\quad\text{Dihydropteridine}\quad\text{NADPH} \\ \text{reductase}}]{\substack{\text{BH}_4\quad\quad\quad\quad\quad\text{BH}_2}}$$

Iron Metabolism. The concentration of iron in the human body is normally about 40 to 50 mg Fe per kilogram of body weight; women typically have lower and men higher amounts. Most of this iron, about 30 mg Fe per kilogram, is contained within circulating red cells as hemoglobin; an additional 5 to 6 mg Fe per kilogram is present in tissues throughout the body in functional form in a variety of heme compounds (myoglobin, cytochrome), enzymes with iron-sulfur complexes, and other iron-dependent enzymes. The remainder of the iron (5 mg Fe per kilogram in women; 10 to 12 mg Fe per kilogram in men) is stored as ferritin and hemosiderin in the liver, bone marrow, spleen, and muscle, serving as a readily available reserve in the event of blood loss. Only a small fraction of the body iron (about 3 mg) circulates in plasma carried by the iron transport protein, transferrin.[460]

Iron balance is normally regulated by controlling iron absorption; iron stores and iron absorption are reciprocally related so that as stores decline, absorption increases. In some chemically sensitive patients, this regulation appears to be disturbed with a deficiency resulting, but not being severe enough to affect hemoglobulin synthesis. The rate of erythropoiesis is also a major determinant of iron absorption, with increased erythropoietic activity linked to enhanced iron absorption. Iron is absorbed through the upper intestine under the regulation of the intestinal mucosa. The amount and bioavailability of dietary iron, lumenal pH and motility, and other factors influence but do not regulate absorption. The means whereby iron is transferred from the lumen to mucosal cells remain unknown. In some animals a lumenal apotransferrin seems to act as a shuttle for iron, but there is no evidence of a physiological role for transferrin in the absorption of iron in humans.[461] The body lacks any effective means to excrete excess iron. Iron exchange is limited so that an adult man absorbs and loses only about 0.01 mg Fe per kilogram per day. A woman of child-bearing age loses slightly more iron because of menstrual flood flow, and in balance absorbs a larger amount so that iron exchange is about 0.015 mg per kilogram per day.[462] This balance is often disturbed by malabsorption or inability to transport, as evidenced by low serum ferritin without anemia in the chemically sensitive.

Stored iron is usually present in roughly equal amounts in the macrophages of the reticuloendothelial system, in hepatic parenchymal cells, and in skeletal muscle. The concentration of storage iron in skeletal muscle is low (15 to 30 $\mu g/g$), but because of the large mass present (about 30,000 g in an adult man), about one third of the body reserve is normally found in muscle. With iron overload, muscle storage iron is increased, but not as much as are hepatic or bone marrow stores; the muscle storage iron seems relatively nonmiscible.[463] Reticuloendothelial cell iron is derived almost entirely from phagocytosis of senescent erythrocytes or defective developing red cells, with the exception of the iatrogenic source of parenteral iron preparations. The phagocytized iron is recycled to plasma transferrin and then transported to the erythroid marrow for use in hemoglobin production. Normally, only a small portion of the transferrin iron enters hepatocytes, which can also derive iron from methemalbumin, hemoglobin-haptoglobin, and heme-hemopexin complexes formed after intra-vascular hemolysis. Ferritin is also taken up by liver parenchymal cells. In the normal individual, the overall extent of iron exchange by hepatocytes is much less (about one fifth) than that by the reticuloendothelial cells.[464,465]

Deficiency. Iron-deficiency anemia is a major nutritional concern worldwide, and its prevalence may range as high as 50% in some segments of the population. About one in every four college women in America has depleted iron stores. Iron may be a major influence on the development of chemical sensitivity as it is essential to the function of P-450, without which the body cannot detoxify. The reduced iron supply to tissues results in diminished energy production and the characteristic symptoms of lethargy, tiredness, apathy, reduced brain function, pallor, headache, heart enlargement, spoon-shaped nails, depleted iron stores, and a plasma iron of less than 40 meg/100 ml.[466] These symptoms are often seen in the chemically sensitive.

Epidemiological, clinical, and experimental studies provide conflicting evidence for the relationship between iron deficiency and immunocompetence.[467] Clearly, lack of iron affects the chemically sensitive both in immune competence and the cytochrome P-450 system. Lovric[468] found that 20% of the children with infections admitted to hospitals had anemia from iron deficiency, compared to the 3% rate of anemia in age-matched children. A retrospective analysis of hospital records showed that about 66.8% of iron-deficient children had first sought medical advice on account of intercurrent infections.[469] Recent clinical studies demonstrate the occurrence of opportunistic infections among iron-deficient subjects and support the hypothesis of altered immunocompetence in these subjects as a cause for increased susceptibility to infection. This vulnerability to infections is seen in a subset of the chemically sensitive.

Mucocutaneous candidiasis and recurrent herpetic lesions are typical infections[470] in iron-deficient subjects, which signifies the role of iron in cellular immune responses.[471-473]

Iron deficiency during the initial development of lymphoid organs may cause permanent damage to the tissues, limiting the ability of the host to mount an effective immune response. Pathogenesis of immunological changes in iron deficiency most likely includes several factors that compromise cellular and humoral immune functions. A number of iron-requiring enzymes are involved in bactericidal function of phagocytic cells.[474] Leukocytes of iron-deficient patients have less myeloperoxidase activity than those of normal individuals.[475] Lysozyme, a bacteriolytic enzyme, is another nonspecific host defense mechanism. Changes in its activity during iron deficiency may also have an effect on susceptibility to bacterial infections. Total DNA was decreased by iron deficiency during gestation and lactation or during the post-weaning period. The cellular basis of both B and T cell function can be dramatically altered during neonatal iron deficiency and may provide a mechanism for the impairments in both CMI and HI that have been reported in iron deficiencies.

By value of its role in cell division, iron may also play a more direct role in CMI as well.[476]

Probably iron is present in superoxide dismutase. Because iron is necessary for the cytochrome P-450 system to function, it is essential in chemically sensitive patients. Therefore, its absence or depletion may negatively influence them. In our experience, use of iron supplementation to restore serum ferritin levels without anemia to normal often decreases chemical sensitivity. Cytochrome P-450 is discussed in detail in Chapter 4.

Causes of Iron Overload. Iron overload can result from increased absorption, excess parenteral administration, fecal sequestration, and hereditary disease in young.[477]

Iron overload, long considered a rarity, is now recognized as a common disorder of iron metabolism and is a form of pollutant injury. In the U.S., a genetically determined form of iron loading affects as much as 0.5% of the population, or a million individuals or more.[478,479]

In adults and children, iron overload may be produced by an increased absorption of dietary iron, by parenteral administration of iron, or both. An increased absorption of iron may be the result (1) of an inappropriately elevated uptake from a diet with normal amounts of iron, as occurs in hereditary hemochromatosis, the iron-loading anemia, and other conditions; or (2) of consumption of large amounts of bioavailable iron, as is found in some regions in Africa and, possibly, with prolonged ingestion of medicinal iron. Parenteral iron loading is produced by repeated blood transfusion or, less often, by injections of therapeutic iron preparations. Iron overload has also been recognized in the neonate and infant, presumably as the result of a disturbance in the regulation of maternal-fetal iron balance, but the pathogenesis of such disorders is still uncertain. Rarely, sequestration of iron, as occurs in pulmonary hemosiderosis, may lead to a focal iron overload.[480,481]

The distribution of the excess iron between relatively benign reticuloendothelial sites and potentially toxic parenchymal locations, internal redistribution of iron, amounts of circulating nontransferrin-bound iron, ascorbate status, and other factors also influence the extent of tissue damage.

Mechanisms of Iron Toxicity. Several mechanisms have been proposed whereby excess hepatic iron could cause cellular injury with resultant fibrosis and cirrhosis. One hypothesis suggests that increased iron-induced lipid peroxidation results in lysosomal and membrane fragility with cerebral rupture, which is responsible for cellular injury in iron overload.[482-489] Peroxidative injury to isolated hepatic lysosomes by iron salts has been demonstrated,[490,491] but the relevance of laboratory observations of increased lysosomal fragility to hepatic fibrosis is still uncertain.[492]

Iron-induced peroxidation of membrane lipids of subcellular organelles other than lysosomes, such as mitochondria and microsomes, leading to functional insufficiency with subsequent cell injury and death is an alternative possibility.

Iron-stimulated collagen biosynthesis has been proposed as a possible explanation for fibrosis that does not necessarily require prior iron-mediated cellular damage.[493] Iron-induced damage to nucleic acids could produce cell injury and be a cofactor or cause of neoplasia.[494] The toxic effects of non-transferrin-bound iron (plasma iron not complexed with transferrin) have also been considered as potentially responsible for both hepatic damage[495] and iron loading[496] as well as for injury to other tissues.[497,498] All of these proposed mechanisms could play a role in exacerbation of clinical sensitivity.

Several clinical reports have suggested that an increased availability of iron might be pathogenetically related to infections with certain organisms, including *Vibrio vulnificus, Listeria monocytogenes, Yersinia enterocolitica, Escherichia coli,* and *Candida* species.[499-503] Also, ascorbic acid deficiency and osteoporosis have been reported.

Sources. Excellent sources of iron are liver and other organ meats, beef, dried fruits, lima beans, ham, legumes, dark green, leafy vegetables, sardines, prune juice, and oysters. Good sources are whole-grain cereals and breads, tuna, green peas, chicken, strawberries, egg, tomato juice, enriched grains, brussels sprouts, winter squash, blackberries, pumpkin, nuts, canned salmon, and broccoli. Small amounts are supplied by potatoes, applesauce, corn muffins, peanut butter, watermelon, corn, pears, and peaches. Cooking acidic foods in cast iron pots can increase the iron content 30-fold. Iron intake can also be increased by combining heme and nonheme food sources, consuming vitamin C-containing foods with each meal and selecting iron-rich foods.[504]

Diagnosis. Low levels of serum ferritin, iron-binding capacity, or total serum iron indicate iron deficiency whereas high levels indicate iron excess. Either can result in chemical sensitivity, one from low antipollutant enzyme (cytochrome P-450 system), the other by giving toxic pollutant overload.

Molybdenum

Molybdenum (Mo) is important in the chemically sensitive in four and possibly five different enzyme reactions. Several studies[505-507] have shown that molybdenum gives a growth response in chickens and other poultry and probably humans. Tungsten, a known Mo antagonist, is required to antagonize an equivalent amount of Mo.[507-509]

Other minerals, such as tungstate, fluoride, sulfate, and Cu, can effect the Mo requirement as well as Cd, Mn, Zn, and Fe which can interact indirectly.[510]

Molybdenum is essential in the function of four and possibly five enzyme systems, all of which effect normal function in the chemically sensitive (Table 18). These systems are the catalytic role of xanthine oxidase in uric acid

Table 18. Molybdenum

Sources
Hardwater, meats, whole grains, legumes, leafy vegetables

Functions
Mo counterbalances silicon

Molybdenum-Containing Enzymes

Enzyme	Substrate	Electron Donor or Acceptor
Aldehyde oxidase	Aldehydes	O_2
Sulfite oxidase	SO_3^{2-}	O_2
Xanthine dehydrogenase	Purines	NAD
Xanthine dehydrogenase	Purines	Ferredoxin
Xanthine oxidase	Purines	O_2

formation and the aldehyde oxidase role in the oxidation of various aldehydes, the sulfide oxidase and nitrate reduction xanthine dehydrogenase.

$$\text{Acetaldehyde} \xrightarrow[\text{Mo, Fe}]{\text{Aldehyde dehydrogenase}} \text{Acetic acid}$$

NAD → NADH + H$^+$; H$_2$O

$$\text{Sulfinylpyruvic acid} \xrightarrow{\text{(Enzyme uncertain)}} \text{Pyruvic acid (cytosol)}$$

Both enzymes contain FAD (a riboflavin enzyme) and are important in electron transport. Xanthine oxidase is also important in converting iron from the ferrous to the ferric form. Therefore, molybdenum, like copper, is necessary in iron metabolism. Sulfide oxidase and xanthine dehydrogenase are the other molybdenum-dependent reactions.

Molybdenum is sensitive to sulfur metabolism. Inorganic sulfate or endogenous sulfur from amino acids can affect the tissue concentration of the mineral. An increased sulfur intake causes a decline in molybdenum status.

$$SO_3 = \xrightarrow[Mo^{6+}]{\text{Sulfite oxidase}} SO_4 =$$

H_2O ; $O_2 + H_2O$; H_2O_2

Molybdenum can interfere with copper absorption, as the two minerals compete for similar absorption sites in the intestines. It is excreted in the urine and bile.[511]

Deficiency. Molybdenum-deficient kids weighed 6% less than the controls and had growth depression and reproduction disorders under the conditions of molybdenum intake of 24 μg/kg ratio of dry matter.[512]

In two recent human studies, a lethal inborn error in metabolism that markedly deranged cysteine metabolism in a patient was determined to be a sulfite oxidase deficiency. Another patient on prolonged total parenteral nutrition (TPN) therapy acquired a syndrome described as acquired molybdenum deficiency. This syndrome, exacerbated by methionine administration, was characterized by hypermethioninemia, hypouricemia, hyperoxypurinemia, hypouricosura, and very low urinary sulfate excretion. In addition, the patient suffered mental disturbances that progressed to coma. The symptoms were indicative of a defect in sulfur amino acid metabolism at the level of sulfite transformation to sulfate (sulfite oxidase deficiency). Supplementation of the patient with ammonium molybdate improved the clinical condition, reversed the sulfur handling defect, and normalized uric acid production.[513] Defects in sulfur amino-requiring metabolism Mo supplementation have been seen in some patients with chemical sensitivity at the EHC-Dallas.

Sources. Dietary intake of molybdenum depends on the status of the soil on which grains and vegetables are raised. Plants grown on molybdenum-rich soil may contain 500 times as much of the minerals as plants grown on depleted soil. Hard water can provide up to 41% of daily intake. Meats, whole grains, legumes, leafy vegetables, and organ meats are other good sources of molybdenum.[514]

Diagnosis. Diagnosis is by empirical challenge and intracellular mineral levels.

Treatment. Some chemically sensitive patients have responded to Mo supplementation.

Cobalt

Cobalt (Co) is essential to erythropoiesis in the human body because it is a constituent of cobalamin (vitamin B_{12}). Cobalt has several functions (Table 19) including acting as a substitute for manganese in the activation of several enzymes (such as glycylglycine dipeptidase); replacing zinc in some enzymes (such as carboxypeptidase A and B and bovine, human, and monkey carbonic arrhydrase); activating phosphotransferase and other enzymes (even though these enzymes are activated in the presence of other metals or in the absence

Table 19. Cobalt

Sources	Figs, cabbage, beet greens, buckwheat, lettuce, watercress
Function	Component of B_{12} Can be substituted for Mn, Zn in enzymes Activities phosphotransferase and other enzymes In biotin-dependent oxaloacetic transcarboxylase Erythropoesis

of any metal); and participating in the biotin-dependent oxalacetate transcarboxylase.

Because of its relationship with vitamin B_{12}, cobalt must be absorbed as a component of B_{12}. The amount absorbed is stored in the liver and kidney, with a reserve of 0.2 ppm of dry weight. The major route of excretion of cobalt in humans is the urine, and small amounts of cobalt are also lost by way of the feces.[515]

Excess cobalt is a potential pollutant in humans. Turk and Kratzer[516] have found that Co is more toxic to chickens when added to a purified diet than to a practical diet of conventional feedstuffs. Toxicity symptoms were emaciation, debility, inanition, sensitivity to being touched, poor growth, and mortality. Alleviation of the toxicity of Co was obtained by the addition of EDTA, cystine, or cysteine, but not methionine. This alleviation indicates that a chelation phenomenon binds the Co to reduce its toxicity for the chick. It is probable that the partial protection from the diet of conventional feedstuffs was due to a component of the diet which has Co-binding properties.[517]

Deficiency. Low cobalt levels create different reactions in animals. Cattle and sheep grazed on cobalt-deficient lands become emaciated and anemic, whereas horses raised on the same land show no deficiency symptoms. Any deficiency is ultimately a vitamin B_{12} deficiency, and administration of the vitamin alleviates the condition. An excess intake of molybdenum may interfere with vitamin B_{12} synthesis in the rumen of cattle. At this time, the effects of low cobalt on the chemically sensitive are undetermined.

Sources. Foods containing about 0.2 ppm cobalt are figs, cabbage, spinach, beet greens, buckwheat, lettuce, and watercress.[518]

Vanadium

The role of vanadium (V) is unknown in chemical sensitivity, but as with many trace elements, knowledge of its functions may throw some light on the problem (Table 20).

Table 20. Vanadium

Sources	Dill, black pepper, spinach, parsley, mushrooms, shellfish, oysters
Function	Regulation of Na, K, Ca, ATPase, and Na pump
	Regulation of phosphoryl transfer enzymes
	Regulation of adenylcyclase — stimulates synthesis of ATP
	Enzyme cofactor
	Glucose metabolism — mimics actions of insulin
	NADH oxidation $\rightarrow$ NADH — vanadate oxireductase
	Enhances DNA synthesis
	Inhibits the rate of receptor hormone complex
	Anticarcinogenic

It is clear that the toxicity of V is greatly influenced by the diet in which it is fed. A purified diet or the addition of skim milk, sucrose, or lactose to the diet increases the toxicity of V. It is probable that the effect of skim milk is due to its lactose component. Ascorbic acid and EDTA, on the other hand, reduce the toxicity of V. While these are distinct differences in the availability of the element, it is difficult to explain these positive and negative effects of chelation. The effect of Cr in reducing the toxicity of V is predictable from the Hill and Matrone[519] proposal that "those elements whose physical and chemical properties are similar will act antagonistically to each other biologically."[520]

Vanadium is active in vitro, and in vivo, physiologic functions for this element have been suggested. These functions include regulation of (Na,K)-ATPase and the sodium pump. In vitro, vanadate inhibits (Na,K)-ATPase activity, and reduction of vanadate to vanaolyl reverses that inhibition. Because of this action, it has been hypothesized that in vivo vanadium might function as a physiologic regulator of sodium pump activity. ATPases (Ca-ATPase, gastric mucosa and colon epithelium H^+, K^+-ATPase, and myosin ATPase are inhibited by physiologic concentrations of vanadium in vitro), phosphoryl transfer enzymes (glucose-6-phosphatase, alkaline, phosphatase, acid phosphatase, and phosphoglucomutase are inhibited by physiologic concentrations of vanadate in vitro), and adenylate cyclase (vanadium stimulates the synthesis of cyclic AMP in various cell membranes through the activation of adenylate cyclase) are inhibited and use vanadyl as an enzyme cofactor. Other than a possible regulatory role through vanadate and redox mechanisms, vanadium as the vanadylcation might be a cofactor for some enzymes. When vanadyl replaces other metals in metalloproteins, the metals replaced include Zn, Cu, and Fe; thus, it is possible that vanadyl has a role similar to that of these cations. In vitro studies have shown that vanadium might affect glucose metabolism by altering or mimicking the action of insulin or by altering the activity of the multifunctional enzyme glucose-6-phosphates. Pharmacologic levels of dietary vanadium improved oral glucose tolerance in studies on guinea pigs.

Pharmacologic levels of dietary vanadium also prevented an increase in glucose in blood, despite low insulin, and prevented the decline in cardiac performance of rats made diabetic with streptozocin.[521] Erdmann et al.[522] reported that cardiac and erythrocyte cell membranes contain a NADH-vanadate oxidoreductase, which reversibly converts vanadate to vanadyl.[523] Jones and Reid[524] and McKeehan et al.[525] reported vanadium can enhance DNA synthesis; vanadate inhibited the rate of transformation of receptor-hormone complex to the activated form. In rats, vanadium metabolism is disturbed by endocrine deficiency induced by hypophysectomy or thyroidectomy-parathyroidectomy.[526] Vanadium has been reported to inhibit cholesterol biosynthesis in human and animal organs.[527] This inhibition was accompanied by decreased plasma phospholipid and cholesterol levels and by reduced aortic cholesterol concentrations and anticarcinogenic action. The induction of murine mammary carcinogenesis in rats by 1-methyl-1-nitrosourea was blocked by feeding 25 μg of vanadyl per gram of diet during the postinitiation stages of the neoplastic process. The vanadyl sulfate treatment reduced both tumor incidence and the average number of tumors per rat and prolonged the median cancer-free time without inhibiting overall growth of the animals.[528]

Deficiency. Vanadium deficiency in rats and chicks produced reduced growth, poor reproductive performance, changes in hematological parameters, bone defects, and alterations of lipid metabolism. Basing his conclusions on the effects seen in animals, Czajka-Narins has suggested that the most important effect on humans may be the influence on lipid metabolism.[529]

Sources. Myron et al.[530] used atomic absorption to show that beverages, fat and oil, fresh fruits, and vegetables contained the least vanadium, ranging from <1 to 5 ng/g. Whole grains, seafood, meats, and dairy products were generally within a range of 5 to 30 ng/g. Prepared foods ranged from 11 to 93 ng/g, while dill seed and black pepper contained 431 and 987 ng/g, respectively. Spinach, parsley, mushrooms, and oysters contained relatively high amounts of vanadium. Shellfish apparently are a rich source of vanadium because several types were found to contain >100 ng/g vanadium on a fresh basis.[531]

Silicon

Although the role of silicon (Si) is unknown in the chemically sensitive, it is deficient in 33% of the patients. Therefore, attempts to understand its function are necessary (Table 21).

Silicon is widely distributed in nature and is the second most abundant element in the earth's crust. Silica $(SiO_2)_n$ occurs in nature in several different

Table 21. Silicon

Sources	Oats, legumes
Function	Essential for bone calcification in prolyl hydroxylase Molybdate counteracts silicon and vice versa Phytolith in lymph nodes, lung, liver, spleen Present in diatomaceous earth Aluminum relationship in Alzheimer's

crystalline forms (quartz, cristibalite, and tridymite) and as a supercooled amorphous liquid glass.[532,533] The sodium and potassium salts of silicic acid (H_4SiO_4) are soluble and can be precipitated with acid to yield SiO_2 $[X]H_2O$, which can be partially dehydrated to a silica gel of importance in chromatography. Magnesium trisilicate $(Mg_2Si_3O_8)$, an insoluble hygroscopic salt of a partially condensed silicic acid, has been used as an antacid in man for many years.[534]

The highest levels of silicon are found in the epidermis and its appendages and in connective tissues in general. The eggs of birds, milk, and the fetuses of mammals have small quantities. The blood of humans and other mammalian species averages about 0.5 mg/L of blood plasma, a level that is not significantly increased by the inhalation of silica dust. Dietary silicon supplements have been reported[535] to have little effect on the silicon concentration of cow milk. Moderate increases have been obtained in rat blood, however, after feeding silicon as sodium metasilicate, and much higher levels have been reached after feeding organic silicates. The consistently low concentrations of silica in most organs do not appear to vary appreciably during life. Parenchymal tissues such as liver, heart, and muscle, for example, range from 2 to 10 mg/kg. The lungs are an exception. Varying amounts of silica entering the respiratory tract normally cross the barrier of the lung as silicic acid, which is eventually eliminated. Nevertheless, the lungs ordinarily accumulate large amounts of silicon from long-continued inhalation of finely particulate silica.

Silicic acid in foods and beverages is readily absorbed across the intestinal wall and is rapidly excreted in the urine. Silicon absorption studies using intestinal ligature techniques showed that the level of silicon in blood and intestinal tissues of male and female rats is affected by age, sex, castration, adrenalectomy, and thyroidectomy.

Silicon is essential in bone formation, cartilage formation, connective tissue matrix, collagen, and glycosaminoglycans.[536-543]

In humans, it was shown in an earlier study[544] that silicon levels gradually increase with age in the human peribronchial lymph nodes, even in subjects who have no history of unusual exposure to dust. More recently, in Alzheimer's disease,[545] a presenile condition characterized pathologically by the presence of glial plaques in the brain, an unexpectedly high increase in silicon has been reported in the cores and rims of the senile plaques. The precise relationship of silicon with the ageing process remains to be determined.

Although a copper-molybdenum-sulfate interrelationship previously has been shown in animal species, there is work demonstrating a silicon-molybdenum interaction in humans.[546] Plasma silicon levels were strongly and inversely affected by molybdenum intake; silicon-supplemented chicks on a liver-based diet (Mo 3 ppm) had a plasma silicon level 348% lower than chicks on a casein diet (Mo 1 ppm). Molybdenum supplementation also reduced silicon levels in those tissues examined. Conversely, plasma molybdenum levels are also markedly and inversely affected by the inorganic silicon intake. Silicon also reduced molybdenum retention in tissues. The interaction occurs within normal dietary levels of these elements. Although a copper-molybdenum-sulfate interrelationship has been shown in animal species, this is the first work demonstrating a silicon-molybdenum interaction.

In addition to possibly producing pathological effects, the presence of phytoliths throughout many tissues, especially lungs, liver, lymph nodes, and spleen, makes it difficult to determine the amount of silicon in animal tissues that is actually contributing to physiological mechanisms. The gut wall of humans is also permeable to particles the size of diatoms. Carlisle[547] showed that diatomaceous earth particles are absorbed through the intact intestinal mucosa, pass through the lymphatic and circulatory systems, and reach other tissues supplied by arterial blood via the alveolar region of the lung. Examination of human organs has revealed diatoms in lungs, liver, and kidneys as a consequence of their presence in atmospheric dust and their movement from the respiratory tract. The capacity of these particles to travel in the blood and the penetrate membranes, including the placenta, is illustrated further by their presence in the organs of stillborn and premature infants.[548]

Silicon is found to be freely diffusible throughout tissue fluid.[549-551]

A dependency on silicon for maximal prolyl hydroxylase activity has been demonstrated.[552] The results support the in vivo and in vitro findings of a requirement for silicon in collagen biosynthesis, the activity of prolyl hydroxylase being a measure of the rate of collagen biosynthesis.

The minimum dietary silicon requirements compatible with satisfactory growth and health are largely unknown.[553-557]

The demonstration of the essentiality of silicon for higher animals is recent, so that reliable data on the silicon content of human foods and dietaries are meager. Furthermore, little is known of the extent of silicon absorption from various sources; for example, some forms of silicon are very insoluble. Also, since silicon is ubiquitous in the environment, the likelihood of a silicon deficiency arising under natural conditions in humans or domestic animals might be questioned. Of possible significance here is the suggestion that silicon absorption might be under hormonal regulation,[558] and if so, a decline in hormonal activity in senescence might result in decreased silicon absorption.

Sources. Foods of plant origin are normally much richer in silicon than those of animal origin. Whole grasses and cereals may contain 3 to 4% of the whole dry plants, as SiO_2, with levels up to 6% silica in some range grasses.[559] In leguminous plants, total silicon concentrations are appreciably lower, the levels approximating those found in animal tissues, with a high proportion of the relatively low amounts of silicon present as monosilicic acid. Solid silica is only sparsely deposited in these species.[560,561] Cereal grains high in fiber such as oats are much richer in silicon that low-fiber grains such as wheat or maize.[562] This finding suggests that the silicon content of patent white flour is significantly lower than that of the whole wheat from which it is made. Substantial losses of silicon occur in the refining of sugar.[563]

In mature gramineous plants, most of the silicon present is in the form of solid mineral particles, known as opal phytolithis (SiO_2 H_2O).[564] In another study, prairie grass hay (mainly *Festuca scabrella*) averaged 6 to 27% total silicon (dry basis) compared with only 0.39% silicon in alfalfa hay.[565]

This source of adventitious silica has been implicated as a significant source of wear of teeth in grazing sheep.[566,567]

One preliminary study[568] was undertaken to investigate the effect of dietary silicon and aluminum on levels of these elements in the brain and is undetermined.

Deficiency. Most of the signs of silicon deficiency in chickens and rats show aberrant metabolism of connective tissue and bone. Chicks fed a semisynthetic, silicon-deficient diet exhibit skull structure abnormalities associated with depressed collagen content in bone and long-bone abnormalities characterized by small, poorly formed joints and defective endochondral bone growth. Tibias of silicon-deficient chicks exhibit depressed contents of anticular, certilage, water, hexosamine, and collagen.[569] Silicon deficiency is seen in 33% of the chemically sensitive patients measured at the EHC-Dallas, but the significance of this finding is not known.

Sulfur

Sulfur (S), which comprises 0.25% of the total body weight, is found in all tissues, especially those of high protein content (Table 22). It is extremely important in protecting the chemically sensitive patient against chemical pollutant damage by its antioxidant and conjugating properties of toxic chemicals, yet 33% of the chemically sensitive have sulfur deficiencies. Most of the sulfur is found in the sulfur-containing amino acids methionine, cystine, glutathione, taurine, and cysteine. These are the antipollutants necessary for methylation and sulfur conjugation detoxification reactions as well as noncatalytic detoxification reactions. Sulfur also occurs in organic sulfates and sulfide in smaller amounts and in the two B vitamins thiamine and biotin. Their depletion will

Table 22. Sulfur

Source	Meat, poultry, eggs, fish, legumes, milk
	↓
Function	In methionine cysteine, cystine, taurine and glutathione amino acids for detoxication
	In thiamin (B_1) and biotin vitamins
	In inorganic sulfates used for detoxication
	In all tissues, especially hair, nails, skin
	Stabilize protein molecule
	In mucopolysacchride

result in decreases in detoxification function and an exacerbation of chemical sensitivity.

Sulfur compounds are metabolically important because of their ability to interconvert disulfide (RSSR) and sulfhydryl groups in oxidation-reduction reactions. As an example, cystine (a disulfide — RSSR) can be reduced to cysteine (a sulfhydryl, –SH). Cystine incorporated into keratin in human hair is responsible for the sulfur smell when hair is burned. Nails, fur, feathers, and skin also contain substantial amounts of sulfur-containing amino acids. Disulfide and sulfhydryl bonds provide the configuration and stabilization for protein molecules (for example, the permanent wave in hair or the biologically active shape of enzymes).

Glutathione activity in oxidation-reduction reactions is also dependent on the sulfhydryl group of cysteine. The active sites of CoASH and lipoic acid are the sulfhydryl portions. Besides its role in oxidation-reduction reactions, sulfur is important in many other compounds, reactions, and metabolites. Taurine, the precursor for the bile acid taurocholic acid, is synthesized from cystine by way of cysteine (see the section in this chapter on amino acids). The mucopolysacchrides (especially chondroid sulfate and collagen) contain sulfur. Sulfur, in the presence of magnesium, is important in detoxifying metabolic sulfuric acid. The esters produced are excreted through the kidneys. Sulfolipids are found in the liver, brain, and kidneys.

Most dietary sulfur is ingested in the amino acid forms, and excesses are excreted in the urine.[570] For a discussion of the sulfur pool, see Chapter 4.

Deficiency. All symptoms of sulfur deficiency are unknown, although it is conceivable that a diet severely lacking in protein could produce a deficiency as well as poorly functioning detoxification systems.

Sources. The ultratrace elements are presented in Table 23. Protein-containing foods such as meats, poultry, eggs, fish, legumes, and milk are good sources of the minerals.

Table 23. Classification Summary of the Ultratrace Elements

Element	Evaluation of Essentiality[a]	Selected Major or Reported Deficiency Sign[b]	Possible Need in Normal Function Of/In
Arsenic	E	Depressed growth (C, R, P, G)	Taurine or sulfate production from methionine
Baron	E	Myocardial damage (G) Depressed growth (C)	Major mineral (Ca, P, Mg) metabolism via parathormone
Bromine	NE	Insomnia? (G, H)	?
Cadmium	NE	Depressed growth (G, R)	?
Fluorine	NE	Depressed growth? (M, R) Depressed hematopoiesis? (M)	Calcified tissue structure
Lead	NE	Hypochromic microcytic anemia? (R) Disturbed iron metabolism? (R)	Iron absorption
Lithium	PE	Depressed growth (G) Depressed fertility (R, G)	Endocrine regulation
Nickel	E	Depressed growth (G, P, R, S, B) Depressed hematopoiesis (C, R, S, G)	Iron absorption; some metalloenzyme
Tin	NE	Depressed growth? (R)	?

Source: Nielsen.[509]

[a] E = essential; PE = probably essential, but further study required to establish essentiality; NE = should not be considered essential at present because evidence for essentiality is weak, has shortcomings, or is questionable.

[b] Letter in parentheses indicates species: B = bovine-cow; C = chicken; g = goat; H = human; M = mice; P = minipig; R = rat; S = sheep. For citations to original reports describing signs of deficiency, refer to other sources.

[c] Fluorine should be recognized as an element with beneficial pharmacologic properties (i.e., anticariogenic property).

Lithium

Lithium is an essential mineral in stabilization of the membranes in chemical sensitivity. Supplementation has been noted to stabilize membranes in many chemically sensitive patients. Lithium is probably the regulator of homeostatic mechanisms. It influences the ion transport systems, cyclic AMP-dependent systems, anedylcyclase, and phosphoinositide (via inhibition of enzymes). Lithium has been reported to influence autoimmune states such as arthritis, psoriases, and myosthenic gravis. It has been reported to influence the immune system. It enhances leukocyte (PMN) production and natural killer cell development. Lithium also increases phagocytic activity, which has been found to be impaired in a subset of the chemically sensitive.

Amino Acids

Understanding protein and amino acid changes is extremely important in understanding pollutant injury in the chemically sensitive because alterations in both can produce changes in the immune and nonimmune detoxification systems that can adversely effect the chemically sensitive.

Protein

The average American adult consumes between 80 and 125 g of protein daily. Protein provides 12% of the daily energy needs. This intake has been constant since 1900, but the animal protein portion is almost 70% which has more than doubled vs. that from vegetable protein. Protein is essential as a buffer against pollutant injury and is essential in many immune functions. Altered proteins are seen in the chemically sensitive, which can result in IGE responses as haptens and nonimmulogic metabolic dysfunctions.

Introduction to Pollutant Overload and Amino Acids

Amino acids are chemical building blocks that plants and animals use to form protein. There are many different amino acids, and distinctions can be made by examining their chemical structures as well as their biological source markers. Study of these varied aspects are only beginning in the chemically sensitive. However, both the EHC-Dallas and Bionostics have accumulated a significant amount of information on the effects of pollutants on amino acid metabolism in the chemically sensitive. We have studied the metabolism of several thousand patients, and data on them will be presented in this and other chapters. Differences in proteins result from variations in the way that amino acids are arranged in protein molecules. For humans there are eight amino acids (valine, leucine, isoleucine, methionine, lysine, phenylalanine, tryptophan, and threonine) that are nutritionally essential. These eight are needed to grow and repair tissue, but humans do not have the capability to form them. They must be obtained in adequate amounts from the diet. This acquisition is sometimes difficult in the chemically sensitive because of pollutant damage which results in difficulty with intake, absorption, transport, or renal reabsorption. In addition to these eight, there are other nonessential amino acids, which may be formed within the body and which also are used for protein formation. Pollutant damage in the chemically sensitive may affect the status of any individual amino acid by altering intake, absorption, transportation, incorporation, and elimination, making it difficult for proper nutrition and detoxication in some cases.

Proper digestion (often disturbed in the chemically sensitive) in the stomach and small intestine breaks down food protein into individual amino acids or to small peptides that may be two or three amino acids long. These individual amino acids and the small peptides then are absorbed from the small intestine into the bloodstream. Improper digestion and breakdown are often seen in the chemically sensitive, resulting in altered peptides, and thus, malabsorption or abnormal absorption with organ malfunction. Hypochlorhydria, which often occurs in the chemically sensitive, will decrease the breakdown of protein. Transport may be damaged by pollutants, making it difficult for amino acids to transfer from organ to organ and across membranes. Formation of human proteins and other products from these amino acids occurs in the liver and in other organs and tissues. Pollutant injury may cause major dysfunction in metabolism because it may disturb any of the following which proteins become: (1) catalysts (enzymes) for metabolism; (2) antibody or immunoglobulin formation; (3) neurotransmitters to biologically activate chemicals that influence nervous system functioning; (4) hormone-like functions, such as leucine and arginine, which stimulate insulin release from the pancreas; and (5) reduce inflammation by combining oxidizing agents or with inflammatory radicals or ions that normally are produced as part of the body's oxidant-response chemistry (Table 24) (see Chapter 4).

Pollutant injury may not only alter amino acid absorption; it may also interfere with or cause renal reabsorption. These conditions are seen in 60% of the more severely disturbed, chemically sensitive patient. Amino acids that are needed (or are present in relative excess) and waste products of metabolism are still excreted in the urine with early pollutant injury. The body's capability for removing the relative excess of an amino acid is limited by the kidney's capability for filtering from the bloodstream. Differently structured amino acids have different clearance rates in the kidneys. They have different reabsorption routes or mechanisms in kidney tubules, and sometimes an excess level of one amino acid can interfere with renal transport of other amino acids. Also, pollutant damage to the tubules may then selectively alter the amino acid

Table 24. Pollutant Injury to Amino Acids and Proteins May Cause Malfunction of the Following in the Chemically Sensitive

Catalysts — enzymes
Antibodies — immunoglobulin formation
Neurotransmitters
Hormone-like functions — leucine and arginine stimulate insulin
Anti-inflammatory by combining with oxidizing agents

output or reabsorption, thus exacerbating an existing problem with chemical sensitivity.

Natural Configuration of Protein Amino Acids

All naturally occurring amino acids that are used for protein formation in animals and plants have a particular configuration with respect to the position of the amino group in the molecule. Except for glycine, all amino acids have at least one carbon that can be considered asymmetric. The amino group ($-NH_2$) in naturally occurring amino acids from protein always is oriented in a certain way relative to the carbon atom to which it is attached, and that carbon atom is linked directly to the acid groups [the carboxyl group $-COOH$]. This is the "alpha carbon," and protein amino acids are alpha-form amino acids. The alpha carbon is the "number two carbon" in an amino acid.

Besides being attached to the alpha carbon, the $-NH_2$ can be oriented in a specific way relative to the carbon atom and the entire amino acid molecule. The simplistic way to depict this is by drawing the chemical structure and observing that the NH_2 can be located on one side or the other of the alpha carbon. In stereochemistry, the one orientation is called the "L-configuration;" the other is the "D-configuration." Naturally occurring protein amino acids all are L-configured with respect to the absolute configuration of the "reference compound," which is glyceraldehyde, the smallest sugar (3 carbons) that has an asymmetric carbon atom in its molecule. For amino acid configurations that capital L or capital D does not mean levorotatory (l or –) or dextrototory (d or +), which refers to the direction in which a plane of polarized light is rotated as it goes through an aqueous solution of the amino acid. Some natural L-configured amino acids are levorotatory (isoleucine, alanine), and some are dextrototory (leucine, phenylalanine).[571]

One of the problems with synthetically derived foods, nutrient supplementation, and toxic chemical additives in the chemically sensitive is that the L-configurated amino acids may be attempted to be substituted as D or DL forms. The notation "L,D-amino acid" means that a mixture of the two configurations is present being a blend of L- and D-isomers. This substitution will not work and can cause metabolic problems. This substitution problem is also seen in synthetically derived medications and vitamins which usually are of D-configuration. All of these D-form substances are particularly difficult for the chemically sensitive to tolerate.

Some amino acids contain more than one –NH$_2$; for example, the essential amino acid lysine contains two amino groups. Natural lysine from protein is L-configured with respect to the –NH$_2$ on the alpha carbon. All naturally occurring protein amino acids are alpha-L-configured although some may contain more than one –NH$_2$. This fact may be a critical issue when attempting to correct individual problems with synthetic substances and the patient fails to respond. Also, some amino acids may contain nitrogen atoms within the carbon chain in addition to their amino group. Citrulline, an intermediate of the urea cycle and not a protein amino acid, contains two NH$_2$ groups and also includes a C-N-C structure. Some amino acids may have more than one –COOH, some may have ring structures in their carbon chains, and a few include sulfur as well as carbon, hydrogen, oxygen, and nitrogen in their molecules (Figure 28 A and B).

Animals and plants do not include D-configured amino acids in protein formation. For the body there is limited enzymatic capacity to remove D-configured amino groups from amino acid molecules. (Racemizing: switching a D-configuration to an L-configuration.) Clinical evidence shows that most D-configured amino acids, whether produced endogenously (such as D-beta-aminoisobutyric acid) or whether taken orally (such as D-methionine might be), end up in the urine or are metabolized to unused products (methionine sulfoxide) that also end up in the urine. D-configured amino acids can be enzyme inhibitors (D-phenylalanine is), and they may induce aminoaciduria patterns identical to that of pathological conditions[572] seen in the chemically sensitive individual.

Essential Amino Acids

The nutritionally essential amino acids (those that cannot be synthesized endogenously) in humans are isoleucine, leucine, valine, lysine, methionine, phenylalanine, threonine, and tryptophan.

Endogenous formation of these eight is insignificant; therefore, intake and absorption are critical to balance at appropriate levels. Often, however, these are disturbed in the chemically sensitive. In biochemical terms, the catabolism sequences for essential amino acids include some chemical steps that are made irreversible or nearly irreversible by the lack of enzymes or pathways for reversing the chemistry. This prevents formation of the essential amino acids from other amino acids or catabolic products that might be available. Pollutant injury can be selective or generalized and cause malfunction of the enzymes for catabolism of the essential amino acids or just incorporation and utilization (Table 25).

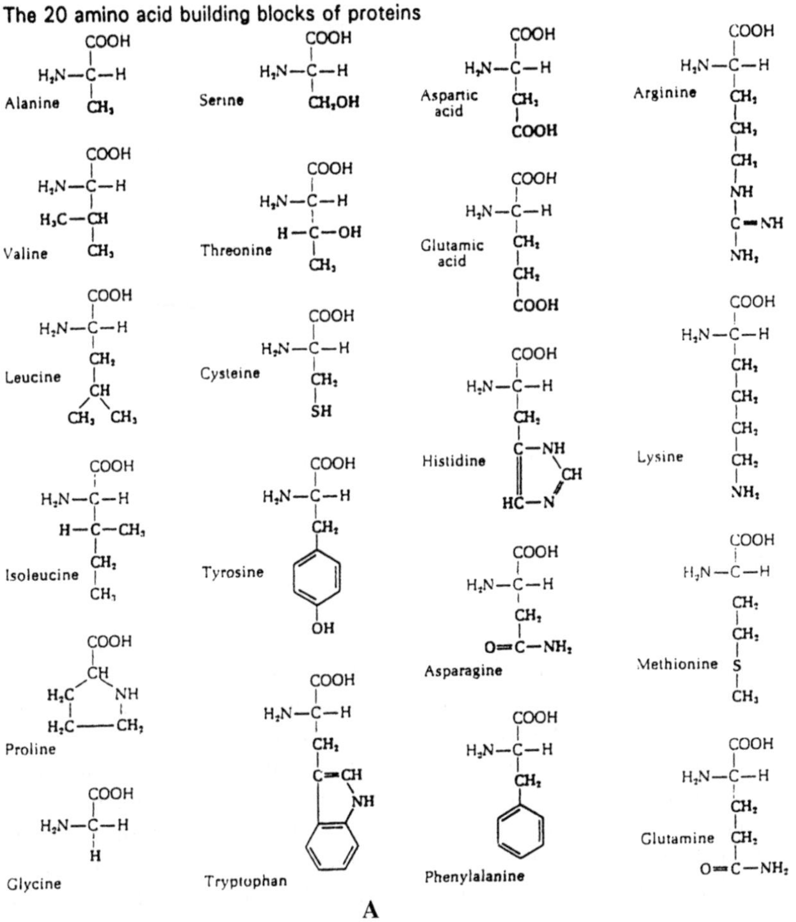

(A) The 20 amino acid building blocks of proteins

Figure 28. **Amino acids potentially injured in chemical sensitivity.**

The major functions of the essential amino acids are shown in Table 26, emphasizing a potential for pollutant damage causing dysfunction. Each will be discussed separately except for valine, leucine, and isoleucine, which will be treated together because of their close proximity in formulation and function.

Valine, Leucine, and Isoleucine. Valine, leucine, and isoleucine (Figure 29) are branched-chain-structured essential amino acids, and their presence in adequate amounts in the chemically sensitive is necessary because they are major components of connective protein tissues (collagen) and are abundant in the structure of elastin, the yellow elastic protein found in ligaments. Clinically, leucine is observed to promote wound healing. It can be used to lower blood sugar levels in hyperglycemia because it stimulates the pancreas to release insulin.

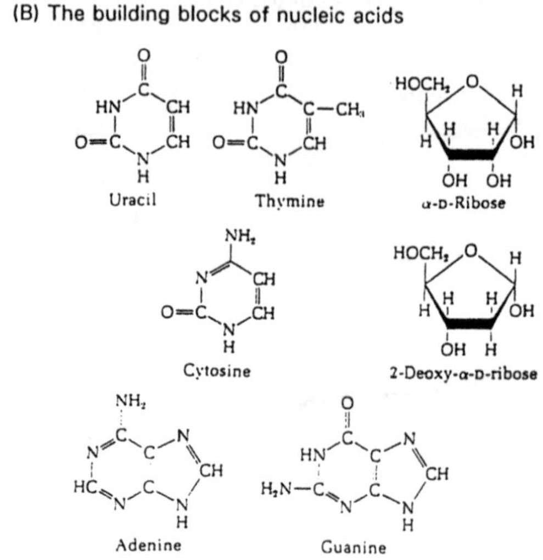

(B) The building blocks of nucleic acids

Uracil

Thymine

α-D-Ribose

Cytosine

2-Deoxy-α-D-ribose

Adenine

Guanine

Figure 28B.

Foods that are rich in these amino acids are soybeans and soy protein, sunflower seeds and meal, swiss cheddar and other cheeses, common beans (navy, pea beans, white), peanuts and peanut butter, lima beans and lentils, cereal grain flours and wheat germ, and cashews.

Uptake of valine, leucine, and isoleucine is particularly subject to pancreatic function and conditions in the mucosa of the small intestine and, therefore, may vary in chemical sensitivity since these functions are often disturbed. When disturbance occurs in the digestive tract, aminopeptidase enzymes (activated by zinc) hydrolyze small peptides to release these branched-chain amino acids. Enzymatic activity is dependent upon pH. Bicarbonate from the pancreas is needed to bring the pH of the small intestine up to 7 or higher so that peptidase enzymes can function properly. Pollutant-damaged individuals with pancreatic dysfunction, duodenal acidity, and food allergies often present with deficient urine levels and marginal or low blood levels of valine, leucine, and isoleucine. Chemically sensitive individuals often have difficulty in maintaining adequate pH and need to have copious bicarbonate supplementation to allow the enzymatic activity to be adequate.

Case experience at Bionostics[573] indicates a syndrome in approximately 20% of patients from all practices that treat "food sensitivity" patients. This syndrome includes various degrees of subnormal coenzyme activity of pyridoxal phosphate, hyper-alpha- and beta-aminoacidurias (under normal protein diet conditions), and excess urine taurine. There is both protein and fat intolerance; bile formation appears disordered (per stool characterization), and plasma vitamin A is often low or deficient. An intolerance to the essential amino acid valine may develop if beta-aminoisobutyric aciduria is present.

Table 25. Pollutant Injury May Cause Change in the Formation of the Following Glycoproteins

Threonine
Asparagine
Serine
Hydroxyproline

Table 26. Major Functions of Amino Acids Possibly Altered by Pollutant Injury as Seen in Some Chemically Sensitive Patients

Valine Leucine Isoleucine	Connective tissue; elastin; wound healing; leucine stimulates pancreas to release insulin
Lysine	Hormones; structural proteins; connecting linkage pyridoxial precursor for carnitine which is phosphate for transamination, the transporting agent for lipid metabolism; reduces arginase activity
Methionine	Component of physiologically active peptides, e.g., encephalin; adenysl methione for methylation; metabolism to cysteine to form CoA; protection against cancer, then propogation of it
Threonine	Precursor of serine, glycine; slowest absorbed; therefore, deficient in malabsorption; link CHO to proteins to produce glycoproteins (agglutinins); copper transporter
Tryptophan	Precursor of niotinic acid (B3); precursor of picolinic acid (for zinc absorption and transport); precursor of seratonin in argintaffin cells and platelets; generates indol in intestine with excessive bacterial action
Phenylalanine	Precursor of catechol amines; main metabolite is tyrosine; therefore, influences thyroid metabolism, phenylketonuria

This beta-acid catabolizes to methylmalonic semialdehyde, which is a component of the catabolism pathway of valine (Figure 30).

Recent studies also have demonstrated that urinary beta-aminoisobutyric acid (beta-AIBA) consists of primarily the D-configured amino acid and very little L-beta-AIBA (L- and D-isomers relative to glyceraldehyde).[574] Only L-form amino acids are found in natural protein. However, some D-form intermediate amino acids are naturally formed within cells.

Lysine (Lys). Lysine (Figure 31) is an extremely important component of hormonal and structural proteins and is therefore important in the chemically sensitive. The lysine residue in an enzyme protein can also provide the

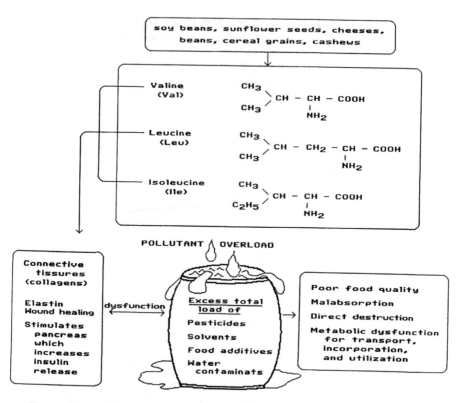

Figure 29.   Valine, leucine, and isoleucine.

connecting linkage for coenzyme pyridoxal phosphate, which catalyzes amino group transfer (transamination) reactions (often disturbed in the chemically sensitive). The lysyl residue in a certain protein is the precursor of carnitine, which is a carrier or transporting agent needed for fatty acid metabolism which is sometimes disturbed in the chemically sensitive. A deficient intake of lysine can lead to anemia, loss of appetite, and weight loss.

Foods that are high in lysine are soybeans and soy protein, sardines, herring, other fish, and shellfish, swiss, cheddar, and other cheeses, poultry, beef, and veal, common beans (navy, pea beans, and white beans), and lentils.

A common therapeutic use of lysine is to reduce an infection by the herpes simplex virus, type 1 (herpes labialis) or "cold sores" type and sometimes type 2 (herpes genitalis) or genital herpes which is a recurrent problem in many chemically sensitive patients. Therapeutic doses of lysine (2 to 5 g/day or more) should not be used continuously without performing a urine or plasma amino acid analysis to check adequacy of the arginase enzyme. This is the urea-forming enzyme in the liver. Lysine can reduce arginase activity via competitive inhibition because arginase has a very high affinity for lysine.

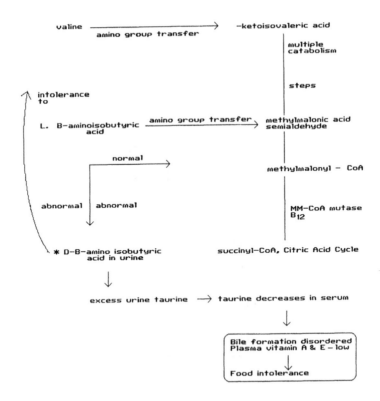

Figure 30. **Pollutant injury to valine pathway (pyridoxal phosphate sub-normal activity). *D-form is an intermediate of natural protein breakdown of L-BAIB and cannot break down further, causing the aminoaciduria, or is more likely an unnatural product.**

Methionine (Met). Methionine (Figure 32) is important in the chemically sensitive for many reasons, including its involvement in xenobiotic methylation conjugation reactions.

In a study of the nutritional and metabolic data for over 1500 individuals presenting with food and chemical intolerance, degenerative diseases, neuro-muscular dysfunction, and mental diseases, Bionostics[575] found that the metabolism of methionine is the *most frequently impaired or disordered amino acid metabolism.* This disorder leads not only to the lack of detoxification of xenobiotics, but also to endogenous epinephrine, norepinephrine, and serotonin.

Some methionine becomes adenosylmethionine, which provides methyl groups for methylation of DNA (controls gene expression), methylation of adrenal catecholamine (helps control levels by inactivation of epinephrine through formation of metanephrine), synthesis of choline (precursor to

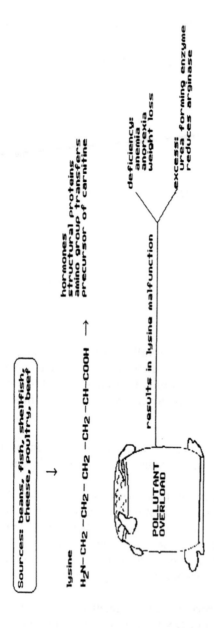

Figure 31.   Potential pollutant damage to lysine.

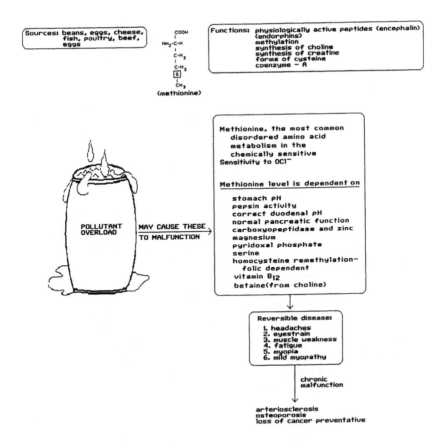

Figure 32. **Pollutant overload causes malfunction of methionine metabolism.**

neurotransmitter acetylcholine), and formation of creatine (precursor of creatinine), etc. Methionine can be metabolized to form cysteine, which becomes part of coenzyme-A, which is required for Krebs or citric acid cycle function, and part of glutathione, which is a biologically active peptide involved in control of insulin levels, prostaglandin synthesis, enzymatic detoxification, and control of oxidizing or inflammatory substances and many xenobiotics. Cysteine, in turn, is a precursor of taurine and contributes to the sulfur pool.

The sulfur-containing essential amino acid, methionine, is a component of physiologically active peptides such as methionine enkephalin and various endorphins, the body's natural analgesics. They are produced in the pituitary, and they bind to the same central nervous system receptors as do morphine opiates. These peptides control the perception of pain and are many times more potent than morphine. Often the chemically sensitive are very sensitive to pain stimuli, perhaps due to malfunction of this pathway.

Assimilation, often disturbed in the chemically sensitive, requires proper stomach pH for pepsin activity, proper duodenal pH, normal pancreatic function, and proper activity of carboxypeptidase enzymes (some require zinc for activation). The nutritional cofactors on which methionine is dependent are magnesium, pyridoxal phosphate from vitamin B_6, serine (a nonessential amino acid), and remethylation of homocystine, which involves folic acid, vitamin B_{12}, and betaine (which comes from choline).

Nutritionally induced, mild impairments in methionine metabolism usually are nonfixed disease conditions as far as amino acid metabolism is concerned and often can be reversed with intelligent manipulation in the chemically sensitive. However, chronically impaired methionine metabolism is associated with the development of named clinical disorders over a period of time and are much more difficult to correct once they are entrenched. Cardiovascular disorders including arteriosclerosis and skeletal disorders including osteoporosis are associated with impaired methionine metabolism. Many variable clinical manifestations of impaired methionine metabolism have been documented. Symptoms may be present, including headaches, susceptibility to eye strain, muscle weakness, fatigue, mild myopathy, and myopia; or there may be no evident symptoms in mildly impaired methionine metabolism. See also Cysteine in this chapter for possible symptoms if cysteine formation is impaired. *Sensitivity to the hypochlorite molecule is present with impaired methionine metabolism and widespread chemical sensitivity develops* (see Taurine in this chapter and Chapter 4).

Hoffman[576] showed that cancer cells contain undermethylated DNA. Methionine is "protective" against cancer from the aspect of being a dietary source of the methyl group that is needed for methylation. However, once cancer has developed, many strains become methionine dependent. Controlled studies with cancer tumors in animals show that removal of methionine from the diet (and substitution of homocysteine) prevents tumor growth and metastasis. To our knowledge, no such studies have been performed on humans.

For the next step for sulfur metabolism, see the sections on cysteine, cystine, homocysteine, and taurine in this chapter.

Foods rich in methionine include brazil nuts, swiss, cheddar, and other cheeses, sardines and halibut, herring and other fish, shellfish, chicken, turkey, soybeans, beef, and eggs.

Phenylalanine (Phe). As an essential amino acid, phenylalanine (Figure 33) is the direct precursor of tyrosine, which leads to synthesis of dopamine, epinephrine, and other adrenal catecholamines, and also to the iodated thyroid hormones via thyroglobulin. These pathways and functions are extremely important in the chemically sensitive since there frequently appear to be excess epinephrine-type responses in a large subset and impaired thyroid function and

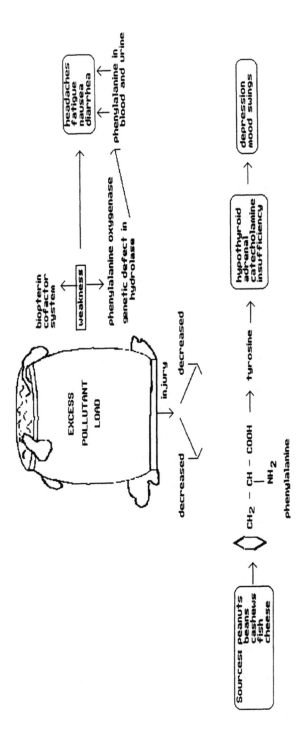

Figure 33. Pollutant overload causes the malfunction of phenylalanine metabolism.

thyroiditis in another. Phenylalanine is a ring-structured (aromatic) amino acid; so is its principal metabolite, tyrosine. Excess and insufficient catechol amines have been seen in some chemically sensitive patients, the imbalance apparently due to pollutant injury. Chronically low phenylalanine and tyrosine can also cause adrenal catecholamine insufficiency, which can be seen in some chemically sensitive patients. The predominant symptom of this is mental depression. Mood swings may occur with varying catecholamine levels, as seen in a subset of chemically sensitive patients.

Tyrosine leads to iodated thyroid hormones (T_1, T_2, T_3, T_4) via thyroglobulin. Chronically low phenylalanine and/or tyrosine may lead to or be coincident with a hypothyroid condition often seen in pollutant injury.[577,578] Recorsinols (mutiphenols), cyanogens, and halogen are the pollutants which are known to give injury to the thyroid (see Chapter 4 and "thyroid" in Chapter 38).

The biochemistry of the phenylalanine-to-tyrosine step is rather complex and involves a hydroxylase enzyme (phenylalanine monooxygenase), with direct and indirect cofactors: oxygen, reduced biopterin, iron, reduced and phosphorylated NAD, and the enzyme dihydrobiopterin reductase, all of which are prone to pollutant injury.

Some individuals have pollutant or genetically induced (subclinical) weakness in phenylalanine monooxygenase or the biopterin cofactor system such that a diet abnormally enhanced in phenylalanine will provoke hyperphenylalaninuria. Consequent symptoms may include headaches, fatigue, nausea, or diarrhea. Phenylalanine-containing artificial sweeteners such as aspartame should be avoided by those individuals and often exacerbate symptoms in the chemically sensitive.

Phenylketonuria (PKU) is an inheritable inborn error of metabolism. Acute hyperphenylalaninuria and hyperphenylalaninemia can result from the hydroxylase enzyme being defective, which occurs when a recessive mutation alters the controlling gene, which occurs in about 1 in 10,000 humans. A secondary phenylalanine catabolism sequence then is overused with transamination of phenylalanine to its keto analogue, phenylpyruvic acid. Deficient biopterin synthesis also can cause PKU. If not treated medically and nutritionally, PKU can result in mental retardation.[579]

Some phenylalanine-rich foods are peanuts, peanut butter, swiss, cheddar, and other cheeses, other dairy products, soybeans and soy protein, fish and shellfish, lima beans, lentils, common beans (navy, white, pea beans), and cashews.

Tyrosine (Tyr). Tyrosine (Figure 34) is an important ring-structured nonessential amino acid that is a component of human protein. Its nutritionally essential precursor is phenylalanine, and tyrosine also can come directly from digestion of dietary protein. Impaired digestion as seen in the chemically sensitive may cause inappropriate levels of tyrosine.

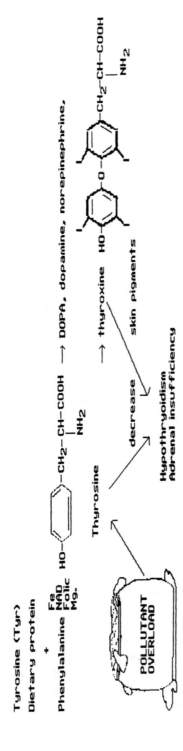

Figure 34. Pollutant injury to tyrosine metabolism.

The importance of tyrosine derives from the fact that it is a precursor to several metabolism sequences, whose components or products are biologically active. Tyrosine begins the chemistry of adrenal catecholamine formation that leads to DOPA, DOPamine, norepinephrine, epinephrine, etc. Epinephrine, for example, functions as a hormone for rapid response and provides sympathetic stimulation of organ activity following various stimuli (heat, cold, realization of danger). This function has been found to be abnormal in some chemically sensitive patients.

Tyrosine participates in thyroid function via thyroglobulin. Iodine atoms are attached to the tyrosine residue in thyroglobulin. Di-iodotyrosine bonds with a double-iodated tyrosine residue to form thyroxine (T_4).

Deficient levels of blood and/or urine tyrosine may indicate hypothyroidism, especially if the deficiency is a chronic condition.[580,581] Elevated levels of tyrosine have been clinically observed in hyperthyroidism.[582,583]

Tyrosine also initiates the chemical sequence that leads to skin pigments, melanin, but linking abnormal tyrosine to abnormal pigmentation is tenuous. Hyperpigmentation can be a symptom of adrenal insufficiency (as in Addison's disease).[584] Also, inborn defects in melanocytes, albinism, and (acquired) damage to melanocytes can cause abnormal pigmentation, but measured blood tyrosine levels are normal. An error in a tyrosine metabolism enzyme that might contribute to elevated tyrosine (deficient homogentisic acid dioxygenase) can cause "alkaptonuria" and abnormal pigmentation of collagen tissue. Supplementing phenylalanine or tyrosine would worsen this condition.

Tyrosine deficiency symptoms often seen in chemically sensitive patients are variable, depending upon affected metabolism and the degree of deficiency. Adrenal catecholamine insufficiency may result in mental symptoms such as apathy, indifference, and depression; and it may also result in reduced physical responses, or low blood pressure. The patient may experience hypothyroid symptoms including fatigue, weight gain, and subnormal basal temperature. The symptom of "restless legs" is reported by some clinicians to be relieved by tyrosine supplementation. This condition is seen in a large subset of chemically sensitive patients. In the experience of the EHC-Dallas, avoidance of the offending pollutants and foods usually solves the restless leg problem. Bionostics[585] described a case of increased blood pressure following use of tyrosine. If tyrosine increases adrenal catecholamine metabolism, increased blood pressure is the expected result.

Tyrosine can be subnormal if phenylalanine is low or if there is protein malnutrition or gastrointestinal dysfunction, as seen in the chemically sensitive. Tyrosine might also be low if its formation step from phenylalanine is impaired. This impairment may occur in iron or NAD deficiency, folic acid deficiency, disordered assimilation of magnesium, or hypoinsulinemia. Tyrosine deficiency is a result of most types of PKU (which is a rare inborn metabolism error, about 1 in 10,000 births).[586] The most commonly observed cause of elevated tyrosine in blood and/or urine is pyridoxal phosphate deficiency or dysfunction (observation based on Bionostics case experience). Also, two steps in tyrosine metabolism require vitamin C for maximum activity in vivo.[587] Adequacy of tyrosine is best determined by a quantitative 24-h urine or by fasting blood plasma amino acid analysis.

Threonine (Thr). Threonine (Figure 35) can serve as a precursor to the nonessential amino acids serine and glycine, and threonine is an endogenous source of acetaldehyde and acetic acid in human metabolism when it is used for glycine formation (an essential amino acid for acyl [peptide] conjugation of xenobiotics).

In the small intestine threonine is the most slowly absorbed essential amino acid, and it can be particularly deficient in cases of intestinal malabsorption seen in the chemically sensitive. Threonine also is one of the five amino acids that link with carbohydrates to form glycoprotein, and it is the only essential amino acid among these five; the other amino acids are asparagine, serine, hydroxyproline, and hydroxylysine.[588,589] Glycoprotein are required for proper immune response, which appears to be lacking in many chemically sensitive patients. A major fraction of plasma protein really is glycoprotein, including the agglutinins that establish blood type. Threonine (also histidine, glutamine, and threonine-histidine-albumin combinations) acts as a copper transporter in the bloodstream and is primarily responsible for the transport of copper to cells.[590] Copper transport appears to be altered in some chemically sensitive.

Some threonine-rich foods are meats, poultry, fish, most cheeses and other dairy products, legumes, cashews, sunflower seeds and meal, and sesame seeds.

Tryptophan (Trp). The structure of tryptophan is somewhat like that of phenylalanine or tyrosine in that it includes a carbon ring, but adjacent to that ring in another ring that contains a nitrogen atom. The amino group attached to the alpha carbon contains the second nitrogen atom. Understanding tryptophan metabolism is essential in understanding some problems that occur with chemical sensitivity, because many chemically sensitive patients have problems with sleep and niacin deficiency and serotonin, to all of which tryptophan is a precursor (Figure 36).

Tryptophan is an endogenous precursor of nicotinic acid, a form of vitamin B_3 that is extremely important in detoxification reactions and in mobilizing xenobiotics from fat. Cofactors, coreactants, and mineral activators of enzymes required for tryptophan to form nicotinic acid are pyridoxal phosphate, ascorbic acid (stimulates activity of tryptophan oxygenase and may be considered as an assist rather than as a requirement), oxygen, NADPH, ATP, phosphorylated ribose, glutamine, iron, and magnesium. Approximately 1.5 to 2% of absorbed tryptophan normally becomes nicotinic acid, while most is formed in the liver and kidney.[591] In the pancreas, a small amount of the body's tryptophan normally becomes picolinic acid via a portion of the same biochemical

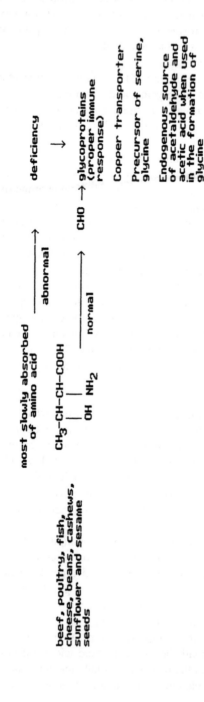

Figure 35.   Potential pollutant injury to threonine.

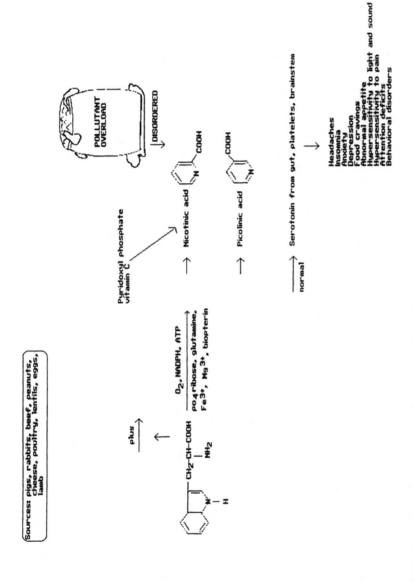

Figure 36.   Potential pollutant injury to tryptophan (Trp).

sequence that produces nicotinic acid. Picolinic acid probably is a minor constituent of the zinc absorption and transport mechanism[592,593] (Figure 38). The complicated pathway by which a small portion of tryptophan is transformed into nicotinic and picolinic acids leaves many opportunities for nutritional deficits or imbalances to impair or impede the chemistry. In some animals (dogs, pigs, rabbits, rats) tryptophan conversion to nicotinic acid provides a nutritionally adequate supply of vitamin B_3. In humans, tryptophan does not, and other dietary sources of B_3 are required.

Tryptophan also is the precursor of the vasoconstrictor and neurotransmitter, 5-hydroxytryptamine (5-HT or serotonin). This is an extremely important neurotransmitter in the chemically sensitive due to its alteration of blood vessel function and is often found to be abnormal. Provocation and neutralization injections with serotonin often aids these patients (see Chapter 36). Subnormal serotonin due to low tryptophan uptake or to retarded metabolism of tryptophan may cause disordered sleep patterns or insomnia and can be coincident with abnormal states of anxiety or depression, which are often seen in some chemically sensitive individuals.[594] Abnormal appetite and food cravings may occur. Acutely deficient tryptophan and serotonin also may lead to hypersensitivity to various external stimuli (light, sound) and to excessive sensitivity to pain, which is frequently observed in chemically sensitive individuals. Disordered tryptophan-to-serotonin metabolism and abnormal blood serotonin levels have been clinically associated with deficits in attention and behavior disorders[595] seen in some of the chemically sensitive.

Metabolism of tryptophan to serotonin depends upon iron, oxygen, biopterin, and pyridoxal phosphate (for decarboxylation). Biopterin synthesis is extremely complicated in humans; it starts with guanosine triphosphate and may be abnormal in cases of disordered purine metabolism or in some types of PKU. These forms of PKU may also include deficient serotonin. Over 95% of the serotonin measured in blood or blood platelets is formed in the mucosa of the intestines. Brain serotonin originates from tryptophan transported across the blood-brain barrier, and formation of this serotonin begins in the brainstem. Supposedly, very little, if any, intestinally produced serotonin crosses the blood-brain barrier.[596,597] However, this observation may not be true in the chemically sensitive individual, since the blood-brain barrier appears to be disturbed and may account for some of the vessel deregulation seen in the chemically sensitive.

When tryptophan is measured to be deficient in blood or urine, one prospect is intestinal malabsorption with excessive bacterial action on unabsorbed tryptophan in the intestine. This malabsorption causes elevated levels of indole compounds including indoxylsulfate (indican) as seen in some chemically sensitive. A urine indican test can be given to confirm or rule out this possibility. An individual must discontinue vitamin C supplements for at least 5 days prior to urine sampling for a valid indican test.

Table 27. Nonessential Amino Acid Functions May Be Altered by Pollutant Injury

Metabolic intermediates — without function, biological activity, or part of building protein
Metabolic intermediates — with special function and activities, but not part of protein
Metabolic intermediates or products that are part of protein

Hartnup disease is an uncommon inherited disease where renal conservation is deficient for tryptophan and other amino acids. Pellagra-like symptoms are presented, and indicanuria and B_3 deficiency can be observed in this disease.

Foods that are relatively high in tryptophan content are soybeans and soy protein, sunflower seeds and meal, cashews, swiss, cheddar, and other cheeses, peanuts, peanut butter, veal, poultry, egg yolk, beef, lamb, pork, lentils, and eggs.

Semiessential and Nonessential Acids

The eight strictly essential amino acids lead via many different chemical sequences or metabolic pathways to other amino acids that are, by definition, nonessential. Also, 2 semiessential and 10 nonessential amino acids are gotten directly via digestion of dietary protein and these are used together with the essential amino acids to build human protein (Table 27). They may be disordered in the chemically sensitive.

Two amino acids (arginine and histidine) can be formed in humans, but not in amounts sufficient to satisfy physiological needs, particularly in children. Arginine is the only urea-cycle intermediate that is also a protein amino acid and, along with leucine, also stimulates the pancreas to release insulin. Histidine is the precursor of histamine, which is functionally active in stimulating capillary dilation, smooth muscle contraction, and gastric secretions as well as mediating hypersensitivity or allergic response. These may be disordered in the chemically sensitive, as evidenced by the continued vasoconstriction seen in many chemically sensitive.

Nonessential amino acids may be grouped into three general classifications. The first is metabolic intermediates without special function or biological activity and not part of protein (Table 28). There are many such amino acids and levels of concentrations. Some of them are measured in blood or urine for the purpose of assessing intermediary metabolism. Examples are saccharopine and alpha-aminoadipic acid (from methionine, the latter also may be derived from threonine) and ornithine, citrulline, and argininosuccinic acid, which are urea-cycle intermediates. Although these amino acids are not used to form protein and they do not possess (known) activities or functions other than being

Table 28. Biologically Active Amino Acids Disturbed by Pollutant Injury

Leucine Arginine	Stimulate the release of insulin
Arginine	Stimulates the release of human growth hormone
Glycine	Inhibits neurotransmitter activity for spinal cord interneurons
Aspartic acid	An excitatory neurotransmitter in the CNS
Glutamic acid	An excitatory neurotransmitter in CNS and spinal sensory nerves

chemical stepping stones along metabolic pathways, their rates of formation and subsequent change into something else are important to metabolism. Inordinate accumulation or deficiency of these intermediates may cause or be the result of errors in amino acid metabolism as seen in some chemically sensitive patients.

The second group is metabolic intermediates or products that do have special functions or activities but are not part of protein. Examples are GABA, a neurotransmitter; taurine, a sulfonic amino acid derived from methionine via cysteine; and the thyroid hormone, thyroxine.[598,599] These are extremely important in the chemically sensitive.

The third group consists of metabolic intermediates or products that are part of protein, such as arginine and histidine. This type of metabolism may be altered in the chemically sensitive.

The 10 nonessential amino acids are alanine, asparagine, aspartic acid, cysteine (cystine is two cysteines joined together), glutamic acid, glutamine, glycine, proline, serine, and tyrosine.

Two more amino acids sometimes are included in the (nonessential) protein amino acid list—hydroxyproline and hydroxylysine. Humans do not build protein using hydroxyproline or hydroxylysine; rather, the hydroxyl group is added after the proline or lysine residue is in the peptide or protein.[600,601]

Thus, hydroxyproline and hydroxylysine are in dietary and human protein, especially collagen (the protein in connective tissue), but they are not among the chemical building blocks used to form protein. They are, however, along with threonine, serine, and asparagine, found in glycoproteins, which may be altered in some chemically sensitive patients.

Some protein amino acids are known to be biologically active or to have hormone-like functions (they fit into several classifications). Abnormal triggering of these occurs quite often in the chemically sensitive, often resulting in a diverse set of symptoms depending upon the type produced. For example, leucine (essential) and arginine (semiessential) both stimulate the pancreas to release insulin.[602] Excess or underproduction might explain the pollutant-induced blood sugar swings seen in some chemically sensitive. Arginine also stimulates release of human growth hormone.[603] Glycine (nonessential) is an inhibiting neurotransmitter for spinal cord interneurons; aspartic acid (nonessential) is an excitatory neurotransmitter in the CNS; glutamic acid (nonessential) is an excitatory neurotransmitter in the brain and spinal sensory neurons. (More information on neurotransmitter functions of amino acids may be found in Reference 604; also, see Table 28). Pollutant injury frequently disrupts these functions, resulting in over- or underactivity in their specialized area. This malfunction may then spread to other areas of the body, causing further dysfunction with fully developed chemical sensitivity.

The following paragraphs describe some of the functions of the semiessential and nonessential amino acids that are used for human protein formation (classification 3 above) and aspects of certain nonprotein amino acids. Pollutants may injure these, exacerbating chemical sensitivity (Table 29).

Alanine (Ala). The nonessential protein amino acid, alanine, has several sources in humans. Addition of an amino group to pyruvic acid forms alanine and is the major endogenous source of this amino acid. Amino group transfer to pyruvic acid (usually from glutamic acid) requires a transaminase enzyme with pyridoxal phosphate (PyrPO$_4$) (B$_6$) as the coenzyme. Often this vitamin is deficient in the chemically sensitive, resulting in a slow ability to conjugate; a relatively small amount of alanine is produced in the metabolism of tryptophan; the immediate precursor is 3-hydroxykynurenine. Alanine also is obtained directly from digested food protein.

Alanine has no known special biologic or hormone-like activity. It does participate in an energy transport process between the liver (where chemical energy, glucose, is regenerated) and muscle tissue (where glucose is expanded to form pyruvic or lactic acids) (Figure 37). Alanine formed from pyruvate goes back via the bloodstream to the liver, where more chemical energy (glucose) is generated (gluconeogenesis). This intraorgan transfer is selectively changed with fasting in the chemically sensitive (see Chapter 34). Inappropriate transfer may lead to loss of energy as seen in the chemically sensitive.

The alanine-forming enzyme, alanine transaminase (EC 2.6.1.2), is stimulated to have high activity by glucocorticoid. This activity accounts for the fact that alanine is often measured to be higher than normal in the blood or urine of individuals presenting with Cushing's Syndrome.[605]

β-Alanine (β-Ala). β-Alanine (Figure 38) is a nonprotein amino acid with endogenous sources (catabolism of DNA and RNA), dietary sources (the dipeptides carnosine and anserine), and bacterial sources (usually intestinal flora that may be foreign). In humans, essentially all β-alanine undergoes amino group transfer to form malonic semialdehyde. The transamination coenzyme requires pyridoxal phosphate (PyrPO$_4$) and coreactant α-ketoglutaric acid.[606,607]

Table 29. Pollutant Injury to Semi- and Nonessential Amino Acids and Metabolism May Result in Chemical Sensitivity

Alanine	Energy transport process from liver to muscle for glucose
Arginine	Protein, amino acid and urea cycle — cannot produce from creatinine and ornithine; stimulates insulin and growth hormone; component of vasopressin; only amino acid in urea cycle that produces protein
Histidine	Precursors of histamine; transport of copper; maintenance and growth of tissue
Homocysteine	With serine → cystathione
Tyrosine	Precursor of DOPA, dopamine, norepinephrine, epinephrine, thyroxin, skin pigments
Asparagine	Back-up carrier of nitrogen NH_3; link to glycoprotein antigenic determinants and virus receptors on the cell
Aspartic acid	With glutamine forms asparagine; with citrulline forms arginosuccinate; allows oxidative phosphorylation and chemical energy conversion; urea cycle using ATP; pyrimidines nucleotides
Glutamine	Carrier and disposer of (NH_4) for CNS
Glycine	With cysteine, glutamine → glutathione; building blocks to physiologically active purines, porphyrins, hemoproteins, bile salts; forms creatine
Proline or (hydroxy)	Allows peptides and proteins to have a sharp bend or turn in shape
Serine acid	Synthesis of proteins; proper metabolism of methionine; neurotransmitter; energy production
Glutamic acid	Protein formation; NH_4 transporter to liver; keto analogue is α-ketoglutaric acid; easily crosses blood-brain barrier; with cysteine and glycine → glutathione
Cysteine	Increases rates of insulin degeneration; holds insulin protein and many others with S-S bond; + pantothenic acid → CoA; + glycine, glutamic acid → glutathione

Healthy individuals without tissue necrosis, without bacterial infection, and with adequate pyridoxal phosphate have very low levels of β-alanine in their urine (less than 5 μmol/24 hr) and less than detectable β-alanine in their blood (less than 0.1 μmol/100 mL plasma).[608] β-Alanine abnormalities are seen in some chemically sensitive patients' blood and urine.

β-Alanine can be detrimental when present at elevated levels because it impairs renal conservation of other amino acids that are needed and beneficial. Renal conservation of taurine is especially impaired by β-alanine in the urine and can account for a severe exacerbation of chemical sensitivity.

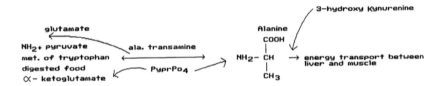

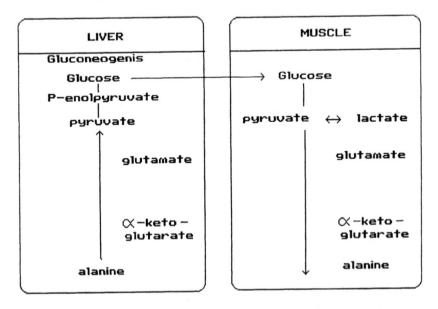

Figure 37.   Potential areas of pollutant injury to alanine transfer between the liver and muscle in the chemically sensitive.

Bacteria produce pantothenic acid in the human gut by combining β-alanine and pantoic acid. Because bacteria is essential to this process and humans cannot manufacture pantothenic acid in a sterile gut, a proper balance of intestinal flora is necessary in the chemically sensitive.

A clinical observation is that many chemically sensitive behavior-disordered children, some with seizures, some diagnosed as having Tourette syndrome, and some that are aggressive and combative, present elevated levels of β-alanine along with elevated levels of the (precursor) dietary peptides anserine and carnosine. Supplementation of pyridoxal phosphate together with reduction of dietary anserine (poultry, tuna, and salmon) and dietary carnosine (pork, beef, tuna, and salmon) are observed to normalize β-alanine, anserine, and carnosine. Coincident improvements in behavior are reported or behavior modification efforts are then successful.

Chemically sensitive food reactors presenting this amino acid abnormality may regress to the category of "universal reactor" if not treated effectively. One strategy for treatment is to reduce β-alanine sources in the diet.

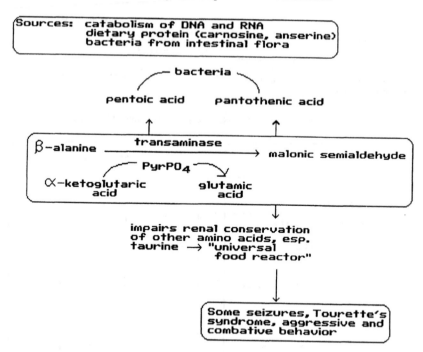

Figure 38. β-Alanine.

These sources are the anserine and carnosine peptide foods: chicken, turkey, duck, rabbit, beef, pork, tuna, and salmon. Chemical overload may alter intestinal absorption or bacterial flora giving more β-alanine.

Arginine (Arg). Arginine (Figure 39) is considered to be a "semiessential" amino acid. It is a protein amino acid, and physiological requirements exceed those which are produced in the urea cycle from its precursor, argininosuccinic acid. Hence, dietary protein sources of arginine are required to supplement that which is formed endogenously.

In the urea cycle, arginine leads to formation of urea, an excreted form of nitrogen, and to the next intermediate of the urea cycle, ornithine. The arginase enzyme which catalyzes formation of urea and ornithine is activated by manganese.

Arginine also combines with glycine to produce the precursor of creatine; ornithine is a side product. Creatine is present in muscle and brain tissue, and its level in the blood or urine can be monitored to assess the possibilities of muscle-wasting disease (muscular dystrophies, amyotrophic lateral sclerosis, poliomyelitis), hyperthyroidism, or hypothyroidism. There is increased utilization of arginine to form increased levels of creatine in muscle-wasting diseases and in hyperthyroidism. There is decreased utilization of arginine and decreased creatine in hypothyroidism, which is seen in some chemically

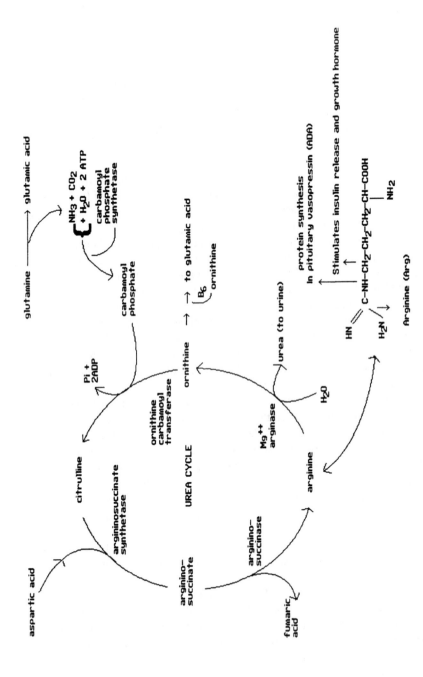

Figure 39.   Arginine.

sensitive patients. Creatine is the precursor of creatinine, an excretant form of nitrogen. Creatinine levels can be used to judge renal clearance or kidney functions.

Arginine is known to have two hormone-like functions: (1) stimulation of insulin release from the pancreas and (2) stimulation of growth hormone release from the pituitary gland.[609,610] Arginine is a component of the pituitary hormone vasopressin (blood pressure, antidiuretic functions), and it is the only amino acid in the urea cycle that is also needed for synthesis of human protein. Many problems with edema and carbohydrate metabolism that may be due to pollutant injuries to these pathways have been observed in the chemically sensitive.

Asparagine (Asn). Asparagine (Figure 40) is formed in humans from aspartic acid and glutamine. Asparagine also is obtained from digestion of food protein. The formation enzyme, asparagine synthetase, is activated by magnesium; glutamic acid is a side product of asparagine synthesis.

Asparagine formed in this manner can be thought of as a "back-up" (for glutamine) as a carrier of nitrogen or ammonia, and, of course, some of this asparagine could be used for formation of protein. Asparagine sometimes is elevated in blood or urine in cases of impaired nitrogen detoxification or hyperammonemia in the chemically sensitive; glutamine usually is elevated in these cases. (See subsequent section on glutamine.) In the kidney, asparaginase is a minor source of urinary ammonia (glutamine decomposition via glutaminase is the major source of prompt urinary ammonia). Renal acidosis favors decomposition of asparagine in the distal tubule of the kidney; renal alkalosis inhibits decomposition of asparagine.

Asparagine, like threonine, serine, hydroxyproline, and hydroxylysine, is used to form the link between the protein and the carbohydrate parts of some glycoproteins. Glycoprotein on the outer surface of the cell membranes are considered to be partly responsible for recognition of cell individuality and identity. They may serve as antigenic determinants and as virus receptors for cells.[611] These may be damaged by pollutant overexposure, causing loss of antigen recognition, as frequently seen in the chemically sensitive patient with a spreading phenomenon. Also, recurrent infections may occur.

Aspartic Acid (ASP). Aspartic acid (Figure 41) is involved in the Krebs cycle, which is essential to intermediate metabolism. It is therefore important in the chemically sensitive, and its metabolism must be understood.

Aspartic acid, a nonessential protein amino acid, is an acidic amino acid with two carboxyl groups, one at each end of the molecule. Aspartic acid can be obtained directly from dietary protein via digestion, and there are numerous aspartic acid-forming steps in human metabolism. A major route for formation of aspartic acid uses oxaloacetic acid (Krebs or citric acid cycle intermediate) and glutamic acid.

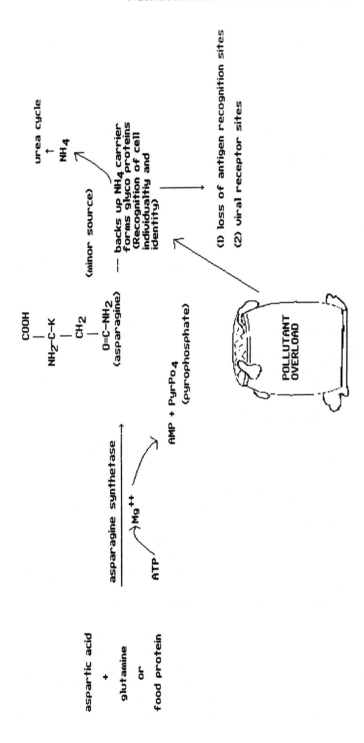

Figure 40. Potential pollutant injury to asparagine.

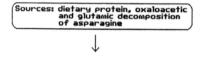

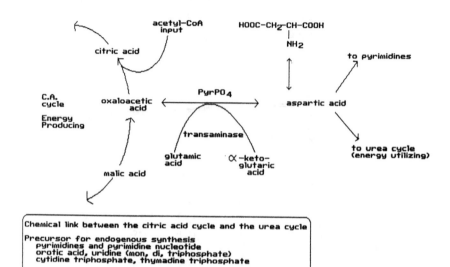

Figure 41.   Aspartic acid (ASP).

Oxaloacetic acid and glutamic acid form aspartic acid using a transaminase enzyme and pyridoxal phosphate as a coenzyme. Aspartic acid and glutamine can form asparagine; decomposition of asparagine produces ammonia and aspartic acid. Aspartic acid also combines with citrulline in the urea cycle (a liver function) to produce argininosuccinic acid. Via oxaloacetic acid and argininosuccinic acid, aspartic acid forms a chemical link between the energy-producing Krebs cycle (allows oxidative phosphorylation and chemical energy conversion to occur) and the energy-consuming urea cycle (which consumes some chemical energy in the form of ATP and which fixes nitrogen in a nontoxic form: urea). Aspartic acid is a precursor for endogenous synthesis of pyrimidine and pyrimidine nucleotides, orotic acid, uridine (mono-, di-, and triphosphates), cytidien triphosphate, and thymidine triphosphate.

In spite of its multiple routes of formation, aspartic acid has been measured to be deficient in blood plasma in some chemically sensitive individuals. Consequent or coincidental conditions reported to Bionostics include fatigue, lack of physical endurance, transient hyper/hypoglycemia, and occasionally, nitrogen imbalances with high ammonia (venous and/or arterial hyper-ammonemia). At the EHC-Dallas, we have seen these conditions occur in the chemically sensitive.

Cysteine (Cys). Cysteine (Figure 42) is a sulfur-containing amino acid that is extremely important in aiding control of chemical sensitivity because of its widespread use in metabolism. It is an important source for the sulfur pool (see Chapter 4). Cystine and cysteine come directly from dietary protein, and cysteine is formed metabolically from methionine. Cystine is the oxidized dimer form of cysteine; it is two cysteines attached by a sulfur-to-sulfur bond. A cystine molecule can be converted to two cysteine molecules by enzymatic processes in the body; two hydrogen atoms are needed to accomplish the chemical reduction of cystine to cysteine.[612,613] Cysteine is a chemically reactive amino acid bearing a sulfhydryl group (–SH), which becomes a point of attachment or chemical bonding.

Cysteine bonds with itself (S-S bond) to hold the insulin protein and many other protein molecules together. Cysteine also reacts with phosphorylated pantothenic acid in a sequence of chemical steps that leads to the formation of coenzyme A, which is required in many metabolism steps, including those of the citric acid (Krebs) cycle. Also, some xenobiotics are rendered less toxic by acetylation.

A cysteine deficiency or limited concentrations of cysteine can result in allergic-like chemical sensitivities and in abnormal glucose metabolism. Cysteine, glutamic acid, and glycine form glutathione, a tripeptide. When cysteine is low, this tripeptide can be in limited supply. Glutathione is needed in the synthesis of prostaglandin, and also it is a detoxifying agent and is involved in the enzymatic conjugation of foreign chemicals such as chlorinated benzene derivatives (organochlorine pesticides)[614] (see Chapter 4). Glutathione functions as a chemical reducing agent that combines with and deactivates inflammatory oxidizing agents such as hydrogen peroxide and other peroxides[615] both by itself and through its glutathione peroxidase enzyme. Glutathione acts as a coenzyme (to glutathione-insulin transhydrogenase) for degrading (deactivating) insulin in the liver and kidney.[616] This deactivation is desired in cases of hypoglycemia where increased glucose levels are beneficial. However, cysteine and glutathione can be detrimental in insulin-dependent diabetes and hyperglycemia because they increase the rate of insulin degeneration. As shown in Chapter 4, glutathionine is one of the prime sulfur conjugators, being essential for detoxification of many chemicals. Chemically sensitive patients are frequently deficient in GSH and glutathione peroxidases due to pollutant injury directly by the heavy metals at the replenishing cycle site and indirectly by overuse when excess xenobiotic load occurs reducing its nutrient fuel. These patients frequently need supplementation.

Within cells, cysteine is the primary precursor of taurine (see Taurine in this chapter); therefore, its pollutant injury will have a more widespread amplification effect in the chemically sensitive because of the many functions of taurine, including peptide conjugation of xenobiotics, membrane stabilization, and leukocyte respiratory burst in phagocytosis.

Cystine and cysteine are contraindicated in cystinuria, which is a renal transport disorder in which conservation of cystine is impaired and cystine

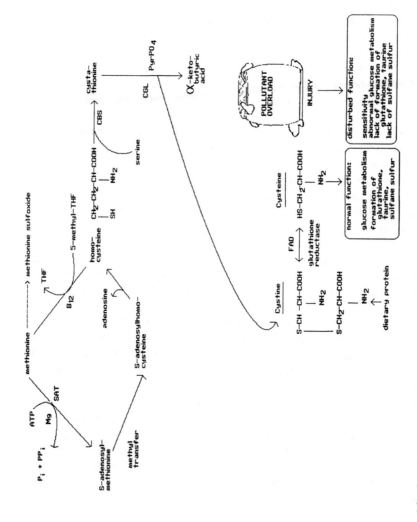

Figure 42.   Methionine, homocysteine, cysteine, cystine formation which can be in chemical sensitivity.

stones may form in the kidneys if urine cystine concentrations are high enough. Clinically, cystine and cysteine are observed to worsen symptoms of intestinal candidiasis, probably because cysteine (like glucose) is a particularly easily metabolized nutrient for *Candida albicans*.[617,618] Temporary use of amino acid blends containing methionine but excluding both cystine and cysteine may be beneficial while the acute phase of candidiasis is treated.

Homocysteine (Hcy). Homocysteine is a nonprotein, nonessential intermediate in metabolism sequence of methionine. It is the "branch point" in methionine metabolism. Some homocystine is methylated with a methyl group ($-CH_3$) donated by methylated folic acid (needs vitamin B_{12} as coenzyme) or by betaine ("trimethylglycine"). The remainder of the homocysteine combines with serine in a vitamin B_6-dependent step to form cystathionine (precursor to cysteine) (Figure 42).

Homocystine is measured (particularly in the urine) to assess whether or not all the nutritional factors are adequate (serine, pyridoxal phosphate, folic acid, B_{12}, betaine, or choline) and to rule out the several inborn and pollutant-acquired errors of metabolism that might cause homocystinuria. Frequently homocysteine is altered in the chemically sensitive.

An interesting research finding on mice with implanted cancerous tumors is that metastasis does not occur if methionine is removed from their diet, while homocysteine, a methyl source, folic acid and B_{12} are present.[619] Of academic interest is the fact that both humans and mice can thrive if the nutritionally essential amino acid methionine is replaced by homocysteine and adequate methyl group sources (with folic acid and B_{12}). This phenomenon might be considered an exception to one aspect of nutritionally essential amino acids — except for the fact that there is negligible homocysteine in food protein.

Taurine (Taur). Taurine (Figure 43) is an unusually structured amino acid being 2-aminoethane-sulfonic acid. It is like an amino acid in that it carries an amino group ($-NH_2$), but it is different because it carries a sulfonic acid group ($-SO_3H$) instead of a carboxyl group ($-CO_2$). It got that way by starting with cysteine, losing CO_2 (which requires pyridoxal phosphate for the decarboxylation), and adding oxygen at the sulfhydryl location ($-SH$). Its function is essential in understanding and treating chemical sensitivity since taurine is used in peptide conjugation of many xenobiotics.

Disordered Metabolism of Taurine Found in the Chemically Sensitive. Taurine is a metabolite of the nutritionally essential amino acid methionine via cysteine or via dietary cystine or cysteine. Taurine levels are particularly susceptible to pollutant injury because they can be affected by incomplete digestive proteolysis or malabsorption, by any of the numerous disorders that can affect methionine metabolism, or by several modes of impaired renal conservation includ-

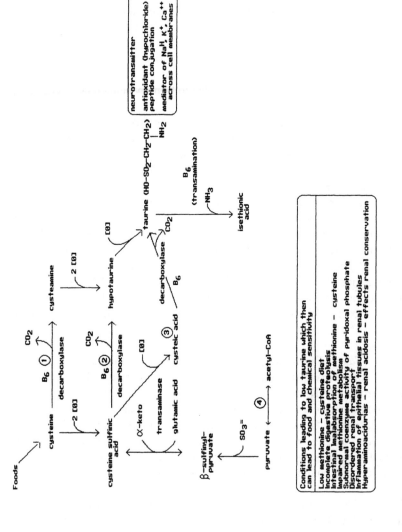

Figure 43.   Pathways for taurine formation.

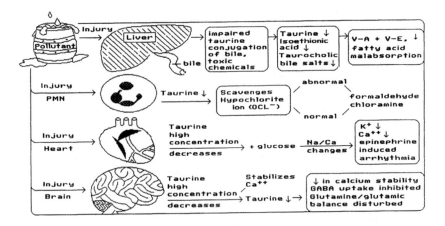

Figure 44.   **The effect of pollutant injury to taurine metabolism usually resulting in chemical sensitivity.**

ing β-aminoaciduria or acidosis with general hyperaminoacidurias. Taurine is not a protein amino acid. It is biologically active with neurotransmitter-, antioxidant- (for the hypochlorite ion), biliary- (peptide conjugation of bile acids) (consumes cholesterol via formation of taurocholic acid), and electrolyte mineral-regulating functions[620,621] (mediator of Na^+, K^+, and Ca^+ ion flux across cell membranes).

Taurine is important in the liver (in bile toxic chemical peptide conjugation) and PMNs, heart, and brain. Each will be discussed individually (Figure 44).

Studies at EHC-Dallas reveal disordered intracellular electrolytes in many patients with chemical sensitivity. Low taurine plays a major part in some of these imbalances.

Taurine and Heart Electrolytes. Low taurine can be coincident with electrolyte mineral imbalance at the cellular level. Potassium, calcium, and sodium are most affected. The effect of taurine on cellular levels of potassium, sodium, and calcium has been studied in mice under conditions of close experimental control. In heart muscle, when taurine is low and when glucose is introduced, potassium and sodium leave the cell and the Na/Ca ratio rises within the cell.[622] This is the reason for observed cardiac arrhythmia when taurine is subnormal in heart muscle cells. Taurine reverses the calcium and potassium depletion.

Taurine is present in relatively high concentrations in the heart and is reported to modulate cardiac activity. It inhibits epinephrine-induced cardiac arrhythmia while at the same time inhibiting the reduction of myocardial potassium concentrations caused by epinephrine. This effect may be due to taurine itself or to its metabolite, isethionic acid.[623,624] Some cases of environmentally induced arrhythmias in the chemically sensitive have been controlled by taurine administration at the EHC-Dallas.

Taurine and Brain Electrolytes. Taurine is also present in relatively high concentrations in brain tissue, but it passes the blood-brain barrier at a notably slow rate. However, changes may occur to the barrier after pollutant injury in the chemically sensitive. In brain tissue, only the calcium-stabilizing effect of taurine has been demonstrated. Fasting and induced vitamin B$_6$ deficiency cannot be shown to lower brain taurine levels significantly.[625] Clinically, taurine administration is observed to alleviate some seizure conditions, and the regulation of membrane transport of electrolyte cations by taurine might be one effect.[626,627] However, low taurine does not appear to be an essential abnormality in epilepsy. Many studies indicate that the ratio of the relative concentrations of the excitatory amino acid glutamic acid to the inhibitory amino acid GABA is much more important in epilepsy. It seems that taurine has a role in keeping this ratio more or less constant, or at least in returning it to normal.[628] Taurine can inhibit GABA uptake but it does not appear to modify the concentrations of the enzymes that form or decompose GABA.

Taurine and Bile. Taurine is important in natural and xenobiotic conjugation in the chemically sensitive via the acylation (peptide conjugation) route and coenzyme A. Herbivores use more glycine and carnivores use more taurine. Baby's milk is high in taurine. Taurine, with glycine, is a key component of bile acid.[629,630] If bile synthesis is disordered, as seen in some cases of chemical sensitivity, then it is possible that assimilation of vitamins A, D, and E (and other lipid-soluble vitamins) is disordered since the intestinal absorption of lipid-soluble vitamins is bile sensitive.[631] Vitamin D "resistance" and chronically disordered assimilation of calcium and phosphorus are observed in some cases of taurine deficiency with chemical sensitivity. Assimilation of essential and dietary fatty acids also may be affected. Bile synthesis utilizes cholesterol; one possible reason for elevated cholesterol can be disordered or deficient bile formation. Pollutant damage in the chemically sensitive enhances exposure.

Beginning with the essential precursor, methionine, a series of metabolic steps leads to cysteine. These steps are sensitive to enzyme assimilation of magnesium for activity of S-adenosyltransferase, coenzyme activity of pyridoxal phosphate for activity of the enzymes cystathionine β-synthase and γ-lyase, and to serine for the formation of cystathionine[632] (Figure 43).

Actually, there are three proposed pathways to taurine from cysteine; all depend upon pyridoxal phosphate for coenzyme activity in a decarboxylation step where CO_2 is released. The major pathway in humans is the one involving cysteine sulfinic acid and hypotaurine.[633] Coenzyme A also has been proposed as a precursor to taurine.

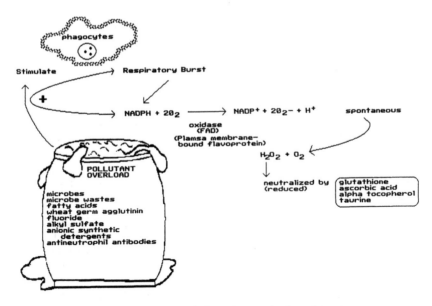

Figure 45.   Potential pollutant injury to respiratory bursts.

When the metabolism of methionine is impaired for some reason, the supply of taurine is dependent upon the dietary uptake of cysteine, cystine (two cysteines linked by a sulfur-to-sulfur bond), and upon the enzymatic reduction of cystine to cysteine. Glutathione reductase (GR) catalyzes the reduction of the sulfur-to-sulfur bond (cystine form) in oxidized glutathione to form GSH (cysteine form). The needed coenzyme is FAD, which is synthesized endogenously from riboflavin (vitamin B_2).[634,635]

Taurine is not part of animal protein. Direct dietary taurine is inadequate for physiological needs, and the above metabolism sequences are required for adequate taurine (Figure 44).

The Antioxidant Role of Taurine. Taurine is an essential amino acid counteracting the hypochlorite reaction in the chemically sensitive. Often dysfunction of its metabolism will lead to a severe aggravation of the sensitivity. The following scenario occurs: phagocytes have the ability to change or destroy foreign chemicals or bacteria by chemical oxidation. In phagocytes, the oxidizing agents are formed in a metabolic sequence called the "respiratory burst." The respiratory burst is initiated when an oxidase enzyme (a plasma membrane-bound flavoprotein) is activated to catalyze the reaction (Figure 45).

Besides microbes and microbial waste products, many different substances have been found to stimulate phagocytic cells to initiate the chemical sequence that results in formation of various oxidizing agents. Included among the substances are certain fatty acids, wheat germ agglutinin, fluoride, anionic detergents (the negatively charged alkyl sulfate or alkanesulfonate part of typical synthetic detergents are examples), and antineutrophil antibodies.[636] Ultimately, various chemical reducing agents are required to keep the oxidizing agents at appropriate levels. Such reducing agents include glutathione, ascorbic acid, α-tocopherol, and taurine.

The $H_2O_2^-$ products are only weakly microbicidal, but they are precursors of more powerful oxidants. The enzyme myeloperoxidase (present in neutrophils and immature mononuclear phagocytes) catalyzes the oxidation of phenols and aromatic amines, and most importantly, this enzyme catalyzes the oxidation of chloride ions[637] (Figure 46). OCl^- is a powerful oxidizing agent. It oxidizes amino acids to chloroamines and to monochloroamine (Figure 47).

Excess pollutant load yields ammonia chloride, CO_2, and aldehydes. Aldehydes (such as formaldehyde) can be handled in very low concentrations by the body's normal metabolic abilities of oxidation; however, higher concentrations and volumes produce notably toxic effects, exacerbating the chemical sensitivity (see Chapter 4).

When taurine is present, the above sequence of reactions proceeds only to the formation of chloroamines (NH_2Cl); aldehydes are not formed. The chloramine is relatively stable and nonreactive. Neutrophils contain relatively high concentrations of taurine, and this taurine is thought to be a control agent for the OCl^-. Houpert et al.[638] report the taurine concentration in PMNs to be 10 times higher than that of any other free-form amino acid. The taurine scavenges excess hypochlorite ion, OCl^- (Figure 48). Once pollutant overload occurs, this scavenging ability decreases and then chemical sensitivity is exacerbated.

Of course, taurine, a unique amino acid carrying SO_3H, instead of COOH, cannot lose CO_2, and it does not form an aldehyde; hence the reason that taurine is an antioxidant that specifically controls the chloride ion and OCl^- concentration. Individuals who are subnormal in taurine at the cellular level may become very sensitive to aldehydes, chlorine, chlorite (bleach), possibly to phenols, and certain amines, resulting in chemical sensitivity or its exacerbation.

Taurine Excretion. Taurine is excreted as taurine or as a catabolite, such as isethionic acid, or it is secreted in the bile as bile salts. The normal amount of urine taurine for the adult female is about 220 to 1300 μmol/24 h, and 350 to 1850 μmol/24 h for adult males.[639,640]

Urine levels can be abnormal in chemical sensitivity due to several causes or dysfunction. For example, a low methionine/cystine diet or incomplete digestive proteolysis or malabsorption often seen in the chemically sensitive can cause reduced levels of taurine. Impaired metabolism of methionine tends to lower taurine levels, and urine taurine can be reduced as a result. Subnormal coenzyme activity of pyridoxal phosphate often effected by pollutant injury may lead to subnormal taurine. Disordered renal transport also can lead to

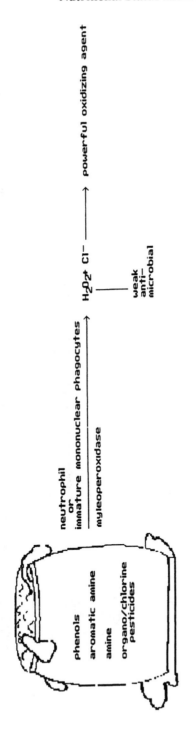

neutrophil
or
immature mononuclear phagocytes

myleoperoxidase

phenols
aromatic amine
amine
organo/chlorine
pesticides

H₂O₂ + Cl⁻ ⟶ powerful oxidizing agent

weak
anti-
microbial

Toxic chemicals may trigger the production of the hypochlorite ion (OCl⁻) which is a powerful oxidizing agent and will damage the surrounding tissues if not neutralized.

Figure 46.   Hypochlorite reaction resulting in chemical sensitivity.

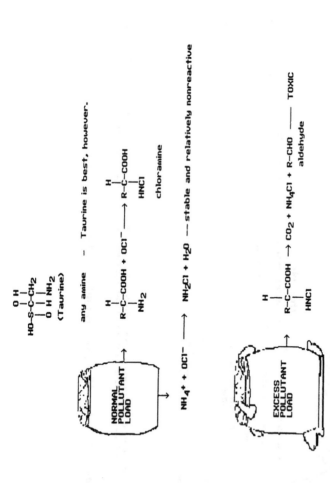

Figure 47.   Scavenging effect of amino acids upon the hypochlorite ion.

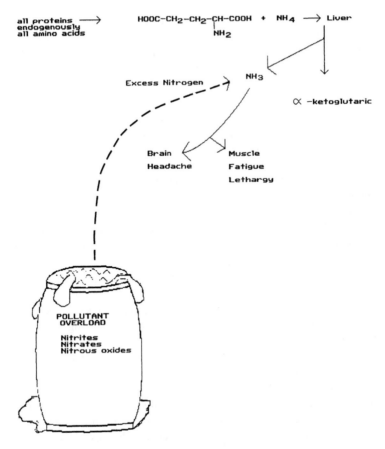

all proteins ⟶
endogenously
all amino acids

$HOOC-CH_2-CH_2-CH-COOH$ + NH_4 ⟶ Liver
                      |
                      NH_2

Excess Nitrogen ⟶ NH_3

α -ketoglutaric

Brain   Muscle
Headache   Fatigue
       Lethargy

POLLUTANT
OVERLOAD

Nitrites
Nitrates
Nitrous oxides

Figure 48.   Glutamic acid.

subnormal taurine in liver cells and in the bloodstream. Inflammation of the epithelial tissue in kidney tubules and hyperaminoacidurias due to pollutant injury in general (renal acidosis) also appear to affect renal conservation of taurine (clinical observation) in the chemically sensitive. β-Aminoaciduria occurring with disordered metabolism of β-aminoisobutyric acid, β-alanine, and methylhistidine can block tubular reabsorption of taurine in the kidneys.[641] Poor renal conservation of taurine can cause elevated urine levels and subnormal cellular levels. Subnormal or abnormal bile formation can result.

Glutamic Acid (Glu). Glutamic acid (Figure 48) is available from all protein foods and is the amino acid that "started" the business of synthetically producing amino acids on a large commercial scale.[642] In the form of monosodium glutamate, the sodium salt of glutamic acid is used as a flavor-enhancing food additive and certainly causes many exacerbations of chemical sensitivity when ingested.

The major endogenous formation route for glutamic acid results from the addition of an amino group ($-NH_2$) to α-ketoglutaric acid. The amino group may come from any one of a number of essential or nonessential amino acids, and the amino group transfer ("transamination") requires pyridoxal 5-phosphate as a coenzyme.

Amino acids that transaminate and lead to endogenous formation of glutamic acid include alanine, β-alanine, aspartic acid, leucine, isoleucine, valine, tyrosine, and ornithine. Glutamic acid formed in this manner becomes a carrier of nitrogen for body tissue (see Chapter 4).

Glutamic acid can be excreted in the urine, but it usually is well conserved in the kidneys, and urine levels are relatively low. For most of the human body (not the brain or CNS) glutamic acid acts as an ammonia transporter by bringing the amino group to the liver, where it gives up the amino group as ammonium ions. The nitrogen from this ammonia then is chemically bound as urea by the urea cycle. In the process of giving up its nitrogen, glutamic acid reverts to its keto analogue, α-ketoglutaric acid.

We have seen a subset of chemically sensitive patients who emit a strong ammonia odor. This odor usually increases episodically with pollutant injury, causing periods of brain dysfunction. Other patients have been seen who have had a strong ammonia odor emanating from the vagina following a food or chemical reaction.

In the CNS, glutamic acid adds ammonia to form glutamine, which acts as the ammonia carrier. Glutamine crosses the blood-brain barrier much more rapidly than does glutamate.[643,644]

Glutamic acid can be detrimental as a food additive or nutritional supplement if ammonia detoxification metabolism or urea cycle impairments are present, as seen in some chemically sensitive individuals, because the ingestion of glutamic acid directly adds transferable nitrogen as an amino group to the body. The most commonly observed symptom of excess nitrogen in the CNS of the chemically sensitive is headache, while fatigue and lethargy may occur when nitrogen is elevated in muscle tissue. Unfortunately, much the same symptoms are observed in nitrogen-deficit conditions such as malnutrition resulting from gastrointestinal dysfunction, protein-deficient diet, or alcoholism. Glutamic acid supplementation can be beneficial in nitrogen-deficit conditions, but it does not (directly) raise brain levels of nitrogen as glutamine might. Glutamic acid is contraindicated in gout, where blood levels may be notably elevated, possibly because of reduced renal clearance for glutamine and other amino acids.

Glutamine (GLN). Glutamine (Figure 49) is glutamic acid with the carboxyl group on one end replaced by an "amide" group. During endogenous formation of glutamine, glutamic acid receives amide structure in the process of adding on "ammonia" (refer to the previous section for information on glutamic acid). Besides endogenous formation from glutamic acid, especially in the CNS, glutamine is an amino acid obtained from dietary protein via digestive prote-olysis, which may be impaired in the chemically sensitive.

Glutamine serves as the carrier and disposer of nitrogen (ammonia) for the CNS.[644] The major pathway for removal of ammonia in the brain is addition of ammonia to α-ketoglutaric acid to form glutamic acid; next, the addition of more ammonia and amide formation to make glutamine; then, passage of glutamine out of the brain. After crossing the blood-brain barrier, glutamine can participate in one of four processes: (1) urinary excre-tion; (2) reaction with aspartic acid to form asparagine and glutamine acid (asparagine synthetase enzyme, needs Mg^{++} and ATP); (3) addition of water and release of ammonia (ammonium ion) which replaces the amide group with a carboxyl group to revert back to glutamic acid (glutaminase enzyme) (in the liver the ammonia may then be processed into urea; the major route for ammonia detoxification for the rest of the human body is via the liver's urea cycle); and (4) various metabolic sequences, including formation of protein.

Glutamine supplementation is contraindicated in hyperammonemia, ammonia intoxication, or nitrogen excess. Glutamine can be beneficial in case of CNS nitrogen depletion (as in alcoholism) or other xenobiotic alcohol excess as seen in some chemically sensitive patients. Both excesses and deficiencies of nitrogen in the brain can lead to headaches. Glutamine along with glycine and taurine is part of the peptide conjugation of xenobiotics with acyl-CoA.

Glycine (Gly). Glycine (Figure 50) is the most simply structured amino acid; it does not have an asymmetric carbon, and the D- or L-configuration issue does not apply. This nonessential amino acid is abundant in protein foods and is obtained via digestive proteolysis. It can be formed endogenously from the essential amino acid threonine or from the nonessential amino acid serine. Transamination from glutamic acid to glyoxylate (requires pyridoxal phos-phate) also leads to the formation of glycine. Glycine is a precursor or building block for physiologically active compounds such as purine, porphyrin, hemo-protein, bile salts (glycine and taurine conjugates; glycine and choleic acid form glycocholic acid), glutathione (an anti-inflammatory or antioxidant and detoxifying peptide composed of glycine, cysteine, and glutamic acid), and creatine. Glycine, along with taurine and glutamine, are utilized in the xenobiotic peptide conjugation reactions using acyl CoA.

Glycine metabolism can be impaired in the chemically sensitive with pyri-doxal phosphate dysfunction, and elevated blood and urine glycine levels can result. Hyperglycinuria caused by pollutant injury with defective reabsorption in kidney tubules can be coincident with chronic or severe vitamin D defi-ciency. Renal transport disorders that cause poor conservation of glycine and lowered blood levels are much less common than the B_6 dysfunction condition (observation based on Bionostics case experience).

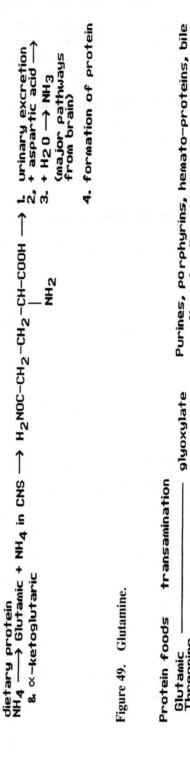

dietary protein

$NH_4 \xrightarrow{} $ Glutamic + NH_4 in CNS $\longrightarrow$ $H_2NOC-CH_2-CH_2-CH-COOH \longrightarrow$
& α–ketoglutaric

NH_2

1. urinary excretion
2. + aspartic acid $\longrightarrow$
3. + $H_2O \longrightarrow NH_3$
(major pathways from brain)

4. formation of protein

Figure 49.   Glutamine.

Protein foods   transamination

Glutamic \
Threonine $\xrightarrow{B_6}$ glyoxylate

Serine \
Glutamic acid $\xrightarrow{B_6}$ NH_2-CH_2-COOH $\xrightarrow{B_6}$

glycine

Purines, porphyrins, hemato-proteins, bile
   salts, glycolic acid
Glutathione
Creatine

Figure 50.   Glycine.

Serine (Ser). Serine contains one carboxyl group, one amino group and one hydroxyl group (–OH) (Figure 51). Serine is a nutritionally nonessential amino acid that is required for the synthesis of human protein and is needed for the proper metabolism of methionine. Serine is considered nonessential because there are several possible routes for its endogenous synthesis from dietary precursors. It can be formed from the essential amino acid threonine via glycine (also from dietary glycine directly), or by dephosphorylation of phosphoserine. By reacting with tetrahydrofolate, serine forms glycine (and methylene-tetrahydrofolate); by reacting with methylene-tetrahydrofolate, glycine forms serine. Serine is formed as a product of glycolysis from phospho-glycerate via transamination of 3-phosphohydroxypyruvate and subsequent dephosphorylation of phosphoserine. In most cases, pyridoxal phosphate activity is required for the endogenous synthesis of serine, and the glycolysis route requires a number of other co-factors, including magnesium. Of course, digestion of dietary protein provides a direct source of serine as well. Serine, along with threonine, asparagine, hydroproline, and hydroxylysine, forms glycoproteins which are often disturbed in the chemically sensitive.

In addition to its use for protein formation, serine is an intermediate or precursor for many metabolism sequences, some of which are indicated in the previous diagram. Ethanolamine and phosphoethanolamine lead to choline and the neurotransmitter, acetylcholine. Pyruvic acid and oxaloacetic acid are involved in the body's energy conversion chemistry (Krebs or citric acid cycle). Serine also combines with homocysteine in the methionine metabolism sequence to form cystathionine; pyridoxal phosphate is required as the coenzyme. Subnormal serine in this chemistry can result in homocystinuria.

In spite of its several routes for formation or supply, serine deficiency has been occasionally demonstrated through laboratory testing (from Bionostics and EHC-Dallas case experience). The reported symptoms for those individuals are essentially those of cysteine deficiency: fatigue, enhanced inflammatory responses, allergic-like sensitivities (or enhancement of true allergic responses), and extreme chemical sensitivities or susceptibility to chemical toxicity. One case of sexual dysfunction was reported with normalization of function several days after beginning supplementation of serine (500 mg/day).

Histidine (His). Histidine (Figure 52) is a semiessential amino acid that contains three nitrogen atoms; two are included in the imidazole ring, and one nitrogen is in the amino group attached to the α-carbon.

Histidine is needed for maintenance and growth of tissue, and it is the precursor of histamine, a vasodilator secreted by cells in the liver, lungs, stomach, and other tissues. Histamine secretion may occur as a result of allergic hypersensitivity or inflammation. Histamine also stimulates gastric secretions and is required for proper functioning of the stomach. Histamine release often occurs in chemical sensitivity, exacerbating the problem; however, many patients with chemical sensitivity have peripheral vasoconstriction

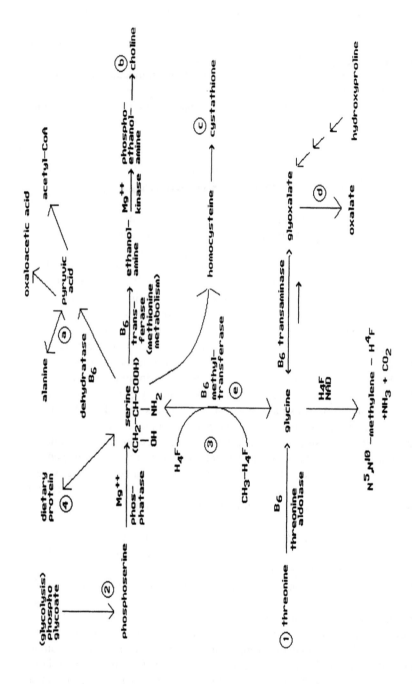

Figure 51.   Ways of forming serine 1, 2, 3, 4, — its action as a precursor in metabolic sequences a, b, c, d.

and hypochlorohydria, which suggests a block or retardation of the conversion of histamine from histidine. In addition, observations of the chemically sensitive at EHC-Dallas have revealed that the proper intradermal dose of histamine will stop reactions (see Chapter 36).

Histidine is catabolized to form glutamic acid in a chemical sequence that requires proper activity of tetrahydrofolate. The condition of elevated histidine in the urine, histidinuria, may be the result of (1) poor renal conservation, a kidney disorder that may or may not be clinically consequential; (2) nutritionally induced by folic acid deficiency (formiminoglutamic acid, or "FIGlu," then is high in the urine); and (3) an enzymatic dysfunction such as with histidine-ammonia lyase. Conditions (2) and (3) also cause histidinemia.

A clinical finding in some cases of rheumatoid arthritis (RA) is that plasma histidine is subnormal (not observed in all cases of RA).[645,646] Also to be noted is the role of histidine in the transport of copper,[647] and the clinical finding of a low copper to zinc ratio in SOD in arthritis.[648] Bionostics has received reports of the remission of RA following supplementation of histidine and copper.

Proline (Pro), Hydroxyproline (Hyp). Proline (Figure 53) is a ring-structured, nonessential amino acid that includes a nitrogen atom in the ring. With this structure it is actually an imino acid rather than an amino acid. Hydroxyproline (Figure 53) occurs in two forms, both are proline with an –OH attached to one of the carbon atoms in the ring. 3-Hydroxyproline (3-hydroxy-pyrrolidinecarboxylic acid) and 4-hydroxyproline (4-hydroxy-pyrrolidinecarboxylic acid) are found in human collagen, as is proline. The structure of proline allows peptides and proteins to have a sharp bend or turn in their shape because these are the linkage points where other amino acids are attached to form peptides and proteins. They also form glycoproteins which are so important in chemical sensitivity, along with asparagine, serine, threonine, and hydroxylysine.

When collagen is formed, the hydroxy group is added after proline is in the peptide structure.[649,650] Hydroxyproline itself is not a building block for protein and is somewhat extraneous when it is included in amino acid blends where proline also is present in adequate amounts.

Normally, human urine contains negligible amounts of proline; it is well conserved in the kidneys. Metabolically, proline exchanges with glutamic acid or glutamic acid semialdehyde; the ring opens or closes for this exchange. Slight elevations of proline in the urine usually indicate a slight problem in renal conservation with little or no clinical significance. Slight elevations in hydroxyproline in the blood usually mean enhanced intestinal absorption (or malabsorption) of dietary peptides that contain 3- or 4-hydroxyproline. Notable excesses of proline and/or hydroxyproline can indicate a disorder in collagen metabolism. Slight or moderate excretion of hydroxyproline may occur following burns, fractures, or surgery, or it can be coincidental with arthritis, lupus, osteoporosis, and other collagen-degenerative diseases with a chemical sensitivity component.

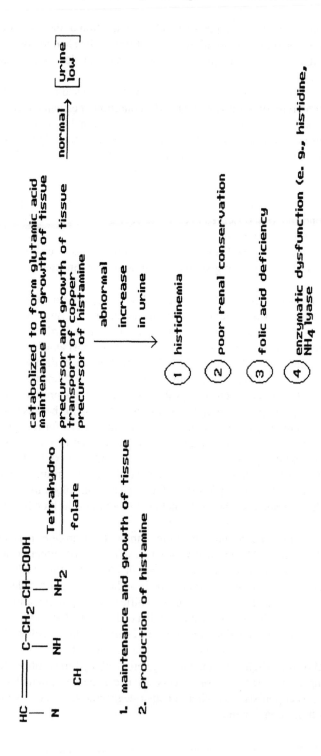

Figure 52.   Histidine.

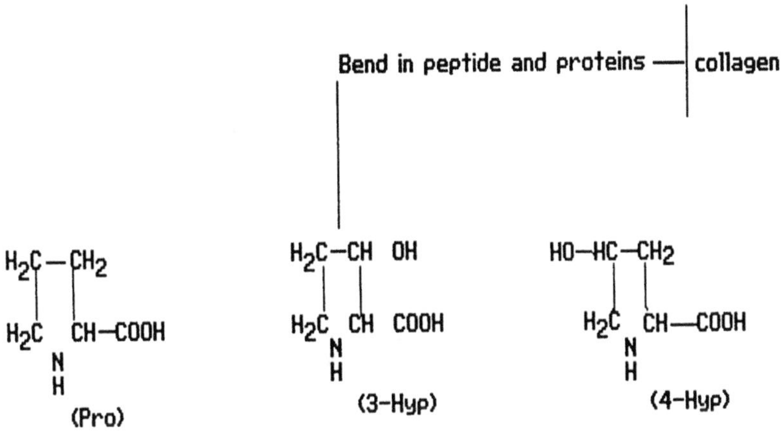

Bend in peptide and proteins —| collagen

H₂C—CH₂

H₂C   CH—COOH
     N
     H
   (Pro)

H₂C—CH OH

H₂C CH COOH
     N
     H
   (3-Hyp)

HO—HC—CH₂

H₂C CH—COOH
     N
     H
   (4-Hyp)

Figure 53.   Proline.

Effects of Pollutants on Fat and Fat Metabolism

The effect of quantity and quality of dietary fat on the metabolism of drugs and xenobiotics has drawn the attention of investigators (Figure 54). Dietary fat may have a profound effect on the toxicity of foreign compounds and on the action of drugs through altering microsomal drug metabolism or by changing the bioavailability of drug or toxicants to the site of action. Phosphatidylcholine is known as an essential component of the MFO system,[651] and its fatty acid composition in the membrane can be altered through dietary fat treatments[652,653] as well as xenobiotic insult. The requirement for quality and quantity of dietary fat, for optimal activity and induction of cytochrome P-450 has been investigated. Caster et al.[654] and Lam and Wade[655] have reported an increase in hexabarbitol metabolism (increase in cytochrome P-450) in the liver of rats fed 3% plus 10% corn oil. Similar findings for phenobarbital were found in animals fed herring oil or linoleic acid.[656] Linseed and menhaden oils showed similar responses while coconut oil, beef fat, and low levels of corn oil had the lowest stimulation of in vitro drug metabolism.[657] Andersen et al.[658] showed a high fat diet (70%) increased the half-life of antipyrine and theophylline and low fat (30%) resulted in a lower half life for both drugs.

Tumor enhancement by a high fat diet was reported by Watson and Mellanby.[659] Tumor promoting action of dietary fat through microsomal enzyme induction has been suggested for aflatoxin-induced liver tumor. Newberne et al.[660] found that rats fed with beef fat had depressed p-nitroanisol demethylase activity as compared to corn oil-fed rats. Cheng et al.[661] found the concentration of nuclear envelope P-450 from 10% corn oil-fed rats was higher than rats fed fat-free diets.

Fatty acid composition of phospholipid in the endoplasmic reticulum may exert a major effect on the antibody of the cytochrome P-450 systems. Davison

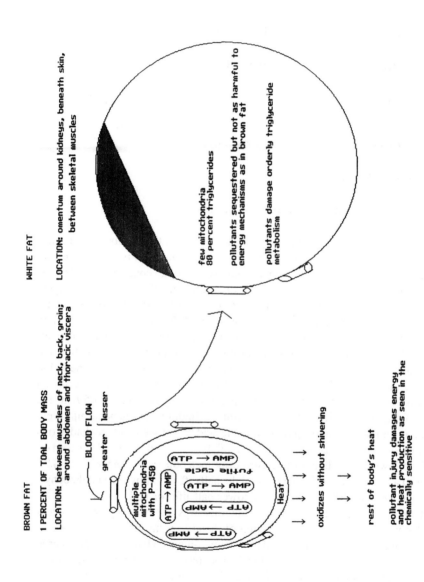

Figure 54.   Pollutant effects on brown and white fat.

and Wills[662] demonstrated that the incorporation of linocate into phospholipid of microsomes increases when phenobarbital or 3-methylcholanthrene (3-MC) inducers are used.

Lambert and Wills[663] reported a maximal in vitro demethylation in both induced and controlled rats fed herring oil, which was high in polyunsaturated fatty acid, but contained 1.17% lineolate. Lang[664] has shown that o-demethylation was inhibited with high levels of dietary linolate whereas arylhydrocarbon hydroxyls (AHK) activity was depressed with low levels of dietary fat. Pollutant injury as well as drug injury will be influenced by these factors.

The effects of pollutants on fat and its metabolism are extremely complex in the chemically sensitive. Toxic chemicals tend to be lipophilic and, therefore, accumulate in fats, especially lipid membranes. They disturb lipid metabolism and also dietary intake.

The average American consumes 38 to 42% of his dietary calories in fat. This consumption is almost the same proportion that comes from carbohydrates, which results in doubling the calories over the carbohydrates. This percentage of fat to carbohydrate intake is much less in the chemically sensitive and probably accounts for some of the weight loss often seen in the chemically sensitive, especially in more advanced cases. Cholesterol consumption in the general population has remained constant over the years, but the proportion of unsaturated fats has decreased. However, 15% of these unsaturated fats are *trans*- rather than *cis*-forms and are probably unavailable for proper nutrition. They do still carry toxic chemicals which are delivered to intestinal flora and villa for absorption. Increased ingestion of fats has been linked to increased cancer and arteriosclerosis in rats and humans and may be linked to some of the initial stages of chemical sensitivity. As shown throughout this text, many toxic chemicals are lipophilic and, therefore, are attracted to and grab onto fat molecules. This attachment to fat not only causes local membrane damage and disturbance of fatty acid metabolism, it also allows a pool for rapid transfusion of toxic chemicals into the blood when fat is metabolized. This fact is demonstrated in Chapters 34 and 35. The chemically sensitive person is very aware of the reactions that occur from these processes.

There are two types of adipose tissue: white and brown. These have defined functions and understanding them is crucial when treating the chemically sensitive because brown fat is high in microsomes containing high energy phosphate bonds which supply energy fuels for the body, and thus heat which is critical to the chemically sensitive who are often cold. White fat is full of triglycerides, which can act as a reservoir for toxic chemical sequestration.

Brown Fat

Brown tissue is located between the muscles of the neck and back, in the groin, around the abdominal viscera, and the thorax. This tissue makes up 1%

of body mass and when "revved up" can produce an amount of heat equal to what the rest of the body can produce. The cells of brown fat are only about 10% the size of white adipose cells but have larger and more numerous mitochondria with the cytochrome P-450 system, which apparently gives them their brown color.

Brown fat is well developed in newborn infants and will oxidize its fat during exposure to cold in order to produce heat without shivering. The thermogenesis is apparently accomplished via a futile cycle in which triglycerides are hydrolyzed to fatty acids, then to their coenzyme derivatives, and back to triglycerides. In this process, high energy phosphate is expended. ATP is hydrolyzed to AMP and to inorganic phosphate and then is regenerated. The net result is the combustion of fuel that generates heat but accomplishes nothing else. Since toxic chemicals are attracted to fat and it is clear that pollutant injury to the mitochondria frequently occurs, one can understand how and why the chemically sensitive patient is usually hypothermic with temperatures running in the 95° to 97°F range. Once this mechanism is disturbed by pollutants, the chemically sensitive individual may not function well and may not generate heat well. As shown in multiple areas in this book, blood flow is altered due to pollutant-triggered vascular malfunction. Normally there is a greater blood flow to brown fat, which can be disturbed by the vascular malfunction resulting in inappropriate dissipation. This disturbed blood flow, in addition to direct pollutant damage to the heat production process, may be highly significant in the chemically sensitive.

White Fat

White fat differentiates at 3 to 4 months' gestation from mesoderm and is a form of connective tissue. It is found in the abdominal cavity (omentum), around the kidneys, underneath the skin, and between skeletal muscle fibers. Its primary function is to store energy.

A 70-kg man contains on an average 9 to 13 kg of adipose tissue of which 80% is triglyceride. The total weight of adipose tissue is capable of varying 50-fold, a change greater than any other organ while still being capable of life. An increase in adipose tissue can occur by either an increase in the number or size of cells and is seen in many patients in the early stage of chemical sensitivity prior to development of the severe catabolic phase. Toxic overload appears to trigger obesity in some chemically sensitive, perhaps as an adaptive mechanism to sequester harmful chemicals.

Decrease in adipose organ size is probably due to only a decrease in the size of the cells. Each cell of white adipose tissue contains a thin layer of cytoplasm surrounding the triglyceride in a single droplet. The cell nucleus lies in the layer of cytoplasm so that a cross-section looks like a signet ring.

Adipose tissue is in a continuous state of flux and is important in many metabolic processes. These cells form and store triglycerides and release fatty acids for energy. Pollutant injury can alter this state of flux, depleting reserves in the chemically sensitive, which may be one reason for their loss of not only immediate energy but also stamina. Adipose tissue can acquire fatty acids from synthesis within the cell of excess glucose, acquisition from those circulating in the plasma bound to albumin, or following release from lipoproteins by the lipase enzymes. Pollutant injury may imbalance these routes of fatty acid acquisition, resulting in excess or lack of fatty acids. The two lipase enzymes may be directly damaged by pollutant injury. Again, release of toxic chemicals may occur with this exchange.

Pollutant Injury to Triglycerides in the Chemically Sensitive

Triglycerides are the most efficient form of fat (in fat cells 99% of volume) for storing the calories essential to energy-requiring processes in the body. When utilized, they are about 9 kcal/g vs. 4.5 kcal/g of carbohydrate or protein. Besides energy use, triglycerides can be converted to cholesterol, phospholipid, and other lipids. They are also used as insulation and padding (Figure 55).

Most of the lipid that accumulates in the tissue in response to a toxic agent is composed primarily of triglycerides, followed by fatty acids. Six general mechanisms can account for the accumulation of lipid material in the liver tissue.[665,666] Triglyceride synthesis increases, release of triglycerides from the liver into the bloodstream decreases, triglyceride synthesis shifts to nonsecretory or slow secreting pools, a combination of increased synthesis and decreased release of triglycerides and fatty acids occurs, and an increased supply of fatty acids and a decreased synthesis of glycoprotein is possible. Some patients with chemical sensitivity have been observed to experience disturbance of one or more of these mechanisms resulting in deranged metabolism.

Some effects that lead to liposes are mediated via changes in transport, synthetic, or catabolic mechanisms. Some could be indirectly mediated via alterations in hormonal-regulating mechanisms. Catecholamine regulation of body fat mobilization, which is often disturbed in the chemically sensitive, can be altered by toxic chemicals to result in increased availability of fatty acids. Increased hepatic triglyceride synthesis could occur because of the increased availability of fatty acids and glycerol phosphate. This triglyceride synthesis can directly cause increased accumulation of triglycerides in the liver. Increased synthesis of fatty acids has been demonstrated after ethanol exposure in a variety of in vitro and in vivo models. This increased synthesis may be related to the increased NADPH/NAD ratios as a result of alcohol metabolism. Alcohol metabolism catalyzed by alcohol and aldehyde dehydrogenases results in increased generation of NADPH and acetyl fragments which facilitate the availability of fatty acids. Increased supply of fatty acids to the liver can also result from alcohol-evoked mobilization of peripheral fat. Increased production of reducing equivalents may also produce a block in the utilization of acetyl fragments that can be shunted to elongation of fatty acids. Thus,

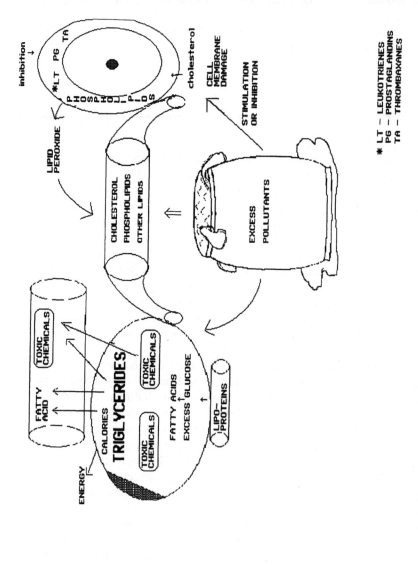

Figure 55. Pollutant injury may imbalance these routes of fatty acid acquisition → excess or lack.

ingestion of alcohol can result in a dual effect of decreased oxidation of fatty acids and increased synthesis, as well as contributing to the elevation of fatty acids. The chemically sensitive individual is often overloaded by excess xenobiotic alcohols from many pollutant sources.[667]

Mere accumulation of fat in the tissue does not necessarily represent a toxic response of the liver tissue to injury inflicted by toxic chemicals. In most cases, high accumulation of fat in the liver or other tissue should be taken to mean that many metabolic pathways have been altered from the norm by the toxic chemical. This would signal the high-risk status of the liver for hepatotoxic mechanisms to operate, for example, the free-radical mechanisms that may result in lipid peroxidation, as exemplified by CCl_4 hepatotoxicity. Accumulation of a high level of fat in the liver or other tissues must add to the lipid peroxidative damage that ensues. Thus, in most instances, it is clear that accumulation of fat by itself may not represent a toxic response but may in fact represent a high-risk status of the liver tissue for damage by other agents.[667]

Pollutant Injury to Phospholipids and Cholesterol in the Chemically Sensitive

The principal function of phospholipids and cholesterol is the formation of all interior and exterior cell membranes, all or part of which may be damaged in the chemically sensitive. These lipids provide a semifluid matrix within which float various forms of membrane protein. The lipid matrix consists of a bilayer of interior cholesterol and two layers of phospholipid with polar groups facing the exterior and interior aqueous environments. The presence of specific polyunsaturated fatty acids on membrane phospholipids is important, not only for cell membrane structure, but also as a source of substrate for the formation of prostaglandins, leukotrienes, and thromboxanes, which are essential for normal functioning. The ratio of formation of these substances appears to be altered in some chemically sensitive patients. Pollutant injury to these areas may not only cause membrane inhibition or leak but damage may also result in altered responses to the normal formation of prostaglandins, leukotrienes, and thromboxanes. Free radicals also can be generated in the form of lipid peroxides (Ro Roo⁻) and their presence seems inordinately high in the chemically sensitive. Some pollutants such as DDT and chlordane have long half-lives and may remain on these membranes for years, thus propagating chemical sensitivity.

Pollutant Injury to Lipoproteins and Apolipoproteins

Pollutant injury to lipid proteins may be a special problem in the chemically sensitive due to the unique features of the lipid and protein components.[668]

Lipoproteins are dynamic particles that transport lipids and proteins in the circulation. Both lipoproteins and apolipoproteins are strong predictors of coronary heart disease (CHD) risk. Many dietary factors affect plasma lipid and lipoprotein levels. It is clear that the diet modifications can beneficially affect plasma lipid (specifically plasma total-C) levels and in turn lower the risk/incidence of CHD in both chemically sensitive individuals and the population at large.[668]

There are five major classes of lipoproteins: chylomicrons, very low-density lipoproteins (VLDL), intermediate-density lipoproteins (IDL), low-density lipoproteins (LDL), and high-density lipoproteins (HDL), which have to be considered when evaluating the chemically sensitive. HDL are further divided into HDL_2 and HDL_3 and LDL into small and large particles. Lipoproteins are separated on the basis of their density, which is determined by the relative proportion of lipid to protein in the particles.[668]

Lipoproteins have a hydrophobic core consisting of nonpolar lipids (triglyceride and cholesterol esters) coated with surfactants (phospholipids and unesterified or free cholesterol) and proteins that are referred to as apolipoproteins. Toxic chemicals may insert in these areas, causing disturbances in the chemically sensitive (Figure 55). The proportions of triglyceride, cholesterol, phospholipid, and protein differ among the specific lipoprotein classes. Chylomicrons and VLDL are triglyceride-rich lipoproteins. LDL, the major cholesterol transport particle in plasma, carries 70% of plasma total cholesterol. HDL is a protein-rich particle; 50% of its mass is protein.[668]

Lipoprotein metabolism is complex and is a function of the relative rates of synthesis, degradation, and exchange among the lipoprotein classes, all of which may be altered by toxic chemical exposures. Many factors such as total body load, genetics, hormones, diet, age, and gender affect lipoprotein metabolism and, hence, concentration. Conceptually, there are three lipoprotein metabolic systems which may be disturbed by toxic exposure. These include exogenous fat transport, endogenous fat transport, and reverse cholesterol transport (see Havel[669] and Breslow[670] for recent reviews). A brief description of each pathway follows, and a summary is given in Figure 56.

Dietary triglycerides are converted into chylomicrons in the intestine. Nascent chylomicrons containing apolipoproteins B-48, A-I, and A-IV on the surface of the particles enter lacteals in the intestinal villus and are transported through the thoracic duct into the blood. Chylomicrons acquire apolipoproteins C and E in the blood from HDL. Apolipoprotein C11, a required co-factor for lipoprotein lipase (LPL), is an enzyme that resides on the capillary endothelium and hydrolyzes chylomicron triglycerides. During the depletion of the triglyceride core of the chylomicron, some surface material (primarily phospholipids, cholesterol, and apolipoproteins) is transferred to HDL. Apolipoprotein E is transferred from HDL to chylomicron remnants, which are then efficiently removed by the hepatic apolipoprotein E receptor. The receptor-bound remnant is then internalized via endocytosis and translocated to the lysosymes, where

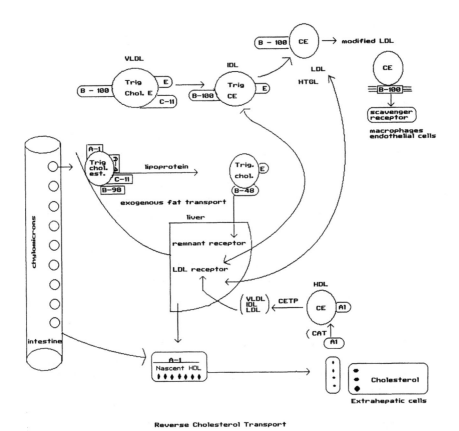

Figure 56.   Lipid transport — pathways endogenous fat transport.

the remnant lipid and protein constituents are catabolized. Lipophilic toxic chemicals in the gastrointestinal tract will be incorporated in the chylomicrons and transported to the liver, probably entering and possibly interfering with the orderly disturbance of the process in pollutant injury.

Endogenous fat transport refers to the synthesis and catabolism of VLDL and LDL (Figure 56). VLDL is a triglyceride-rich lipoprotein that is hydrolyzed to IDL by LPL. Surface material from IDL is transferred to HDL. Unesterified cholesterol is esterified in HDL via the action of the enzyme lecithin cholesterol acyltransferase (LCAT) which may be damaged by pollutants. The esterified cholesterol is transferred back to IDL, leading to a cholesterol ester enrichment of IDL. Some IDL is cleared from plasma by hepatic LDL receptors that bind apolipoprotein E. The remainder undergo rapid continued lipolysis, probably by hepatic triglyceride lipase, during which time all apolipoproteins except B-100 are transferred to other lipoproteins. The result is the formation of LDL with apolipoprotein B-100 on its surface. That form of apolipoprotein B is recognized by the hepatic and extrahepatic high affinity LDL receptors. Approximately 70% of LDL is removed from plasma via this mechanism, principally by the liver. The remaining LDL is modified in plasma and removed by scavenger receptors on macrophages and endothelial cells. Plasma LDL levels are influenced by the LDL receptor number, which in turn is regulated by the need for cholesterol by the cells. When the need is low, cells make fewer receptors and remove LDL at a reduced rate. As a result of those cellular events, the rate at which LDL is removed from the plasma decreases, and there is a corresponding rise in plasma LDL.[671] An elevated LDL leads to an accelerated rate of atherogenesis.

HDL functions in reverse cholesterol transport. This system provides a mechanism by which cholesterol in peripheral tissues can be excreted. Nascent HDL, produced in the liver and intestine, accepts unesterified cholesterol from extrahepatic cells. The free cholesterol is esterified via the enzyme LCAT, utilizing apolipoprotein A-1 as a cofactor, and moves to the core of HDL. Mature HDL is formed by the addition of surface components derived from chylomicron and VLDL catabolism (i.e., HDL_3 is converted to HDL_2). One can see how in the subset of chemical sensitivity how toxic chemicals could deregulate the LDL and HDL mechanism causing increased cholesterol in the blood.

The apolipoproteins play an important role in lipoprotein metabolism. The plasma transport lipoprotein for each apolipoprotein is identified in Figure 56, and the major characteristics and functions are clinically relevant in CHD and familial lipidemia.

Plasma total-C, LDL cholesterol, and HDL cholesterol (LDL-C, HDL-C) are strong predictors of CHD risk.[672] Plasma total-C and LDL-C are positively and HDL-C is negatively related to CHD risk. HDL_2 is a better predictor of CHD risk than either total HDL-C or HDL_3.[673] Smaller LDL are thought to be more atherogenic than larger LDL.[674]

The LDL-C and total-C:HDL-C ratios are frequently examined as an assessment of CHD risk. The LDL-C:HDL-C ratio provides information about the relative proportion of cholesterol transported in the undesirable vs. desirable lipoprotein fraction. A high LDL-C:HDL-D ratio is strongly associated with an increased risk of ischemic heart disease.[675] The total-C:HDL-C ratio provides information about the proportion of total-C carried in the HDL fraction. This ratio has been shown to be a consistent and important indicator of CHD risk.[676] Since LDL-C parallels plasma total-C, the ratios of LDL-C:HDL-C and total-C:HDL-C provide similar information. Gordon and associates[677] advise against using these ratios because some individuals may have similar ratios, yet show large differences in the concentration of plasma total-C, LDL-C, and HDL-C. The ratios, presented alone, do not provide information on the LDL-C level, a significant risk factor for CHD. Presently, the National Cholesterol Education Program's Adult Treatment Panel recommends that plasma total-C be assessed first. If it is higher than 240 mg/dL, then a lipoprotein cholesterol analysis is indicated. A LDL-C level of 130 to 159 mg/dL places an individual at risk, and a level of $\geq$160 mg/dL is indicative of high risk for CHD. An HDL-C level of <35 mg/dL is a risk factor for CHD.[678]

There is increasing evidence that serum apolipoproteins may be better discriminators of CHD risk than plasma lipid and lipoprotein levels.[679] Lipoprotein metabolism is regulated by physiological events that affect synthesis and catabolism of the various classes. Effectors (e.g., diet, gender, and hormones) of lipoprotein metabolism, therefore, play an important role in the development of atherosclerosis.

Arteriosclerosis

The endoplasmic reticulum (ER) is fundamentally involved in the etiology of cardiovascular disease primarily through its function in oxidative reactions. Arteriosclerotic plaques have a monoclonal origin, and focal proliferation of smooth muscle cells in the intima is required for its development. Environmental mutagens can be activated by enzymes bound to the ER of cells in the arterial wall, and consequently, trigger local somatic mutations which lead to subsequent arteriosclerotic lesions, similar to the actions of environmental toxic chemicals on the function of the hepatic ER.

It is clear that pollutant injury can alter fat and its metabolism, having far-reaching effects in the chemically sensitive patients.

Laseter[680] has demonstrated an increase in toxic chemicals in arteriosclerotic plaques. Chronic alteration of methionine metabolism has also been shown to be a part of the pathogenesis of arteriosclerosis. Many environmental chemicals such as the presence of nitrous oxide or chlorinated pesticides have been shown to alter methionine metabolism.

Essential Fatty Acids

The body can manufacture some fatty acids from various ingested fats eaten while others cannot be manufactured by the body. These essential fatty acids (Figure 57) are linolenic acid, which converts to eicosapentanoic acid (EPA); linoleic acid, which converts to arachidonic acid (AA); and gamma-linoleic acid (GLA) which is metabolized to PG_1. It has been observed that appropriate proportions of these substances are altered in some chemically sensitive patients. Fatty acids have two major uses in the body. They form cell walls and manufacture prostaglandin molecules. The cell wall serves as a place to store essential fatty acids, which can then be made into prostaglandins. It also is a place to capture and store lipophilic toxic chemicals as seen on the biopsies of many chemically sensitive patients. It is a very dynamic substance because of nutrient transport across it. (Pollutant injury will damage these membranes,

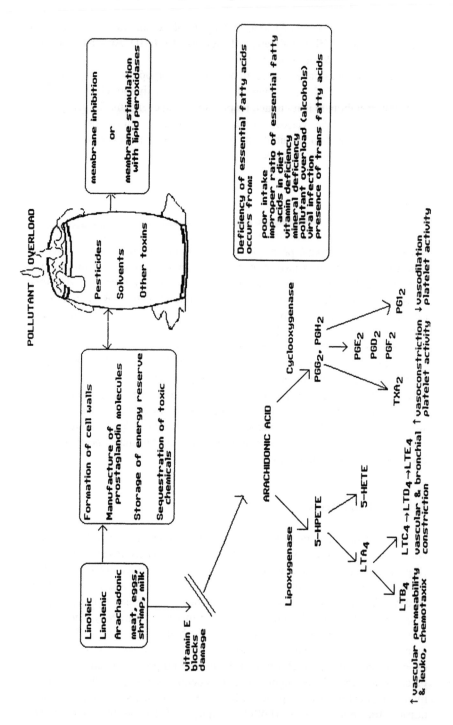

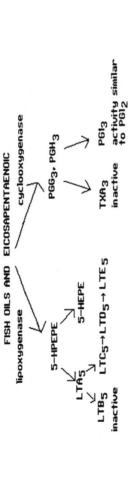

Arachidonic acid metabolism results in formation of thromboxane A$_2$ (TXA$_2$) in platelets, prostacyclin (PGI$_2$) in the blood vessels, and leukotrienes (LTs) of 4-series and hydroxy eicosateiraenoic acid (HETE) primarily in the leukocytes. Eicosapentaenoic acid metabolism results in formation of inactive thromboxane A$_3$ (TXA$_3$) and leukotrienes (LTs) of the 5-series. PGI$_3$ has biological activity similar to that of PGI$_2$.

Figure 57.   Essential fatty acids.

causing lipid peroxidation or inhibition. Pollutant injury to cell membranes is discussed in Chapter 4). Prostaglandin types can be divided into three classes: PG_1, PG_2, and PG_3.

PG_1 is derived from GLA. Evening primrose oil is a rich source of GLA, which can benefit people in whom the conversion of linoleic acid into GLA is deficient because of a lack of the converting enzymes, delta-6-desaturase. The deficiency of this enzyme may be due to dietary deficiency or excess or toxic chemical injury. Diabetes, atopia, and high saturated fat intake seem to inhibit this enzyme. PG_1 is also a protective factor against heart attacks caused by blood clots. It keeps blood platelets from sticking together, whereas PG_2 (from AA) promote platelet aggregation. PG_1 is also necessary for normal functioning of T cells of the immune system; they reduce the release of inflammation causing lysosomes and block the release of AA from storage (Figure 58). Alterations of all of these have been seen in some chemically sensitive patients.

AA is metabolized to PG_2 and comes primarily from meat, milk, egg yolks, and shrimp. PG_2 helps control the body's response to injuries, such as triggering clot formation, or helping to form mucous to protect respiratory membranes. PG_2 responses are kept in balance by adequate amounts of PG_1 and/or PG_3.

PG_3 is derived from EPA,[681] and fish oils are the primary source of EPA. Linolenic acid, found primarily in dark green leafy vegetables and linseed oil, can be converted by humans into EPA and dehydroascorbic acid (DHA), but the conversion is slow. Clinical studies indicate that PG_3 lowers blood viscosity and platelet stickiness and can, therefore, lessen the tendency to form clots. PG_3 also lowers plasma cholesterol and triglycerides (Figure 59).

Over the years Westerners have been gradually reducing the consumption of unsaturated fats such as linseed oil, fish oil, and walnut oil, which are converted to PG_1 and PG_3, and at the same time, dramatically increasing saturated fats such as those found in beef, pork, and lamb which convert to PG_2. Our society consumes large amounts of dairy products, eggs, and meats. While most people also consume plenty of corn or soy oil, high heat extraction methods generally used to cook changes the EFA from *cis* to *trans*, which cannot be used by humans to make prostaglandins. We then have an excess of PG_2, but an insufficient amount of PG_1 and PG_3 and, therefore, AA is not kept in balance. The various fatty acids compete with each other for enzyme systems. The excessive AA converts to leukotrienes which are proinflammatory agents, and end results may be seen in some chemically sensitive vasculitis patients. Research has shown increased PG_2 products in many chronic diseases such as asthma, arthritis, premenstrual syndrome, etc.[682] Pollutant injury may also damage the various enzyme steps resulting in blocks to the different steps. This blocking is sometimes seen in the chemically sensitive.

Essential fatty acid deficiency can occur as a result of vitamin and mineral deficiency, as a result of pollutant injury as seen in the chemically sensitive, and an improper ratio of essential fatty acids in the diet. The occurrence of

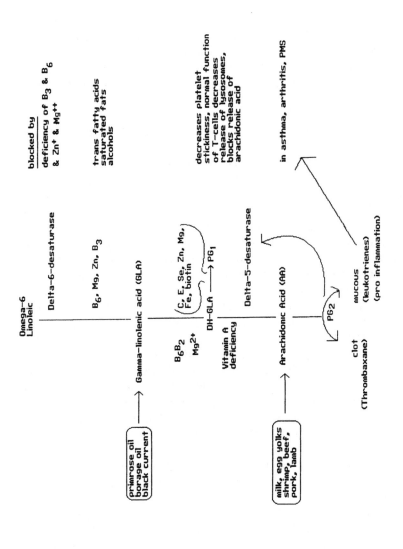

Figure 58. Omega-6 linoleic acid metabolism.

these deficiencies and dysfunction are discussed throughout this book. Metabolizing linoleic acid to PG_1 requires the enzyme delta-6-desaturase, which can be blocked by the deficiency of vitamin B_3, B_6, zinc, or magnesium, or the presence of *trans*-fatty acids (Figure 58) saturated fats, alcohol, some viruses, and some toxic chemicals. Other reasons PG_1 may be inadequate include old age, radiation, and diabetes.

The GLA to DGLA step needs pyridoxal 5-phosphate that the body makes from vitamin B_6 (pyridoxine), which requires B_2 and magnesium. The conversion of DGLA to PG_1 requires adequate amounts of vitamin E, C, selenium, zinc, magnesium, iron, and biotin. Deficiencies of these nutrients often seen in the chemically sensitive result in inadequate production of PG_1 and PG_3. Some research indicates that vitamin A deficiency shunts DGLA into AA and away from the production of PG_1.

Metabolism of α-linolenic to EPA also requires delta-6-desaturase (Figure 59).

Epidemiologic studies show that Greenland Eskimos consume a diet rich in animal fats from fish, yet are virtually free of coronary artery disease and diabetes mellitus;[683-685] however, they bruise more easily, have increased bleeding time, and may have more of a tendency toward cerebral vascular accidents.[689] The Japanese also have a low death rate from coronary artery disease and also consume fish as a large percentage of their diet.[687] The fish get EPA/DHA from a marine plant, chlorella, which is part of oceanic plankton.[688] Bang and Dyerberg[687] proposed that the relative lack of atherosclerotic disease among Greenland Eskimos is the result of their high intake of fish oils, which contain omega-3 fatty acids. Although their dietary lipids are higher, they have lower levels of triglycerides, cholesterol, β-lipoprotein (LDL), and pre β–lipoprotein (VLDL). In addition, they have a lower concentration of linoleic acid and AA, and a higher concentration of EPA and DHA.

Phillipson[690] studied the treatment of hypertriglyceridemia and found the same results as the Greenland Eskimo study for omega-3 fatty acids. These studies used 10 patients with type-IIIa hyperlipidemia (increased levels of both VLDL and LDL) and 10 patients with type-V hyperlipidemia (increases fasting levels of chylomicrons and VLDL) who were placed successively on three isocaloric diets, each maintained for 4 weeks: a diet high in fish oil (including fish and omega-3 fish oil supplements); a diet high in vegetable oil (including corn and safflower oil); and a control diet consisting of cereal, fruit, vegetable, and milk.

In type-IIb patients, the fish oil diet decreased plasma cholesterol by 27%. The decrease was significant for cholesterol in both the VLDL and LDL fractions, but was highly significant for VLDL fractions. The vegetable diet produced less striking decreases in plasma cholesterol and triglycerides, although it was associated with slightly increased HDL levels as compared to the fish oil diet. In type-V patients on fish oil diets, fasting chylomicronemia disappeared and the triglyceride content of the chylomicron fraction fell by

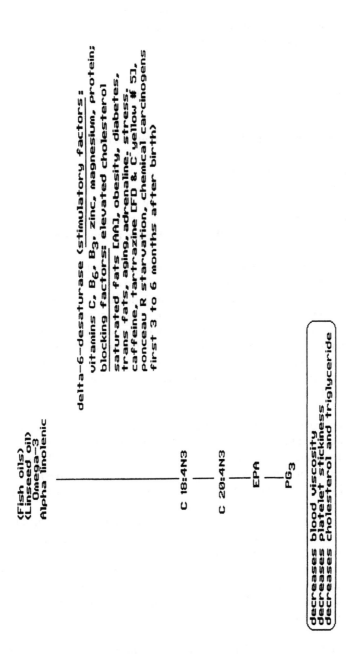

Figure 59.   α–linoleic metabolism omega-3.

95%. Total plasma triglyceride decreased by 79%, VLDL triglycerides decreased by 85%, and plasma cholesterol decreased into the normal range by 44%. Fish oil is often contaminated with polychlorinated biphenyls (PCBs), polyaromatic hydrocarbons, and other toxic chemicals. Therefore, there may be detrimental as well as beneficial effects in the chemically sensitive patient.

It is clear that pollutant injury and diet can alter fat and its metabolism, having far-reaching effects in chemically sensitive patients.

Pollutant Injury and the Enzymes of Biotransformation

Humans have a number of enzymes that catalyze oxidation, reduction, degradation, and conjugation reactions in order to eliminate foreign and endogenous compounds. The presence or absence of some enzymes appears to play a crucial role in propagating and perhaps creating chemical sensitivity. Partial supply may be the rule rather than total absence in most cases of chemical sensitivity.

All pure enzymes are proteins and exist as large convoluted molecules with molecular weights that range from about 12,000 to over 1 million. By international agreement, these known enzymes, which number over 2000, are classified into 6 main categories according to their basic action.[691] Their integrity depends entirely on the molecular spacial configuration, and processes such as the denaturing caused by heat or inhibition by the union of some other molecule to the active site by chemical injury will completely change the enzymatic activity,[692] as seen in the chemically sensitive. Enzymes are classified basically as follows:

1. Hydrolases — splitting of compounds into fragments, by the enzymatic addition of water ($H^+ + OH$) (degradation reactions).
2. Transferases — group-transfer reactions such as the addition or removal of methyl or acetyl groups (conjugation reactions).
3. Lyases — addition or removal of groups at double bonds (examples are carboxylases, decarboxylases, aldolases, synthetases).
4. Oxidoreductases — transfer of electrons, such as seen with dehydrogenases, reductases, oxidases, transhydrogenases, peroxidases, and oxygenases (oxidation reduction reactions).
5. Isomerases — produces isomeric forms of molecules by group transfers, producing *cis*- and *trans*-forms and vice versa.
6. Ligases — forms C–O, C–N, C–C, and C–S bonds by the breakdown of a pyrophosphate bond, such as in ATP.

The complex and very large enzyme molecule has an active site on the surface that is usually an extremely small percentage of the total area or volume and is easily damaged by pollutants.

Table 30. Examples of Enzymes Requiring Inorganic Ions for Function

Fe^{2+}	Cytochrome oxidase
	Catalase
	Peroxidase
	Aldehyde oxidase
	Superoxide dismutase
Se	Glutathione peroxidase
Mo	Nitrate reductase, aldehyde oxidase, xanthine oxidase, aldehyde dehydrogenase, sulfur oxidase controversial
Mg^{2+}	Hexokinase
	Glucose-6-phosphatase
Mn^{2+}	Arginase
	Superoxide dismutase, glutamine synthetase
Ni^{2+}	Urease pyridoxine linase
Zn^{2+}	DNA polymerase, RNA polymerase, thymidine kinase
	Alcohol dehydrogenase carboxypeptidase
	Superoxide dismutase alkalic phosphatase
	Carbonic anhydrase
Cu^{2+}	Superoxide dismutase, tyrosine hydroxylase
	Cytochrome c oxidase, dopamine hydroxylase, lysyl oxidase
K^+	Pyruvate kinase
Co	Component of vitamin B_{12}
Va	Cofactor for nitrate reductase
Cr^{2+}	Insulin, trypsin

Many enzymes require cofactors for their catalytic action, usually in the form of an inorganic ion, or as a complex organic molecule, or coenzyme. Some coenzymes are TPP, which uses thiamine (B_1), FAD, which utilizes vitamin B_2 (riboflavin), NAD, which uses nicotinic acid (B_3), coenzyme A of pyridoxal phosphate, which uses pyridoxine (B_6). Occasionally, some enzymes require both inorganic and organic co-factors. Iron may play an essential role at the active site on the surface of the enzyme, e.g., nickel, as a co-factor for ureas. Or it may serve as an integral part of the internal structure, like iron in the heme complex of catalase or cytochrome oxidase. Some examples of enzymes that require inorganic ions for their function are shown in Table 30.

As shown previously, many chemically sensitive patients have vitamin and mineral deficiencies and these have an impact on depollutant enzyme function, either exacerbating or triggering the chemical sensitivity.

Two interrelated general properties of the enzymes considered to be active in chemical detoxication and detoxification appear to apply: the general lipophilic nature of the substrates that were converted to more hydrophilic products and the loose specificity of enzymes for toxic substances. Despite their occasional participation in anabolic reactions the evolutionary development of the enzymes suggests xenobiotics as their natural substrates. These enzymes are characterized by simplicity rather than efficiency on the removal of a multitude of xenobiotics to which the individual is repeatedly exposed. *Among the mechanisms responsible for toxicity, sensitivity, and detoxication there is a growing appreciation that a delicate balance exists in each tissue between enzymes that form toxic intermediates and those that detoxify these highly reactive intermediates.* An imbalance appears to be occurring in chemical sensitivity. Factors contributing to an individual's susceptibility to drug or chemical toxicity or sensitivity include genetic predisposition, age, state of nutrition, compartmentalization of the enzyme, efficiency of DNA repair, RNA synthesis, total body load, and immunological competence. When these do not occur effectively, chemical sensitivity may occur.

Genetic variation in the face of excess pollutant load or malnutrition in any cellular process involving foreign substance metabolism might lead to atypical responses through: (1) transport (absorption, plasma protein binding); (2) transducer mechanisms (receptors, enzyme induction or inhibition); (3) biotransformation; and (4) excretory mechanisms (renal and biliary transport) (Figure 60), thus resulting in chemical sensitivity.

The bulk of understood pharmacogenetic disorders reflects alterations in biotransformation. These await environmental triggers for their genetic time bombs. There are a few under genetic transport. Individual differences (of twoto threefold) exist in displacing effects of drugs such as sulfisoxazole, salicylic acid, and salicyluric on bilirubin plasma binding.[693] Genetically inherited structural differences in the albumin molecule (one binding vs. two binding sites) may affect the transport of certain drugs such as warfarin[694,695] and other toxic chemicals which could result in a tendency toward chemical sensitivity.

According to Nebert,[696] pharmacogenetic subgroups can be identified in two ways. (1) The specific aspect of drug (or toxic chemical) absorption assumes paramount importance in the overall host response to that drug or other toxic chemical. Genetic variation in that process may be identified as a discontinuous response in the population, e.g., atypical pseudocholinesterase, first detected as a prolonged apneic response to succinylcholine.[697] (2) One can select a drug or toxic chemical or enzyme activity to test rather than looking for the more obvious atypical drug or chemical response, e.g., clinical survey of polymorphism of dopamine B hydroxylase;[698] catechol *o*-methyl transferase activity;[699] and barbiturate metabolites in the urine.[700-702] The drug acetylation polymorphism[703] was uncovered following documentation that genetic variability exists on the rate of isoniazid metabolism in man.[704-708] Studies show that there are five possible phenotypes for acid phosphate that exist in humans and are that are controlled by three alleles at a single genetic locus.[709,710] Glucose-6-phosphate dehydrogenase activity in red cells is genetically heterogenous.[711] All of these reports emphasize the observations by environmental physicians of the varied nature of chemical sensitivity in a group of individuals.

According to Nebert,[712] studies in twins have established that individual differences in drug metabolism are largely under genetic control for isoniazid,[713]

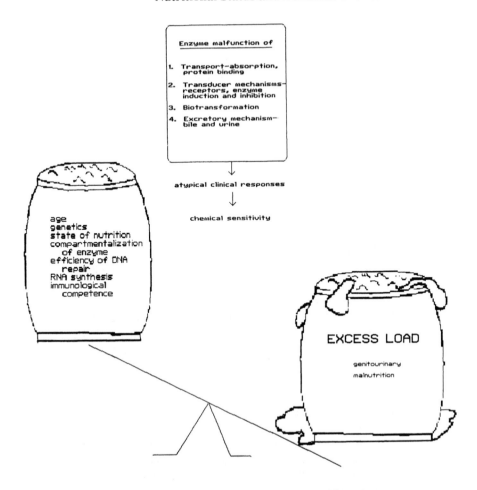

Figure 60.   Excess pollutant load causes enzyme malfunction.

phenylbutazone,[714] antipyrine dicoumeral,[714] nortryptyline,[715] ethanol,[716] and halothane.[717] Evidence also exists[718] that differences in response to inducers of drug metabolism such as phenobarbital are genetically controlled.[719] However, pollutant injury may create genetic damage or suppress enzyme function, resulting in a similar metabolic situation, both of which are seen in the chemically sensitive.

Of randomly selected gene products, 30% may exhibit polymorphic variations.[720] This high percentage of polymorphism implies that the hundreds of genes in the body of each individual will likely have one or more polymorphisms. These variations with a small volume of response enzymes are time bombs awaiting to be triggered by the right type and quantity of environmental pollutants, thus exacerbating or resulting in chemical sensitivity. For example, caffeine-sensitive type human malignant hyperthermia may reflect two or more genes. Low catechol o-methyltransferase activity is an inherited autosomal recessive trait.[721,722] Chemically sensitive patients appear to have found their triggers for these time bombs.

According to Jakoby,[723] an arbitrary classification of human pharmacogenetic disorders is as follows: given the right set-up circumstances, a spreading phenomenon may occur resulting in chemical sensitivity. This classification will be listed so the environmentally oriented physician may develop a general understanding of the possibility of genetic time bombs.

Disorders with Increased Drug Sensitivity

Diminished detoxification resulting from enzyme deficiencies can occur and are caused by partial or a total absence of enzymatic function.[724]

Succinylcholine apnea is an autosomal recessive characteristic involving pseudocholinesterase,[725] of which the incidence is about 1 in 2500.

Serum Paroxonase Polymorphism. Paraoxon is an organophosphate pesticide with potent cholinesterase inhibiting properties. People who are homozygous (autosomal recessive) may not be able to metabolize this chemical like the bulk of the population.[726] Chemical sensitivity may result similar to that seen after exposures to other organophosphates.

Sulfate Oxidase Deficiency. Sulfate oxidase deficiency will produce urinary metabolites of sulfur metabolism in large amounts, due to the enzyme deficiency. This autosomal recessive disorder may be a serious condition in the presence of sulfur-containing medications[727] and sulfur-inhalant exposures, as shown in Chapters 4, 9, and 10, which discuss nonimmune mechanisms, and outdoor and indoor air pollution. Some patients with chemical sensitivity have been seen with this defect.

The antihypertensive debrisoquine is known to produce profound orthostatic hypotension in people who have the autosomal recessive trait for slow hydroxylation. About 9% of the U.K. population appears to have this phenotype.[728] Monro[729] has shown a high proportion of her chemically sensitive patients to have this defect.

Acetylation of xenobiotics may also be significantly impaired by an autosomal recessive trait.[730,731] It has been estimated that half of the European population exhibits this characteristic,[732] which is especially important in the metabolism of isoniazid. Slow acetylators are more prone to develop the toxic peripheral neuropathy, which can be prevented by the simultaneous supplementation with pyridoxine (B_6).[733] Similarly, positive antinuclear antibody tests are seen more frequently in slow acetylators receiving procainamide anti-arrhythmia therapy.[734] For more information, see Chapter 4.

In addition, the slow acetylation phenotype appears predominately in systemic lupus erythematosus patients by a wide margin (p <0.01),[735] suggesting that inadequate handling of environmental factors may play a role in this disease. Monro[736] has reported that this defect occurs in some chemically sensitive patients who were studied in an environmental care unit.

Extremely high blood levels of phenytoin have been produced by low doses of the drug in a family reported to have symptoms of toxicity. The presumed defect is in a cytochrome P-450-dependent monoxygenase[737] and appears to result in the spreading phenomenon seen in some chemically sensitive patients.

Dicoumarol sensitivity has also been reported, whereby plasma half-life of the drug was three times longer than normal, apparently because of an inability to metabolize.[738]

Decreased levels of catalase is transmitted as an autosomal recessive characteristic,[739] reported in only 54 families in Japan, Korea, Switzerland, and Israel. The only established consequence is an abnormal response to topical hydrogen peroxide.

The Crigler-Najjar syndrome is a severe form of congenital nonhemolytic jaundice, which is autosomal recessive.[740] Liver formation of glucuronide conjugates of bilirubin are undetectable.

Conditions such as malignant hyperthermia with anesthetics is about 1 in 20,000, and 5% of the U.S. population[741] appears susceptible to steroid-induced glaucoma, which seems to be mediated by a recessive allele.

Disorders of Increased Drug Resistance

Defective Receptor. The autosomal dominant trait of coumarin resistance has been described, whereby patients may require much larger doses than usual.[742] We have seen this phenomenon in a chemically sensitive patient who was sensitive to both coumarin and heparin. She had high levels of antiheparin antibodies requiring 8400 units of heparin every 4 h intravenously and could not tolerate dicoumeral in any form.

Of all Caucasians, 30% are homozygous for the recessive allele that causes the inability to taste phenylthiourea or phenylthiocarbamide.[743,744] This characteristic is thought to represent an altered receptor for these phenylthio analogues.

Familial hypercholesterolemia may be linked to genetic differences in the LDL receptor system.[745] Inheritance appears to be autosomally autonomous; whereby hetero- and homozygotes have significantly increased risks of myocardial infarctions.

Defective absorption is typified by juvenile pernicious anemia, whereby ideal absorption of vitamin B_{12} is defective because of a congenital lack of intrinsic factor.[746]

Disorders Characterized by Increased Metabolism[747]

Succinylcholine resistance has been seen in one family, whereby the succinylcholine is hydrolyzed at three times the normal rate.[748]

Alcohol dehydrogenase will convert ethanol to acetaldehyde, and atypical alcohol dehydrogenase is associated with a five- to sixfold increase in the conversion rate in vitro. This atypical alcohol dehydrogenase may produce a significant buildup of the more toxic acetaldehyde.[749] As shown in Chapter 4, many chemically overloaded patients have trouble with the buildup of aldehydes. Some of the problems experienced by the chemically sensitive may result from this enzyme variation.

Resistance to the analgesic pentazocine has been noted and thought to be related to the inducement of cytochrome P-450.

Enzyme Induction Influencing Genetic Disorders[750]

Hepatic porphyria can be worsened substantially by barbiturates and steroids, apparently by inducing aminolevulinic acid synthetase, the rate-limiting step in heme synthesis.[751] Porphyria appears to be an autosomal dominant genetic characteristic. Many people with a chemical overload have been seen to develop a transient porphyria (see Chapter 21). This condition probably is relative to the quantity of enzyme deficiency vs. the total pollutant load presented. Chemically sensitive patients are seen in these porphyria cases.

The autosomal recessive trait *pentosuria* appears to result from a marked decrease in xylitol dehydrogenase activity, an enzyme in the glucuronic acid oxidation pathway.[752]

Polycyclic hydrocarbons are known to induce various forms of cytochrome P-450 and others[753] (see Chapter 4). Genetic differences in the ability to produce these detoxifying enzymes may account for some increased susceptibility to the carcinogenic effect of environmental pollutants as well as many cases of chemical sensitivity. The majority of chemically sensitive patients are extremely susceptible to polycyclic aromatic hydrocarbons (PAHs) found in food, air, and water. They are especially susceptible to coal smoke.

Variations in Metabolic Potentiation

Vitamin D resistant rickets is the result of a partial or complete inability to respond to large doses of vitamin D. One severely chemically sensitive, pesticide-damaged, magnesium renal-wasting patient who was almost entirely resistant to vitamin D supplementation was seen at EHC-Dallas. When vitamin D was administered in too high a dose, the patient would become comatose but still give no positive metabolic response to vitamin D.

The catecholamine neurotransmitter exhibits polymorphisms in biosynthesis and degradation, which has important implication for depression, schizophrenia, Parkinson's disease, hypertension, and chemical sensitivity.[754] Pollutant overload may outstrip the ability to detoxify catecholamines, resulting in excess quantities of circulating catecholamines and chemical sensitivity.

Abnormalities of plasma dopamine B hydroxylase have been demonstrated in patients with Down's syndrome, torsion dystonia, and familial dysautonomia.[755]

Monoamine oxidase (MAO) inhibitors are widely used clinically, and there appear to be genetic differences in platelet MAO activity[756] which may play a role in the variability of effects. Pollutant overload has been seen to stress this system in chemical sensitivity.

Human genetic defects have been found in sex steroid or glucocorticoid biosynthesis or degradation.[757,758]

Rarely, aplastic anemia induced by chloramphenicol may not be dose-related, suggesting a pharmacogenetic disorder.

Hepatitis may be induced by isoniazid in fast acetylators,[759] whereas the slow acetylators of this drug are more susceptible to toxic neuropathy. Similar responses in the chemically sensitive patient may be seen to environmental chemicals requiring acetylation for detoxification (see Chapter 4).

G-6-PD deficiency was discovered because the tendency of the antimalarial primaquine to cause acute hemolysis in 10% of black Americans.[760]

One hundred million people are affected by the more than 80 G-6-PD variants which are X-linked. Amine derivatives, nitrofuran derivatives, quinoline derivatives, naphthalene, and other drugs are known agents that will induce the hemolysis in susceptible individuals as well as in chemical sensitivity.

Inherited disorders of GSH production and utilization have been counted with spontaneous and drug-induced hemolysis.[761]

Unstable hemoglobins are easily denatured, and are phenotypically inherited as autosomal dominant traits. These individuals are susceptible to sulfonamide-induced hemolysis (see Chapter 4). Hereditary methemoglobinemia may predispose to drug-induced hemolysis as well.[761,762]

Polymorphism of aryl hydrocarbon hydroxylase and cytochrome P-450 inducibility has been seen in mice[763] and evidence for the same genetic Ah locus exists in man.[764,765] PAHs are inducers of some chemical sensitivity and are also potent environmental contaminants, as shown throughout this book. Cytochrome P-450 is an important rate-limiting step in the metabolic pathway of these xenobiotics with overload resulting in inadequate detoxication and chemical sensitivity. This genetic polymorphism may play a pivotal role in the striking differences in chemical carcinogenesis, mutagenesis, toxicity, and sensitivity that is seen in these chemicals.[766-768]

The expanding knowledge of these genetic influences on the metabolism of medication and environmental pollutants will have a profound impact on our understanding of chemical sensitivity (see Chapter 3).

The idea of "threshold" concentrations and the "average" response to so-called "low levels" of pollutants must be reevaluated when one considers how these genetic variations complicate and ally with total pollutant load, pollutant-created defects, and nutritional status resulting in individual susceptibility. These are truly time bombs waiting for environmental triggers.

Without a doubt, some individuals, while "normal" in many parameters, are extraordinarily sensitive or susceptible to certain environmental factors. No doubt our knowledge of the vital genetic factors surrounding this susceptibility will only expand as research progresses.

Enzymes of Detoxication

Several enzymes of detoxication and those of detoxification are essential in metabolism of the chemically sensitive, and exert a profound influence on cellular chemistry and the transformation and elimination of toxins of whatever source. Understanding these processes and the effects of pollutants upon them is paramount in understanding chemical sensitivity.

Cytochrome P-450-Mixed-Function Oxidase. Microsomal cytochrome P-450 is a terminal oxidase of the microsomal electron transport system which is responsible for the metabolism of xenobiotics in the chemically sensitive. It is a hemoprotein. There are multiple forms: P-448, P-450, P-450b, P-450c, etc.[769] Each may have general and specific activity. These are the prime oxidases for enzyme detoxication of xenobiotics. The types of chemical transformations in the hepatic microsomes in cascading oxidative reactions are seen. Here, the molecular oxygen is inserted into the organic molecule such as aliphatic and aromatic hydroxylations, N-oxidation, sulfoxidation, epoxidation, N-, S-, and O-dealkylations, -desulfurations, and -deaminations as shown in Chapter 4. Reductive reactivators involving direct electron transfer such as reductases of AZO; nitro-, N-oxide, and epoxides; and the not-so-well-understood dehalogenations are also catalyzed by the mixed function oxidases. P-450 brings about chemical changes in fatty acids, steroid prostaglandins, petroleum products, drugs, pesticides, and anesthetics. As one can see, malfunction would lead to a build up of toxic substances with resultant overload and the propagation and possibly the initiation of chemical sensitivity. Finally, total tissue injury may occur, allowing small doses of toxic and at times even nontoxic chemicals to then trigger the responses in the chemically sensitive.

This enzyme system of liver microsomes is probably one of the most versatile of human detoxication and detoxification mechanisms, utilizing a spectrum of chemical reactions from oxidations to conjugations including acetylation, acylation, methylation, gluconation, and sulfur conjugation and sulfoxidation, through dealkylations to deaminations, and others (Table 31).

Variable enzyme activities which had been seen in animals suggested that there might be numerous forms of the cytochrome enzyme,[770-773] although the

Table 31. Reactions Catalyzed by Cytochrome P-450

Oxidation

Hydroxylation	Addition of (–OH) group to form alcohols
Epoxidation	Adds O_2 → highly reactive
N^-, O^-, S^-, Dealkalation	Adds H to N, O_2, or S
Deamination	Removes $-NH_3$
Desulfuration and sulfoxidation	Removes sulfur; adds O_2 to sulfur
Dehalogenation	Removes chlorine

Reduction

Azos, aromatic nitros, disulfides, double-bond compounds, dehydroxylations, valence, dehalogenation

Conjugation

Acetylation, acylation (peptide conjugation), sulfuration, methylation, gluconation

Pollutant damage may disturb any or all of these reactions.

latest evidence indicates five or six distinct forms. The general reaction uses NADPH during the reaction, and molecular oxygen is also required. The cytochrome P-450 activity seems mostly associated with the ER, where various reactions take place on a wide variety of substrates. The level of microsomal liver cytochrome P-450 can be increased to a point by the presence of certain inducers such as phenobarbital, benzopyrene, dioxin, and others. After a certain point, they become overloaded and then depleted, and antibody reactions occur causing increased dysfunction. Damage to this system is probably how chemical sensitivity occurs in many people.

The liver cytochrome P-450 system is the primary site for xenobiotic metabolism, but this system is also found in other tissues that may have primary contact with potential toxins, such as gastrointestinal tract, heart, skin, lungs, and kidneys. It is also found in the placenta, lymphocytes, monocytes, pulmonary alveolar macrophages, adrenal, and brain.[774]

Because of the many isoenzymes of cytochrome P-450, there is a whole family of potential reactions, but this family is characterized by an overlapping specificity that will permit a very wide array of specific enzyme activity.[775]

Liver microsomal cytochrome P-450 has been further identified as having three components, that being the cytochrome P-450, NADPH-cytochrome P-450 reductase, and diacylglyceryl-3-phosphorylcholine (diacyl-GPC).[776,777] Under the influence of vitamin B_3-containing NADPH and oxygen, these three can be unified into the active functioning system. The NADPH cytochrome P-450 reductase utilizes both the vitamin B_2 FMN and FAD. It has been isolated from rat and rabbit liver after induction with phenobarbital.[778-781] Iron is another essential mineral co-factor for the cytochrome oxidase system, and on stressed, chemically sensitive individuals, supplementation with iron often produces clinical improvement and well-being in spite of the lack of evidence of anemia or overt iron deficiency. The necessity of iron for the proper functioning of this enzyme and other systems may partially explain this observation. Copper also is a catalyst in the cytochrome oxidase system and its deficiency may lead to inadequate function with an exacerbation of chemical sensitivity.

Aromatic (Aryl) Hydrocarbon Hydroxylase (AHK). AHK acts on aromatic ring compounds and exhibits polymorphism in its inducibility. Polycyclic aromatic inducers will cause this induction in mice, and evidence for the same genetic locus also exists in man. The hydroxylation of aromatic compounds produces arene oxides as important intermediates, which may be the mechanism of carcinogenesis, mutagenicity,[782-784] DNA effects, drug toxicity, and teratogenesis[785-787] (see Chapter 4). This production of intermediates may also result in chemical sensitivity.

Microsomal Flavin-Containing Monooxygenase. This enzyme will oxidize xenobiotics which contain nitrogen or sulfur atoms, like amines, organic sulfur compounds, such as sulfides, thiols, disulfides, and thiones.[788-790] Some chemically sensitive individuals have been specifically sensitive to these substances.

Alcohol Dehydrogenase/Aldehyde Oxidizing Enzymes. Aldehydes are produced from several sources, including the oxidation of biogenic amines and the peroxidation of polyunsaturated lipids. The most common pathway, however, is the production of acetaldehyde from alcohol. The aldehydes are generally much more toxic than their precursors and will interact readily with cellular material. The two major oxidizing systems are aldehyde dehydrogenase, which is NAD-dependent, and aldehyde oxidase, which requires flavin. These pathways convert the aldehydes to acids. Aldehyde dehydrogenases are found in every organ system.[791,792] There is also a specific type of aldehyde dehydrogenase that oxidizes only formaldehyde.[793-796]

Alcohol dehydrogenase is the major enzyme mechanism acting on many alcohols, oxidizing them to aldehydes, which are then converted to carboxylic acids. Individual differences in alcohol metabolism and sensitivity may be partly accounted for because of the genetic variations in isoenzyme distribution as well as in environmental overload damage.[797-803] Specific end-organ levels may be present causing varied rates of detoxification which would result in different symptoms in chemically sensitive individuals. The liver is the principal site of ethanol oxidation in man, where at least 75% of an alcohol dose is metabolized.[804,805] This enzyme is NADPH-dependent, and its overuse in alcoholism or xenobiotic alcohol excess may explain why these patients are often niacin (B_3) deficient. Chemical sensitivity may be accentuated or perhaps even induced in people who have a small amount of this enzyme. The chemically sensitive can tolerate little or no alcohols. This fact has been emphasized by the thousands of double-blind inhaled challenges of petroleum-derived ethanol as well as oral and intradermal challenges performed in chemically sensitive patients at EHC-Dallas (see Chapter 31).

Aldehyde Reductase. Aldehyde reductases have wide-ranging biologic functions in the reduction of carbonyl compounds. NADPH is an essential coenzyme, and these reductases also act in xenobiotic and naturally occurring substances. These enzymes are found mainly in kidney, liver, and brain.[806-811] The aldehyde reductases are inhibited by a variety of medications, including barbiturates, hydantoins, and other anticonvulsants.[812,813]

Ketone Reductases. Ketones may be reduced to their corresponding alcohols by the ketone reductases.[814] Xenobiotic ketones of unsaturated aromatic, aliphatic, and alicyclic varieties are metabolized to free or conjugated alcohols. Closely related to the aldehyde reductases, ketone reductases have been found in almost all mammalian tissues, and appear to use NADPH as a cosubstrate. The chemically sensitive have problems with xenobiotic ketones.

Xanthine Oxidase and Aldehyde Oxidase. These enzymes are similar to one another, and are classified as metalloflavoproteins, containing molybdenum, FAD, and iron. They catalyze the oxidative hydroxylation of their substrate, usually xanthine, hypoxanthine, and purine by the xanthine oxidase, and hypoxanthine and purine for aldehyde oxidase. The genetic deficiency of xanthine oxidase produces xanthinuria.

Monoamine Oxidase. (FAD-Cysteine Compound). MAO catalyzes the oxidative deamination of a wide variety of amines, including the endogenous neurotransmitters, amines, and adrenaline. It is widely found in most mammalian tissues. This enzyme is in high concentration in nerve tissue, where it plays an essential role in the breakdown of noradrenaline, dopamine, and serotonin, when it is released into the synaptic cleft, but it is also found in high concentration in the heart, liver, and intestines. Detoxication of exogenous amines in the intestine and liver may account for the higher levels in these organs. MAO is usually tightly bound to the mitochondrial outer membrane.[815,816] Detergents or organic solvents will bring it into solution and may dislodge it from the mitochondrial membrane, resulting in some aspects of chemical sensitivity. Several antidepressants, such as amitryptyline and imipramine, are reversible MAO inhibitors, and apparently exert their effect by the subsequent rise in the neurotransmitters. These substances are generally unsuccessful in the chemically sensitive who have brain dysfunction.

   The principal way in which the activity of the transmitter amines is terminated is via the energy-dependent high affinity transport system into presynaptic terminals where it either may be taken up into storage vesicles for reuse or degraded by MAO. Many toxic chemicals inhibit or stimulate MAO. Some chemically sensitive patients are seen to have problems with MAOs.

Superoxide Dismutase (SOD). SOD catalyzes the dismutation reaction which produces hydrogen peroxide from two superoxide radicals:

$$SOD$$
$$(B_2NADH)$$
$$2O_2^- + 2H^+ \rightarrow H_2O_2 + O_2$$

It is the first line of the body's defense against free radical damage. The resultant hydrogen peroxide is far less toxic (especially to cell membrane components) than is the superoxide. It appears this reaction is the sole biological function of SOD. SOD, which is present in all cells, but markedly depleted in many chemically sensitive patients (see Chapter 30) is essential for the maintenance of cellular integrity. Variation in the levels of SOD may have an important influence on a host of other reactions in the chemically sensitive. Significantly lowered blood levels of SOD have been found among some environmentally sensitive patients, a finding that may be either a cause or an effect of their illness, but after it occurs certainly exacerbates the sensitivity. Many of these sensitive patients are found to be deficient in zinc or copper, the essential metal co-factors of the cellular zinc/copper SOD of eukaryotic cells. Mitochondrial or manganese SOD, however, has manganese as its metal co-factor and also has been found to be deficient as well as excessive in some of the chemically sensitive. A fourth type, iron SOD, is found mostly in prokaryotic cells with no nucleus, and in some higher plants.[817]

Other patients have no currently measurable intracellular mineral deficiency. They may have direct damage to the enzyme due to environmental toxins. This is evidenced by the fact that studies done at EHC-Dallas reveal that as the total load is reduced by good environmental control, nutrition, and heat depuration, the levels of SOD often improve. These may eventually return to control levels or even attain the ability to give an induction response to a new toxic chemical assault.

Some toxic xenobiotics, such as the pesticide paraquat[818,819] mediate the toxic effects by the reaction with oxygen to produce higher levels of the superoxide radical (O_2^-), eventually depleting the enzymes and thereby overwhelming the ability of the SOD to cope with this cell-damaging free radical.

This toxic overload and enzyme depletion phenomenon may then produce a vicious cycle of exaggerated pollutant response with increasing tissue injury until the process and damage become autonomous. Numerous clinical studies have now been done of conditions such as RA,[820] septic shock,[821] and ischemic states,[822] suggesting that SOD may markedly decrease the tissue damaging effects of free radicals.

Substances known to stimulate SOD are paraquat, pyocyanine, phenazine meth-sulfate, streptonigrin, juglone, menadione, plumbin, methylene blue, and azurea. As shown in Chapter 30, these enzymes are altered in some chemically sensitive patients.

Glutathione peroxidase. Glutathione peroxidase enzymatically converts moderately toxic hydrogen peroxide (H_2O_2) to oxygen and water, thereby reducing the rate of lipid peroxidation of cellular membrane surfaces (Figure 61). Glutathione peroxidase and catalase are the second line of defense after SOD in the chemically sensitive. Glutathione peroxidase is found in erythrocytes and protects them from oxygen damage. This enzyme contains the iron porphyrin heme group, acts with the tripeptide glutathione to protect cell membranes from peroxidation, and protects hemoglobulin from the formation of methemoglobin by the action of hydrogen peroxide. Glutathione itself contains L-glutamic acid, L-cysteine, and glycine and is present in all animal cells in high concentrations. It acts as the reductant of the toxic peroxides, via the catalytic action of glutathione peroxidase, as shown in Figure 61.

The active site of the glutathione peroxidase molecule contains selenocysteine, an analogue of cysteine, in which the sulfur atom normally contained in the molecule is replaced by selenium. The enzymatic action is thereby dependent on adequate available stores of this essential mineral. Selenium is the essential co-factor of glutathione peroxidase. It is not surprising that this enzyme is sometimes inducible by oral supplementation with selenium, as reported by Levine and Kidd[823] and seen at EHC-Dallas, a finding that has obvious therapeutic importance. In addition, some chemically sensitive patients seen at EHC-Dallas have low levels of intracellular sulfur. This intracellular level occurs in 36% of the patients, suggesting among other things that there is a low total body sulfur pool. A low intracellular sulfur pool might decrease the amount of sulfur-containing amino acids available for the generation of the glutathione peroxidase enzymes, thus decreasing the ability to counteract pollutant damage to cell membranes.

Glutathione peroxidase is unstable at pH values approaching the isoelectric point and especially below pH 6.0. It appears stable up to pH 9.0. Optimum function is probably around pH 7.0 and is another reason that bicarbonate supplementation helps many chemically sensitive.

Inhibitors of GSH peroxidase are idioacetate, chloroacetate, and KCN. Its activity is influenced by insufficient amounts of dietary selenium, amino acids, and peroxidized lipids, and by the estrous cycle.

Glutathione peroxidase will also convert aliphatic and aromatic peroxides to their less toxic alcohols. This enzyme is one of the body's major defenses against the destructive effects of peroxides, which are constantly formed in cells by metabolic activity.[824,825] Variations in the levels of this enzyme, which

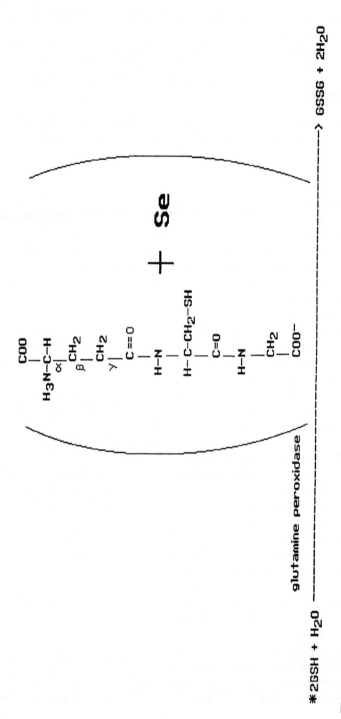

Figure 61. Glutathione peroxidase. *Glutathione (GSH), a tripeptide containing L-glutamic acid, L-cysteine, and glycine.

is known to occur (with environmentally sensitive patients often being deficient), and will logically have profound metabolic consequences. Ethane and pentane are known to be produced by the action of peroxides on lipid-containing cell membranes. Ethane content of expired air has been used as a measurement of lipid peroxidation in selenium and/or vitamin E-deficient rats. This harmful lipid peroxidation can be greatly reduced by a normal glutathione peroxidase activity[826,827] as well as by supplementation. Pentane levels have been used in measuring the ability of individuals to tolerate more hypoxia resulting from mountain climbing and marathon racing. As shown in our Laboratory chapter, glutathione peroxidase is frequently altered in the chemically sensitive.

Catalase

Catalase acts like glutathione peroxidase in that it converts hydrogen peroxide to oxygen and water, and therefore also protects against the toxic lipid peroxidation effects of hydrogen peroxide. It also contains the heme group, and is thereby dependent on the presence of adequate iron. Many environmentally sensitive patients will improve clinically with the addition of more available iron in the diet, in spite of the lack of overt signs of classical iron deficiency, such as anemia or deficient iron stores. The fact that the iron-containing heme group is essential for the formation of cytochrome oxidase and peroxidases, as well as catalase, may be a partial explanation for this observation.

It has been observed that there is an enhancement of antipollutant enzyme activity if a combination of catalase and SOD are used together intravenously. Catalase is present in all mammalian cells and in many peroxisomes and microperoxisomes. Some chemically sensitive patients have been shown to have altered catalases.

A genetic defect characterized by a decreased level or absence of catalase has been identified[828-830] that produces an abnormal response to topical hydrogen peroxide.

Glutamine Synthetase

Brain glutamine synthetase consists with four Mn(II) ions very tightly bound. The native enzyme Mn_4 accounts for 80% of all Mn(II) in the brain. Both are essential for normal brain functioning and may be abnormal in some chemically sensitive patients.

There may be other enzymes that are not yet clearly described that have influenced pollutant load in the chemically sensitive. However, present knowledge of those already-described possible genetic pollutant-induced defects is enough to understand some aspects of chemical sensitivity.

REFERENCES

1. Newsholme, E. A., and B. Crabtree. "Substrate cycles in metabolic regulation and in heat generation," *Biochem. Soc. Symp.* 41:61–109 (1976).
2. Murphy, S. D. "Toxic effects of pesticides," in *Casarett and Doull's Toxicology. The Basic Science of Poisons*, 3rd ed., C. D. Klassen, M. O. Amdur, and J. Doull, Eds. (New York: Macmillan, 1986), p. 537.
3. Anderson, K. E., and A. Kappas. "Dietary regulation of cytochrome P-450," in *Annual Review of Nutrition* (Palo Alto, CA: Annual Review Inc., 1991), 11:42.
4. Merliss, R. R. "Phenol marasmus," *J. Occup. Med.* 14:55–56 (1972).
5. Claudio, T. "Molecular genetics of acetyl choline receptor channels," in *Molecular Neurobiology*, D. M. Glover and B. D. Hames, Eds. (Oxford: IRL Press at Oxford University Press, 1990), pp. 63–79.
6. Calabrese, E. J. *Nutrition and Environmental Health*, Vol. I, (New York: John Wiley & Sons, 1980).
7. Calabrese, E. J. *Nutrition and Environmental Health*, Vol. I, (New York: John Wiley & Sons, 1980), pp. 163–164 and 348–355.
8. Calabrese, E. J. *Nutrition and Environmental Health*, Vol. I, (New York: John Wiley & Sons, 1980), p. 434.
9. Bendich, A., and C. March. *Pesticide Toxicities: Protective Effects of Antioxidant Vitamins C and E and Beta-Carotene*, ROCHE Dept. of Clinical Nutrition Manuscript No. N-121727 (1986).
10. Calabrese, E. J. *Nutrition and Environmental Health*, Vol. I, (New York: John Wiley & Sons, 1980).
11. "Dietary Levels of Households in the United States," Agriculture Research Service, U.S. Dept. of Agriculture (1968), pp. 12–17.
12. "Ten State Nutrition Survey," U.S. Dept. of Health, Education, and Welfare, Health Services and Mental Health Administration Center for Disease Control, DHEW Publ. No. (HSM) 72-8130-8134, Atlanta, GA (1972).
13. "First Health and Nutrition Examination Survey," Public Health Service, Health Resources Administration, U.S. Dept. of Health (1971–1972).
14. "Nationwide Food Consumption Survey," U.S. Dept. of Agriculture, Science and Education Administration, Beltsville, MD (Spring 1980).
15. "Dietary Intake Source Data: U.S. 1976-1980," Data from the National Health Survey, Series 11, No. 231, DHHS Publ. No. (PHS) 83-1681 (March 1983).
16. Pao, E. M., and S. J. Mickle. "Problem nutrients in the United States," *Food Technol.* 35(9):58 (September 1981).

17. "Nationwide Food Consumption Survey — Continuing Survey of Food Intakes by Individuals. 1986. Low Income Women 19–50 and Their Children 1–5, 1-day," USDA Human Nut Inf. Serv. Nutrition Monitoring Div. NCS, CSFII Report #86-1, U.S. Dept. of Agriculture, Science and Education Administration, Beltsville, MD (January 1987).
18. Hunter, B. T. *Consumer Beware*, (New York: Simon and Schuster 1971), pp. 79–107.
19. Calabrese, E. J. *Nutrition and Environmental Health*, Vol. I, (New York: John Wiley & Sons, 1980).
20. Leach, J. F., A. R. Pingstone, K. A. Hall, F. A. Ensell, and J. L. Burton. "Interrelation of atmospheric ozone and cholecalciferol (vitamin D_3) production in man," *Aviat. Space Environ. Med.* 47(6):620–633 (1976).
21. Krinsky, N. I. "Singlet oxygen in biological systems," *Trends Biochem. Sci.* 2:35–38 (1977).
22. Lui, N. S. T., and O. A. Roels. "The vitamins: A. Vitamin A and carotene," in *Modern Nutrition in Health and Disease*, 6th ed., R. S. Goodhart and M. E. Shils, Eds. (Philadelphia: Lea and Febiger, 1980), p. 142.
23. Calabrese, E. J. *Nutrition and Environmental Health*, Vol. I, (New York: John Wiley & Sons, 1980).
24. Brabec, M. J., and I. A. Bernstein. "Cellular, subcellular, and molecular targets of foreign compounds," in *Toxicology: Principles and Practice*, Vol. 1, A. L. Reeves, Ed. (New York: John Wiley & Sons, 1981), pp. 33–34.
25. Biehl, J. P., and R. W. Vilter. "Effect of isoniazid on vitamin B_6 metabolism: Its possible significance in producing isoniazid neuritis," *Proc. Soc. Exp. Biol. Med.* 85:389–394 (1954).
26. Bender, D. A., and R. Russell-Jones. "Isoniazid-induced pellagra despite vitamin-B_6 supplementation," *Lancet* 2:1125–1126 (1979).
27. Heller, C. A., and P. A. Friedman. "Pyridoxine deficiency and peripheral neuropathy associated with long-term phenelzine therapy," *Am. J. Med.* 75:887 (1983).
28. Calabrese, E. J. *Pollutants and High Risk Groups: The Biological Basis of Increased Human Susceptibility to Environmental and Occupational Pollutants* (New York: John Wiley & Sons, 1978).
29. Calabrese, E. J. *Ecogenetics: Genetic Variation in Susceptibility to Environmental Agents* (New York: John Wiley & Sons, 1984).
30. Evans, D. A. P., M. F. Bullen, J. Housto, C. A. Hopkins, and J. M. Vetters. "Antinuclear factor in rapid and slow acetylator patients treated with isoniazid," *J. Med. Genet.* 9:53–56 (1972).
31. Burrell, R. "Immunology of occupational lung disease," in *Occupational Lung Disease*, 2nd ed., W. K. C. Morgan and A. Seaton, Eds. (New York: W. B. Saunders, 1984), pp. 196–211.

32. Brin, M. "Marginal Deficiency and Immunocompetence," presented at American Chemical Society Symposium, Las Vegas, NV (1980).

33. Schumann, I. "Preoperative measures to promote wound healing," *Nurs. Clin. North Am.* 14(4):683 (1979).

34. Hook, G. E. R., and J. R. Bend. "Pulmonary metabolism of xenobiotics," *Life Sci.* 18:279 (1976).

35. Calabrese, E. J. *Pollutants and High Risk Groups: The Biological Basis of Increased Human Susceptibility to Environmental and Occupational Pollutants* (New York: John Wiley & Sons, 1978).

36. Kirkpatrick, P. Personal communication, Kirkpatrick Medical Laboratory, Houston, TX (1983).

37. Baker, H. Personal communication, UMD, New Jersey Medical School. Newark, NJ 07107–3306 (1983).

38. Miroff, G. Personal communication, Monroe Medical Research Laboratory. Southfields, NY (1983).

39. Sauberlich, H. E., J. H. Skala, and R. P. Dowdy. *Laboratory Tests for the Assessment of Nutritional Status*, 6th ed. (Boca Raton, FL: CRC Press, 1981).

40. See References 36 and 37.

41. Miroff, G. Personal communication, Monroe Medical Research Laboratory. Southfields, NY (1983).

42. Sauberlich, H. E., J. H. Skala, and R. P. Dowdy. *Laboratory Tests for the Assessment of Nutritional Status*, 6th ed. (Boca Raton, FL: CRC Press, 1981).

43. Ross, G. H., W. J. Rea, A. R. Johnson, B. J. Maynard, and L. Carlisle. "Evidence for vitamin deficiencies in environmentally-sensitive patients," *Clin. Ecol.* 6(2):60–66 (1989).

44. Krinsky, N. I. "Singlet oxygen in biological systems," *Trends Biochem. Sci.* 2:35–38 (1977).

45. Lui, N. S. T., and O. A. Roels. "The vitamins: A. Vitamin A and carotene," in *Modern Nutrition in Health and Disease*, 6th ed., R. S. Goodhart and M. E. Shils, Eds. (Philadelphia: Lea and Febiger, 1980), p. 142.

46. Ongsakul, M., S. Sirisinha, and A. J. Lamb. "Impaired blood clearance of bacteria and phagocytic activity in vitamin A-deficient rats 41999," *Proc. Soc. Exp. Biol. Med.* 178 (2):204–208 (1985).

47. Hof, H., and C. H. Wirsing. "Anti-infective properties of vitamin A," *Z. Ernaehrungswiss.* 18(4):221–232 (1979).

48. Barclay, A. J. G., A. Foster, and A. Sommer. "Vitamin A supplements and mortality related to measles: A random clinical trial," *Br. Med. J.* 294:294–296 (1987).

49. Sommer, A., G. Hussaini, I. Tarvotjo, and D. Susanto. "Increased mortality in children with mild vitamin A deficiency," *Lancet* 2:585–588 (1983).

50. Colby, H. D., R. E. Kramer, J. W. Greiner, D. A. Robinson, R. F. Krause, and W. J. Canady. "Hepatic drug metabolism in retinol-deficient rats," *Biochem. Pharmacol.* 24:1644–1646 (1975).

51. Coats, J. R. "Toxicology of pesticide residues in foods," in *Nutritional Toxicology*, Vol. 2, J. H. Hathcock, Ed. (Orlando, FL: Academic Press, 1987), p. 249.

52. Wedzisz, A., T. Sloczynska, and J. Szwejda. "Wplyw preparatu gardona na niektone skladniki warzyw," *Bromatol. Chem. Toksykol.* 10:29–34 (1977a).

53. Engst, R., M. Blazovich, and R. Knoll. "Über das vorkommen von lindan," in *Möhren and Seinen Einfluss auf dem Carotengehalt Nahrung,* 11:389–399 (1967).

54. Mlodecki, H., W. Lasota, and J. Dutkiewicz. "Wplyw fenitrotionu na zawartoxc β-Karotenu i sumy karotenoidow w marchw bromacol," *Chem. Toksykol.* 6:221–228 (1973).

55. Zolnierz-Piotrowska, M. "Wplyw Wybranych Fungicydow na Zawartosc Witaminy c i Cukrow Redukujacych w Owocach Czarnej Porzeczki Oraz na Zawartosc β-Karotenu i Sumy Karotenoidow w Pomidorach," Ph.D thesis, Library of National Institute of Hygiene, Warsaw, Poland (unpublished, 1975).

56. Wilkinson, R. E. "Physiological response of lipid components to thiocarbamates and antidotes," in *Chemistry and Action of Herbicide Antidotes*, F. M. Pallos and J. E. Casida, Eds. (New York: Academic Press 1978), p. 85.

57. Wilkinson, R. E. "Physiological response of lipid components to thiocarbamates and antidotes," in *Chemistry and Action of Herbicide Antidotes*, F. M. Pallos and J. E. Casida, Eds. (New York: Academic Press 1978), pp. 105–106.

58. Wilkinson, R. E. "Physiological response of lipid components to thiocarbamates and antidotes," in *Chemistry and Action of Herbicide Antidotes*, F. M. Pallos and J. E. Casida, Eds. (New York: Academic Press 1978), p. 105.

59. Wilkinson, R. E. "Physiological response of lipid components to thiocarbamates and antidotes," in *Chemistry and Action of Herbicide Antidotes*, F. M. Pallos and J. E. Casida, Eds. (New York: Academic Press 1978), p. 106.

60. Krause, M. V., and L. K. Mahan. *Food, Nutrition and Diet Therapy* (Philadelphia: W.B. Saunders, 1979), p. 153.

61. Rodriguez, M. S., and M. I. Irwin. "A conspectus of research and Vitamin A," *J. Nutr.* 102:75 (1972).

62. "Vitamin A and the thyroid," *Nutr. Rev.* 37:90 (1979).

63. Rodriguez, M. S., and M. I. Irwin. "A conspectus of research and Vitamin A," *J. Nutr.* 102:75 (1972).

64. "The role of growth hormone in the action of vitamin B_6 on cellular transfer of amino acids," *Nutr. Rev.* 37:300 (1979b).

65. Linder, M. C., Ed. *Nutritional Biochemistry and Metabolism with Clinical Applications* (New York: Elsevier, 1985), pp. 109–110.
66. Gaitan, E. "Endomic goiter in Western Clombia," *Ecol. Dis.* 2(4):295–308 (1983).
67. Pfeiffer, C. C. *Mental and Elemental Nutrients: A Physicians' Guide to Nutrition and Health Care* (Princeton, NJ: Ketas Publishers, 1975).
68. Seshadry, S. P., and J. Ganguly. "Studies on vitamin A enterase 5. A comparative study of vitamin A enterase and cholesterol esterase of rat and chicken liver," *Biochem. J.* 80:397–406 (1961).
69. Khan, M. A. "Systemic pesticides for use on animals," *Annu. Rev. Entomol.* 14:369–386 (1969).
70. Ember, M., L. Mindszenty, B. Rengei, and G. Y. Gal. "Changes of vitamin A in blood and liver organophosphorous poisoned suicides," *Res. Commun. Chem. Pathol. Pharmacol.* 3:561–571 (1970).
71. Ember, M., L. Mindszenty, B. Rengei, L. Csaszar, and L. Crizna. "Secondary vitamin A deficiency in organophosphate formulators and spray workers," *Res. Commun. Chem. Pathol. Pharmacol.* 3:145–154 (1972).
72. Menzel, D. B., and M. O. Amdur. "Toxic responses of the respiratory system," in *Casarett and Doull's Toxicology: The Basic Science of Poisons*, 3rd ed., C. D. Klaassen, M. O. Amdur, and J. Doull, Eds. (New York: Macmillan, 1986), p. 335.
73. Villeneuve, D. C., D. L. Grant, W. E. J. Phillips, M. L. Clark, and D. J. Clegg. "Effects of PCB administration of microsomal enzyme activity in pregnant rabbits," *Bull. Environ. Contam. Toxicol.* 6:120 (1971).
74. Cecil, H. C., S. J. Harris, J. Bitman, and G. F. Fries. "Polychlorinated biphenyl-induced decrease in liver vitamin A in Japanese quail and rate," *Bull. Environ. Contam. Toxicol.* 93:179 (1973).
75. Linder, M. C. "Nutrition and metabolism of vitamins," in *Nutritional Biochemistry and Metabolism with Clinical Applications*, M. C. Linder, Ed. (New York: Elsevier 1985), p. 109.
76. Runner, M. N. "Rescue from disproportionate dwarfism in mice by means of caffeine modulation of the 4-hour early effect of excessive vitamin A," *Prog. Clin. Biol. Res.* 110(Pt A):345–353 (1983).
77. Linder, M. C. "Nutrition and metabolism of vitamins," in *Nutritional Biochemistry and Metabolism with Clinical Applications*, M. C. Linder, Ed. (New York: Elsevier, 1985), p. 110.
78. Grangaud, R., M. Nichol, and D. Desplanques. "Effect of vitamin A in enzymatic conversion of the Δ^5-3-β-Hydroxy-into Δ^4-3-oxosteroids by adrenals of the rat," *Am. J. Clin. Nutr.* 22:991–1002 (1969).
79. Seifter, E., G. Rettura, J. Padawer, F. Stratford, D. Kambosos, and S. M. Levenson. "Impaired wound healing in streptozotocin diabetes. Prevention by supplemental vitamin," *Ann. Surg.* 194:42–50 (1981).
80. "Vitamin A deficiency and anemia," *Nutr. Rev.* 37:38–40 (1979).

81. Bazai, R. "Vitamin "A" and infection," *JPMA* 37(5):115–116 (1987).
82. Frame, B., H. L. Guiang, A. M. Frost, and W. A. Reynolds. "Osteomalacia induced by laxative (phenolphthalein) ingestion," *Arch. Intern. Med.* 128:794–796 (1971).
83. Linder, M. C. "Nutrition and metabolism of vitamins," in *Nutritional Biochemistry and Metabolism with Clinical Applications*, M. C. Linder, Ed. (New York: Elsevier 1985), p. 109.
84. Burrows, M. T., and W. K. Farr. "The action of mineral oil per os on the organism," *Proc. Soc. Exp. Biol. Med.* 24:719–723 (1927).
85. Ahotupa, M., V. Bussacchini-Griot, J. C. Béréziat, A. M. Camus, and H. Bartsch. "Rapid oxidative stress induced by N-nitrosamines," *Biochem. Biophys. Res. Commun.* 146(3):1047–1054 (1987).
86. Guilbert, H. R., and G. H. Hart. "Storage of vitamin A in cattle," *J. Nutr.* 8:25 (1934).
87. Morabia, A., A. Sorenson, S. K. Kumanyika, H. Abbey, B. H. Cohen, and E. Chee. "Vitamin A, cigarette smoking, and airway obstruction," *Am. Rev. Respir. Dis.* 140(5):1312–1316 (1989).
88. "Present knowledge of naturally occurring toxicants in foods," *Nutr. Rev.* 26(5):129–33 (1968).
89. Simpson, K. L., and C. O. Chichester. "Metabolism and nutritional significance of caronoids," *Annu. Rev. Nutr.* 1:351–374 (1981).
90. Krause, M. V., and L. K. Mahan. *Food, Nutrition and Diet Therapy* (Philadelphia: W.B. Saunders, 1979), p. 153.
91. Mandani, K. A., and M. B. Elmongy. "Role of vitamin A in cancer," *Nutr. Res.* 6:863–875 (1986).
92. Innami, S., H. Tojo, S. Utsuja, A. Nakamur, and S. Nagazama. "PCB toxicity and nutrition I. PCB toxicity and vitamin A," *Jpn. J. Nutr.* 32:58–66 (1974).
93. Buist, R. A. "Anxiety neurosis: The lactate connection," *Int. Clin. Nutr. Rev.* 5:1–4 (1985).
94. Garrison, R. H., Jr., and E. Somer. *The Nutrition Desk Reference* (New Canaan, CN: Keats, 1985), pp. 44–46.
95. Garrison, R. H., Jr., and E. Somer. *The Nutrition Desk Reference* (New Canaan, CN: Keats, 1985), pp. 44–46.
96. Singer, T. P., R. W. VonKoff, and D. L. Murphy. *Monoamine Oxidase: Structure Function and Altered Function* (New York: Academic Press, 1979).
97. Kulkarni, A. P., and E. Hodgson. "The metabolism of insecticides: The role of mono-oxygenase enzymes," *Annu. Rev. Pharmacol. Toxicol.* 24:19–42 (1984).
98. Garrison, R. H., Jr., and E. Somer. *The Nutrition Desk Reference* (New Canaan, CN: Keats, 1985), pp. 44–46.
99. Garrison, R. H., Jr., and E. Somer. *The Nutrition Desk Reference* (New Canaan, CN: Keats, 1985), p. 45.

100. Garrison, R. H., Jr., and E. Somer. *The Nutrition Desk Reference* (New Canaan, CN: Keats, 1985), p. 46.
101. Garrison, R. H., Jr., and E. Somer. *The Nutrition Desk Reference* (New Canaan, CN: Keats, 1985), p. 47.
102. Singer, T. P., R. W. VonKoff, and D. L. Murphy. *Monoamine Oxidase: Structure Function and Altered Function* (New York: Academic Press, 1979).
103. Garrison, R. H., Jr., and E. Somer. *The Nutrition Desk Reference* (New Canaan, CN: Keats, 1985), p. 53.
104. Garrison, R. H., Jr., and E. Somer. *The Nutrition Desk Reference* (New Canaan, CN: Keats, 1985), p. 53.
105. Toth, B. "Synthetic and naturally occurring hydrazines as possible carcinogens," *Cancer Res.* 35:3693 (1975).
106. Toth, B., S. Erickson. "Reversal of toxicity of hydrazine analogue by pyrioxine," *Toxicology* 7:31 (1984).
107. Cunnane, S. C., M. S. Manku, and D. F. Horrobin. "Accumulation of linoleic and γ-linoleic acid in tissue lipids of pyridoxine-deficiency in rats," *J. Nutr.* 114(10):1754–1761 (1984).
108. Waldinger, C., and R. B. Berg. "Signs of pyridoxine dependency manifest at birth in siblings," *Pediatrics* 32:161 (1963).
109. Mehansho, H., D. D. Buss, M. W. Hamm, and L. V. M. Henderson. "Transport and metabolism of pyridoxine in rat liver," *Biochim. Biophys. Acta* 631:112 (1980).
110. Ellis, J., K. Folkers, T. Watanabe, M. Kaji, S. Saji, J. W. Caldwell, C. A. Temple, and F. S. Wood. "Clinical results of a cross-over treatment with pyridoxine and placebo of the carpal tunnel syndrome," *Am. J. Clin. Nutr.* 32:2040 (1979).
111. Linder, M. C. *Nutritional Biochemistry and Metabolism with Clinical Applications* (New York: Elsevier, 1985), p. 88.
112. Garrison, R. H., Jr., and E. Somer. *The Nutrition Desk Reference* (New Canaan, CN: Keats, 1985), p. 49.
113. Garrison, R. H., Jr., and E. Somer. *The Nutrition Desk Reference* (New Canaan, CN: Keats, 1985), p. 48.
114. Garrison, R. H., Jr., and E. Somer. *The Nutrition Desk Reference* (New Canaan, CN: Keats, 1985), p. 52.
115. Butterfield, S., and D. H. Calloway. "Folacin in wheat and selected foods," *J. Am. Diet. Assoc.* 60:310 (1972).
116. Garrison, R. H., Jr., and E. Somer. *The Nutrition Desk Reference* (New Canaan, CN: Keats, 1985), p. 49.
117. Garrison, R. H., Jr., and E. Somer. *The Nutrition Desk Reference* (New Canaan, CN: Keats, 1985), p. 49.
118. Garrison, R. H., Jr., and E. Somer. *The Nutrition Desk Reference* (New Canaan, CN: Keats, 1985), p. 49.

119. Levine, M. R., R. Cuzzi-Spada, R. Cardenas, R. D. Buckley, and O. J. Bolchan. "P-aminobenzoic acid as a protective agent in ozone toxicity," *Arch. Environ. Health* 24:243–247 (1972).

120. Levine, M. R., R. Cuzzi-Spada, R. Cardenas, R. D. Buckley, and O. J. Bolchan. "P-aminobenzoic acid as a protective agent in ozone toxicity," *Arch. Environ. Health* 24:243–247 (1972).

121. Barnhart, E. R. *Physicians' Desk Reference*, 40th ed., (Oradell, NJ: Medical Economic Co., Inc., 1988), pp. 1035–1036.

122. Sieve, B.F. "Further investigation in the treatment of vitiligo," *Va. Med. Mon.* 72:6–17 (1945).

123. Hughes, C.G., "Oral PABA and vitiligo," *J. Am. Acad. Dermatol.* 9:770 (1983).

124. Zarafonetis, C. J. D., E. B. Johnwick, L. W. Kirkman, and A. C. Curtis. "Para amino benzoic acid in dermatitis herpetiformis," *Arch. Dermatol. Syph.* 63:115–132 (1951).

125. Barnhart, E. R. *Physicians' Desk Reference*, 40th ed., (Oradell, NJ: Medical Economic Co., Inc., 1988), pp. 1035–1036.

126. Zarafonetis, C. J. D. "The treatment of scleroderma: results of potassium para aminobenzoate therapy in 104 cases," in *Inflammation and Disease of Connective Tissue*, L. C. Mills and T. H. Moyer, Eds. (Philadelphia: W.B. Saunders, 1961), pp. 688–696.

127. Grace, W. J., R. J. Kennedy, and A. Formato. "Therapy of scleroderma and dermatomyositis," *N.Y. State J. Med.* 63:140–144 (1963).

128. Barnhart, E. R. *Physicians' Desk Reference*, 40th ed., (Oradell, NJ: Medical Economic Co., Inc., 1988), pp. 1035–1036.

129. Barnhart, E. R. *Physicians' Desk Reference*, 40th ed., (Oradell, NJ: Medical Economic Co., Inc., 1988), pp. 1035–1036.

130. Kantor, G. K., and J. L Ratz. "Letter to the Editor. Liver toxicity from potassium para-aminobezoate," *J. Am. Acad. Dermatol.* 13(4):671–672 (1985).

131. Werbach, M. R. *Nutritional Influences on Illness. A Sourcebook of Clinical Research* (Tarzana, CA: Third Line Press, Inc., 1988), p. 491.

132. Garrison, R. H., Jr., and E. Somer. *The Nutrition Desk Reference* (New Canaan, CN: Keats, 1985), p. 52.

133. Williams, S. R. *Essentials of Nutrition and Diet Therapy* (Times Mirror/Mosby College Publishing, 1986), p. 118.

134. Machlin, L. J., and A. Bendich. "Free radical tissue damage: Protective role of antioxidant nutrients," *Fed. Am. Soc. Exp. Biol.* 1:441–445 (1987).

135. Sprince, H., C. M. Parker, and G. G. Smith. "Comparison of protection by L-ascorbic acid, L-cysteine, and adrenergi-blocking agents against acetaldehyde, acrolein, and formaldehyde toxicity: Implications in smoking," *Agents Actions* 9(4):407–414 (1979).

136. Pelletier, O. "Cigarette smoking and vitamin C," *Nutrition Today* 5:12 (1970).

137. Thorn, N. A., F. S. Nielsen, C. K. Zeppesen, B. L. Christensen, and O. Farver. "Uptake of dehydroascorbic acid and ascorbic acid to isolated nerve terminals and secretor granules from OX neurohypohyses," *Acta. Physiol. Scand.* 128:629–638 (1986).

138. Cullen, E. I., V. May, and B. A. Eipper. "Transport and stability of ascorbic acid in pituitary cultures," *Mol. Cell. Endocrinol.* 48(2-3):239–250 (1986).

139. Ragoebar, M., J. A. Huisman, W. B. van den Berg, and C. A. van Ginneken. "Characteristics of the transport of ascorbic acid into leukocytes," *Life Sci.* 40:499–510 (1987).

140. Rose, R. C. "Ascorbic acid transport in mammalian kidney," *Am. J. Pathol.* 120:244–247 (1986).

141. Bianchi, J., and R. C. Rose. "National independent dehydro-L-ascorbic acid uptake in renal brush-border membrane vesicles," *Biochim. Biophys. Acta* 819:75–82 (1986).

142. Beers, M. F., R. G. Johnson, and A. Scarpa. "Evidence for an ascorbate shuttle for the transfer of reducing equivalents across chromaffin granule membranes," *J. Biol. Chem.* 261:2524–2535 (1986).

143. Thorn, N. A., F. S. Nielsen, C. K. Zeppesen, B. L. Christensen, and O. Farver. "Uptake of dehydroascorbic acid and ascorbic acid to isolated nerve terminals and secretor granules from OX neurohypohyses," *Acta. Physiol. Scand.* 128:629–638 (1986).

144. Glembotski, C. C. "The characterization of the ascorbic acid mediated α-amidation of α-melanotrophic cultured intermediate pituitary lobe cells," *Endocrinology* 118:1461–1468 (1986).

145. Shields, P. P., T. R. Gibson, and C. C. Glembotski. "Ascorbate transport by AtT_{20} mouse pituitary corticotropic tumor cells uptake and secretion studies," *Endocrinology* 118:1452–1460 (1986).

146. Mueller, H. W., and S. Tannert. "The significance of electron spin resonance of the ascorbic acid radical in freeze-dried human brain tumours and oldematons or normal periphery," *Brain J. Cancer* 53:385–391 (1986).

147. Pietronigro, D. D., M. Hovsepain, W. B. Demopoulas, and E. S. Flamon. "Reductive metabolism of ascorbic acid in the central nervous system," *Brain Res.* 333(1):161–164 (1985).

148. Overbeck, G. A. "Hormonal regulation of ascorbic acid in the adrenal of the rat," *Acta Endocrinol. (Copenhagen)* 109(3): 483–489 (1986).

149. Lam, D. K., and P. M. Daniel. "The influx of ascorbic acid into the rat's brain," *Q. J. Exp. Physiol.* 71(3):483–489 (1986).

150. Kamata, K., R. L. Wilson, K. D. Alloway, and G. V. Rebec. "Multiple amphetamine injections reduce the release of ascorbic acid in the meostriatum of the rat," *Brain Res.* 362:331–338 (1986).

151. Wilson, R. L., K. Kamata, J. C. Bigelow, G. V. Rebec, and R. M. Wightman. "Crus cerebri lesions abolish amphetamine-induced ascorbate release in the rat neostriatum," *Brain Res.* 370:393–396 (1986).

152. Pietronigro, D. D., V. DeCrescito, J. J. Tomasula, H. B. Demopoulos, and E. S. Flamm. "Ascorbic acid: a putative biochemical marker of irreversible neurologic functional loss following spinal cord injury," *Cent. Nerv. Syst. Trauma* 2:85–92 (1985).

153. Lovilot A., F. Gonon, M. Buda, H. Simon, M. LeMoal, and J. F. Pajol. "Effects of D- and L-amphetamine on dopamine metabolism and ascorbic acid levels in nucleus accumens and olfactory tubercle as studied by in vivo differential puls voltammetry," *Brain Res.* 336(2):253–263 (1985).

154. Kratzing, C. C., J. D. Kelly, and J. E. Kratzing. "Ascorbic acid in fetal rat brain," *J. Neurochem.* 44(5):1623–1624 (1985).

155. Obrenovitch, T. P., and J. L. Gillard. "Decreased brain levels of ascorbic acid in rats exposed to high pressures," *J. Appl. Physiol.* 58(3):839–843 (1985).

156. Subaticanec, K., V. Folnegofic-Smale, R. Turcin, B. Mestrovic, and R. Buzina. "Plasma levels and urinary vitamin C excretion in schizophrenic patients," *Hum. Nutr. Clin. Nutr.* 40(6):421–428 (1986).

157. Rose, R. C. "Ascorbic acid transport in mammalian kidney," *Am. J. Pathol.* 120:244–247 (1986).

158. Bianchi, J., and R. C. Rose. "National independent dehydro-L-ascorbic acid uptake in renal brush-border membrane vesicles," *Biochim. Biophys. Acta* 819:75–82 (1986).

159. Cowley, D. M., B. C. McWhinney, J. M. Brown, and A. H. Chalmers. "Chemical factors important to calcium nephrolithiasis: Evidence for impaired hydroxicarboxylic acid absorption causing hyperoxaluria," *Clin. Chem.* 33(2):243–247 (1987).

160. Chalmers, A. H., D. M. Cowley, and J. M. Brown. "A possible etiological role for ascorbate in calculi formation," *Clin. Chem.* 32(2):333–336 (1986).

161. Faizallah, R., A. I. Morriss, N. Krasner, and R. J. Walker. "Alcohol enhances vitamin C excretion in the urine," *Alcohol Alcoholism* 21(1):81–84 (1986).

162. Wakefield, L. M., A. E. Cass, and G. K. Radda. "Electron transfer across the chromaffin granule membrane. Use of EPR to demonstrate reduction of intravesicular ascorbate radical by the extravesicular mitochondrial NADH: ascorbate radical oxidoreductase," *J. Biol. Chem.* 261(21):9746–9752 (1986).

163. Beers, M. F., R. G. Johnson, and A. Scarpa. "Evidence for an ascorbate shuttle for the transfer of reducing equivalents across chromaffin granule membranes," *J. Biol. Chem.* 261:2524–2535 (1986).

164. Morita, K., M. Levine, E. Heldman, and H. B. Pollard. "Ascorbic acid and catecholamine release from digitonin-treated chromaffin cells," *J. Biol. Chem.* 260(28):15112–15116 (1985).

165. Boidin, M. P., W. E. Erdmann, and N. S. Faithfull. "The role of ascorbic acid in etomidate toxicity," *Eur. J. Anaesthesiol.* 3(5):417–422 (1986).

166. Arts, T., J. T. Kuikka, R. S. Reneman, and J. B. Bassingthwaighte. "Polarographic measurement of ascorbate washout in isolated perfused rabbit hearts," *Am. J. Physiol.* 249(1):H150–154 (1985).

167. Dubick, M. A., H. Heng, and R. B. Rucker. "Effects of protein deficiency and food restriction on a lung ascorbic acid and glutathione in rats exposed to ozone," *J. Nutr.* 115(8):1050–1056 (1985).

168. Brenes, A., L. S. Jensen, K. Takahashi, and S. L. Bolden. "Dietary effects on content of hepatic lipid, plasma minerals and tissue ascorbic acid in hens and estrogenized chicks," *Poultry Sci.* 64(5):947–954 (1985).

169. Strnadova, E., and J. Prokopic. "Changes in ascorbic acid content in various organs and serum of mice experimentally infected with Taenia crassicepts (Zeder, 1800) cysticerci," *Folia Parsitol. (Progue)* 32(2):185–188 (1985).

170. Dubick, M. A., H. Heng, and R. B. Rucker. "Effects of protein deficiency and food restriction on a lung ascorbic acid and glutathione in rats exposed to ozone," *J. Nutr.* 115(8):1050–1056 (1985).

171. Shah, S., and N. Nath. "Effect of castration on the metabolism of L-ascorbic acid in rat prostate," *J. Nutr. Sci. Vitaminol. (Tokyo)* 31(1):107–113 (1985).

172. Goldie, R. G., D. Spina, P. J. Rigby, and J. W. Paterson. "Autodradiographic localization of ascorbic acid-dependent binding sites for [125I] iodocyanopindolol in guinea pig trachea," *Eur. J. Pharmacol.* 124(1-2):179–182 (1986).

173. Basu, T. K. "Effects of estrogen and progestogen on the ascorbic acid status of female guinea pigs," *J. Nutr.* 116(4): 570–577 (1986).

174. Tadera, K., M. Arima, S. Yoshino, F. Yagi, and A. Kobayashi. "Conversion of pyridoxine into 6-hydrosypyridoxine by food components, especially ascorbic acid," *J. Nutr. Sci. Vitaminol.* 32(3):267–277 (1986).

175. Lambelet, P., F. Saucy, and J. Loliger. "Chemical evidence for interactions between vitamins E and C," *Experientia* 41(11):1384–1388 (1985).

176. Marx, G., and M. Chevion. "Site-specific modification of albuin by free radicals. Reaction with copper (II) and ascorbate," *Biochem. J.* 236(2):397–400 (1986).

177. Chiou, S. H., W. C. Chang, Y. S. Jon, H. M. Chung, and T. B. Lo. "Specific cleavages of DNA by ascorbate in the presence of copper ion or copper chelates," *J. Biochem. (Tokyo)* 98(6):1723–1726 (1985).

178. Marx, G., and M. Chevion. "Site-specific modification of albumin by free radicals. Reaction with copper (II) and ascorbate," *Biochem. J.* 236(2):397–400 (1986).

179. Majamaa, K., V. Giunzler, H. M. Hanauske-Abel, R. Myllyl:a, and K. I. Kivirikko. "Partial identity of the 2-oxoglutarate and ascorbate binding sites of prolyl 4-hydroxylase," *J. Biol. Chem.* 261(17):7819–7823 (1986).

180. Marx, G., and M. Chevion. "Site-specific modification of albumin by free radicals. Reaction with copper (II) and ascorbate," *Biochem. J.* 236(2):397–400 (1986).

181. Pethig, R., P. R. Gascoyne, J. A. McLaughlin, and A. Szert-Gyorgyi. "Enzyme-controlled scavenging of ascorbyl and 2,6-dimethoxy-semiquinone free radicals in Ehrlich ascites tumor cells," *Proc. Natl. Acad. Sci. U.S.A.* 82(5):1439–1442 (1985).

182. Bhattacharjee, J., A. S. Chakraborty, N. K. Sarkar, A. Basu, and S. Mitra. "Study of ascorbate status in murine and human leukemias," *J. Comp. Pathol.* 95(1):87–91 (1985).

183. Ghosh, J., and S. Das. "Evaluation of vitamin A and C status in normal and malignant conditions and their possible role in cancer prevention," *Jpn. Cancer Res.* 76(12):1174–1178 (1985).

184. Costamagna, L., I. Rosi, I. Garuccio, and O. Arrigoni. "Ascorbic acid specific utilization by some yeasts," *Can. J. Microbiol.* 32(9):756–758 (1986).

185. Moriya, Y., M. Goto, T. Nakamura, and J. Koyama. "Hemolysis of sheep erythrocytes with the cell membrane of liquid paraffin-induced guinea pig macrophages," *J. Biochem. (Tokyo)* 100(3):521–529 (1986).

186. Ohel, G., R. Kisselevitz, E. J. Margalisth, and J. G. Schenker. "Ascorbate-dependent lipid peroxidation in the human placenta and fetal membrane," *Gynecol. Obstet. Invest.* 19(2):73–77 (1985).

187. Devasagayam, T.P., and C. K. Pashpendran. "Changes in ascorbate-induced lipid peroxidation of hepatic rough and smooth microsomes during postnatal development and aging of rats," *Mech. Ageing Dev.* 34:13–21 (1986).

188. Shah, S., and N. Nath. "Metabolism of L-ascorbic acid in the prostate of normal and scorbatic guinea pigs," *Metabolism* 34(10):912–916 (1985).

189. Ingermann, R. L., L. Stankova, and R. H. Bigley. "Role of monosaccharide transporter in vitamin C uptake by placental membrane vesicles," *Am. J. Physiol.* 250(4 Pt 1):C637–C641 (1986).

190. Forsman, S., and K. O. Frykholm. "Benzene poisoning. II. Examination of workers exposed to benzene with reference to the presence of estersulfate, muconic acid, urochrome A, and polyphenols in the urine together with vitamin C deficiency. Prophylacti-measures," *Acta Med. Scand.* 128:256 (1947).

191. Gabovich, R. D., and P. N. Maistruk. "On the therapeutic and prophylacti-diet in the flourine manufacturing industry," *Vopr. Pitan.* 22:323–338 (1963).

192. Centi, R., and O. Zaffini. "Notes from an extremely serious case of acute oxycaronism successfully proven to high doses of ascorbic acid and with L-dopa," *Minerva Anestesiol.* 37:406–414 (1971).

193. Klenner, F. R. "The role of ascorbic acid in therapeutics," *Tri-State Med. J.* 1977:51 (1977).

194. Zaffini, O., G. Cala, R. Centi, and A. Salicone. "A new therapeutic method for acute oxycarbonism (with the method of Cala-Zaffini) with intravenous infusions of high doses of ascorbic acid," *Minerva Anestesiolgica* 37:332–339 (1971).

195. Yunice, A. A., and R. D. Linderman. "Effect of ascorbic acid and zinc sulfate on ethanol toxicity and metabolism," *Proc. Soc. Exp. Biol. Med.* 154:146–150 (1979).

196. Pelletier, O. "Vitamin C and cigarette smokers," *Ann. N. Y. Acad. Sci.* 258:156–167 (1975).

197. Pelletier, O. "Smoking and vitamin C levels in humans," *Am. J. Clin. Nutr.* 2:1259–1267 (1968).

198. Pelletier, O. "Vitamin C and cigarette smokers," *Ann. N. Y. Acad. Sci.* 258:156–167 (1975).

199. Varghese, A. J., P. C. Land, R. Furrer, and W. R. Bruce. "Non-volatile N-nitroso compounds in human feces," in *International Agency for Research on Cancer, Environmental Aspects of N-Nitroso Compounds*, E. A. Walker, L. Griciute, M. Castegnaro, and R. E. Lyle, Eds. (Lyon: IARC Scientific Publications, 1978), 19:257–264.

200. Samita, M. H., D. M. Scheiner, and S. A. Katz. "Ascorbic acid in the prevention of chrome dermatitis: Mechanism of inactivation of chromium," *Arch. Environ. Health* 17:44–45 (1968).

201. Chakraborty, D., A. Bhattacharyya, K. Majumdar, G. C. Chatterjee. "Studies on L-ascorbic acid metabolism in rats under chronic toxicity due to organophosphorus insecticides and effects of supplementation of L-ascorbic acid in high doses," *J. Nutr.* 108:973–980 (1978).

202. Rea, W. J., I. R. Bell, C. W. Strust, and R. E. Smiley. "Food and chemical overexposure: Case histories," *Ann. Allergy* 41:2 (1978).

203. Rea, W. J., J. R. Butler, J. L. Laseter, and I. R. DeLeon. "Pesticides and brain-function changes in a controlled environment," *Clin. Ecol.* 2(3):145–150 (1984).

204. Krause, M. V., and L. K. Mahan. *Food, Nutrition and Diet Therapy* (Philadelphia: W.B. Saunders, 1979), p. 153.

205. Halliday, J. W., and L. W. Powell. "Iron overload," *Semin. Hematol.* 19:42 (1982).

206. Herbert, V. D. "Megavitamin therapy," in *Contemporary Nutrition Controversies*, T. P. Labuza and A. E. Sloan, Ed. (Los Angeles: West Publishing, 1979), p. 223.

207. Finley, E., and F. Cerklewski. "Influence of ascorbic acid supplementation on copper status in young adult men," *Am. J. Clin. Nutr.* 37(4):553–556 (1983).

208. Leach, J. F., A. R. Pingstone, K. A. Hall, F. A. Ensell, and J. L. Burton. "Interrelation of atmospheric ozone and cholecalciferol (vitamin D_3) production in man," *Aviat. Space Environ. Med.* 47(6):620–633 (1976).

209. Chicago National Dairy Council. "Recent developments in vitamin D," *Dairy Counc. Dig.* 47(3):13–17 (1976).

210. Garrison, R. H., Jr., and E. Somer. *The Nutrition Desk Reference* (New Canaan, CN: Keats, 1985), pp. 39–40.

211. Packer, L., and S. Landvik. "Vitamin E: Introduction to biochemistry and health benefits," in *Annals of the New York Academy of Sciences: Vitamin E Biochemistry and Health Implications*, Vol. 570, A. T. Diplock, L. J. Machlin, L. Packer, and W. Pryor, Eds. (New York: New York Academy of Sciences, 1989), pp. 1–6.

212. Blumberg, J. B., and S. N. Meydani. "Role of dietary antioxidants in aging," in *Nutrition and Aging*, Vol. 5., H. Munro and M. Hutchinson, Eds. (New York: Academic Press, 1986), pp. 85–97.

213. Harman, D. "The free-radical theory of aging," in *Free Radicals in Biology*, Vol. 5, W. A. Pryor, Ed. (New York: Academic Press, 1982), pp. 255–273.

214. Meydani, S. N., M. Meydani, P. N. Barklund, S. Liu, R. A. Miller, J. G. Cannon, R. Rocklin, and J. B. Blumberg. "Effect of vitamin E supplementation on immune responsiveness of the aged," in *Annals of the New York Academy of Science: Vitamin E Biochemistry and Health Implications*, Vol. 570, A. T. Diplock, L. J. Machlin, L. Packer, and W. A. Pryor, Eds. (New York: New York Academy of Sciences, 1989), pp. 283–290.

215. Wang, Y. M., M. Purewal, B. Nixon, D. H. Li, and D. Soltysiak-Pawluczuk. "Vitamin E and cancer prevention in an animal model," in *New York Academy of Science: Vitamin E Biochemistry and Health Implications*, Vol. 570, A. T. Diplock, L. J. Machlin, L. Packer, and W. A. Pryor, Eds. (New York: New York Academy of Sciences, 1989), pp. 383–390.

216. Pryor, W. A. "Views on the wisdom of using antioxidant vitamin supplements," *Free Rad. Biol. Med.* 3(3):189–191 (1987).

217. Pryor, W. A. "The free-radical theory of aging revisited: A critique and a suggested disease-specific theory," in *Modern Biological Theories of Aging*, H. R. Warner, R. N. Butler, and R. L. Sprott, Eds. (New York: Raven Press, 1987), pp. 89–112.

218. Karpen, C. W., A. J. Merola, R. V. Trewyn, D. G. Cornwell, and R. V. Panganamala. "Modulation of platelet thromboxane A_2 and arterial prostacyclin by dietary vitamin E," *Prostaglandins* 22:651–661 (1981).

219. Okuma, T., H. Takahasi, and H. Uchino. "Generation of prostacyclin-like substance and lipid peroxidation in vitamin E-deficient rats," *Prostaglandins* 19:527–536 (1980).

220. Steiner, M., and J. Anastasi. "Vitamin E. An inhibition of the platelet release reaction," *J. Clin. Invest.* 57:732–737 (1976).

221. Sonneveld, P. "Effect of α-tocopherol on the cardiotoxicity of Adrianmycin in the rat," *Cancer Treat. Rep.* 62:1033–1036 (1978).

222. Varma, S. D., N. A. Beachy, and R. D. Richards. "Photooxidation of lens lipids prevention by vitamin E," *Photochem. Photobiol.* 36:623–626 (1982).

223. Varma, S. D., D. Chand, Y. R. Sharma, J. F. Kuck, and R. D. Richards. "Oxidative stress on lens and cataract formation: role of light and oxygen," *Curr. Eye Res.* 3:35–57 (1984).

224. Viguie, C. A., L. Packer, and G. A. Brooks. "Muscle trauma does not change blood antioxidant status," *Med. Sci. Sports* 20(2):510 (1988).

225. Gohil, K., L. Packer, B. De Lumen, G. A. Brooks, and S. E. Terblanche. "Vitamin E deficiency and vitamin C supplements: Exercise and mitochondrial oxidation," *J. Appl. Physiol.* 60(6):1986–1991 (1986).

226. Roehm, J. N., J. G. Hadley, and D. B. Menzel. "Antioxidants vs. lung disease," *Arch. Intern. Med.* 128:88–93 (1971).

227. Roehm, J. N., J. G. Hadley, and D. Menzel. "The influence of vitamin E on the lung fatty acids of rats exposed to ozone," *Arch. Environ. Health* 24:237–242 (1972).

228. Chow, C. K., and A. L. Tappel. "An enzymatic protective mechanism against lipid peroxidation damage to lung of ozone exposed rats," *Lipids* 7:518–524 (1972).

229. Kagan, V. E. "Tocopherol stabilizes membrane against phospholipase A, free fatty acids, and lysophospholipids," in *Annals of the New York Academy of Science: Vitamin E Biochemistry and Health Implications,* Vol. 570 (part II), A. T. Diplock, L. J. Machlin, L. Packer, and W. A. Pryor, Eds. (New York: New York Academy of Sciences, 1989), pp. 121–135.

230. Fletcher, B. L., and A. L. Tappel. "Protective effects of dietary α-tocopherol in rats exposed to toxic levels of ozone and nitrogen dioxide," *Environ. Res.* 6:165–175 (1973).

231. Goldstein, B. D., O. S. Balchum, H. B. Demopoulos, and H. B. Duke. "Electron paramagnetic resonance spectroscopy. Free radical signals associated with ozonization of linoleic acid," *Arch. Environ. Health* 17:46–49 (1968).

232. Levander, O. A., V. C. Morriss, and R. S. Ferretti. "Comparative effects of selenium and vitamin E in lead poisoned rats," *J. Nutr.* 107:378–382 (1977).

233. Garrison, R. H., Jr., and E. Somer. *The Nutrition Desk Reference* (New Canaan, CN: Keats, 1985), p. 40.

234. Menzel, D. B. "The role of free radicals in the toxicity of air pollutants (nitrogen oxygen oxides and ozone)," in *Free Radicals in Biology,* Vol. 2, W. A. Pryor, Ed. (New York: Academic Press, 1976), pp. 181–200.

235. Guggenheim, M. A., S. P. Ringel, A. Silverman, and B. E. Grabert. "Progressive neuromuscular disease in children with chronic cholestasis and vitamin E deficiency: diagnosis and treatment with alpha tocopherol," *J. Pediatr.* 100:51–58 (1982).

236. Rosenblum, J. L., J. P. Keating, A. L Prinsky, and J. S. Nelson. "A progressive neurologic syndrome in children with chronic liver disease," *N. Engl. J. Med.* 304:503–508 (1981).

237. Sokol, R. J., J. E. Heubi, S. Iannaccone, K. E. Bove, and W. F. Balistreri. "Mechanism causing vitamin E deficiency during chronic childhood cholestasis," *Gastroenterology* 85:1172–1182 (1983).

238. Brin, M. F., T. A. Pedley, R. E. Lovelace, R. G. Emerson, P. Gouras, C. Mackay, H. J. Kayden, J. Levy, and H. Baker. "Electrophysiologic features of abetalipoproteinemia: Functional consequences of vitamin E deficiency," *Neurology* 36:669–673 (1986).

239. Brin, M. F., M. R. Fetell, P. H. A. Green, H. J. Kayden, A. P. Hays, M. M. Behrens, and H. Baker. "Blind loop syndrome, vitamin E malabsorption, and spinocerebellar degeneration," *Neurology* 35:338–342 (1985).

240. Elias, E., D. P. R. Muller, and J. Scott. "Association of spinocerebellar disorders with cystic fibrosis on chronic childhood cholestastis and very low serum vitamin E," *Lancet* 2:1319–1321 (1981).

241. Traber, M. G., R. J. Sokol, S. P. Ringel, H. E. Neville, C. A. Thellman, and H. J. Kayden. "Lack of tocopherol in peripheral nerves of vitamin E-deficient patients with peripheral neurophathy," *N. Engl. J. Med.* 317:262–265 (1987).

242. Bus, J. S., S. D. Aust, and J. E. Gibson. "Lipid peroxidation: A possible mechanism for paraquat toxicity," *Res. Commun. Chem. Pathol. Pharmacol.* 11:31–38 (1975).

243. Block, E. R. "Potentiation of acute paraquat toxicity by vitamin E deficiency," *Lung* 156(3):195–203 (1979).

244. Wasserman, B., and E. R. Block. "Prevention of acute paraquat toxicity in rats by superoxide dismutase," *Aviat Space Environ. Med.* 49(6):805–809 (1978).

245. Guggenheim, M. A., S. P. Ringel, A. Silverman, and B. E. Grabert. "Progressive neuromuscular disease in children with chronic cholestasis and vitamin E deficiency: Diagnosis and treatment with alpha tocopherol," *J. Pediatr.* 100(1):51–58 (1982).

246. Burck, U., H. H. Goebel, H. D. Kuhlendahl, C. Meir, and K. M. Goebel. "Neuromyopathy and vitamin E deficiency in man," *Neuropediatrics* 12:267–278 (1981).

247. Landrieu, P., J. Selva, F. Alvarez, A. Ropert, and S. Metral. "Peripheral nerve involvement in children with chronic cholestasis and vitamin E deficiency: a clinical electrophysiological and morphological study," *Neuropediatrics* 6:194–201 (1985).

248. Nelson, J. S. "Neuropathological studies of chronic vitamin E deficiency in mammals including humans," *Ciba Found. Symp.* 101:92–105 (1983).

249. Kayden, H.J. "Tocopherol content of adipose tissue from vitamin E-deficient humans," *Ciba Found. Symp.* 101:70–91 (1983).

250. Kayden, H.J., L. J. Hatam, and M. G. Traber. "The measurement of manograms of tocopherol from needle aspiration biopsies of adipose tissue: normal and abetalipoproteinemic subjects," *J. Lipid Res.* 24:652–656 (1982).

251. Traber, M. G., R. J. Sokol, S. P. Ringel, H. E. Neville, C. A. Thellman, and H. J. Kayden. "Lack of tocopherol in peripheral nerves of vitamin E-deficient patients with peripheral neurophathy," *N. Engl. J. Med.* 317:262–265 (1987).

252. Traber, M. G., R. J. Sokol, S. P. Ringel, H. E. Neville, C. A. Thellman, and H. J. Kayden. "Lack of tocopherol in peripheral nerves of vitamin E-deficient patients with peripheral neurophathy," *N. Engl. J. Med.* 317:262–265 (1987).

253. Bus, J. S., S. D. Aust, and J. E. Gibson. "Lipid peroxidation: A possible mechanism for paraquat toxicity," *Res. Commun. Chem. Pathol. Pharmacol.* 11:31–38 (1975).

254. Traber, M. G., R. J. Sokol, S. P. Ringel, H. E. Neville, C. A. Thellman, and H. J. Kayden. "Lack of tocopherol in peripheral nerves of vitamin E-deficient patients with peripheral neurophathy," *N. Engl. J. Med.* 317:262–265 (1987).

255. Burck, U., H. H. Goebel, H. D. Kuhlendahl, C. Meir, and K. M. Goebel. "Neuromyopathy and vitamin E deficiency in man," *Neuropediatrics* 12:267–278 (1981).

256. Landrieu, P., J. Selva, F. Alvarez, A. Ropert, and S. Metral. "Peripheral nerve involvement in children with chronic cholestasis and vitamin E deficiency: a clinical electrophysiological and morphological study," *Neuropediatrics* 6:194–201 (1985).

257. Wichman, A., F. Buchthal, G. H. Pezeshkpour, and R. E. Grigg. "Peripheral neuropathy in abetalipoproteinemia," *Neurology* 35:1270–1289 (1985).

258. McCoy, P. B. "Vitamin E: Interactions with free radicals and ascorbate," *Annu. Rev. Nutr.* 5:323–340 (1985).

259. Fariss, M. W. "Oxygen toxicity: Unique cyto-protective properties of vitamin E succinate," *Free Radical Biol. Med.* 9:333–343 (1985).

260. Pascoe, G. A., M. N. Farris, R. K. Olatsdotti, and D. J. Reed. "The role of vitamin E in protection against cytotoxicity. Maintenance of intracellular glutathione precursors and biosynthesis," *Eur. J. Biochem.* 166:241–247 (1985).

261. Kayden, H. J., and M. G. Traber. "Clinical, nutritional and biochemical consequences of apolipoprotein B deficiency," in *Proc. Int. Symp. Lipoprotein Deficiency Syndromes: Vancouver British Columbia, Canada, May 1985*, A. Angel and J. Frohlick, Eds. (New York: Plenum Publishing Corporation, 1986), pp. 67–81.

262. Brin, M. F., T. A. Pedley, R. E. Lovelace, R. G. Emerson, P. Gouras, C. Mackay, H. J. Kayden, J. Levy, and H. Baker. "Electrophysiologic features of abetalipoproteinemia: Functional consequences of vitamin E deficiency," *Neurology* 36:669–673 (1986).

263. Guggenheim, M. A., S. P. Ringel, A. Silverman, and B. E. Grabert. "Progressive neuromuscular disease in children with chronic cholestasis and vitamin E deficiency: diagnosis and treatment with alpha tocopherol," *J. Pediatr.* 100:51–58 (1982).

264. Kayden, H. J., and M. G. Traber. "Clinical, nutritional and biochemical consequences of apolipoprotein B deficiency," in *Proc. Int. Symp. Lipoprotein Deficiency Syndromes: Vancouver British Columbia, Canada, May 1985*, A. Angel and J. Frohlick, Eds. (New York: Plenum Publishing Corporation, 1986), pp. 67–81.

265. Sokol, R. J., M. Guggenheim, S. T. Iannaccone, P. E. Barkhaus, C. Miller, A. Silverman, W. F. Balistreri, and J. E. Heubi. "Improved neurologic function after long-term correction of vitamin E in children with chronic cholestasis," *N. Engl. J. Med.* 313:1580–1586 (1985).

266. Kayden, H. J., and M. G. Traber. "Clinical, nutritional and biochemical consequences of apolipoprotein B deficiency," in *Proc. Int. Symp. Lipoprotein Deficiency Syndromes: Vancouver British Columbia, Canada, May 1985*, A. Angel and J. Frohlick, Eds. (New York: Plenum Publishing Corporation, 1986), pp. 67–81.

267. Sokol, R. J., M. Guggenheim, S. T. Iannaccone, P. E. Barkhaus, C. Miller, A. Silverman, W. F. Balistreri, and J. E. Heubi. "Improved neurologic function after long-term correction of vitamin E in chidlren with chronic cholestasis," *N. Engl. J. Med.* 313:1580–1586 (1985).

268. Kayden, H. J., R. J. Sokol, and M. G. Traber. "Familial vitamin E deficiency with neurologic abnormalities in the absence of fat malabsorption," in *Clinical and Nutritional Aspects of Vitamin E*, O. Hayaishi and M. Min, Eds. (Amsterdam: Elsevier Science Publishers, 1987), p. 193.

269. Burck, U., H. H. Goebel, H. D. Kuhlendahl, C. Meir, and K. M. Goebel. "Neuromyopathy and vitamin E deficiency in man," *Neuropediatrics* 12:267–278 (1981).

270. Traber, M. G., R. J. Sokol, S. P. Ringel, H. E. Neville, C. A. Thellman, and H. J. Kayden. "Lack of tocopherol in peripheral nerves of vitamin E-deficient patients with peripheral neurophathy," *N. Engl. J. Med.* 317:262–265 (1987).

271. "Recommended Dietary Allowances," 8th ed. (Washington, DC: Food and Nutrition Board, National Research Council, National Academy of Science, 1974).

272. Garrison, R. H., Jr., and E. Somer. *The Nutrition Desk Reference* (New Canaan, CN: Keats, 1985), p. 42.
273. Ketola, H. G., and M. C. Nesheim. "Influence of dietary protein and methionine levels on the requirement for choline by chickens," *J. Nutr.* 104:1484–1489 (1974).
274. Aoyama, H. J., H. Yasui, and K. Ashida. "Effect of dietary protein and amino acids in a choline deficient diet on lipid accumulation in rat liver," *J. Nutr.* 101:730–734 (1971).
275. Zeisel, S. H. " 'Vitamin-like' molecules. (A) choline," in *Modern Nutrition in Health and Disease*, 7th ed., M. E. Shils and V. R. Young, Eds. (Philadelphia: Lea and Febiger, 1988), p. 444.
276. Olson, R. E. "Vitamin K," in *Modern Nutrition in Health and Disease*, 7th ed., M. E. Shils and V. R. Young, Eds. (Philadelphia: Lea and Febiger, 1988), pp. 328 and 335.
277. Rogers, A. E., and P. M. Newberne. "Lipotrope deficiency in experimental carcinogenesis," *Nutr. Cancer* 2:104–112 (1980).
278. Michael, U. F., S. L. Cookson, R. Chavez, and V. Pardo. "Renal function in the choline deficient rat (39103)," *Proc. Soc. Exp. Biol. Med.* 150:672–676 (1975).
279. Chang, C. H., and L. S. Jensen. "Inefficiency of carnitine as a substitute for choline for normal reproduction in Japanese quail," *Poultry Sci.* 54:1718–1720 (1975).
280. Kratzing, C.C., and Perry, J.J. "Hypertension in young rats following choline deficiency in maternal diets," *J. Nutr.* 101: 1657–1662 (1971).
281. Davis, K. L., P. A. Berger, and L. E. Hollister. "Choline for tardive dyskinesia," *N. Engl. J. Med.* 293:152 (1975).
282. Growdin, J. H., M. J. Hirsch, R. J. Wurtman, and W. Wiener. "Oral choline administration to patients with tardive dyskinesia," *N. Engl. J. Med.* 297:524–527 (1977).
283. Drachman, D. A. "Memory and cognitive function in man: Does the cholinergic system have a specific role?," *Neurology* 27:783–790 (1977).
284. Bartus, R.T., R. L. Dean, A. J. Goas, and A. S. Lippa. "Age-related changes in passive avoidance retention: Modulation with dietary choline," *Science* 209:301–303 (1980).
285. Sitaram, N., H. Weingartner, and J. C. Gillin. "Human serial learning: Enhancement with arecholine and choline and impairment with scopolamine," *Science* 201:274–276 (1978).
286. Zeisel, S., D. Reinstein, S. Corkin, J. Grondon, and R. Wurtman. "Cholinergic neurones and memory," *Nature* 293:187–188 (1981).
287. Fisman, M., H. Mersky, E. Helmes, J. McCready, E. H. Colhoun, and B. J. Rylett. "Double-blind study of lecithin in patients with Alzheimer's disease," *Can. J. Psychiatry* 26:426–428 (1981).
288. Wurtman, R. J., M. J. Hirsch, and J. H. Growdon. "Lecithin consumption raises serum free choline levels," *Lancet* 2:68–69 (1977).

289. Cohen, E. L., and R. J. Wurtman. "Brain acetylcholine: increase after systemic choline administration," *Life Sci.* 16:1095–1102 (1975).

290. Cohen, E. L., and R. J. Wurtman. "Brain acetylcholine: Control by dietary choline," *Science* 191:561–562 (1976).

291. Boyd, W. D., J. Graham-White, G. Blackwood, I. Glen, and J. McQueen. "Clinical effects of choline in Alzheimer senile dementia," *Lancet* 2:711 (1977).

292. Singh, G. S. "Letter. Action of choline on the rat blood pressure," *Indian J. Physiol. Pharmacol.* 17:125–131 (1973).

293. Zeisel, S. H. " 'Vitamin-like' molecules. (A) choline," in *Modern Nutrition in Health and Disease*, 7th ed., M. E. Shils and V. R. Young, Eds. (Philadelphia: Lea and Febiger, 1988), p. 447.

294. Broquist, H. P. "Vitamin-like moleculars, (c) inositol," in *Modern Nutrition in Health and Disease*, 7th ed., M. E. Shils and V. R. Young, Eds. (Philadelphia: Lea and Febiger, 1988), p. 459.

295. Eisenberg, F., Jr. "D-Myoinositol L-phosphate as product of cyclization of glucose 6-phosphate and substrate for a specific phosphatase in rat testis," *J. Biol. Chem.* 242:1375–1382 (1967).

296. Chen, C. H.-J., and F. Eisenberg, Jr. "Myoinosose-2 L-phosphate: An intermediate in the myoinositol L-phosphate synthase reaction," *J. Biol. Chem.* 250:2963–2967 (1975).

297. Paulus, H., and E. P. Kennedy. "The enzymatic synthesis of inositol monophosplatide," *J. Biol. Chem.* 235:1303–1311 (1960).

298. Neya, S., and I. Morishima. "Interaction of methemoglobin with inositol hexaphosphate. Presence of the T state in human adult methemoglobin in the low spin state," *J. Biol. Chem.* 256:793–798 (1981).

299. Mitchell, R. H. "Inositol phospholipids in membrane function," *Trends Biol. Sci.* 4:128–131 (1979).

300. Holub, B. J. "Metabolism and function of myo-inositol and inositol phospholipids," *Annu. Rev. Nutr.* 6:563–597 (1986).

301. Michell, B. "A link between lithium, lipids and receptors?," *Trends Biol. Sci.* 7:387–388 (1982).

302. Holub, B. J. "The nutritional significiance, metabolism, and function of myo-inositol and phosphatidylinositol in health and disease," *Adv. Nutr. Res.* 4:107–141 (1982).

303. Clements, R. S., Jr., and R. Reynertson. "Myoinositol metabolism in diabetes mellitus: Effect of insulin treatment," *Diabetes* 26:215–221 (1977).

304. Broquist, H. P. "Vitamin-like moleculars, (c) inositol," in *Modern Nutrition in Health and Disease*, 7th ed., M. E. Shils and V. R. Young, Eds. (Philadelphia: Lea and Febiger, 1988), p. 459.

305. Salway, J. G., J. A. Finnegan, D. Barrett, L. Whitehead, A. Karunanayaka, and R. B. Payne. "Effect of myo-inositol on peripheral-nerve function in diabetes," *Lancet* 2:1282–1284 (1978).

306. Broquist, H. P. "Vitamin-like moleculars, (c) inositol," in *Modern Nutrition in Health and Disease*, 7th ed., M. E. Shils and V. R. Young, Eds. (Philadelphia: Lea and Febiger, 1988), p. 459.

307. Broquist, H. P. "Vitamin-like moleculars, (c) inositol," in *Modern Nutrition in Health and Disease*, 7th ed., M. E. Shils and V. R. Young, Eds. (Philadelphia: Lea and Febiger, 1988), p. 459.

308. Kuhnau, J. "The flavonoids. A class of semi-essential food components: their role in human nutrition," *World Rev. Nutr. Diet.* 24:117–191 (1976).

309. Herrmann, K. "Flavonols and flaornes in food plants: A review," *J. Food Technol.* 11:433–448 (1976).

310. Robbins, R. C. "Stabilization of flow properties of blood with phenylbenzo-gama-pyrone derivatives (flavonoids)," *Int. J. Vitamin Nutr. Rev.* 47:373 (1977).

311. Miner, R. W. "Bioflavinoids and the capillary," *Ann. N.Y. Acad. Sci.* 61:637 (1955).

312. Harborne, J., and H. Mabry, Eds. *The Flavoenoids, Part 1 and Part 2* (New York: Academic Press, 1975).

313. Jacobs, W. M. "Citrus bioflavonoid compounds in pH-immunized gravidas. Results of a 10-year study," *Obstet. Gynecol.* 25:648 (1965).

314. Redman, J. C. "Letter to the editor," *Med. Trib.* 21(15):11 (April 16, 1980).

315. Linder, M. C., Ed. *Nutritional Biochemistry and Metabolism with Clinical Applications* (New York: Elsevier, 1985), pp. 109–110.

316. Cody, M. M. "Substances without vitamin status," *Food Sci. Technol.* 13:571 (1984).

317. Hill, Y. H. "Vitamins, structural formulas," in *Human Nutrition and Diet Therapy*, Y. H. Hui, Ed. (Monterey, CA: Wadsworth Health Sciences, 1983), p. 155.

318. Kratzer, F. H., and P. Vohra. *Chelates in Nutrition* (Boca Raton, FL: CRC Press, 1986), p. 157.

319. Punyasingh, K., and J. M. Navia. "Biochemical role of fluoride on in vitro bone resorption," *J. Dent. Res.* 63:432 (1984).

320. Faelten, S., and the Editors of *Prevention* Magazine, Eds. *The Complete Book of Minerals for Health* (Emmaus, PA: Rodale Press, 1981), p. 19.

321. Garrison, R. H., Jr., and E. Somer. *The Nutrition Desk Reference* (New Canaan, CN: Keats, 1985), p. 59.

322. Lehninger, A. L. *Principles of Biochemistry* (New York: Worth Publishers, Inc., 1982), pp. 269 and 892.

323. Lehninger, A. L. *Principles of Biochemistry* (New York: Worth Publishers, Inc., 1982), pp. 27, 381–382, 698–699.

324. Linder, M. C. *Nutritional Biochemistry and Metabolism with Clinical Applications* (New York: Elsevier Science Publishing Co., Inc., 1985), p. 141.

325. Lehninger, A. L. *Principles of Biochemistry* (New York: Worth Publishers, Inc., 1982), pp. 27, 381–382, 698–699.

326. Linder, M. C. *Nutritional Biochemistry and Metabolism with Clinical Applications* (New York: Elsevier Science Publishing Co., Inc., 1985), p. 142.

327. Garrison, R. H., Jr., and E. Somer. *The Nutrition Desk Reference* (New Canaan, CN: Keats, 1985), p. 58.

328. Garrison, R. H., Jr., and E. Somer. *The Nutrition Desk Reference* (New Canaan, CN: Keats, 1985), p. 59.

329. Rawn, J. D., Ed. *Biochemistry* (Burlington, NC: N. Patterson Publishers, 1989), p. 391.

330. Claudio, T. "Molecular genetics of acetyl choline receptor channels," in *Molecular Neurobiology*, D. M. Glover and B. D. Hames, Ed. (Oxford: IRL Press at Oxford University Press, 1990), pp. 63–79.

331. Patel, K. Personal communication (1990).

332. Rodale, J. I. *Complete Book of Minerals for Health* (Emmaus, PA: Rodale Press, 1972), p. 37.

333. Gallagher, J. C., B. L. Riggs, J. Eisman, A. Hamstra, S. B. Arnaud, and H. F. Deluca. "Intestinal calcium absorption and serum vitamin D metabolites in normal subjects and osteoporotic patients," *J. Clin. Invest.* 64:729–736 (1979).

334. Faelten, S., and the Editors of *Prevention* Magazine, Eds. *The Complete Book of Minerals for Health* (Emmaus, PA: Rodale Press, 1981), p. 22.

335. Fabio, A. "Rheumatoid Diseases Cured at Last," A Publication of the Rheumatoid Disease Foundation. Rt. 4, Box 137, Franklin, TN 37064 (1985), p. 98.

336. Rea, W. J. "Inter-relationships between the environment and premenstrual syndrome," in *Functional Disorders of the Menstrual Cycle*, M. G. Brush and E. M. Goudsmit, Eds. (New York: John Wiley & Sons, 1988), p. 154.

337. Krane, S. M., and J. T. Potts, Jr. "Skeletal remodeling and factors influencing bone and bone mineral metabolism," in *Harrison's Principles of Internal Medicine*, 8th ed., G. W. Thorn, R. D. Adames, E. Braunwald, K. J. Isselbacher, and R. G. Petersdorf, Eds. (New York: McGraw Hill, 1977), p. 2007.

338. Goto, S., and T. Sawamura. "Effect of excess calcium intake on absorption of nitrogen, fat, phophorus, and calcium in young rats. The use of organic calcium salt," *J. Nutr. Sci. Vitaminol.* 19:355–360 (1973).

339. McCance, R. A., and E. M. Widdowson. "Mineral metabolism of healthy adults on white and brown bread dietaries," *J. Physiol.* 101:44 (1942).

340. Reinhold, J. G., K. Nasr, R. Lahimgarzadeh, and H. Hedayati. "Effects of purified phytate and pytate-rich bread upon metabolism of zinc, calcium, phosphorus and nitrogen in man," *Lancet* 1:283 (1973).

341. Slavin, J. L., and J. A. Marlett. "Influence of refined cellulose on human small bowel function and calcium and magnesium balance," *Am. J. Clin. Nutr.* 33:1932 (1980).

342. James, W. P. T., W. J. Branch, and D. A. T. Southgate. "Calcium binding by dietary fiber," *Lancet* 1:638 (1978).

343. James, W. P. T., W. J. Branch, and D. A. T. Southgate. "Calcium binding by dietary fiber," *Lancet* 1:638 (1978).

344. Harmuth-Hoene, A. S., and R. Schelenz. "Effect of dietary fiber on mineral absorption in growing rats," *J. Nutr.* 110:1774 (1980).

345. Norris, L. C., F. H. Kratzer, H. J. Lin, A. B. Hellewell, and J. R. Beljan. "Effect of quantity of dietary calcium on maintenance of bone integrity in mature white leghorn male chickens," *J. Nutr.* 120:1085 (1972).

346. Cummings, J. H., D. A. T. Southgate, W. J. Branch, H. S. Wiggins, H. Houston, and D. J. A. Jenkins. "The digestion of pectin in the human gut and its effect on calcium absorption and large bowel function," *Br. J. Nutr.* 41:477 (1979).

347. Zarembski, P. M., and A. Hodgkinson. "The oxalic acid content of English diets," *Br. J. Nutr.* 16:627 (1962).

348. Lengemann, F. W., R. H. Wasserman, and C. L. Comar. "Studies on the enhancement of radiocalcium and radiostrontium absorption by lactose in the rat," *J. Nutr.* 68:443–456 (1959).

349. Armbrecht, H. J., and R. H. Wasserman. "Enhancement of Ca^{++} uptake by lactose in the rat small intestine," *J. Nutr.* 106:1265 (1976).

350. Kobayashi, A., S. Kawai, Y. Ohbe, and Y. Nagashima. "Effects of dietary lactose and a lactase preparation on the intestinal absoroption of calcium and magnesium in normal infants," *Am. J. Clin. Nutr.* 28:681–683 (1975).

351. Vaughn, O. W., and L. J. Filer, Jr. "The enhancing action of certain carbohydrates on the intestinal absorption of calcium in the rat," *J. Nutr.* 71:10 (1960).

352. Weiner, I. M. "Diuretics and other agents employed in the mobilization of edema fluid," in *Goodman and Gilman's The Pharmacological Basis of Therapeutics*, 8th ed., A. G. Gilman, T. W. Rall, A. S. Nies, and P. Taylor, Eds. (New York: Pergamon Press, 1990), pp. 713–731.

353. Kobayashi, A., S. Kawai, Y. Ohbe, and Y. Nagashima. "Effect of dietary lactose and a lactase preparation on the intestinal absorption of calcium and magnesium in normal infants," *Am. J. Clin. Nutr.* 28:681 (1975).

354. Westrich, B. H. "Effect of physical activity on skeletal integrity and its implications for calcium requirement studies," *Nutr. Health* 5(1/2):53–60 (1987).

355. Delgado-Escueta, A. V., A. A. Ward, Jr., D. M. Woodbury, and P. J. Porter. "Basic mechanisms of the epilepsies: Molecular and cellular approaches," in *Advanced Neurology* (New York: Raven Press, 1986) pp. 1–20.

356. Jowsey, J. "Osteoporosis postgraduate," *Med.* 60:75–79 (1976).

357. Carlson, R. J. "Longitudinal observations of two cases of organic anxiety syndrome," *Psychosomatics* 27(7):529–531 (1986).

358. Cohlan, S. Q., G. Bevelander, and T. Tiamsic. "Growth inhibition of prematures receiving tetracycline: Clinical and laboratory investigation," *Am. J. Dis. Child.* 105:453–461 (1963).
359. Allen, L. H. "Calcium bioavailability and absorption: A review," *Am. J. Clin. Nutr.* 35:783 (1982).
360. Mellanby, E. "The rickets-producing and anti-calcifying action of phytate," *J. Physiol. (London)* 109:488 (1949).
361. Yamaguchi, T., F. Matsumura, and A. A. Kadons. "Inhibition of synaptic ATPases by heptachlorepoxide in rat brain," *Pest. Biochem. Physiol.* 11:285–293 (1979).
362. Haynes, R. C. Jr. "Agents affecting calcification: Calcium, parathyroid hormone, calcitonin, vitamin D, and other compounds," in *Goodman and Gilman's The Pharmacological Basis of Therapeutics*, 8th ed., A. G. Gilman, T. W. Rall, A. S. Nies, and P. Taylor, Eds. (New York: Pergamon Press, 1990), p. 1514.
363. Haynes, R. C. Jr. "Agents affecting calcification: Calcium, parathyroid hormone, calcitonin, vitamin D, and other compounds," in *Goodman and Gilman's The Pharmacological Basis of Therapeutics*, 8th ed., A. G. Gilman, T. W. Rall, A. S. Nies, and P. Taylor, Eds. (New York: Pergamon Press, 1990), p. 1512.
364. Garrison, R. H., Jr., and E. Somer. *The Nutrition Desk Reference* (New Canaan, CN: Keats, 1985), p. 60.
365. Haynes, R. C. Jr. "Agents affecting calcification: Calcium, parathyroid hormone, calcitonin, vitamin D, and other compounds," in *Goodman and Gilman's The Pharmacological Basis of Therapeutics*, 8th ed., A. G. Gilman, T. W. Rall, A. S. Nies, and P. Taylor, Eds. (New York: Pergamon Press, 1990), p. 1512.
366. Garrison, R. H., Jr., and E. Somer. *The Nutrition Desk Reference* (New Canaan, CN: Keats, 1985), p. 60.
367. Agriculture Handbook No. 456, U.S. Dept. of Agriculture, (1975).
368. Pangborn, J. B. Personal communication on magnesium and bionostics amino acids evaluations (1978).
369. Itokawa, Y., C. Tanaka, and M. Fujiwara. "Changes in body temperature and blood pressure in rats with calcium and magnesium deficiencies," *J. Appl. Physiol.* 37:835–839 (1974).
370. Bloom, B. "Cardiomyopathy of magnesium deficiency and ischemia," Paper presented at the Fourth International Symposium on Magnesium. Abstract number 19, *J. Am. Coll. Nutr.* 4:314 (1985).
371. James M. F., and G. A. Wright. "Tetany and myocardial arrhythmia due to hypomagnesaemia: A case report," *S. Afr. Med. J.* 69:48–49 (1985).
372. James M. F., and G. A. Wright. "Tetany and myocardial arrhythmia due to hypomagnesaemia: A case report," *S. Afr. Med. J.* 69:48–49 (1985).
373. Seelig, M. S., A. R. Berger, and N. Spielholz. "Latent tetany and anxiety. Marginal magnesium deficit and normocalcemia," *Dis. Nerv. Syst.* 36:461–465 (1975).

374. Rude, R. of USC qtd. in "How come hypomagnesemia in 50 percent of CCU admitters?" *Medical Tribune*, R. McGuire. Aug. 1986.(No. 24):3.
375. Whang, R. "Routine serum magnesium determination — a continuing unrecognized need," *Magnesium* 6:1–4 (1987).
376. Rea, W. J., A. R. Johnson, R. E. Smiley, B. Maynard, and O. D. Brown. "Magnesium deficiency in patients with chemical sensitivity," *Clin. Ecol.* 5(1):17–20 (1987).
377. Whang, R. "The need for routine serum magnesium: Clinical observations," Paper presented at the Fourth International Symposium on Magnesium, *J. Am. Coll. Nutr.* 4:330 (Abstr.) (1985).
378. Kobayashi, J. "On geographical relationship between the chemical nature of river water and death race from apoplexy," *Ber. Ohara Inst.* 11:12–21 (1957).
379. Seelig, M. S. *Magnesium Deficiency in the Pathogenesis of Disease. Early Roots of Cardiovascular, Skeletal, and Renal Abnormalities* (New York: Plenum Medical Book, 1980), pp. 1–26.
380. McGuire, R. "Mg in the water may mediate CHD protection," *Med. Trib.* 49 (April 24, 1985).
381. McGuire, R. "Mg in the water may mediate CHD protection," *Med. Trib.* 49 (April 24, 1985).
382. Rude, R. of USC qtd. in "How come hypomagnesemia in 50 percent of CCU admitters?" *Medical Tribune*, R. McGuire. Aug. 1986.(No. 24):3.
383. Sjogren, A., L. Edvinsson, and A. Ottosson. "Enhanced vasoconstrictor responses to potassium, 5-hydroxytryptamine, and prostaglandin F2 alpha of isolated coronary arteries from magnesium-deficient rats. Comparison with vasomotor activity of aorta," *J. Am. Coll. Nutr.* 7(6):461–469 (1988).
384. Barch, G. E., and T. D. Giles. "The importance of magnsium deficiency in cardiovascular disease," *Am. Health J.* 94(5):649–657 (1977).
385. Garrison, Robert H., Jr., and Somer, Elizabeth. *The Nutrition Desk Reference* (New Canaan, CN: Keats, 1985), p. 69.
386. Wacker, W. C. "Biochemistry of zinc-role in wound healing," in *Zinc and Copper in Clinical Medicine*, K. M. Hambidge and B. L. Nichols, Jr., Eds. (New York: SP Medical and Scientific Books, Inc., 1978), p. 15.
387. Solomons, N. W. *Nutrition in the 1980s: Constraints on our Knowledge* (New York: Alan R. Liss, 1981), p. 97.
388. Linder, M. C., Ed. *Nutritional Biochemistry and Metabolism with Clinical Applications* (New York: Elsevier, 1985), p. 161.
389. Quarterman, J., C. F. Mills, and W. R. Humphries. "The reduced secretion of, and sensitivity to insulin in zinc-deficient rats," *Biochem. Biophys. Res. Commun.* 25:354 (1966).
390. Slater, J. P., A. S. Mildvan, and L. A. Loeb. "Zinc in DNA polymerases," *Biochem. Biophys. Res. Commun.* 44:37 (1971).
391. Linder, M. C., Ed. *Nutritional Biochemistry and Metabolism with Clinical Applications* (New York: Elsevier, 1985), p. 161.

392. Chapril, M., J. N. Ryan, and C. F. Zukaski. "The effect of zinc and other metals on the stability of lysosomes," *Proc. Soc. Exp. Biol. Med.* 140:642–646 (1972).

393. Chapril, M., S. L. Elias, J. N. Ryan, and C. F. Zukoski. "Considerations on the biological effects of zinc," in *International Review of Neurobiology*, C. C. Pfeiffer, Ed. (New York: Academic Press, 1973), pp. 115–173.

394. Chrapil, M., J. N. Ryan, and C. F. Zukoski. "The effect of zinc and other metals on the stability of lysosomes," *Proc. Soc. Exp. Biol. Med.* 140:642–646 (1972).

395. Chapril, M., S. L. Elias, J. N. Ryan, and C. F. Zukoski. "Considerations on the biological effects of zinc," in *International Review of Neurobiology*, C. C. Pfeiffer, Ed. (New York: Academic Press, 1973), pp. 115–173.

396. McClintock, B. "1. Spread of mutational change along the chromosome. 2. A case of $\underline{Ac}$-induced instability at the Bronze locus in chromosome 9. 3. Transposition sequences of $\underline{AC}$. 4. A suppressor-mutator system of control of gene action and mutational change. 5. System responsible for mutations at a_1-m2," *Maize Genet. Coop. News Lett.* 29:9–13 (1955).

397. Henkin, R. I., and F. R. Smith. "Zinc and copper metabolism in acute viral hepatitis," *Am. J. Med. Sci.* 264:401–409 (1972).

398. Niell, H. B., E. Leach, and A. P. Kraus. "Zinc metabolism in sickle cell anemia," *JAMA* 242(24):2686–2690 (1979).

399. McClain, C. J., and L. C. Su. "Zinc deficiency in the alcoholic: A review," *Alcoholism: Clin. Exp.* 7:5 (1983).

400. Wu, C. T. et al. "Serum zinc, copper, and ceruloplasmin levels in male alcoholics," *Biol. Psychiatr.* 19:1333–1338 (1982).

401. Linder, M. C., Ed. *Nutritional Biochemistry and Metabolism with Clinical Applications* (New York: Elsevier, 1985), pp. 109–110. p. 165.

402. Spencer, J. C. "Letter to the editor. Direct relationship between the body's copper/zinc ratio, ventricular premature beats and sudden cardiac death," *Am. J. Clin. Nutr.* 32:1184–1185 (1979).

403. Whelen, P., B. E. Walker, and J. Kelleher. "Zinc, vitamin A and prostatic cancer," *Br. J. Urol.* 55(5):525–528 (1983).

404. Williams, S. R., Ed. *Essentials of Nutrition and Diet Therapy*, 4th ed., (St. Louis: Times Mirror/Mosby College Publishing, 1986), p. 164.

405. Kurkus, J., N. W. Alcock, and M. E. Shils. "Manganese content of large-volume parenteral solutions and of nutrient additives," *J. Parenteral Enteral Nutr.* 8:254–257 (1984).

406. Garrison, R. H., Jr., and E. Somer. *The Nutrition Desk Reference* (New Canaan, CN: Keats, 1985), p. 78.

407. Kaplan, J., J. W. Hess, and A. S. Prasad. "Impairment of immune function in the elderly: Association with mild zinc deficiency," in *Current Topics in Nutrition and Disease*, Vol. 18, A. S. Prasad, Ed. (New York: Alan R. Liss, Inc., 1988), pp. 309–317.

408. Chandra, R. K. "Nutrition and immunity — Basic considerations. 1," *Contemp. Nutr.* 11(11) (1986).

409. Chandra, R. K. "Trace element regulation of immunity and infection," *J. Am. Coll. Nutr.* 4(1):5–16 (1985).

410. Rogers, S. A. "Zinc deficiency as a model for developing chemical sensitivity," *Int. Clin. Nutr. Rev.* 10(1):253–259 (1990).

411. Garrison, R. H., Jr., and E. Somer. *The Nutrition Desk Reference* (New Canaan, CN: Keats, 1985), p. 78.

412. Brem, S. S., D. Zag Zag, A. M. Tsanaclis, S. Gately, M. P. Elkowby, and S. E. Brien. "Inhibition of angiogeneses and tumor growth in the brain suppression of endothelial cell turnover by penicilliamine and depletion of copper, an angiogenic cofactor," *Am. J. Pathol.* 137(5):1121–1142 (1990).

413. Garrison, R. H., Jr., and E. Somer. *The Nutrition Desk Reference* (New Canaan, CN: Keats, 1985), p. 63.

414. Menkes, J. H., M. Alter, G. K. Steigleder, D. R. Weakley, and J. H. Sung. "A sex-linked recessive disorder with retardation of growth, peculiar hair, and focal cerebral and cerebellar degeneration," *Pediatrics* 29:764 (1962).

415. Garrison, R. H., Jr., and E. Somer. *The Nutrition Desk Reference* (New Canaan, CN: Keats, 1985), p. 64.

416. Avioli, L. V. "Major minerals. A. calcium and phosphorus," in *Modern Nutrition in Health and Disease*, 6th ed., R. S. Goodhart and M. E. Shils, Eds. (Philadelphia: Lea and Febiger, 1980), p. 294.

417. Schwarz, K., and W. Mertz. "Chromium (IV) and the glucose tolerance factor," *Arch. Biochem. Biophys.* 85:292–295 (1959).

418. Newman, H. A. I., R. F. Leighton, R. R. Lanese, and N. A. Freedland. "Serum chromium and angiographically determined coronary artery disease," *Clin. Chem.* 24:541 (1978).

419. Roginski, E. E., and W. Mertz. "Effects of chromium (III) supplementation on glucose and amino acid metabolism in rats fed a low protein diet," *J. Nutr.* 97:525–530 (1969).

420. Mertz, W., E. W. Toepfer, E. E. Roginski, and M. M. Polansky. "Present knowledge of the role of chromium," *Fed. Proc. Fed. Am. Soc. Exp. Biol.* 33:2275 (1974).

421. Farkas, T. G., and S. L. Robertson. "The effect of Cr^{3+} on the glucose utilization of isolated lenses," *Exp. Eye Res.* 4:124–126 (1965).

422. Schwarz, K., and W. Mertz. "A glucose tolerance factor and its differentiation from factor 3," *Arch. Biochem. Biophys.* 72:515–518 (1957).

423. Mertz, W., E. E. Roginski, and K. Schwarz. "Effect of trivalent chromium complexes on glucose uptake by epidiolymal fat tissue of rats," *J. Biol. Chem.* 236:318–322 (1961).

424. Mertz, W., E. E. Roginski, and H. A. Schroeder. "Some aspects of glucose metabolism of chromium-deficient rats raised in a strictly controlled environment," *J. Nutr.* 86:107–112 (1965).

425. Schroeder, H. A. "Chromium deficiency in rats: A syndrome simulating diabetes mellitus with retarded growth," *J. Nutr.* 88:43-445 (1966).

426. Mertz, W. "The newer essential trace elements, chromium, tin, vanadium, nickel and silicon," *Proc. Nutr. Soc.* 33:307–313 (1974).

427. Baig, H. A., and J. C. Edezien. "Carbohydrate metabolism in kwashiorkor," *Lancet* 2:662–665 (1965).

428. Gürson, C. T., and G. Saner. "Effect of chromium on glucose utilization in marasmic protein-calorie malnutrition," *Am. J. Clin. Nutr.* 24:1313–1319 (1971).

429. Hopkins, L. L., Jr., O. Ransome-Kuti, and A. S. Majaz. "Improvement of impaired carbohydrate metabolism by chromium (III) in malnourished infants," *Am. J. Clin. Nutr.* 21:203–211 (1968).

430. Garrison, R. H., Jr., and E. Somer. *The Nutrition Desk Reference* (New Canaan, CN: Keats, 1985), p. 62.

431. Wallach, Stanley. "Clinical and biochemical aspects of chromium deficiency," *J. Am. Coll. Nutr.* 4:107–120 (1985).

432. Kumpulainen, J., J. Lehto, and P. Kowistoinen. "Determination of chromium in human milk, serum, and urine by electrothermal atomic absorption spectometry without preliminary ashing," *Sci. Total Environ.* 31:71–80 (1985).

433. Baig, H. A., and J. C. Edezien. "Carbohydrate metabolism in kwashiorkor," *Lancet* 2:662–665 (1965).

434. Kayne, F. J., G. Komar, H. Laboda, and R. E. Vanderlinde. "Atomic absorption spectrophotometry of chromium in scrum and urine with a modified Perkin-Elmer 603 atomic absorption spectrophotometer," *Clin. Chem.* 24:2151–2154 (1978).

435. Versicck, J., J. DeRudder, and F. Barbier. "Serum chromium levels," *JAMA* 242:1613 (1979).

436. Guthrie, B. E., W. R. Wolf, and C. Veillon. "Background correction and related problems in the determination of chromium in urine by graphite furnace atomic absorption spectometry," *Anal. Chem.* 50:1900–1902 (1978).

437. Veillon, C., K. Y. Patterson, and N. A. Bryden. "Direct determination of chromium in human urine by electrothermal atomic absorption spectometry," *Anal. Chim. Acta* 136:233–241 (1982).

438. Veillon, C., K. Y. Patterson, and N. A. Bryden. "Chromium in urine as measured by atomic absorption spectrometry," *Clin. Chem.* 28:2309–2311 (1982).

439. Anderson, R. A., M. M. Polansky, N. A. Bryden, E. E. Roginski, K. Y. Patterson, C. Veillon, and W. Glinsmann. "Urinary chromium excretion of human subjects: Effects of chromium supplementation and glucose loading," *Am. J. Clin. Nutr.* 36:1184–1193 (1982).

440. Veillon, C. W., R. Wolf, and B. E. Guthrie. "Determination of chromium in biological materials by stable isotope dilution," *Anal. Chem.* 51:1022–1024 (1979).

441. Anderson, R. A., M. M. Polansky, N. A. Bryden, K. Y. Patterson, C. Veillon, and W. H. Glinsmann. "Effects of chromium supplementation on urinary Cr excretion of human subjects and correlation of Cr excretion with selected clinical parameters," *J. Nutr.* 113:276–281 (1983).

442. Roginski, E. E., R. A. Anderson, M. M. Polansky, N. A. Bryden, W. H. Glinsmann, K. Y. Patterson, C. Veillon, and W. Mertz. "Urinary chromium excretion in human subjects following a glucose load," *Fed. Proc. Fed. Am. Soc. Exp. Biol.* 40:886 (1981).

443. Anderson, R. A., M. M. Polansky, N. A. Bryden, E. E. Roginski, K. Y. Patterson, and D. C. Reamer. "Effects of exercise (running) on serum glucose insulin, glucagon, and chromium excretion," *Diabetes* 31:212–216 (1982).

444. Bunker, V. W., M. S. Lawson, H. T. Kieves, and B. E. Clayton. "The uptake and excretion of chromium by the elderly," *Am. J. Clin. Nutr.* 39:797–802 (1984).

445. Wallach, S. "Clinical and biochemical aspects of chromium deficiency," *J. Am. Coll. Nutr.* 4:107–120 (1985).

446. Beisel, W. R., R. Edelman, K. Nauss, and R. M. Suskind. "Single-nutrient effects on immunologic functions," *JAMA* 245:53 (1981).

447. Chow, C. K. "Effect of dietary selenium and vitamin E on the biochemical responses in the lungs of ozone-exposed rats," *Fed. Proc. Fed. Am. Soc. Exp. Biol.* 36:1094 (1977).

448. Stampfer, M. J., G. A. Colditz, and W. C. Willett. "The epidemiology of selenium and cancer," *Cancer Surv.* 6:623 (1987).

449. Garrison, R. H., Jr., and E. Somer. *The Nutrition Desk Reference* (New Canaan, CN: Keats, 1985), p. 74.

450. Wilgus, H. S., Jr., L. C. Norris, and G. F. Heuser. "The role of certain inorganic elements in the cause and prevention of perosis," *Science* 84:232 (1936).

451. Davis, P. N., L. C. Norriss, and F. H. Kratzer. "Interference of soy protein with the utilization of trace minerals," *J. Nutr.* 77:217 (1962).

452. Suso, F. A., and H. M. Edwards, Jr. "Influences of various chelating agents on absorption of Co, Mn and Zn by chickens," *Poultry Sci.* 47:1417 (1968).

453. Chan, W. Y., J. M. Bates, Jr., and O. M. Reussert. "Comparative studies of manganese binding in human breast milk, bovine milk, and infant formula," *J. Nutr.* 112:642 (1982).

454. Lonnerdal, B., C. L. Kein, and L. S. Hurley. "Manganese binding in human milk and cow's milk — an effect on bioavailability," *Fed. Proc. Fed. Am. Soc. Exp. Biol.* 42:926 (1983).

455. Garcia-Aranda, J. A., R. A. Wapnise, and F. Lifshitz. "In vivo intestinal absorption of manganese in the rat," *J. Nutr.* 113:2601 (1983).

456. Kratzer, F. H., and P. Vohra. *Chelates in Nutrition* (Boca Raton, FL: CRC Press, 1986), p. 124.

457. Garrison, R. H., Jr., and E. Somer. *The Nutrition Desk Reference* (New Canaan, CN: Keats, 1985), p. 70.

458. Cavalieri, R. R. "Trace elements. A. Iodine," in *Modern Nutrition in Health and Disease*, 6th ed., R. S. Goodhart and M. E. Shils, Eds. (Philadelphia: Lea and Febiger, 1980), p. 395.

459. Garrison, R. H., Jr., and E. Somer. *The Nutrition Desk Reference* (New Canaan, CN: Keats, 1985), pp. 70-71.

460. Gordeuk, V. R., B. R. Bacon, and G. M. Brittenham. "Iron overload: Causes and consequences," *Am. Rev. Nutr.* 7:485–508 (1987).

461. Bezwoda, W. R., A. P. MacPhail, T. H. Bothwell, R. D. Baynes, and J. D. Torrance. "Failure of transferrin to enhance iron absorption in achlorhydric human subjects," *Br. J. Haematol.* 63:749–752 (1986).

462. Gordeuk, V. R., B. R. Bacon, and G. M. Brittenham. "Iron overload: Causes and consequences," *Am. Rev. Nutr.* 7:485–508 (1987).

463. Torrance, J. D., R. W. Charlton, A. Schmaman, S. R. Lynch, and T. H. Bothwell. "Storage iron in 'muscle'," *J. Clin. Pathol.* 21:495–500 (1968).

464. Gordeuk, V. R., B. R. Bacon, and G. M. Brittenham. "Iron overload: Causes and consequences," *Am. Rev. Nutr.* 7:485–508 (1987).

465. Brittenham, G. M., E. H. Danish, and J. W. Harris. "Assessment of bone marrow and body iron stores. Old techniques and new technologies," *Semin. Hematol.* 18:194–221 (1981).

466. Garrison, R. H., Jr., and E. Somer. *The Nutrition Desk Reference* (New Canaan, CN: Keats, 1985), p. 66.

467. Bhaskaram, P. "Immunology of iron-deficient subject," in *Contemporary Issues in Clinical Nutrition*, Vol. 11, R. K. Chandras, Ed. (New York: Alan R. Liss, Inc., 1988), pp. 149–168.

468. Lovric, V. "Normal hemotological values in children aged 6–36 months and socio-medical implications," *Med. J. Aust.* 2:366 (1970).

469. Werkman, S. L., L. Shifman, and T. Skelly. "Psycho social correlates of iron deficiency anemia in childhood," *Psychosom. Med.* 26:125 (1964).

470. Bhaskaram, P. "Immunology of iron-deficient subject," in *Contemporary Issues in Clinical Nutrition*, Vol. 11, R. K. Chandras, Ed. (New York: Alan R. Liss, Inc., 1988), pp. 149–168.

471. Higgs, J. M., and R. S. Wells. "Chronic mucocutaneous candidiasis: Associated abnormalities of iron metabolism," *Br. J. Dermatol.* 86(Suppl.):88 (1972).

472. Fletcher, J., J. Mather, M. J. Lewis, and G. Whiting. "Mouth lesions in iron deficient anemia. Relationship to candida albicans in saliva and to impairment of lymphocyte transformation," *J. Infect. Dis.* 131:44 (1975).

473. Chandra, R. K., and A. Grace. "Goldsmith award lecture. Trace elements regulation of immunity and infection," *J. Am. Coll. Nutr.* 4:5 (1985).

474. Sherman, A. R., and L. Helyar. "Iron deficiency, immunity, and disease resistance in early life," in *Contemporary Issues in Clinical Nutrition*, Vol. 11, R. K. Chandra, Ed. (New York: Alan R. Liss, Inc., 1988), p. 169.

475. Prasad, J. S. "Leukocyte function in iron deficiency," *Am. J. Clin. Nutr.* 32:550–552 (1979).
476. Sherman, A. R., and L. Helyar. "Iron deficiency, immunity, and disease resistance in early life," in *Contemporary Issues in Clinical Nutrition,* Vol. 11, R. K. Chandra, Ed. (New York: Alan R. Liss, Inc., 1988), p. 169.
477. Gordeuk, V. R., B. R. Bacon, and G. M. Brittenham. "Iron overload: Causes and consequences," *Am. Rev. Nutr.* 7:485–508 (1987).
478. Dadone, M. M., J. P. Kushner, C. Q. Edwards, D. T. Bishop, and M. H. Skolnick. "Hereditary hemochromatosis: Analysis of laboratory expression of the disease by genotype in 18 pedigrees," *J. Clin. Pathol.* 78:196–207 (1982).
479. Kushner, J., C. Edwards, L. Griffen, M. Dadone, and M. Skolnick. "Incidence of homozygosity for HLA-linked hemochromatosis in healthy young blood donors," *Blood (Suppl. 1)* 64:40 (Abstr.) (1984).
480. Gordeuk, V. R., B. R. Bacon, and G. M. Brittenham. "Iron overload: Causes and consequences," *Am. Rev. Nutr.* 7:485–508 (1987).
481. Suso, F. A., and H. M. Edwards, Jr. "Influences of various chelating agents on absorption of Co, Mn and Zn by chickens," *Poultry Sci.* 47:1417 (1968).
482. Seldon, C., M. Owen, J. M. P. Hopkins, and T. J. Peters. "Studies on the concentration and intracellular localization of iron proteins in liver biopsy specimens from patients with iron overload with special reference to their role in lysosomal disruption," *Br. J. Haematol.* 44:359–603 (1980).
483. Peters, T. J., M. J. O'Connell, and R. J. Ward. "Role of free-radical mediated lipid peroxidation in the pathogenesis of hepatic damage by lysosomal disruption," in *Free Radicals in Liver Injury,* G. Poli, K. H. Cheeseman, M. U. Dianzani, and T. F. Slater, Eds. (Oxford: IRL Press, 1985), pp. 107–115.
484. Dougherty, J. J., W. A. Croft, and W. G. Hoekstra. "Effect of ferrous chloride and iron dextran on lipid peroxidation in vivo in vitamin E and selenium adequate and deficient rats," *J. Nutr.* 3:1784–1796 (1981).
485. Goddard, J. G., D. Basford, and G. D. Sweeney. "Lipid peroxidation stimulated by iron nitrilotriacetate in rat liver," *Biochem. Pharmacol.* 35:2381–2387 (1986).
486. Bacon, B. R., A. S. Tavill, G. M. Brittenham, C. H. Park, and R. O. Recknagel. "Hepatic lipid peroxidation in vivo in rats with chronic iron overload," *J. Clin. Invest.* 71:429–439 (1983).
487. Bacon, B. R., J. F. Healey, G. M. Brittenham, C. H. Park, J. Nunnari, A. S. Tavill. "Hepatic microsomal function in rats with chronic dietary iron overload," *Gastroenterology* 90:1844–1853 (1986).
488. Bacon, B. R., C. H. Park, G. M. Brittenham, R. O'Neill, and A. S. Tavill. "Hepatic mitochondrial oxidative metabolism in rats with chronic dietary iron overlaod," *Hepatology* 5:789–797 (1985).

489. Tavill, A. S. "Hepatic fibrosis in rats with chronic dietary iron overload," *Hepatology* 5:950 (Abstr.) (1985).

490. Mak, T. I., and W. B. Weglicki. "Characterization or iron-mediated peroxidative injury in isolated hepatic lysosomes," *J. Clin. Invest.* 75:58–63 (1985).

491. O'Connell, M. J., R. J. Ward, H. Baum, and T. J. Peters. "The role of iron in ferritin and haemosiderin-mediated lipid peroxidation in 33 lysosomes," *Biochem. J.* 229:135–139 (1985).

492. Gordeuk, V. R., B. R. Bacon, and G. M. Brittenham. "Iron overload: Causes and consequences," *Am. Rev. Nutr.* 7:485–508 (1987).

493. Weintraub, L. R., A. Goral, J. Grasso, C. Franzblau, A. Sullivan, and S. Sullivan. "Pathogenesis of hepatic fibrosis in experimental iron overload," *Br. J. Haematol.* 59:321–331 (1985).

494. Wilson, R. L. "Iron, zinc, free radicals and oxygen in tissue disorders and cancer control," in *Iron Metabolism, Ciba Found. Symp.* (Amsterdam: Elsevier/Excerpta Medica/North Holland, 1977), 51:331–334.

495. Wheby, M. S. "Liver damage in disorders of iron overload. A hypothesis," *Arch. Intern. Med.* 144:621–622 (1984).

496. Brissot, P., T. L. Wright, W. L. Ma, and R. A. Weisiger. "Efficient clearance of non-transferrin-bound iron by rat liver," *J. Clin. Invest.* 76:1463–1470 (1985).

497. Hershko, C., G. Graham, G. W. Bates, and E. A. Rachmilewitz. "Non-specific serum iron in thalassemia-abnormal serum iron fraction of potential toxicity," *Br. J. Haematol.* 40:255–263 (1978).

498. Wang, W. C., N. Ahmet, and M. Hanna. "Non-transferrin-bound iron in long-term transfusion in children with congenital anemias," *J. Pediatr.* 108:522–557 (1986).

499. Boyce, N., C. Wooc, S. Holdsworth, N. M. Thomason, and R. C. Atkins. "Life-threatening sepsis complicating heavy metal chelation therapy with desferrioxamine," *Aust. N.Z. J. Med.* 15:654–655 (1985).

500. Barry, D. M. J., and A. N. Reeve. "Increased incidence of gram-negative neonatal sepsis with intramuscular iron administration," *Pediatrics* 60:908–912 (1977).

501. Murray, M. J., A. B. Murray, N. J. Murray, and M. B. Murray. "Refeeding malaria and hyperferraemia," *Lancet* 1:653–654 (1975).

502. Karp, J. E., and W. G. Mertz. "Association of reduced iron binding capacity and fungal infections in leukemic granulocytopenic patients," *J. Clin. Oncol.* 4:216–220 (1986).

503. Gordeuk, V. R., B. R. Bacon, and G. M. Brittenham. "Iron overload: Causes and consequences," *Am. Rev. Nutr.* 7:485–508 (1987).

504. Garrison, R. H., Jr., and E. Somer. *The Nutrition Desk Reference* (New Canaan, CN: Keats, 1985), p. 68.

505. Reid, B. L., A. A. Kurnick, R. L. Svacha, and J. R. Couch. "The effect of molybdenum on chick and poult growth," *Poultry Sci.* 93:245 (1956).

506. Richert, D. A., and W. W. Westerfield. "Isolation and identification of the xanthine oxidase factor as molybdenum," *J. Biol. Chem.* 203:915 (1953).

507. Leach, R. J., D. E. Turk, T. R. Zeigler, and L. C. Norris. "Studies on the role of molybdenum in chick nutrition," *Poultry Sci.* 41:300 (1962).

508. Payne, C. G. "Involvement of molybdenum in feather growth," *Br. Poult Sci.* 18:427–432 (1977).

509. Nielsen, F. H. "Ultratrace minerals," in *Modern Nutrition in Health and Disease*, 7th ed., M. E. Shils and V. R. Young, Eds. (Philadelphia: Lea and Febiger, 1988), p. 283.

510. Payne, C. G. "Involvement of molybdenum in feather growth," *Br. Boul. Sci.* 18:427 (1977).

511. Garrison, R. H., Jr., and E. Somer. *The Nutrition Desk Reference* (New Canaan, CN: Keats, 1985), p. 71.

512. Anke, M., B. Groppel, W. Arnhold, M. Langer, and U. Krause. "The influence of the ultra trace element deficiency (Mo, Ni, As, Cd, V) on growth, reproduction and life expectancy," in *Trace Elements in Clinical Medicine*, H. Tomita, Ed. (Tokyo: Springer-Verlag, 1990), p. 363.

513. Nielsen, F. H. "Ultratrace minerals," in *Modern Nutrition in Health and Disease*, 7th ed., M. E. Shils and V. R. Young, Eds. (Philadelphia: Lea and Febiger, 1988), p. 284.

514. Garrison, R. H., Jr., and E. Somer. *The Nutrition Desk Reference* (New Canaan, CN: Keats, 1985), p. 71.

515. Valberg, L. S., J. Ludwig, and D. Olatunbosun. "Alteration in cobalt absorption in patients with disorders of iron metabolism," *Gastroenterology* 56(2):241–251 (1969).

516. Turk, J. L., Jr., and F. H. Kratzer. "The effects of cobalt in the diet of the chick," *Poultry Sci.* 39:1302 (1960).

517. Kratzer, F. H., and P. Vohra. *Chelates in Nutrition* (Boca Raton, FL: CRC Press, 1986), p. 126.

518. Garrison, R. H., Jr., and E. Somer. *The Nutrition Desk Reference* (New Canaan, CN: Keats, 1985), p. 62.

519. Hill, C. H., and G. Matrone. "Chemical parameters in the study of in vivo and in vitro interactions of transition elements," *Fed. Proc. Fed. Am. Soc. Exp. Biol.* 29:1474–1481 (1970).

520. Kratzer, F. H., and P. Vohra. *Chelates in Nutrition* (Boca Raton, FL: CRC Press, 1986), p. 127

521. Nielsen, F. H. "Vanadium," in *Trace Elements in Human and Animal Nutrition*, Vol. 1, W. Mertz, Ed. (San Diego: Academic Press, Inc., 1988), pp. 275–300.

522. Erdmann, E., K. Werdan, W. Krawietz, M. Lebuhn, and S. Christl. "Bedentung der NADPH-Vanadat-Oxidoreduktase von Herzmuskel zellmerribranen und erythrozytenmembranen (Significance of NADH-Vanadate oxidoreductase of cardiac and erythrocyte cell membranes)," *Basic Res. Cardiol.* 75:460–465 (1980).

523. Erdmann, E., K. Werdan, W. Krawietz, M. Lebuhn, and S. Christl. "Bedentung der NADPH-Vanadat-Oxidoreduktase von Herzmuskel zellmerribranen und erythrozytenmembranen (Significance of NADH-Vanadate oxidoreductase of cardiac and erythrocyte cell membranes)," *Basic Res. Cardiol.* 75:460–465 (1980).

524. Jones, T. R., and T. W. Reid. "Sodium orthovanadate stimulation of DNA synthesis in Nakano mouse lens epithelial cells in serum-free medium," *J. Cell. Physiol.* 121:199–205 (1984).

525. McKeehan, W. L., K. A. McKeehan, S. L. Hammond, and R. G. Ham. "Improved medium for clonal growth of human diploid fibroblasts at low concentrations of serum protein," *In Vitro* 3(7):399–416 (1977).

526. Peabody, R. A., S. Wallach, R. L. Verch, and M. Lifschitz. "Effect of LH and FSH on vanadium distribution in hypophysectomized rats (40984)," *Proc. Soc. Exp. Biol. Med.* 165:349–353 (1980).

527. Curran, G. L., and R. E. Burch. "Biological and health effects of vanadium," in *Trace Substances in Environmental Health — I* (Proceedings of University of Missouri 1st Annual Conference on Trace Substances Environmental Health), D. D. Hemphill, Ed. (Columbia: University of Missouri Press, 1968), p. 96.

528. Thompson, H. J., N. D. Chasteen, and L. D. Mecker. "Dietary vanadyl (IV) sulfate inhibits chemically-induced mammary carcinogenesis," *Carcinogenesis* 5:849–851 (1984).

529. Czajka-Narins, D. M. "Minerals," in *Food, Nutrition, and Diet Therapy*, 6th ed., M. V. Krause and L. K. Mahan, Eds. (Philadelphia: W. B. Saunders, 1979), p. 114.

530. Myron, D. R., S. H. Givand, and F. H. Nielsen. "Vanadium content of selected foods as determined by flameless atomic absorption spectrometry," *J. Agric. Food Chem.* 25:297 (1977).

531. Ikebe, K., and R. Tanaka. "Determination of vanadium and nickel in marine samples by flameless and flame atomic absorption spectrophotometry," *Bull. Environ. Contam. Toxicol.* 21(4-5):526–532 (1979).

532. Keenan, C. W., J. H. Wood, and D. C. Kleinfelter, Eds. *General College Chemistry*, 5th ed., (New York: Harper & Row, 1976), p. 610.

533. Hicks, J. F. G. "Glass formation and crystal structure," *J. Chem. Educ.* 51:28 (1974).

534. Keeler, R. F., and S. A. Lovelace. "The metabolism of silicon in the rat and its relation to the formation of artificial siliceous calculi," *J. Exp. Med.* 109:601–614 (1959).

535. Kirchgessner, M. Z. *Tierphysiol. Tierernahr Futtermittelkd.* 14:270, 278 (1959).

536. Carlisle, E. M. "Silicon, an essential element for the chick," *Science* 178:619–621 (1972).

537. Carlisle, E. M. "In viro requirement for silicon in articular cartilage and connective tissue formation in the chick," *J. Nutr.* 106:478–484 (1976).

538. Carlisle, E. M. "Silicon as an essential trace element in animal nutrition," in *Silicon Biochemistry*, Ciba Found. Symp. 121 (New York: Wiley Chichester, 1986),..pp. 123–139.

539. Carlisle, E. M. "Silicon: A requirement in bone formation independent of vitamin D," *Calcif. Tissue Int.* 33:27–34 (1981).

540. Schwarz, K., and D. B. Milue. "Growth-promoting effects of silicon in rats," *Nature* 239:333–334 (1972).

541. Jeanloz, R. W. "Mucopolysaccharides of higher animals," in *The Carbohydrates*, Vol. IIB, W. Pigman and D. Horton, Eds. (New York: Academic Press, 1970), pp. 589–625.

542. Carlisle, E. M., and W. Alpenfels. "A requirement for silicon for bone growth in culture," *Fed. Proc. Fed. Am. Soc. Exp. Biol.* 37:1123 (1978).

543. Carlisle, E. M. "Silicon," in *Trace Elements in Human and Animal Nutrition*, Vol. 2, 5th ed., (Orlando, Fl: Academic Press, 1986), p. 382.

544. King, E. J., and T. H. Belt. "The physiological and pathological aspects of silica," *Physiol. Rev.* 18:329–365 (1938).

545. Nikaido, T., J. Austin, L. Trueb, and R. Rinehart. "Studies in aging of brain. II. Microchemical analysis of the nervous system in Alzheimer's patients," *Arch. Neurol.* 27:549–554 (1972).

546. Carlisle, E. M. "A silicon-molybdenum interrelationship in vivo," *Fed. Proc. Fed. Am. Soc. Exp. Biol.* 38:553 (1979).

547. Carlisle, E. M. "Silicon," in *Trace Elements in Human and Animal Nutrition*, Vol. II, E. J. Underwood, Ed. (New York: Academic Press, 1986), p. 378.

548. Geissler, V. U., and J. Gerloff. "Das Vorkommen vo Diatomeen in menschlichen Organen und in der Luft," *Nova Hedwigia Z. Kryptogamenkd.* 10:565 (1965).

549. King, E. J., and M. McGeorge. "LIV. The biochemistry of silicic acid. B. The solution of silica and silicate dusts in body fluids," *Biochem J.* 32:417–425 (1938).

550. Dobbie, J. W., and M. J. B. Smith. "The silicon content of body fluids," *Scott. Med. J.* 27:17–19 (1982).

551. Policard, A., A. Collet, D. H. Moussard, and S. Pragermain. "Deposition of silica in mitochondria: an electron microscopic study," *J. Biophys. Biochem. Cytol.* 9:236 (1961).

552. Carlisle, E. M. "The nutritional essentiality of silicon," *Nutr. Rev.* 40:193–198 (1982).

553. Nielsen, F. H. "Ultratrace minerals," in *Modern Nutrition in Health and Disease*, 7th ed., M. E. Shils and V. R. Young, Eds. (Philadelphia: Lea and Febiger, 1988), p. 286.

554. Schwarz, K., and D. B. Milne. "Growth-promoting effects of silicon in rats," *Nature* 239:333 (1972).

555. Schwarz, K. "Significance and function of silicon in warm-blooded animals," in *Biochemistry of Silicon and Related Problems*, G. Bendz and I. Lindquist, Eds. (New York: Plenum Press, 1978), p. 207.

556. Carlisle, E. M. "Silicon: An essential element for the chick," *Science* 178: 619 (1972).

557. Carlisle, E. M. "A Silicon requirement for normal skull formation in chicks," *J. Nutr.* 110:352–359 (1980).

558. Bernheim, R., G. Thomeret, J. D. Romani, P. Renault, F. Carnot, H. Loo, and M. Albeaux-Fernet. "Gynecomasty revealing a slow-developing leydig cell tumor evolutionary and anato-mopathological peculiarities [French]," *Ann. Endocrinol.* 30(3):480–487 (1969).

559. Carlisle, E. M. "Silicon: An essential element for the chick," *Science* 178: 619 (1972).

560. Kirchgessner, M. Z. *Tierphysiol. Tierernahr Futtermittelkd.* 14:270, 278 (1959).

561. Carlisle, E. M. "Silicon: An essential element for the chick," *Science* 178: 619 (1972).

562. Nottle, M. C. Personal communication (1962).

563. Hamilton, E. I. and M. J. Minski. "Abundance of the chemical elements in man's diet and possible relations with environmental factors," *Sci. Total Environ.* 1:375–394 (1972–1973).

564. Kirchgessner, M. Z. *Tierphysiol. Tierernahr Futtermittelkd.* 14:270, 278 (1959).

565. Bailey, C. H. "Silica excretion in cattle fed a ration predisposing to silica urolithiasis: Total excretion and diurnal variations," *Am. J. Vet. Res.* 28:1743–1749 (1967).

566. Baker, G., L. H. P. Jones, and I. D. Wardrop. "Opal phytoliths and mineral particles in the rumen of the sheep," *Aust. J. Agric. Res.* 12(3):426 (1981).

567. Healy, W. B. and T. G. Ludwig. "Wear of sheep's teeth in the role of ingested soil," *N. Z. J. Agric. Res.* 8:737–752 (1965).

568. Carlisle, E. M., and M. J. Curran. "Effect of dietary silicon and aluminum on silicon and aluminum levels in rat brain," *West. Geriatr. Res. Inst. Alzheimer Dise. Assoc. Disorders* 1(2):83–89 (1987).

569. Carlisle, E. M. "Silicon," *Nutr. Rev.* 33(4):257–260 (1975).

570. Garrison, R. H., Jr., and E. Somer. *The Nutrition Desk Reference* (New Canaan, CN: Keats, 1985), p. 76.

571. Lehninger, A. L. *Biochemistry*, 2nd ed., (New York: Worth Publishers, 1975), pp. 80–88.

572. Mudd, S. H., and H. L. Levy. "Disorders of transulfuration," in *Metabolic Basis of Inherited Disease*, 5th ed., J. B. Stanbury, J. B. Wyngaarden, D. S. Fredrickson, J. L. Goldstein, and M. S. Brown, Eds. (New York: McGraw-Hill, 1983), pp. 522–559.

573. Bionostics, Inc. Technical memorandum 2, "Functions of Amino Acids," Revised Sept. 1986.

574. Bremer, H. J., M. Duran, J. P. Kamerling, H. Przyembel, and S. K. Wadman. *Disturbances of Amino Acid Metabolism: Clinical Chemistry and Diagnosis* (Baltimore, MD: Urban and Schwarzenberg, 1981), p. 21.

575. Bionostics, Inc. Technical memorandum 2, "Functions of Amino Acids," Revised Sept. 1986.

576. Hoffman, R. "Altered methionine metabolism, DNA methylation and oncogene expression in carcinogenesis. A review and synthesis," *Biochim. Biophys. Acta* 738:49–87 (1983).

577. Mudd, S. H., and H. L. Levy. "Disorders of transulfuration," in *Metabolic Basis of Inherited Disease,* 5th ed., J. B. Stanbury, J. B. Wyngaarden, D. S. Fredrickson, J. L. Goldstein, and M. S. Brown, Eds. (New York: McGraw-Hill, 1983), pp. 522–559.

578. Bremer, H. J., M. Duran, J. P. Kamerling, H. Przyembel, and S. K. Wadman. *Disturbances of Amino Acid Metabolism: Clinical Chemistry and Diagnosis* (Baltimore, MD: Urban and Schwarzenberg, 1981), p. 420.

579. Stanburg, J. B. "Dietary treatment of the inborn errors of metabolism," in *Nutritional Biochemistry and Metabolism with Clinical Applications.* M. C. Linder, Ed. (New York: Elsevier, 1985) p. 395.

580. Mudd, S. H., and H. L. Levy. "Disorders of transulfuration," in *Metabolic Basis of Inherited Disease,* 5th ed., J. B. Stanbury, J. B. Wyngaarden, D. S. Fredrickson, J. L. Goldstein, and M. S. Brown, Eds. (New York: McGraw-Hill, 1983), pp. 522–559.

581. Bremer, H. J., M. Duran, J. P. Kamerling, H. Przyembel, and S. K. Wadman. *Disturbances of Amino Acid Metabolism: Clinical Chemistry and Diagnosis* (Baltimore, MD: Urban and Schwarzenberg, 1981), p. 420.

582. Levine, R. J., J. A. Oates, A. Vendsalu, and A. Sjoerdsma. "Studies on the metabolism of aromatic amines in relation to altered thyroid function in man," *J. Clin. Endocronol. Metab.* 22:1242–1250 (1962).

583. Melmon, K. L., R. Rivlin, J. A. Oates, and A. Sjoerdsma. "Further studies of plasma tyrosine in patients with altered thyroid metabolism," *J. Clin. Endocrinol. Metab.* 24:691–698 (1964).

584. Miller, B. F., and C. B. Keane. *Encyclopedia and Dictionary of Medicine, Nursing and Allied Health.* 4th ed., (Philadelphia: W. B. Saunders, 1987), p. 16.

585. Bionostics, Inc. Technical memorandum 2, "Functions of Amino Acids," Revised Sept. 1986.

586. Tourian, A., and J. B. Sidbury. "Phynylketonuria and hyper-phynylalaninuria," in *The Metabolic Basis of Inherited Disease,* 5th ed., J. B. Stanbury, J. B. Wyngaarden, D. S. Fredrickson, J. L. Goldstein, and M. S. Brown, Eds. (New York: McGraw-Hill, 1983), pp. 270–286.

587. Lehninger, A. L. *Biochemistry,* 2nd ed. (New York: Worth Publishers, 1975), p. 570.

588. Harpter, H. A., V. W. Rodwell, and P. A. Mayes. *Review of Physiological Chemistry,* 17th ed. (New York: Lange Medical Publishers, 1979), pp. 660–662.

589. Harpter, H. A., V. W. Rodwell, and P. A. Mayes. *Review of Physiological Chemistry*, 17th ed. (New York: Lange Medical Publishers, 1979), pp. 660–662.
590. Sass-Kortsak, A., and A. G. Bearn. "Hereditary disorders of copper metabolism: [Wilson's disease (Hepatolenticular degeneration) and Menkes' disease (Kiuky-hair or steely-hair syndrome)]," in *The Metabolic Basis of Inherited Disease*, 4th ed., J. B. Stanbury, J. B. Wyngaarden, D. S. Fredrickson, J. L. Goldstein, and M. S. Brown, Eds. (New York: McGraw-Hill, 1978), pp. 1098–1101.
591. Bremer, H. J., M. Duran, J. P. Kamerling, H. Przyembel, and S. K. Wadman. *Disturbances of Amino Acid Metabolism: Clinical Chemistry and Diagnosis* (Baltimore, MD: Urban and Schwarzenberg, 1981), p. 21.
592. Rebello, T., B. Lönnerdal, and L. S. Hurley. "Picolinic acid in milk, pancreatic juice, and intestine: Inadequate for role in zinc absorption," *Am. J. Clin. Nutr.* 35:1–5 (1982).
593. Hurley, L. S., and B. Lönnerdal. "Zinc binding in human milk: Citrate versus picolinate," *Nutr. Rev.* 40:65–71 (1982).
594. Harpter, H. A., V. W. Rodwell, and P. A. Mayes. *Review of Physiological Chemistry*, 17th ed. (New York: Lange Medical Publishers, 1979), pp. 592–593.
595. Irwin, M., K. Belendiuk, K. Mccloskey, and D. X. Freedman. "Tryptophan metabolism in children with attentional deficit disorder," *Am. J. Psychiatry* 138(8):1082–1085 (1981).
596. Rodwell, V. W. "Conversion of amino acids to specialized products," in *Review of Physiological Chemistry*, 17th ed., H. A. Harper, V. W. Rodwell, and P. A. Mayes, Eds. (New York: Lange Medical Publishers, 1979), pp. 434–436.
597. Coleman, M., and C. Gillberg. *The Biology of Autistic Syndromes* (Westport, CN: Praeger, 1985), pp. 76–77.
598. Rodwell, V. W. "Conversion of amino acids to specialized products," in *Review of Physiological Chemistry*, 17th ed., H. A. Harper, V. W. Rodwell, and P. A. Mayes, Eds. (New York: Lange Medical Publishers, 1979), pp. 438–439.
599. Grodesky, G. M. "Chemistry and functions of the hormones. I. Thyroid, pancreas, adrenal, and gastrointestinal tract," in *Review of Physiological Chemistry*, 17th ed., H. A. Harper, V. W. Rodwell, and P. A. Mayes, Eds. (Los Altos, CA: Lange Medical Publishers, 1979), pp. 511–517.
600. Scriver, C. R. "Disorders of proline and hydroxyproline metabolism," in *The Metabolic Basis of Inherited Disease*, 4th ed., J. B. Stanbury, J. B. Wyngaarden, and D. S. Fredrickson, Eds. (New York: McGraw-Hill, 1978), p. 340.
601. Rodwell, V. W. "Biosynthesis of amino acids," in *Review of Physiological Chemistry*, 17th ed., H. A. Harper, V. W. Rodwell, and P. A. Mayes, Eds. (Los Altos, CA: Lange Medical Publishers, 1979), pp. 392 and 661.

602. Grodesky, G. M. "Chemistry and functions of the hormones. I. Thyroid, pancreas, adrenal, and gastrointestinal tract," in *Review of Physiological Chemistry*, 17th ed., H. A. Harper, V. W. Rodwell, and P. A. Mayes, Eds. (Los Altos, CA: Lange Medical Publishers, 1979), pp. 511–517.

603. Grodesky, G. M. "Chemistry and functions of the hormones. II. Pituitary and hypothalamus," in *Review of Physioloical Chemistry*, 17th ed., H. A. Harper, V. W. Rodwell, and P. A. Mayes, Eds. (Los Altos, CA: Lange Medical Publishers, 1979), p. 559.

604. Carlson, N. R. *Physiology of Behavior* (Needham Heights, MA: Allyn & Bacon, Inc., 1977).

605. Rodwell, V. W. "Regulation of enzyme activity," in *Review of Physiological Chemistry*, H. A. Harper, V.W. Rodwell, and P. A. Mayes, Eds. (Los Altos, CA: Lange Medical Publishers, 1979), p. 87.

606. Scriver, C. R., T. L. Perry, and W. Nutzenadel. "Disorders of beta-alanine, carnosine, and homocarnosine metabolism," in *The Metabolic Basis of Inherited Disease*, 5th ed., J. B. Stanbury, J. B. Wyngaarden, D. S. Fredrickson, J. L. Goldstein, and M. S. Brown, Eds. (New York: McGraw Hill Book Company, 1983), pp. 570–579.

607. Bremer, H. J., M. Duran, J. P. Kamerling, H. Przyembel, and S. K. Wadman. *Disturbances of Amino Acid Metabolism: Clinical Chemistry and Diagnosis* (Baltimore, MD: Urban and Schwarzenberg, 1981), p. 10.

608. Bionostics observation based on case records for 1700 individuals with incasurements of urine and/or plasma beta-alanine, carnosine and homocarnosin metabolism.

609. Grodesky, G. M. "Chemistry and functions of the hormones. I. Thyroid, pancreas, adrenal, and gastrointestinal tract," in *Review of Physiological Chemistry*, 17th ed., H. A. Harper, V. W. Rodwell, and P. A. Mayes, Eds. (Los Altos, CA: Lange Medical Publishers, 1979), pp. 511–517.

610. Grodesky, G. M. "Chemistry and functions of the hormones. II. Pituitary and hypothalamus," in *Review of Physioloical Chemistry*, 17th ed., H. A. Harper, V. W. Rodwell, and P. A. Mayes, Eds. (Los Altos, CA: Lange Medical Publishers, 1979), p. 559.

611. Harper, H. A. "Epithelial, connective, and nerve tissues," in *Review of Physiological Chemistry*, H. A. Harper, V.W. Rodwell, and P. A. Mayes, Eds. (Los Altos, CA: Lange Medical Publishers, 1979), p. 660–661.

612. Schneider, J. A., and J. D. Schulman. "Cystinosis," in *The Metabolic Basis of Inherited Disease*, 5th ed., J. B. Stanbury, J. B. Wyngaarden, , D. S. Fredrickson, J. L. Goldstein, and M. S. Brown, Eds. (New York: McGraw Hill, 1983), pp. 1852–1855.

613. Jakoby, W. B. *The Enzymatic Basis of Detoxification*. Vol. II (New York: Academic Press, 1980), pp. 231, 237, and 238.

614. Jakoby, W. B. *The Enzymatic Basis of Detoxification*, Vols. I and II (New York: Academic Press, 1980).

615. Babior, B. M., and C. A. Crowley. "Chronic granulmatous disease and other disorders of oxidative killing by phagocytes," in *The Metabolic Basis of Inherited Disease*, 5th ed., J. B. Stanbury, J. B. Wyngaarden, D. S. Fredrickson, J. L. Goldstein, and M. S. Brown. Eds. (New York: McGraw-Hill, 1983), pp. 1964–1966.

616. Grodesky, G. M. "Chemistry and functions of the hormones. I. Thyroid, pancreas, adrenal, and gastrointestinal tract," in *Review of Physiological Chemistry*, 17th ed., H. A. Harper, V. W. Rodwell, and P. A. Mayes, Eds. (Los Altos, CA: Lange Medical Publishers, 1979), p. 524.

617. Griffin, D. H. *Fungal Physiology* (New York: John Wiley & Sons, Inc., 1981), pp. 124–125.

618. Stone, T. L. Communication, Center for Bio-Ecologic Medicine, Rolling Meadows, IL (1988).

619. Hoffman, R. M. "Altered methionine metabolism, DNA methylation and oncogene expression in carcinogenesis," *Biochim. Biophys. Acta* 738:49–87 (1984).

620. Babior, B. M., and C. A. Crowley. "Chronic granulmatous disease and other disorders of oxidative killing by phagocytes," in *The Metabolic Basis of Inherited Disease*, 5th ed., J. B. Stanbury, J. B. Wyngaarden, D. S. Fredrickson, J. L. Goldstein, and M. S. Brown. Eds. (New York: McGraw-Hill, 1983), pp. 1964–1966.

621. Welty, J. D., M. J. McBroom, A. W. Appelt, M. B. Peterson, and W. O. Read. "Effect of taurine on heart and brain electrolyte imbalances," in *Taurine*, R. Huxtable and A. Barbeau, Eds. (New York: Raven Press, 1976), pp. 155–163.

622. Welty, J. D., M. J. McBroom, A. W. Appelt, M. B. Peterson, and W. O. Read. "Effect of taurine on heart and brain electrolyte imbalances," in *Taurine*, R. Huxtable and A. Barbeau, Eds. (New York: Raven Press, 1976), pp. 155–163.

623. Read, W. O., and J. D. Welty. "Effect of taurine on the action potential of guinea pig papillary muscle," in *Taurine*, R. Huxtable and A. Barbeau, Eds. (New York: Raven Press, 1976), pp. 173–177.

624. Huxtable, R. "Metabolism and function of taurine in the heart," in *Taurine*, R. Huxtable and A. Barbeau, Eds. (New York: Raven Press, 1976), pp. 99–119.

625. Sturman, J. A., and G. F. Gaull. "Taurine in the brain and liver of the developing human and rhesus monkey," in *Taurine*, R. Huxtable and A. Barbeau, Eds. (New York: Raven Press, 1976), pp. 73–84.

626. Barbeau, A., Y. Tsukada, and N. Inoue. "Neuropharmacologic and behavioral effects of taurine," in *Taurine*, R. Huxtable and A. Barbeau, Eds. (New York: Raven Press, 1976), pp. 253–266.

627. Barbeau, A., Y. Tsukada, and N. Inoue. "Neuropharmacologic and behavioral effects of taurine," in *Taurine*, R. Huxtable and A. Barbeau, Eds. (New York: Raven Press, 1976), pp. 253–266.

628. Van Gelder, N. M. "Rectification of abnormal glutamic acid level by taurine," in *Taurine*, R. Huxtable and A. Barbeau, Eds. (New York: Raven Press, 1976), pp. 293–302.

629. Bionostics, Inc. Technical memorandum 2, "Functions of Amino Acids," Revised Sept. 1986.

630. Spaeth, D. G., and D. L. Schneider. "Taurine metabolism: Effects of diet and bile salt metabolism," in *Taurine*, R. Huxtable and A. Barbeau, Eds. (New York: Raven Press, 1976), pp. 35–44.

631. Harpter, H. A., V. W. Rodwell, and P. A. Mayes. *Review of Physiological Chemistry*, 17th ed. (New York: Lange Medical Publishers, 1979), pp. 250–253.

632. Mudd, S. H., and H. L. Levy. "Disorders of transulfuration," in *Metabolic Basis of Inherited Disease*, 5th ed., J. B. Stanbury, J. B. Wyngaarden, D. S. Fredrickson, J. L. Goldstein, and M. S. Brown, Eds. (New York: McGraw-Hill, 1983), pp. 522–559.

633. Schneider, J. A., and J. D. Schulman. "Cystinosis," in *The Metabolic Basis of Inherited Disease*, 5th ed., J. B. Stanbury, J. B. Wyngaarden, , D. S. Fredrickson, J. L. Goldstein, and M. S. Brown, Eds. (New York: McGraw Hill, 1983), pp. 1852–1855.

634. Roe, D. A. *Drug-Induced Nutritional Deficiencies*, 2nd ed., (Westport, CN: AVI Publishing Co., 1985), p. 24.

635. Aurand, L. W., and A. E. Woods. *Food Chemistry* (Westport, CN: AVI Publishing Co., 1973).

636. Babior, B. M., and C. A. Crowley. "Chronic granulmatous disease and other disorders of oxidative killing by phagocytes," in *The Metabolic Basis of Inherited Disease*, 5th ed., J. B. Stanbury, J. B. Wyngaarden, D. S. Fredrickson, J. L. goldstein, and M. S. Brown. Eds. (New York: McGraw-Hill, 1983), pp. 1964–1966.

637. Irwin, M., K. Belendiuk, K. Mccloskey, and D. X. Freedman. "Tryptophan metabolism in children with attentional deficit disorder," *Am. J. Psychiatry* 138(8):1082–1085 (1981).

638. Houpert, Y., P. Tarallo, and G. Siest. "Comparison of procedures for extracting free amino acids from polymorphonuclear leukocytes," *Clin. Chem.* 22:1618–1622 (1976).

639. Soupare, P. "Free amino acids of blood and urine in the human," in *Amino Acid Pools*, J. Holden, Ed. (Amsterdam: Elsevier Publishing Co., 1962), pp. 220–262.

640. Peters, J. H., S. C. Lin, B. J. Berridge, Jr., J. G. Cummings, and W. R. Chao. "Amino acids, including asparagine and glutamine, in plasma and urine of normal human subjects," *Proc. Soc. Exp. Biol. Med.* 131:281–288 (1969).

641. Awapara, J., and M. Bery. "Uptake of taurine by slices of rat heart and kidney," in *Taurine*, R. Huxtable and A. Barbeau, Eds. (New York: Raven Press, 1976), pp. 135–143.

642. Pangborn, J. B. Technical memorandum 2, "Function of Amino Acids," Bionostics Inc. Lisle, IL (1986), p. 22.

643. Rodwell, V. W. "Catabolism of amino acid nitrogen," in *Review of Physiological Chemistry*, H. A. Harper, V. W. Rodwell, and P. A. Mayes, Eds. (Los Altos, CA: Lange Medical Publishers, 1979), pp. 400–401.

644. Pardridge, W. M. "Regulation of amino acid availability to the brain," in *Nutrition and the Brain*, R. J. Wurtman and J. J. Wurtman, Eds. (New York: Raven Press, 1977), pp. 141–190.

645. Sherman, A. R., and L. Helyar. "Iron deficiency, immunity, and disease resistance in early life," in *Contemporary Issues in Clinical Nutrition*, Vol. 11, R. K. Chandra, Ed. (New York: Alan R. Liss, Inc., 1988), pp. 86, 416.

646. Gerber, D. A. "Decreased concentration of free histidine in serum in rheumatoid arthritis," *J. Rhematol.*. 2:384–392 (1975).

647. Sass-Kortsak, A., and A. G. Bearn. "Hereditary disorders of copper metabolism: [Wilson's disease (Hepatolenticular degeneration) and Menkes' disease (Kiuky-hair or steely-hair syndrome)]," in *The Metabolic Basis of Inherited Disease*, 4th ed., J. B. Stanbury, J. B. Wyngaarden, D. S. Fredrickson, J. L. Goldstein, and M. S. Brown, Eds. (New York: McGraw-Hill, 1978), pp. 1098–1101.

648. Banford, J. C., D. H. Brown, R. A. Hazelton, E. J. McNil, R. D. Sturrock, and W. E. Smith. "Serum copper and erythrocyte superoxide dismutase in rhematoid arthritis," *Ann. Rheum. Dis.* 458–462 (1982).

649. Scriver, C. R. "Disorders of proline and hydroxyproline metabolism," in *The Metabolic Basis of Inherited Disease*, 4th ed., J. B. Stanbury, J. B. Wyngaarden, and D. S. Fredrickson, Eds. (New York: McGraw-Hill, 1978), p. 340.

650. Rodwell, V. W. "Biosynthesis of amino acids," in *Review of Physiological Chemistry*, 17th ed., H. A. Harper, V. W. Rodwell, and P. A. Mayes, Eds. (Los Altos, CA: Lange Medical Publishers, 1979), p. 392.

651. Strobel, L., A. H. Y. Lu, S. Heidema, and M. J. Coon. "Phosphatidyl choline requirement in the enzymatic reduction of heam protein P-450 and in fatty acid, hydrocarbon, and drug hydroxylation," *J. Biol. Chem.* 245:4851–4854 (1970).

652. Wade, A. F., Wu, B., and D. Lee. "Nutritional factors affecting drug metaboizing enzymes of the rat," *Biochem. Pharmacol.* 24:785–789 (1975).

653. Hammer, C. T., and E. D. Wills. "The role of lipid components of the diet in regulation of the fatty acid composition of rat liver endoplasmic reticulum and lipid peroxidation," *Biochem. J.* 174:585–593 (1978).

654. Caster, W. O., A. E. Wade, F. E. Greene, and J. S. Meadows. "Effect of different levels of corn oil in the diet upon the rate of hexobarbitol, hetachlor, analine, metabolism in the liver of the male white rat," *Life Sci.* 9:181–190 (1970).

655. Lam, T. C., and A. E. Wade. "Influence fo dietary lipid on the metabolism of hexobarbitol by the ixolated perfused rat liver," *Pharmacology* 21:64–67 (1980).

656. Marshall, W. J., and A. E. M. McLean. "A requirement of dietary lipids for the induction of cytochrome P-450 by phenobarbitone in rat liver microsomal fraction," *Biochemistry* 122:569–573 (1971).

657. Century, B. "A role of dietary lipid in the ability of phenobarbitol to stimulate drug detoxification," *J. Pharmacol. Exp. Ther.* 185:185–194 (1923).

658. Andersen, K. E., A. H. Conney, and A. Kappas. "Nutrition and oxidative drug metabolism in man. Relative influence of dietary protein, carbohydrate, and lipids," *Clin. Pharmacol. Ther.* 26:493–501 (1976).

659. Watson, A. F., and E. Mellanby. "Tar cancer in mia II the/condition of the skin when modified by external treatment or diet, as a factor in influencing the cancerous reaction," *Br. J. Exp. Pathol.* 11:311–322 (1930).

660. Newberne. P. M., J. Weogert., and N. Lula. "Effect of dietary fat on hepatic-mixed function oxidases and hepato cellular carcinoma induced by aflatoxin B, in rats," *Cancer Res.* 39:3986–3991 (1979).

661. Cheng, K. C., Ragland, W. L., and A. E. Wade. "Dietary fat and 3 M.C. induction of hepatic nuclear and microsomal cytochrome P-450," *Drug Nutr. Interact.* 1:163–174 (1981).

662. Davison, S. C., and E. D. Wills. "Studies on the lipid composition of the rat liver endoplasmic reticulum after induction with phenobaritone and 20-methylantheren," *Biochem. J.* 140:461–469 (1974).

663. Lambert, L., and E. D. Wills. "The effect of dietary lipids on 3,4-benzo(a)pyrene metabolism in the hepatic endoplasmic reticulum," *Biochem. Pharmacol.* 26:1423–1427 (1977).

664. Lang, M. "Depression of drug metabolism in liver microsomes after treating rats with unsaturated fatty acids," *Gen. Pharmacol.* 7:415–419 (1976).

665. Reeves, A. L. "The metabolism of foreign compounds," in *Toxicology: Principles and Practice*, Vol. 1, A. L. Reeves, Ed. (New York: John Wiley & Sons, 1981), p. 27.

666. Hayes, W. J., Jr. "Review of the metabolism of chlorinated hydrocarbon insecticides, especially in mammals," *Annu. Rev. Pharmacol.* 5:27 (1965).

667. Levine, S. A., and P. M. Kidd. "Antioxidant Adaptation. Its Role in Free Radical Pathology," Biocurrents Division, Allergy Research Group, San Leandro, CA (1986), p. 221.

668. Kris-Etherton, P. M., D. Krummel, M. E. Russell, D. Dreon, S. Mackey, J. Borchers., and P. D. Wood. "Natl. cholesterol education program: The effect of diet on plasma lipids, lipoproteins, and coronary heart disease," *J. Am. Diet. Assoc.* 88(11):1373–1400 (1988).

669. Havel, R. J. "Approach to the patients with hyperlipemia," *Med. Clin. North Am.* 66:319 (1982).

670. Breslow, J. L. "Apoliprotein genetic variation and human disease," *Physiol. Rev.* 68:85 (1988).

671. Brown, M. S., and J. L. Goldstein. "A receptor-mediated pathway for cholesterol homeostasis," *Science* 232:34 (1986).

672. Gordon, T., W. B. Kannel, W. B. Castelli, and T. R. Dawber. "Lipoproteins, cardiovascular disease and death: The Framingham study," *Arch. Intern. Med.* 141:1128–1131 (1981).

673. Miller, N. E., F. Hammett, S. Sultiss, S. Rao, H. Van Zeller, J. Coltart, and B. Lewis. "Relation of angiographically defined coronary artery disease to plasma lipoprotein subfractions and apoliporpoteins," *Br. Med. J.* 282:1741 (1981).

674. Williams, P. T., R. M. Krauss, S. Kindel-Joyce, D. M. Dreon, K. M. Vranizan, and P. D. Wood. "Relationship of dietary fat, protein, cholesterol, and fiber intake to atherogenic lipoproteins in men," *Am. J. Clin. Nutr.* 44:788–797 (1986).

675. Gordon, T. W., P. Castelli, M. C. Hjortland, W. B. Kannel, and T. R. Dawber. "High density lipoprotein as a protective factor against coronary heart disease: The Framingham study," *Am. J. Med.* 62:707–714 (1977).

676. Castelli, W. P., R. J. Garrison, P. W. F. Wilson, R. D. Abbott, S. Kalousdian, and W. B. Kannel. "Incidence of coronary heart disease and lipoprotein cholesterol levels: The Framingham study," *JAMA* 252:2835 (1986).

677. Gordon, T., W. P. Castelli, M. C. Hjortland, W. B. Kannel, and T. R. Dawber. "High density lipoprotein as a protective factor against coronary heart disease: The Framingham study," *Am. J. Med.* 62:707–714 (1977).

678. "The expert panel: Report of the national cholesterol education program expert panel on detection, evaluation, and treatment of high blood cholesterol in adults," *Arch. Intern. Med.* 148:36–69 (1988).

679. Naito, H. K. "New diagnostic tests for assessing coronary heart disease risk," in *Recent Aspects of Diagnosis and Treatment of Lipoprotein Disorders*, K. Widhalm and H. K. Naito, Eds. (New York: Alan Liss, Inc., 1986), p. 54.

680. Laseter, J. L. Personal communication (1989).

681. Bland, J. S. "Serum lipids, prostaglandins, and marine oils," *J. Int. Acad. Prevent. Med.* 8:3–16 (1983).

682. Crawford, M. "Essential fatty acids and prostaglandins," *Nature* 287:388–389 (1980).

683. Bang, H. O., and J. Dyerberg. "Plasma lipids and lipoproteins in Greenlandic east coast Eskimos," *Acta Med. Scand.* 192:85–94 (1972).

684. Bang, H. O., and J. Dyerberg. "Lipid metabolism and ischemic heart disease in Greenland Eskimos," in *Advanced Nutrition Research*, Vol. 3, H. H. Draper, Ed. (New York: Plenum Press, 1980), pp. 1–22.

685. Arthaud, J. B. "Cause of death in 339 Alaskan natives as determined by autopsy," *Arch. Pathol.* 90:433–438 (1970).

686. Saynor, R., and D. Verel. "Eicosapentaenoic acid, bleeding time and serum lipids," *Lancet* 2:272 (1982).

687. Bang, H. D., and J. Dyerberg. "The composition of food consumed by Greenland Eskimos," *Acta Med. Scand.* 200:69 (1976).

688. Ballard-Barbash, R., and C. W. Callaway. "Marine fish oils," *Mayo Clin. Proc.* 62:113–118 (1987).

689. Dyerberg, J., and H. O. Bang. "Eicosapentanoic acid and prevention of atherosclerosis and thrombosis," *Lancet* 2:117–119 (1978).

690. Phillipson, B. E. "Reduction of plasma lipids, lipoproteins, and apoproteins by dietary fish oils in patients with hypertriglyceridemia," *N. Engl. J. Med.* 312:1210–1216 (1985).

691. Lehninger, A. L. *Principles of Biochemistry* (New York: Worth Publishers, 1982), p. 297.

692. Lehninger, A. L. *Principles of Biochemistry* (New York: Worth Publishers, 1982), p. 209

693. Øie, S., and G. Levy. "Interindividual differences in the effect of drugs on bilirubin plasma binding in newborn infants and in adults," *Clin. Pharmacol. Ther.* 21:627–632 (1977).

694. Wilding, G., B. Paigen, and E. S. Vesell. "Genetic control of interindividual variations in racemic warfarin binding to plasma and albumin of twins," *Clin. Pharmacol. Ther.* 22:831–842 (1977).

695. Wilding, G., B. Paigen, and E. S. Vesell. "Genetic control of interindividual variations in racemic warfarin binding to plasma and albumin of twins," *Clin. Pharmacol. Ther.* 22:831–842 (1977).

696. Nebert, D. W. "Human genetic variation in the enzymes of detoxication," in *Enzymatic Basis of Detoxication*, Vol. I, W. B. Jakoby, Ed. (Orlando, FL: Academic Press, 1980), p. 29.

697. Evans, F. T., P. W. S. Gray, H. Lehmann, and E. Silk. "Sensitivity to succinylcholine in relation to serum cholinesterase," *Lancet* 1:1229–1230 (1952).

698. Weinshilboum, R. M. "Serum dopamine β-hydroxylase," *Pharmacol. Rev.* 30:133–166 (1979).

699. Weinshilboum, R., and F. A. Raymond. "Inheritance of low erythrocyte catechol O-methyltransferase activity in man," *Am. J. Hum. Genet.* 29:125–135 (1977).

700. Kalow, W., D. Kadar, T. Inaba, and B. K. Tang. "A case of deficiency of N-hydroxylation of amobarbital," *Clin. Pharmacol. Ther.* 21:530–535 (1977).

701. Tang, B. K., T. Inaba, and W. Kalow. "N-Hydroxylation of barbiturates," in *Biological Oxidation of Nitrogen*, J. W. Gorrod, Ed. (New York: Elsevier North-Holland, 1978), pp. 151–156.

702. Tang, B. K., W. Kalow, and A. A. Grey. "Amobarbital metabolism in man: N-glucoside formation," *Res. Commun. Chem. Pathol. Pharmacol.* 21:45–53 (1978).

703. Evans, D. A. P., and T. A. White. "Human acetylation polymorphism," *J. Lab. Clin. Med.* 63:394–403 (1964).

704. Bönicke, R., and B. P. Lisboa. "Über die Erbeddingtheit der intraindividuellen Konstanz der Isoniazidausscheidung beim Menschen," *Naturwissenschaften* 44:314 (1957).

705. Harris, H. W., R. A. Knight, and M. J. Selin. "Comparison of isoniazid concentrations in the blood of people of Japanese and European descent — therapeutic and genetic implications," *Am. Rev. Tuberc. Pulm. Dis.* 78:944–948 (1958).

706. Knight, R. A., M. J. Selin, and H. W. Harris. "Genetic factors influencing isoniazid blood levels in humans," *Trans. Conf. Chemother. Tuberc.* 18th Conf. 52–58 (1958).

707. Evans, D. A. P., K. Manley, and V. A. McKusick. "Genetic control of isoniazid metabolism in man," *Br. Med. J.* 2:485–491 (1960).

708. See References 704 to 707.

709. Hopkinson, D. A., N. Spencer, and H. Harris. "Red cell acid phosphatase variants: A New human polymorphism," *Nature* 199:969–971 (1963).

710. Eze, L. C., M. C. K. Tweedie, M. F. Bullen, P. J. J. Wren, and D. A. P. Evans. "Quantitative genetics of human red cell acid phosphatase," *Ann. Hum. Genet.* 37:333–340 (1974).

711. Modiano, G., G. Battistuzzi, G. J. F. Esan, U. Testa, and L. Luzzatto. "Genetic heterogeneity of "normal" human erythrocyte glucose-6-phosphate dehydrogenase: An isoelectrophoretic polymorphism," *Proc. Natl. Acad. Sci. U.S.A.* 76:852–856 (1979).

712. Nebert, D. W. "Human genetic variation in the enzymes of detoxication," in *Enzymatic Basis of Detoxication*, Vol. I, W. B. Jakoby, Ed. (Orlando, FL: Academic Press, 1980), p. 30.

713. Bönicke, R., and B. P. Lisboa. "Über die Erbeddingtheit der intraindividuellen Konstanz der Isoniazidausscheidung beim Menschen," *Naturwissenschaften* 44:314 (1957).

714. Vesell, E. S. "Advances in pharmacogentics," *Prog. Med. Genet.* 9:291–367 (1973).

715. Alexanderson, B., D. A. P. Evans, and F. Sjöqvist. "Steady state plasma levels of nortriptyline in twins: Influence of genetic factors and drug therapy," *Br. Med. J.* 4:764–768 (1969).

716. Vesell, E. S., J. G. Page, and G. T. Passananti. "Genetic and environmental factors affecting ethanol metabolism in man," *Clin. Pharmacol. Ther.* 12:192–201 (1971).

717. Cascorbi, H. F., E. S. Vesell, D. A. Blake, and M. Hebrich. "Genetic and environmental influence on halthane metabolism in twins," *Clin. Pharmacol. Ther.* 12:50–55 (1971).
718. Vesell, E. S., and J. G. Page. "Genetic control of phenobarbital-induced shortening of plasma antipyrine half-lives in man," *J. Clin. Invest.* 48:2202–2209 (1969).
719. See References 715 and 718.
720. Harris, H. *The Principles of Human Biochemical Genetics* (New York: Elsevier North-Holland, 1975).
721. Scanlon, P. D., F. A. Raymond, and R. M. Weinshilboum. "Catechol O-methyltransferase: Thermolabile enzeme in erythrocytes of subjects homozygous for allele for low activity," *Science* 203:63–65 (1979).
722. Weinshilboum, R. M., F. A. Raymond, and M. Frohnauer. "Monogenic inheritance of catechol O-methyltransferase activity in the rat — biochemical and genetic studies," *Biochem. Pharmacol.* 28:1239–1247 (1979).
723. Nebert, D. W. "Human genetic variation in the enzymes of detoxication," in *Enzymatic Basis of Detoxication*, Vol. I, W. B. Jakoby, Ed. (Orlando, FL: Academic Press, 1980), p. 32.
724. Nebert, D. W. "Human genetic variation in the enzymes of detoxication," in *Enzymatic Basis of Detoxication*, Vol. I, W. B. Jakoby, Ed. (Orlando, FL: Academic Press, 1980), p. 32.
725. Kalow, W., and K. Genest. "A method for the detection of atypical forms of human serum cholinesterase. Determination of dibucaine numbers," *Can. J. Biochem. Physiol.* 35:339–346 (1957).
726. Nebert, D. W. "Human genetic variation in the enzymes of detoxication," in *Enzymatic Basis of Detoxication*, Vol. I, W. B. Jakoby, Ed. (Orlando, FL: Academic Press, 1980), p. 33.
727. Shih, V. E., I. F. Abroms, J. L. Johnson, M. Carney, R. Mandell, R. M. Robb, J. P. Cloherty, and K. V. Ragagopalan. "Sulfite oxidase deficiency. Biochemical and clinical investigations of a hereditary metabolic disorder in sulfur metabolism," *N. Engl. J. Med.* 297:1022–1028 (1977).
728. Idle, J. R., A. Mahgoub, R. Lancaster, and R. L. Smith. "Hypotensive response to debrisoquine and hydroxylation phenotype," *Life Sci.* 22:979–984 (1978).
729. Monro, J. Personal communication, The Breakspeare Hospital. London, England (1986).
730. Knight, R. A., M. J. Selin, and H. W. Harris. "Genetic factors influencing isoniazid blood levels in humans," *Trans. Conf. Chemother. Tuberc.* 18th Conf. 52–58 (1958).
731. Evans, D. A. P., K. Manley, and V. A. McKusick. "Genetic control of isoniazid metabolism in man," *Br. Med. J.* 2:485–491 (1960).
732. Motulsky, A. G. "Pharmacogenetics," *Prog. Med. Genet.* 3:49–74 (1964).
733. Evans, D. A. P., and T. A. White. "Human acetylation polymorphism," *J. Lab. Clin. Med.* 63:394–403 (1964).

734. Woolsky, R. L., D. E. Drager, M. M. Reidenberg, A. S. Nies, K. Carrk, and J. A. Oates. "Effect of acetylator phenotype on the rate at which procainamide induces anti-nuclear antibodies and the lupus syndrome," *N. Engl. J. Med.* 298:1157–1159 (1978).

735. Reidenberg, M. M., D. E. Drager, and W. C. Robbins. "Polymorphic drug acetylation and systemic lupus erythematosus," in *Advances in Pharmacology and Therapeutics: Clinical Pharmacology*, Vol. 6, C. P. Duchene-Maru-Ilaz, Ed. (Oxford: Pergamon Press, 1979), pp. 51–56.

736. Monro, J. Personal communication, The Breakspeare Hospital, London, England (1987).

737. Kutt, H., M. Wolk, R. Scherman, and F. McDowell. "Insufficient parahydroxylation as a cause of diphenylhydantion toxicity," *Neurology* 14:542–548 (1964).

738. Solomon, H. M. "Variations in metabolism of coumarin anticoagulant drugs," *Am. N.Y. Acad. Sci.* 151:932–935 (1968).

739. Aebi, H., and H. Suter. "Acatulasemia," *Adv. Hum. Genet.* 2:143–199 (1971).

740. Childs, B., J. B. Sidbury, and C. T. Migeon. "Glucuronic acid conjugation by patients with familial non-hemolytic jaundice and their relatives," *Pediatrics* 23:903–913 (1959).

741. Kalow, W. "Topics in pharmacogenetics," *Ann. N.Y. Acad. Sci.* 179:654–659 (1971).

742. O'Reilly, R. A. "The second reported kindred with hereditary resistance to oral anticoagulant drugs," *N. Engl. J. Med.* 282:1448–1451 (1970).

743. Snyder, L. H. "Studies in human inheritance. IX. The inheritance of taste deficiency in man," *Ohio J. Sci.* 32:436–445 (1932).

744. Blakeslee, A. F. "Genetics of sensory thresholds: Taste for phenylthiocarbamide," *Proc. Natl. Acad. Sci. U.S.A.* 18:120–130 (1932).

745. Brown, M. S., and J. L. Goldstein. "Receptor-mediated endocytosis: Insights from the lipoprotein receptor system," *Proc. Natl. Acad. Sci. U.S.A.* 76:3330–3337 (1979).

746. McIntyre, O. R., L. W. Sullivan, G. H. Jeffries, and R. H. Silver. "Pernicious anemia in childhood" *N. Engl. J. Med.* 272:981–986 (1965).

747. Nebert, D. W. "Human genetic variation in the enzymes of detoxication," in *Enzymatic Basis of Detoxication*, Vol. I, W. B. Jakoby, Ed. (Orlando, FL: Academic Press, 1980), p. 38.

748. Neitlich, H. W. "Increased plasma cholinesterase activity and succinylcholine resistance; a genetic variant," *J. Clin. Invest.* 45:380–387 (1966).

749. Korsten, M. A., S. Matsuzaki, L. Feinman, and C. S. Lieber. "High blood acetaldehyde levels after ethanol administration. Difference between alcoholic and nonalcoholic subjects," *N. Engl. J. Med.* 292:386–389 (1975).

750. Nebert, D. W. "Human genetic variation in the enzymes of detoxication," in *Enzymatic Basis of Detoxication*, Vol. I, W. B. Jakoby, Ed. (Orlando, FL: Academic Press, 1980), p. 39.

751. Brennan, M. J. W., and R. C. Cantrell. "O-Aminolaevulinic acid is a potent agonist for GABA autoreceptors," *Nature* 280:514–515 (1979).

752. Wang, Y. M., and J. van Eys. "The enzymatic defect in essential pentosuria," *N. Engl. J. Med.* 282:892–896 (1970).

753. Nebert, D. W. "The Ah locus. A gene with possible importance in cancer predictability," *Arch. Toxicol.* 43:195–207 (1980).

754. Ciaranello, R. D., H. J. Hoffman, J. G. M. Shire, and J. Axelrod. "Genetic regulation of catecholamine biosynthetic enzymes," *J. Biol. Chem.* 249:4528–4536 (1974).

755. Weinshilboum, R. M. "Catecholamine biochemical genetics in human populations," in *Neurogenetics: Genetic Approaches to the Nervous System*, X. O. Breakfield, Ed. (New York: Elsevier North-Holland, 1979), pp. 257–282.

756. Weinshilboum, R. M. "Catecholamine biochemical genetics in human populations," in *Neurogenetics: Genetic Approaches to the Nervous System*, X. O. Breakfield, Ed. (New York: Elsevier North-Holland, 1979), pp. 257–282.

757. Bercu, B. B., and J. D. Schulman. "Genetics of abnormalities of sexual differentiation and of female reproductive failure," *Obstet. Gynecol. Surv.* 35:1–11 (1980).

758. Lee, P. A., L. P. Plotnick, A. A. Kowarski, and C. J. Migeon. *Congenital Adrenal Hyperplasia* (Baltimore, MD: University Park Press, 1977).

759. Mitchell, J. R., U. P. Thorgeirsson, M. Black, J. A. Timbrell, W. R. Snodgrass, W. Z. Potter, D. J. Jollow, and H. R. Keiser. "Increased incidence of isoniazid hepatitis in rapid acetylators: Possible relation to hydrazine metabolites," *Clin. Pharmacol. Ther.* 18:70–79 (1975).

760. Hockwald, R. S., J. Arnold, C. B. Clayman, and A. S. Alving. "Status of primaquine. IV. Toxicity of primaquine in negroes," *J. Am. Med. Assoc.* 149:1568–1570 (1952).

761. Beutler, E. "Disorders due to enzyme defects in the red blood cell," *Adv. Metab. Disord.* 6:131–160 (1972).

762. Axelrod, J. "Purification and properties of phenylethanolamine N-methyltransferase," *J. Biol. Chem.* 237:1657–1660 (1962).

763. Nebert, D. W., and H. V. Gelboin. "The in vivo and in vitro induction of aryl hydrocarbon hydroxylase in mammalian cells of different species, tissues, strains, and developmental and hormonal states," *Arch. Biochem. Biophys.* 134:76–89 (1969).

764. Atlas, S. A., and D. W. Nebert. "Pharmaogenetics and human disease," in *Drug Metabolism from Microbe to Man*, D. V. Parke and R. L. Smith, Eds. (London: Taylor & Francis, 1976), pp. 393–430.

765. Kärki, N. T., and E. Huhti. "Aryl hydrocarbon hydroxylase activity in cultured lymphocytes from lung carcinoma patients and cigarette smokers," *Abstr. Int. Congr. Pharmacol.*, 7th (644):254 (1978).

766. Kouri, R. E., and D. W. Nebert. "Genetic regulation of susceptibility to polycyclic hydrocarbon-induced tumors in the mouse," in *Origins of Human Cancer*, H. H. Hiatt, J. D. Watson, and J. A. Winsten, Eds. (Cold Spring Harbor, NY: Cold Spring Harbor Laboratory, 1977), pp. 811–835.

767. Nebert, D. W., and N. M. Jensen. "The ah locus: Genetic regulation of the metabolism of carcingens, drugs, and other environmental chemicals by cytochrome P-450-indicated monooxygenases," *CRC Crit. Rev. Biochem.* 6:401–437 (1979).

768. Nebert, D. W. "Human genetic variation in the enzymes of detoxication," in *Enzymatic Basis of Detoxication*, Vol. 1, W. B. Jakoby, Ed. (New York: Academic Press, 1980), pp. 32–54.

769. Thomas, P. E., A.Y.H. Lu, D. Ryan, S. B. West, J. Kawalek, and W. Levin. "Multiple forms of rat liver cytochrome P-450," *J. Biol. Chem.* 251:1385–1391 (1976).

770. Gillette, J. R. "Biochemistry of drug oxidation and reduction by enzymes in hepatic endoplasmic reticulum," *Adv. Pharmacol.* 4:219–261 (1966).

771. Conney, A. H. "Pharmacological implications of microsomal enzyme induction," *Pharmacol. Rev.* 19:317–366 (1967).

772. Gilboin, H. V. "Carcinogens, enzymes induction, and gene action," *Adv. Cancer Res.* 10:1–81 (1967).

773. Gillette, J. R. "Mechanism of oxidation by enzymes in the endoplasmic reticulum," *FEBS Symp.* 16:109–124 (1969).

774. Wislocki, P. G., G. T. Miva, and A. Y. H. Lu. "Reactions catalyzed by the cytochrome P-450 system," in *Enzymatic Basis of Detoxication*, Vol 1, W. B. Jakoby, Ed. (New York: Academic Press, 1980), p. 170.

775. Wislocki, P. G., G. T. Miva, and A. Y. H. Lu. "Reactions catalyzed by the cytochrome P-450 system," in *Enzymatic Basis of Detoxication*, Vol 1, W. B. Jakoby, Ed. (New York: Academic Press, 1980), p. 170.

776. Dean, W. L., and M. J. Coon. "Immunochemical studies on two electrophoretically homogeneous forms of rabbit liver microsomal cytochrome P-450: P-450LM$_2$ and P-450LM$_4$," *J. Biol. Chem.* 252:3255–3261 (1977).

777. Coon, M. J., and A. V. Persson. "Microsomal cytochrome P-450: A central catalyst in detoxication reactions," in *Enzymatic Basis of Detoxication*, Vol. I, W. B. Jakoby, Ed. (New York: Academic Press, 1980), p. 118.

778. Vermilion, J. L., and M. J. Coon. "Highly purified detergent-solubilized NADPH-cytochrome P-450 reductase from phenobarbital-induced rat liver microsomes," *Biochem. Biophys. Res. Commun.* 60:1315–1322 (1980).

476   **Chemical Sensitivity: Principles and Mechanisms**

779. Yasukochi, Y., and B. S. S. Masters. "Some properties of a detergent-solubilized NADPH-cytochrome c (cytochrome P-450) reductase purified by biospecific affinity chromatography," *J. Biol. Chem.* 251:5337–5344 (1976).
780. Dignam, J. D., and H. W. Strobel. "NADPH-cytochrome P-450 reductase from rat liver: Purification by affinity chromatography and characterization," *Biochemistry* 16:1116–1123 (1977).
781. Vermilion, J. L., and M. J. Coon. "Purified liver microsomal NADPH-cytochrome P-450 reductase: Spectral characterization of oxidation-reduction states," *J. Biol. Chem.* 253:2694–2704 (1978).
782. Daly, J. W., D. M. Jerina, and B. Witkop. "Arene oxides and the NIH shift: The metabolism, toxicity and carciongenicity of aromatic compounds," *Experientia* 28:1129–1149 (1972).
783. Jerina, D. M., and J. W. Daly. "Arene oxides: A new aspect of drug metabolism," *Science* 185:573–582 (1974).
784. Wislocki, P. G., G. T. Miwa, and A. Y. H. Lu. "Reactions catalyzed by the cytochrome P-450 system," in *Enzymatic Basis of Detoxication*, Vol. I, W. B. Jakoby, Ed. (New York: Academic Press, 1980), p. 142.
785. Nebert, D. W. "Human genetic variation in the enzymes of detoxication," in *Enzymatic Basis of Detoxication*, Vol. 1, W. B. Jakoby, Ed. (New York: Academic Press, 1980), pp. 32–54.
786. Nebert, D. W. and N. M. Jensen. "The Ah locus: Genetic regulation of the metabolism of carcinogens, drugs, and other environmental chemicals by cytochrome P-450-mediated monooxygenases," *Crit. Rev. Biochem.* 6:401–437 (1979).
787. Kouri, R. E. and D. W. Nebert. "Genetic regulation of susceptibility to polycyclic hydrocarbon-induced tumors in the mouse," in *Origins of Human Cancer*, H. H. Hiatt, J. D. Watson, and J. A. Winsten, Eds. (New York: Cold Spring Harbor, 1977), pp. 811–835.
788. Capuzzi, G., and G. Modena. "Oxidation of thiols," in *The Chemistry of the Thiol Group*, S. Patai, Ed. (New York: John Wiley & Sons, 1974), pp. 785–840.
789. Ziegler, D. M. "Microsomal flavin-containing monooxygenase: oxygenation of nucleophilic nitrogen and sulfur compounds," in *Enzymatic Basis of Detoxication*, Vol. I, W. B. Jakoby, Ed. (Orlando, FL: Academic Press, 1980), p. 204.
790. Sterling, C. J. M. "The sulfinic acids and their derivatives," *Int. J. Sulfur Chem.* 6:277–316 (1974).
791. Tank, A. W. "The Effects of Ethanol on Dopamine Metabolism in the Rat Liver and Brain," Ph.D thesis, Purdue University, West Lafayette, IN (1976).
792. Ziegler, D. M. "Microsomal flavin-containing monooxygenase: oxygenation of nucleophilic nitrogen and sulfur compounds," in *Enzymatic Basis of Detoxication*, Vol. I, W. B. Jakoby, Ed. (New York: Academic Press, 1980), p. 202.

793. Strittmater, P., and E. G. Ball. "Formaldehyde dehydrogenase, a glutathione dependent enzyme system," *J. Biol. Chem.* 213:445–461 (1955).

794. Goodman, J. I., and Tephly, T. R. "A comparison of rat and human liver formaldehyde dehydrogenase," *Biochim. Biophys. Acta* 252:489–505 (1971).

795. Uotiola, L., and M. Koivasalo. "Formaldehyde dehydrogenase from human liver. Purification, properties, and evidence for the formation of glutathione thiol esters by the enzyme," *J. Biol. Chem.* 249:7653–7663 (1974).

796. Weiner, H. "Aldehyde oxidizing enzymes," in *Enzymatic Basis of Detoxication*, Vol. I, W. B. Jakoby, Ed. (New York: Academic Press, 1980), p. 264.

797. Li, T. K. "Enzymology of human alcohol metabolism," *Adv. Enzymol.* 46:427–483 (1977).

798. Partanen, T., K. Brook, and T. Markkanen. "A study on intelligence, personality and use of alcohol of adult twins. Alcohol research in the northern countries," *Finn. Found. Alcohol Stud.* 14:1–159 (1966).

799. Reed, T. E. "Racial comparisons of alcohol metabolism: Background, problems and results," *Alcoholism Clin. Exp. Res.* 2:83–87 (1978).

800. Hanna, J. M. "Metabolic responses of Chinese, Japanese, and Europeans to alcohol," *Alcoholism Clin. Exp. Res.* 2:89–92 (1978).

801. Schaefer, J. M. "Alcohol metabolism and sensitivity reactions among the Reddis of South India," *Alcoholism Clin. Exp. Res.* 2:61–69 (1978).

802. Seto, A., S. Tricomi, D. W. Goodwin, R. Kolodney, and T. Sullivan. "Biochemical correlates of ethanol-induced flushing on orientals," *J. Stud. Alcohol* 39:1–11 (1978).

803. Bosron, W. F., and T.-K. Li. "Alcohol dehydrogenase," in *Enzymatic Basis of Detoxication*, Vol. I, W. B. Jakoby, Ed. (New York: Academic Press, 1980), p. 232.

804. Tygstrup, N., K. Winkler, and T. Lundquist. "The mechanism of the fructose effect of the ethanol metabolism of the human liver," *J. Clin. Invest.* 44:817–883 (1965).

805. Winkler, K., F. Lundquist, and N. Tygstrup. "The hepatic metabolism of ethanol in patients with cirrhosis of the liver," *Scand. J. Clin. Lab. Invest.* 23:59–69 (1969).

806. Bosron, W. F., and R. L. Prairie. "Reduced triphosphopyridine-linked aldehyde reductase. II. Species and tissue distribution," *Arch. Biochem. Biophys.* 154:166–172 (1973).

807. Wermuth, B., J. D. B. Münch, and J. P. von Wartburg. "Purification and properties of NADPH-dependent aldehyde reductase from human liver," *J. Biol. Chem.* 252:3821–3828 (1977).

808. Ahmed, N. K., R. L. Felsted, and N. R. Bachur. "Heterogeneity of anthracycline antibiotic carbonyl reductases in mammalian livers," *Biochem. Pharmacol.* 27:2713–2719 (1978).

809. Ris, M. M., and J. P. von Wartburg. "Heterogeneity of NADPH-dependent aldehyde reductase from human and rat brain," *Eur. J. Biochem.* 37:69–77 (1973).
810. Erwin, V. G. "Oxidative-reductive pathways for metabolism of biogenic aldehydes," *Biochem. Pharmacol.* 23(Suppl. 1):110–115 (1974).
811. Von Wartburg, J. P., and B. Wermuth. "Aldehyde reductase," in *Enzymatic Basis of Detoxication*, Vol. 1, W. B. Jakoby, Ed. (New York: Academic Press, 1980), p. 251.
812. Erwin, V. G., and R. A. Deitrich. "Inhibition of bovine brain aldehyde reductase by anticonvulsant compounds in vitro," *Biochem. Pharmacol.* 22:2615–2626 (1973).
813. Erwin, V. G., B. Tabakoff, and R. L. Bronaugh. "Inhibition of a reduced nicotinamide adenine dinucleotide phosphate-linked aldehyde reductase from bovine brain by barbiturates," *Mol. Pharmacol.* 7:169–176 (1971).
814. Williams, R. T. *Detoxication Mechanisms* (New York: John Wiley & Sons, 1959), pp. 88 and 322.
815. Greenwalt, J. W. "Localization of monoamine oxidase in rat liver mitochondria," *Adv. Biochem. Physchopharmacol.* 5:207:226 (1972).
816. Tipton, K. "Monoamine oxidase," in *Enzymatic Basis of Detoxication*, Vol. I, W. B. Jakoby, Ed. (New York: Academic Press, 1980), p. 357.
817. Bannister, J. V., W. H. Bannister, and G. Rutilio. "Aspects of the structure, function and application of superoxide dismutase," *CRC Crit. Rev. Biochem.* 22(2):111–180 (1987).
818. Hassan, H. M., and I. Fridovich. "Superoxide radical and the oxygen enhancement of the toxicity of paraquat in *Escherichia coli*," *J. Biol. Chem.* 253:8143–8148 (1978).
819. Hassan, H. M., and I. Fridovich. "Superoxide dismutases: Detoxication of a free radical," in *Enzymatic Basis of Detoxication*, Vol. I, W. B. Jakoby. Ed. (New York: Academic Press, 1980), p. 325.
820. Kroker, G. F., R. M. Stroud, R. Marshall, T. Bullock, F. M. Carroll, M. Greenberg, T. G. Randolph, W. J. Rea, and R. E. Smiley. "Fasting and rheumatoid arthritis: A multicenter study," *Clin. Ecol.* 2:137–144 (1984).
821. Robbins, S. L., and R. S. Cotran. "Inflammation and repair," in *Pathologic Basis of Disease*, 2nd ed., S. L. Robbins, Ed. (Philadelphia: W. B. Saunders, 1979), pp. 55–106.
822. Rea, W. J., I. R. Bell, and R. E. Smiley. "Environmentally triggered large vessel vasculitis," in *Allergy: Immunology and Medical Treatment*, J. Johnson and J. T. Spencer, Eds. (Chicago: Symposia Specialists, 1975), pp. 185–198.
823. Levine, S. A., and P. M. Kidd. "Antioxidant Adaptation: Its Role in Free Radical Pathology," (San Leandro, CA: Biocurrents Division, Allergy Research Group, 1986), p. 51.
824. Halliwell, B. "Superoxide dismutase, catalase, and glutathione peroxidase: Solutions to the problem of living with oxygen," *New Phytol* 73:1075–1086 (1974).

825. Wendel, A. "Glutathione peroxidase," in *Enzymatic Basis of Detoxication*, Vol. I, W. B. Jakoby, Ed. (New York: Academic Press, 1980), p. 348.

826. Hafeman, D. G., and W. G. Hoekstra. "Lipid peroxidation in vivo during vitamin E and selenium deficiency in the rat as monitored by ethane evolution," *J. Nutr.* 107:666–672 (1977).

827. Wendel, A. "Glutathione peroxidase," in *Enzymatic Basis of Detoxication*, Vol. I, W. B. Jakoby, Ed. (New York: Academic Press, 1980), p. 336.

828. Takahara, S., H. Sato, M. Doi, and S. Mihara. "Acatalacemia. III. On the heredity of acatalasemia," *Proc. Jpn. Acad.* 28:585–589 (1952).

829. Aebi, H., and H. Suter. "Acatalasemia," *Adv. Hum. Genet.* 2:143–199 (1971).

830. Nebert, D. W. "Human genetic variation in the enzymes of detoxication," in *Enzymatic Basis of Detoxication*, Vol. I, W. B. Jakoby, Ed. (Orlando, FL: Academic Press, 1980), p. 35.

831. Orr, M. L. "Vitamin B_6 and Vitamin B_{12} in Foods," in Home Economics Research Report No. 36, Agricultural Research Service, USDA, Washington, DC (1969).

832. Butterfield, S., and D. H. Calloway. "Folacin in wheat and selected foods," *J. Am. Diet. Assoc.* 60:310 (1972).

833. "Consultation: Improved vitamin D bone therapy," *Med. World News* 14(pt.4):34 (Oct. 19, 1973).

834. Czajka-Narins, D. M., and M. Corics. Agriculture Handbook Number 456, U.S. Dept. of Agriculture, Washington, DC (1979).

835. Murphy, E. W., B. W. Willis, and B. K. Watt. "Provisional tables on the zinc content of foods," *J. Am. Diet. Assoc.* 66:345 (1975).

836. Schroeder, H. A., A. P. Nason, and I. H. Tipton. "Chromium deficiency as a factor in atherosclerosis," *J. Chronic Dis.* 23:123–142 (1970).

837. Diplock, A. T., and F. A. Chaudhry. "The relationship of selenium biochemistry to selenium-responsive disease in man," in *Essential and Toxic Trace Elements in Human Health and Disease*, A. S. Prasad, Ed. (New York: Alan R. Liss, Inc., 1987), p. 218.

838. Yang, G. C., S. Wang, R. Zhou, and S. Sun. "Endemic selenium intoxication of humans in China," *Am. J. Clin. Nutr.* 37:872–881 (1983).

839. Casey, C. E. "Selenophilia," *Proc. Nutr. Soc.* 37:872–881 (1988).

840. Willett, W. C., J. S. Morris, S. Pressel, J. O. Taylor, B. F. Polk, M. J. Stamper, B. Rosner, K. Schneider, and C. G. Hames. "Prediagnostic serum selenium and risk of cancer," *Lancet* 2130–2134 (1983).

841. Salonen, J. T., R. Salonen, B. Lappeteläinen, P. H. Mäenpää, G. Alfthan, and P. Puska. "Risk of cancer in relation to serum concentrations of selenium and vitamin A and E: Matched case-control analysis of prospective data," *Clin. Res.* 290:417–420 (1985).

GLOSSARY

Acetylation A general reaction of intermediary metabolism using acetyl Co-A and the enzyme N-acetyl transferase; also, a chief degradation pathway for foreign compounds like sulfur amines and aromatic amines.

Acute Toxicological Tolerance (Adaptation) (Masking) The adjustment of the body's metabolism including immune and enzymatic detoxification systems to a new set point for acute survival in response to a toxic exposure. Acute toxicological tolerance results in short-term survival at the expense of long-term debit, i.e., lung adaptation to a toxic substance functionally leads to end-organ failure such as emphysema if exposure continues.

Acylation Peptide conjugation using acyl-COA-, acyl transferase enzymes, and amino acids, taurine, glycine, and glutamine.

Adaptation (Masking) (Acute Toxicological Tolerance) A change in homeostasis induced by the internal or external environment followed by accommodation of body function to a new set point. An acute survival mechanism.

Addiction Habituation to some practice considered harmful to an individual. A state of periodic or chronic intoxication produced by repeated exposure to a chemical. This state is characterized by (1) an overwhelming desire or need (compulsion) to continue use of the substance and to obtain it by any means, (2) a tendency to increase the dosage, and (3) a psychological and usually a physical dependence on its effect.

Alarm Stage The first stage of adaptation in which a visible cause and effect relationship betweenpollutant challenge and symptom responses occurs.

Aldehyde Oxidase The major enzyme that is rapid-acting upon aldehyde, converting it to carboxylic acids. A metallo flavoprotein containing FAD, iron, and molybdenum.

Aldehyde Reductase An enzyme that reduces carbonyl compounds. NADPH is an essential coenzyme.

Aliphatic Hydrocarbons A class of carbons derived from fatty tissues or oil including methane, ethane, propane, etc. These are straight chain (C–C–C) carbons.

481

Allergens or Antigen An incitant of altered reactivity.

Allergy An altered reaction induced by an antigen or allergen. Hypersensitivity of the body cells to a specific substance that results in various types of reactions.

Alcohol Dehydrogenase The major enzyme acting on alcohols, oxidizing them to aldehydes. Slow-acting versus the rapid-acting aldehyde oxidase.

Amphipatic Compounds Of or relating to molecules containing groups with characteristically different properties, e.g., both hydrophilic and hydrophobic properties.

Anaphylaxis An immediate, violent, transient, immunological reaction characterized by contraction of smooth muscle and dilation of capillaries due to release of pharmacologically active substances (histamine, bradykinin, serotonin), and slow-reacting substances, classically initiated by the combination of antigen (allergen) with cell-fixed, cytophilic antibody resulting in shock and angioedema.

Aromatic Hydrocarbons Any compound containing only hydrogen and six carbon rings such as benzene and all derivatives which resemble benzenes in chemical behavior.

Aromatic (Aryl) Hydrocarbon Hydroxylase An enzyme that acts primarily on the aromatic rings resulting in detoxification.

Bioaccumulation The absorption and retention of chemical compounds which are stored and not completely metabolized in the body.

Biochemical Individuality Each person's uniqueness; the individual quantity and quality of enzymes, immune systems, fats, proteins, minerals, and vitamins that allow for a unique and different response for each individual to incitants at a given time.

Biopsychological Interaction between the body and the mind. Interface of psychology, biology, physiology, biochemistry, neural sciences, and related areas.

Bipolarity The abnormal fluctuation of the body characterized by a stimulating, withdrawal, and depressive reaction as a result of pollutant stimuli.

Brain Fag A state of fogginess of the brain induced by chemicals, foods, or biological inhalants.

Brittleness A state of extremely easy altering of function in the chemically sensitive by very minute environmental triggers.

Carbamates Class of pesticides which are neutral esters of carbamic acids. They usually have anticholinesterase activity, e.g., "Sevin" (carbyl).

Catalase The antipollutant enzyme which is a heme amino acid that acts like glutathione peroxidase to convert hydrogen peroxide to oxygen and water.

Catalytic Conversion The rendering of foreign chemicals to a less toxic state by virtue of enzymes.

Challenge Tests Tests designed to incite a reaction in the body by any route, i.e., oral, skin, inhalation.

Chemical Sensitivity An adverse reaction to ambient levels of chemicals generally accepted as subtoxic in our environment in air, food, and water.

Chemically Less Contaminated Food Food grown, preserved, packaged, and transported in the relative absence of toxic chemicals.

Chlorinated Hydrocarbons Hydrocarbons charged with chlorine.

Chlorinated Pesticide Organochlorines containing a carbon and chlorine bond, including heptachlor, dieldrin, endrin, mirex, and DDT (TDE, DDE). A contact poison which is insoluble in water but soluble in oils making it readily absorbable through the skin and accumulative in the fatty tissues.

Cofactor A substance, usually a vitamin or mineral, that aids enzymes in their action.

Conjugation The process of detoxifying xenobiotics using nutrient derived enzymes and cofactors and amino acids resulting in acetylation, peptide conjugation, sulfonation, methylation, and gluconation.

Cytochrome P-450—Mixed Function Oxidase System This is a metallo protein system consisting of three components including cytochrome P-450, NADPH-P-450-reductase, and dia-acylglycerl-3-phosphorylcholine. Therefore, it is dependent upon vitamins B_2, B_3, B_6, iron, copper, and several amino acids such as choline. It is the major detoxification system used for antipollution detoxification through oxidation, reduction, and conjugation reactions.

Deadaptation The process of reducing total body burden to the point of unmasking the adaptation mechanism and allowing the response systems to return to a symptom free physiologic state.

Degradation The breaking down of a chemical compound by water and a microsomal enzyme.

Detoxication Refers to the subject matter, rather than "metabolism of foreign compounds," "metabolism of xenobiotics," or "biotransformation." Those chemical changes which foreign organic compounds undergo in the animal body. Emphasis is on the mechanism of conversion rather than on the toxicological aspects. Normal clearing of small amounts of toxic chemicals from the body usually using the oxidation and conjugation systems.

Detoxification Abnormal clearing of excess toxic chemicals from the body usually by oxidation and conjugation reactions. Implies the correction of a state of toxicity as in the efforts made to produce sobriety in one who is inebriated.

Drug Sensitivity Adverse reactivity of the body to medications.

Electromagnetic Sensitivity Adverse reactivity of the body to electric and magnetic stimuli.

Electrophil A chemical compound that has an affinity for bare electrons and will bind easily for detoxification.

End-organ The final target for pollutant injury.

Environmental Control Unit (ECU) A hospital unit whose design facilitates control of common chemical, particulate, biologic, and electromagnetic contaminants, especially those specific to air, food, and water. A therapeutic-diagnostic haven whereby chemically sensitive patients can be diagnosed and treated.

Environmental Illness Adverse response of the body to environmental pollutants including physical phenomena, chemical or biological substances.

Endpoint A specific amount of an antigen which does not elicit an adverse reaction or which can neutralize an occurring adverse reaction.

Enzyme A protein containing organic catalyst classified as one or six types: hydrolases, transferases, lyases, oxi-reductases, isomerases, and ligases.

Extrinsic Coming from or originating from the outside.

Fasting Total abstinence of all food for a limited period of time to deliver a therapeutic effect on the body.

Fixed-named Disease The pathological condition of the body that is generally irreversible and has a name in contrast to nonspecific symptomatology; e.g., systemic lupus or arteriosclerosis or renal failure versus weakness and malaise.

Food Additives Substances added to foods to preserve, enhance color, or taste.

Food Allergy Adverse reactions of the body to foods through the IgE mechanism.

Food Sensitivity Hypersensitivity of the body cells to a specific food that results in various types of reactions due to immune mechanisms other than IgE; often includes food intolerance which suggests and involves nonimmune mechanisms.

Formaldehyde A colorless, pungent hydrocarbon gas used in solution as a strong disinfectant and preservative and used in the manufacturing of synthetic resins, dyes, building materials.

Fragile A condition in the chemically sensitive in which malfunction is easily triggered by an environmental pollutant.

General Adaptation Syndrome A condition in which an animal gets used to a chronic toxic exposure for acute survival. Eventually results in nutrient fuel depletion if the pollutant is not removed.

Gluconation A two-step conjugation process for xenobiotics using glucuronic acid and uridine diphosphenyls and UDP-glucuronic transferase. This system has a high capacity but low affinity for phenol.

Glutathione Glutamyl cysteinyl glycine—a sulfur-containing amino acid that is paramount in xenobiotic detoxification.

Glutathione Peroxidase An antipollutant enzyme which converts moderately toxic hydrogen peroxide to oxygen and water.

Glycoprotein A substance that usually links a carbohydrate and an amino acid together. Usually involves the immune system. The amino acids involved are serine, asparagine, threonine, hydroxy proline, and lysine.

Herbicide Any chemical substance used to check the growth of or destroy plants, especially weeds.

Hydrocarbon Any chemical compound containing only hydrogen and carbon bonds.

Hydrolase A class of enzymes that splits compounds into fragments by the enzymatic addition of water (H^+ and OH^-) through degradation reactions.

Hydrophilic A chemical substance that is easily attributed to water in contrast to one that is hydrophobic.

Hypersensitivity Abnormal or excessive reactivity to any substance.

Incitant A triggering or causative agent; one that induces an allergic or hyper-sensitive reaction.

Individual Adaptation Syndrome Masking, acute toxicological intolerance.

Individual Susceptibility A person's own peculiar sensitivity to a given substance.

Isomerase A class of enzymes that produce isomeric forms of molecules by group transfer producing cis- and trans- forms and vice versa.

Ketone Reductase An enzyme which will reduce carbonyl compounds. *NADPH* is an essential coenzyme.

Ligands Proteins that bind minerals.

Ligases Enzymes that form C–O–, C–N, C–C, and C–S bonds by the breakdown of a pyrophosphate bond such as ATP.

Lyases Enzymes that add or remove double bonds, e.g., carboxylases, aldolases.

Maladaptation Stage IIA of the adaptation syndrome where the body has lost its masking of symptoms from pollutant stimuli.

Masking (Adaptation) (Acute Toxicological Tolerance) Adaptation of the body's metabolism to a new set point in order to survive a toxic exposure. Masking results in short-term survival at the cost of a long-term debit, i.e., long-term adaptation leads to end-organ failure such as lung failure or emphysema.

Methylation A conjugation pathway for detoxifying many synthetics and endogenous toxic compounds as well as nutrients, using the methyl transferase enzymes and the amino acids methionine or ethionine.

Microsomal Flavin Containing Monoxygenase The enzyme that oxidizes xenobiotics containing nitrogen and sulfur atoms like amines and organic sulfur compounds, such as sulfides, thiols, disulfide, thiones.

Monoamine Oxidase An FAD-cysteine containing enzyme that catalyzes oxidative deamination of a variety of amines including neurotransmitters such as epinephrine, histamine, and serotonin.

Natural Gas A mixture of gaseous hydrocarbons, chiefly methane, occurring naturally in the earth, often in association with other petroleum deposits, used as fuel.

Neutralization To change to a neutral state; to render ineffective, to destroy the distinctive or active properties by specific doses of an allergen delivered either sublingually or by injection under the dermal layers.

Nucleophile An electron donor in chemical reactions in which the donated electrons bind other chemical groups easily.

Noncatalytic Conversion The alteration of foreign chemicals to a less toxic state by nonenzymatic chemical neutralization, i.e., combining an acid and base to neutralize.

Nutrients The proteins, minerals, carbohydrates, fats, and vitamins necessary for body function and maintenance of life.

Overload Increase in total body burden to the point of triggering adverse reactions and altering physiology.

Organic Chemicals Substances derived from living organisms; contain carbon.

Organic Foods Less chemically contaminated foods; those grown, stored, preserved, and transported without the use of synthetic chemicals.

Organophosphate Phosphate esterified to organic compounds such as glucose or sorbitol. Many organophosphorus compounds are powerful acetylc holinesterase inhibitors and are used as insecticides. Highly toxic to man at less than 5 mg/kg. A total dose of 2 mg of parathion has been known to kill children.

Oxidation Loss of electrons.

Oxireductases A class of enzymes that transfer electrons such as seen in dehydrogenases, reductases, oxidases, transhdyrogenases, peroxidase, and oxygenases (oxidation-reduction reactions).

Pollutant Injury (Damage) Change in the tissues or physiologic processes due to the intake of noxious substances.

Pollutants Substances which pollute or contaminate the environment, especially harmful chemical or waste material discharged into the atmosphere and water, including gases, particulate matter, pesticides, radioactive isotopes, sewage, organic chemicals and phosphates, solid wastes and many others.

Polycyclic Hydrocarbons A class of carbon containing pollutants that contain multiple benzene rings such as benzapyrene, naphthalene.

Polymorphism A genetic condition in which the same gene is present in varied quantities resulting in varied detoxification responses to the same environmental pollutant.

Reductions Gain electrons.

Rotary Diet A diet in which no food or food family is repeated (ingested) more often than 4 to 7 days.

Safe House, Water, Food, Air Less chemically contaminated air, food, housing, and water which are less apt to trigger reactions in sensitive individuals.

Serial Dilution Precise quantity of challenging doses in decreasing amounts of antigens or incitants, i.e., 1/5, 1/25, 1/625.

Spreading Phenomenon A condition occurring in the body after repeated exposure in which the individual increasingly reacts to more and more new substances which may or may not be reversible upon reducing the total load. Also added involvement of new organs and systems may occur.

Starvation Abstinence from food (not necessarily total abstinence) to the point of harming the body.

Subcutaneous Test Test injection of an incitant under the skin.

Sulfonation A conjugation process for xenobiotics which transfer inorganic sulfate to the hydroxyl groups for detoxification.

Superoxide Dismutase The first antipollutant defense enzyme which comes in contact with pollutants when they initially enter the body, causing a dismutation reaction from superoxide radicals which produces hydrogen peroxide. It contains zinc, copper, and manganese, and is usually dependent upon B_2 and B_3 as cofactors.

Supplements Dietary augmentation of nutrients, especially vitamins, minerals, and amino acids.

Switch Phenomenon Symptoms and conditions changing from one organ to another as a result of pollutant stimuli, e.g., sinusitis to phlebitis to colitis.

Synergisms Synergisms may be additive or potentiative. They are additive when the effects of the pollutants equal the sum of the individual pollutants involved. They are potentiative when the effects of the potentially harmful substances exceed the sum of the individual substances involved.

Terpenes Any hydrocarbon of the formula $C_{10}H_{16}$ derived chiefly from essential oils, resins, and other vegetable aromatic products; often used in perfumes and medicines, etc.

Total Load or Burden The sum total of all of the incitants in the body.

Toxic Load Total body burden.

Transferases A class of enzymes that involves group transfer reactions such as the addition of methyl, acetyl, amino groups (conjugation reactions).

Trigger Any substance which incites a reaction.

Triggering Agent A substance which incites body physiology.

Unmasking A state in which the body becomes clear of accommodation or acute toxicological tolerance.

Volatile Organic Organic compounds with a high vapor pressure; capable of evaporation from a solid or liquid into the air.

Xanthine Oxidase A metallo flavoprotein enzyme containing molybdenum, iron, and FAD which catalyzes the oxidative hydrolation of substances containing xanthin.

Xenobiotics Foreign compounds or chemicals.

INDEX

Acceptor substrates, 118
Accommodation response, 58
Acetaldehyde, 85, 254, 330, 335,
    361, 416, 420
Acetaminophans, 110
Acetanilide, 96
Acetates, 70, 94, 126
Acetic acid, 335, 361
Acetoacetate, 254
Acetone, 79, 93
Acetylation, 99–101, 375, 414,
    418
  defined, 481
  polymorphism of, 412
  vitamin B_5 and, 254
N-Acetylation, 38
Acetylation enzymes, 228, see also
    specific types
Acetylcholine, 72, 229, 230, 294,
    295
  glycine and, 389
  magnesium and, 308
  vitamin B_5 and, 254
Acetylcholinesterase, 66, 98, 230
Acetyl-coenzyme A, 94, 99, 254
Acetyl-coenzyme A carboxylase,
    267
Acetyl-coenzyme A-dependent N-
    acetyl-transferase, 101

N-Acetylcysteine, 108
Acetylcysteines, 104
Acetylcysteyl derivatives, 109
Acetyl radicals, 50
Acetyl salicylic acid, 70, 126
N-Acetyltransferase, 100, 101
Acid activating enzymes, 102, see
    also specific types
Acid aldehyde, 79
Acid phosphatase, 338
Acid phosphate, 412
Acridine orange, 66
ACTH, see Adrenocorticotropin
    hormone
Actinomycin D, 66, 187
Actomyosin, 304
Acute toxicological tolerance, see
    Adaptation; Masking
    phenomenon
Acylacetates, 102
Acylation, 101–104, 380, 418, 481
Acylchlorides, 88, see also
    specific types
Acyl-coenzyme A, 103, 387
Acyl-coenzyme A synthetase, 102,
    103
N-Acyloxyarylamines, 85
Acyltransferases, 102
N-Acyltransferases, 102

ADA, *see* Adenosine deaminase

Adaptation, 17, 21–28, 72, *see*
    also Masking phenomenon
  alarm stage of, 4, 24, 481
  bipolarity in, 33
  cell damage and, 58
  defined, 1–2, 17, 21, 481
  end-organ failure stage of, 26–28,
    32, 229
  general, 484
  masking stage of, *see* Masking
    phenomenon
  metabolic changes during, 22
  nutrient requirements in, 22
  specific, 22
  stages of, 24–28

Addiction, 30, 481

Additives, 39, 484, *see also*
    specific types

Adenoids, 173

Adenosine daminase (ADA), 171

Adenosine diphosphate (ADP),
    106, 230, 308

Adenosine monophosphate (AMP),
    102, 245, 303, 345, 396

Adenosine triphosphatase (ATPase),
    224, 304, 308, 338

Adenosine triphosphate (ATP), 25,
    32, 62, 102, 103, 221, 248
  aspartic acid and, 374
  biotin and, 267
  depletion of, 106
  energy metabolism and, 224
  generation of, 230
  glutamine and, 387
  magnesium and, 308
  PAPS and, 105
  phosphorus and, 322
  synthesis of, 58, 230
  tryptophan and, 361
  vitamin B_2 and, 250

Adenosine triphosphate (ATP)-
    dependent enzymes, 110

Adenosine triphosphate (ATP)
    sulfurase, 106

Adenosine triphosphate (ATP)
    synthetase, 230

S-Adenosylhomocysteine, 120

Adenosylmethionine, 353

S-Adenosyl-L-methionine-depen-
    dent methyltransferases, 119

Adenosyltransferase, 308

S-Adenosyltransferase, 381

Adenylate cyclase, 303

Adipose tissue, 190, 396

ADP, *see* Adenosine diphosphate

Adrenal glands, 192

Adrenaline, 120, 421

Adrenocorticotropin hormone
    (ACTH), 271, 273

Aflatoxins, 60, 88, 393

Agent orange, 9, 55

Aging, 285

AHK, *see* Arylhydrocarbon
    hydroxyls

Alanine, 229, 347, 366, 367, 386

β-Alanine, 229, 254, 367–370,
    385, *see also* Vitamin B_5
  excess, 231
  glutamic acid and, 386
  sources of, 369–370

Alanine-forming enymes, 367, *see*
    also specific types

Alarm stage of adaptation, 4, 24, 481

Albumin, 68, 70, 275, 319, 361

Alcohol dehydrogenase, 94, 317,
    416, 420–421, 482

Alcohols, 74, 77, 98, 126, 408,
    420, *see also* specific types
  aliphatic, 104
  benzyl, 84
  conjugation of, 122
  elimination of, 78
  glutathione peroxidase and, 423
  magnesium deficiency and, 309
  metabolism of, 397, 420
  oxidation of, 78–79
  reduction to, 94
  sulfotransferases and, 108
  toxic, 423

vitamin A and, 243, 245
vitamin B_2 and, 251
vitamin B_6 and, 257, 258
vitamin C and, 272, 278
xenobiotic, 399
zinc and, 317
Aldehyde dehydrogenase, 79, 246,
    250, 330, 420
Aldehyde oxidase, 78, 335, 420,
    421, 481
Aldehyde oxidizing enzymes, 420–
    421, *see also* specific types
Aldehyde reductases, 421, 481
Aldehydes, 74, 83, 85, 126, *see
    also* specific types
  aliphatic, 94
  aromatic, 94
  biogenic, 70, 94
  buildup of, 78
  detoxification of, 243, 247
  excess, 79
  metabolism of, 250
  molybdenum and, 335
  oxidation to, 78, 88
  reduction of, 94–95
  sugar, 94
  taurine and, 381, 382, 384
  vitamin B_2 and, 250
  vitamin B_6 and, 256–258
  vitamin C and, 270
Aldrin, 86, 88, 173, 192
Algae, 183
Alicyclic compounds, 84, 85, 95,
    98, *see also* specific types
Alicyclic ketones, 95
Aliesterase, 98
Aliphatic alcohols, 104, *see also*
    specific types
Aliphatic aldehydes, 94, *see also*
    specific types
Aliphatic amines, 72, 82, 83, 100,
    254, *see also* specific types
Aliphatic compounds, 82, 85, 95,
    229, 481, *see also* specific
    types

Aliphatic disulfides, 108
Aliphatic esters, 98, *see also*
    specific types
Aliphatic hydroxylation, 72
Aliphatic ketones, 95
Aliphatic nitrates, 161
Aliphatic nitrites, 161
Aliphatic steroids, 108
Alkaline phosphatase, 49, 338
Alkalinization, 103
Alkaloids, 62, 85, 230, *see also*
    specific types
Alkanesulfonate, 381
Alkene oxides, 86, *see also*
    specific types
Alkenes, 88, 95, *see also* specific
    types
Alkylating agents, 109, 160, 177,
    see also specific types
Alkylations, 119–120
Alkyl halides, 126, *see also*
    specific types
Alkyl-substituted aromatic com-
    pounds, 83
Alkyl sulfates, 381
Alkyl sulfides, 90
Alkynes, 95, *see also* specific
    types
Allergens, *see* Antigens
Allergic responses, 8–9, 60, 177,
    236, 365, 482, *see also*
    Hypersensitivity
Alloantibodies, 164
Allyl esters, 82
Aluminum, 7, 231, 300
alpha-Amanitine, 62
Amidases, 98, *see also* specific
    types
Amides, 96, 98, *see also* specific
    types
Amine nitrogen, 106
Amine oxidase-catalyzed
    desulfuration, 88
Amines, 70, 82, 83, 98, 417, 421,
    see also specific types

aliphatic, 72, 82, 83, 100, 254
aromatic, *see* Aromatic amines
aryl, 70, 85, 101, 104, 108, 121
biogenic, 79, 420
conjugation of, 122
detoxification of, 82, 119
endogenous, 421
exogenous, 421
phenol, 254
primary, 85
secondary, 85
solubility of, 125
tertiary, 94
transmitter, 421
vasoactive, 34, *see also* specific
  types
Amino acid acylases, 102
Amino acids, 33, 100, 101, 345–
  391, *see also* specific types
absorption of, 346
activation of, 261
bioavailability of, 101
in blood, 222
changes in, 345
in chemical sensitivity, 238
chromium and, 322
deamination of, 253
deficiencies of, 239
defined, 345
degradation of, 261
depletion of, 39, 231, 236
dietary sources of, 350
elimination of, 346
in end-organs, 222
enzyme formation and, 236
essential, 348–349, *see also*
  specific types
excess, 231, 239
generation of, 104
as glutamyl moiety acceptors,
  109, 110
incorporation of, 346
intake of, 346
intracellular levels of, 50
metabolism of, 261

molecular damage and, 65
molybdenum and, 335, 336
natural configuration of, 347–348
nonessential, 365–367, *see also*
  specific types
pollutant overload and, 345–347
pools of, 221, 237
readsorption of, 346
semiessential, 365–367, *see also*
  specific types
sources of, 350
sulfur and, 107, 342
transport of, 112, 346
vitamin B$_3$ and, 253
vitamin B$_6$ and, 258
vitamin B$_{12}$ and, 261
β-Aminoaciduria, 377
Aminoacidurias, 350
Aminoadipic acid, 365
p-Aminobenzoate, 266
4-Aminobiphenyl, 121
4-Aminodiphenyl, 72, 99
2-Aminoethane-sulfonic acid, 377
Aminofluorene, 101
Aminofluorine, 99
Aminoimidazole carboxamide, 265
Aminoisobutyric acid, 351
β-Aminoisobutyric acid, 385
Aminoisobutyric aciduria, 350
Aminolevulinic acid, 259
Aminolevulinic acid dehydrase, 66
Aminolevulinic acid synthetase,
  416
Amino nitrogen, 101
Aminopeptidase, 317
Aminophenols, 114, 161, *see also*
  specific types
Amino-pyridine-4-aminodiphenyl,
  84–85
Aminotransferases, 256
Amitryptyline, 421
Ammonia, 83, 111, 372
aspartic acid and, 374
detoxification of, 386
glutamic acid and, 386

glutamine and, 387
histidine and, 391
magnesium and, 308
taurine and, 381
vitamin B_6 and, 257
zinc and, 316
Ammonium molybdate, 336
AMP, *see* Adenosine monophosphate
Amphetamine, 72, 83, 259, 271
Amphipathic compounds, 101, 482, *see also* specific types
Amphoteric detergents, 67
Amplification systems, 56
Amylnitrate, 114
Amylnitrite, 161
Analines, 67, 95, 121, 161, *see also* specific types
Anaphylactic reactions, 191
Anaphylactoids, 184
Anaphylaxis, 182–184, 482
Anedylcyclase, 345
Anesthetics, 92, 415, 418, *see also* specific types
Angiogenesis, 319
Angiotensin, 297
Aniline, 99, 161
Antagonistic effects, 70
Anthio, 185, 192
Anthocyanins, 298
Anthracene, 104, 108
Antiacetylcholinesterase, 97
Antibiotics, 62, 98, *see also* specific types
Antibodies, 99, 164, 177, 184, 346, *see also* specific types
allo-, 164
antineutrophil, 381
antinuclear, 100, 414
auto-, 58, 158, 164, 184
to cytochrome P-450, 228
DDT and, 191
DNA and, 193
immunoglobulin M, 193
monoclonal, 194

to normal lung tissue, 193
RNA and, 193
synthesis of, 187
thrombocytopenia and, 164
Anticalcium, 306
Anticonvulsants, 94, 421, *see also* specific types
Antidepressants, 421, *see also* specific types
Antifungal compounds, 177, *see also* specific types
Antigen recognition, 32
Antigens, 64, 99, 170, 184, *see also* specific types
defined, 482
injections of, 175
major histocompatibility complex, 171
recognition of, 372
Antimalarials, 156, *see also* specific types
Antimetabolites, 155, 156
Antineutrophil antibodies, 381
Antinuclear antibodies, 100, 414
Antioxidants, 124, 126, 241, 381–384, *see also* specific types
Antipollutant enzymes, 50, *see also* specific types
Antipyrine, 393
Antipyrine dicoumeral, 413
Antivitamins, 258, *see also* specific types
Apolipoproteins, 399–403, *see also* specific types
Appendix, 173
Arachidonic acid, 406
Arene epoxides, 86
Arene oxides, 86, 94, 420
Arginase, 328
Arginine, 346, 365–367, 370–372
Argininosuccinic acid, 365, 370, 374
Aromatic amines, 99, *see also* specific types
Arochlor, 192

Arochlor-1248, 55
Arochlor-1254, 244
Aromatic aldehydes, 94, *see also*
    specific types
Aromatic amines, 84–85, 99–100,
    161
  bladder cancer and, 121
  PABA and, 266
  taurine and, 381
  vitamin B$_5$ and, 254
Aromatic azo groups, 95–96
Aromatic carboxylic acids, 101, 102
Aromatic compounds, 82, 83, 84,
    see also specific types
Aromatic disulfides, 108, *see also*
    specific types
Aromatic epoxide metabolites, 38
Aromatic hydrocarbon hydroxy-
    lase, 420
Aromatic hydrocarbons, 85, 86,
    121, 126, 229
  defined, 482
  halogenated, 177
  polycyclic, 66
Aromatic hydroxylation, 72
Aromatic ketones, 95
Aromatic nitro groups, 95–96
Aromatic polycyclic hydrocarbons,
    see also specific types
Aromatic side chains, 83
Aromatic steroids, 108
Aromatization, 85
Arsenic, 96, 156, 161
Arteriosclerosis, 68, 403
Arylacetates, 103
Arylacetic acid, 101
Arylactates, 102
Arylamides, 98, *see also* specific
    types
Arylamines, 70, 85, 101, 104, 108,
    121, *see also* specific types
Arylesterase, 98
Aryl halides, 126, *see also* specific
    types

Arylhydrocarbon hydroxylase, 25,
    56, 417
Arylhydrocarbon hydroxyls
    (AHK), 395, 420, 482
Arylhydroxamic acids, 84, 85
Arylhydroxamines, 85
Arylhydroxylamines, 70, *see also*
    specific types
Arylhydroxylase, 84
Arylnitro compounds, 70, 161, *see
    also* specific types
Aryl-s-phenols, 82
Aryl sulfates, 106
Aryl sulfides, 90
Aryl sulfotransferases, 108
Asbestos, 175
Ascorbic acid, *see* Vitamin C
L-Ascorbic acid, 106
Asparagine, 308, 366, 372, 387
  aspartic acid and, 374
  glycine and, 389
  proline and, 391
  synthesis of, 372
Asparagine synthetase, 372, 387
Aspartic acid, 308, 366, 367, 372–
    374, 386, 387
Aspirin, 98
ATP, *see* Adenosine triphosphate
ATPase, *see* Adenosine triphos-
    phatase
Atropine, 98
Autoantibodies, 58, 158, 164, 184
Autoimmune diseases, 228, *see
    also* specific types
Autoimmune response, 177, 184
Autonomic nervous system, 32,
    49, 54, 182, 271
Autotranspeptidetion, 108
Azide, 62
Azo reductase, 95
Azurea, 423

Bacterial infections, 9, 59, 60, 124,
    236, 368, 369

Barbital, 90
Barbiturates, 94, 230, 412, 416,
    421, *see also* specific types
Barium, 231, 300
Basophils, 184
B-cells, 8, 173, 187
  evaluation of, 193–195, 198, 201
  iron and, 332
  reduced, 202
  reticuloendothelial system and,
    170
  total count of, 202
Benzaldehydes, 79, 94
Benzapyrenes, 70
Benzene, 74, 84, 85, 156, 198
  degradation of, 86
  3,4-dichloronitro, 108
  immune suppression from, 177
  leukemia and, 169
  stem cells and, 158
  vitamin C and, 278
  vitamin requirement increases
    and, 231
Benzene thiol, 91
Benzidine, 99, 101, 121
Benzilate insecticides, 97
Benzoates, 102, 103, 276
Benzodiazapines, 230
5,6-Benzoflavone, 72
7,8-Benzoflavone, 72
Benzoic acid, 85, 101
Benzo(a)pyrene, 86, 121, 125, 276
Benzopyrenes, 66, 70, 72, 88, 120,
    295, 419
Benzyl alcohol, 84
Benzyl chloride, 108
Berylium, 193
Beryllosis, 193
Beta-carotene, 239, 241, 246, 330,
    see also Vitamin A
Betaine, 356, 377
Bicarbonate, 423
Bile acids, 101
Bile salts, 67, 103, 104

Bilirubin, 109, 122, 415
Bioaccumulation, 13, 14, 19, 39,
    482
Biochemical individuality of
    response, 2, 17, 18
  biotransformation and, 125–126
  defined, 36, 482
  enzyme induction and, 125–126
  genetics and, 2, 8, 36–38, 40,
    222
  to minerals, 300
  to nutrients, 222–223
  nutritional status and, 38–40
  total body load and, 38–40
Biochemical transformation
    systems, 48
Bioflavonoids, 295, 298, *see also*
    specific types
Biogenic acids, 94, *see also*
    specific types
Biogenic aldehydes, 70, 94, *see
    also* specific types
Biogenic amines, 79, 420, *see also*
    specific types
Biological compounds, 7, *see also*
    specific types
Biological detoxification systems,
    20
Biological half-life, 13–14
Biopsychological interaction, 482
Biopterin, 358, 364
Biosynthesizing enzymes, 277, *see
    also* specific types
Biotin, 239, 267–268, 337, 408
Biotinyllysine, 267
Biotransformation, 13, 57, 70, 71,
    86, 99, 100
  of aliphatic amines, 83
  biochemical individuality of
    response and, 125–126
  enzymes of, 410–414, *see also*
    specific types
  factors affecting rates of,
    124–126

nonenzymatic, 70
proteins in, 410
rates of, 124–126
of xenobiotics, 74, 92
Bipolarity, 2, 17, 26, 28–33
  defined, 28, 482
  depression of, 30, 32–33
  stages of, 28–33
Blood, 155, 156, 170
Blood-brain barrier, 364, 380, 386,
  387
Blood-organ barriers, 32
Body load, total, *see* Total body
  load
Bone marrow, 155, 156–160, 223
Borate, 109
Brain electrolytes, 380
Brain fog, 482
Brittleness, 482
Bromobenzenes, 110
Bronchial tree, 53
Brown fat, 224, 394, 395–396
Building materials, 7
Butane, 82
Buteraldehyde, 85

Cadaverine, 83
Cadmium, 58, 67, 177, 193
Caffeine, 245, 258, 413
Calcium, 120, 203, 297, 301–303
  accumulation of, 62
  assimilation of, 380
  cytosolic, 304
  deficiencies of, 305
  depletion of, 306
  dietary sources of, 306
  efflux of, 304
  in elderly, 233
  excess, 231, 235, 309
  intracellular, 308
  loss of, 306
  magnesium and, 308, 312, 314
  metabolism of, 62, 280, 301
  phosphorus ratio to, 301
  plasma, 322

selenium and, 324
serum, 314
sources of, 306
taurine and, 379, 380
transport of, 312
in treatment of chemical sensitivity, 306–307
utilization of, 306
vitamin C and, 272
vitamin D and, 280
vitamins and metabolism of,
  301
in women, 233
Calcium carbonate, 103
Calcium channel blockers, 230,
  see also specific types
Calcium channels, 64
Calcium pump, 235, 303
Calmodulin, 303–307
Caloric deficiency, 228
Caloric metabolism, 221
Cancer chemotherapy, *see* Chemotherapy
Carbamates, 97, 177, 192, *see also*
  specific types
Carbamides, 122, *see also* specific
  types
Carbamyl insecticides, 66, *see also*
  specific types
Carbaryl, 192
Carbohydrates, 25, 36, 68, *see also*
  specific types
  arginine and, 372
  biotin and, 267
  in blood, 222
  catabolism of, 224
  depletion of, 236
  in end-organs, 222
  magnesium and, 312
  metabolism of, 254, 261, 267,
    372
  pools of, 223
  vitamin B_2 and, 248
  vitamin B_5 and, 254
  vitamin B_{12} and, 261

Carbolic acid, 228
Carbomates, 482, *see also* specific
    types
Carbon, 347, 348
Carbon dioxide, 78
Carbon-hydrogen bond, 83
Carbonic arrhydrase, 336
Carbon monoxide, 7, 67, 72, 73,
    88, 92, 231
  amine inhibition and, 94
  dehalogenation and, 96
  red-cell damage from, 161
  vitamin C and, 278
Carbon tetrachloride, 72, 96, 316
Carbonyl compounds, 70, 94, 95,
    421, *see also* specific types
Carboxylases, 101
Carboxylesterases, 82, 96–98
Carboxyl groups, 125, 377
Carboxylic acids, 101, 102, 122
s-Carboxymethyl-L-cysteine,
    36–38
Carboxypeptidases, 336, 356
Carcinogenesis, 84, 85, 121, 417,
    see also Carcinogens;
    Malignancy
Carcinogens, 79, 86, 95, 101, 104,
    188–190, *see also*
    Carcinogenesis; Malignancy;
    specific types
Cardiovascular system damage, 53,
    54
Carnitine, 352
Carnosine, 317
Carotenoids, 243, *see also* specific
    types
Carrageenan, 305
Carrier proteins, 65, 67, *see also*
    specific types
Catalases, 48, 50, 78, 124, 161,
    276, 324, *see also* specific
    types
  decreased levels of, 415
  defense role of, 423
  defined, 482

  functions of, 425
  iron in heme complex of, 411
Catalysis, 66
Catalytic action, 66
Catalytic agents, 177, *see also*
    specific types
Catalytic conversion, 68, 71–74,
    428
Catalytic enzymes, 65, *see also*
    specific types
Catechol, 74, 120
Catecholamines, 353, *see also*
    specific types
  magnesium and, 309
  metabolism of, 360
  phenylalanine and, 356
  polymorphisms of, 417
  regulation of, 397
  tyrosine and, 360
  vitamin B_2 and, 250
  vitamin C and, 270, 273
Catechol O-methyltransferase, 120,
    308, 412, 413
Catechols, 95, *see also* specific
    types
Cell damage, 57–61
Cell dysfunction, 59
Cell leaks, 58
Cell-mediated immune system
    response, 177, 185–187, 192,
    193, 202
Cell membranes, 60, 422
Cell susceptibility, 60
Cellular retinol binding protein
    (CRBP), 243
Ceruloplasmin, 270
Challenge testing, 3, 50, 482
Chemical sensitivity, defined,8,
    482
Chemotherapy, 155, 160, 164, *see
    also* specific types
Childbirth, 9–10
Chloral hydrate, 94
Chloramphenicol, 98, 156
Chlorate salts, 161

Chlordane, 86, 125, 156, 235, 279, 399
Chlorinated compounds, 39, 88, 162, 229, 279, 291, 483, *see also* Organochlorines; specific types
Chlorinated drinking water, 92
Chlorine, 18, 19, 50, 53, 88, 92, 194
  taurine and, 384
  vitamin B and, 257
Chlorite, 384
Chlormycetin, 62
Chloroacetate, 423
Chloroamines, 381, 382, *see also* specific types
Chloroform, 49, 53, 92, 96, 110, 198, 277
Chlorofos, 192
Chloroplasts, 328
Chloroquine, 63
Chlorothiazide, 184
Chlorpromazine, 63, 94, 259
Chlorpromazine dimethyl sulfoxide, 90
Cholecalciferol, 280
Cholesterol, 399, 408
  abnormal, 103
  bile synthesis and, 380
  blood, 68, 402, 406, 408, 410
  cellular need for, 402
  choline and, 294
  consumption of, 395
  eleavated, 168
  free, 400
  high, 239
  impaired, 239
  inhibition of biosynthesis of, 339
  metabolism of, 68, 101, 103
  oxidative degradation of, 103
  plasma, 402, 406, 408, 410
  synthesis of, 254
  taurine and, 380
  transport of, 400

  vanadium and, 339
  vitamin B_5 and, 254
  vitamin E and, 288
Cholesterol esters, 400
Cholesterol sulfate, 106
Choline, 67, 254, 291–295, 353
  deficiencies of, 291–295
  glycine and, 389
  homocysteine and, 377
  magnesium and, 308
  metabolism of, 291
  methionine and, 356
  synthesis of, 261
  vitamin B_{12} and, 261
Cholinesterase inhibitors, 91, 414, *see also* specific types
Cholinesterases, 38, 96, 244, 291, *see also* specific types
Chondratin sulfate, 329
Chondroitin, 105
Chromaffin, 272
Chromaffin cells, 273
Chromatin, 61
Chromium, 193, 203, 279, 299, 322–324
Chylomicrons, 400, 402
Cigarette smoke, 7, 22
Citrate, 254, 272, 328
Citric acid, 301, 355, 372, 375, 389
Citrulline, 348, 374
Cleaning agents, 7, 92, *see also* specific types
Clinical manifestations of chemical sensitivity, 9
Clostridia toxins, 60
Clothing, 7
Coal, 175
Cobalanin, *see* Vitamin B_{12}
Cobalt chloride, 72
Cobaltus chloride, 96
Cocaine, 70, 98
Coenzyme A, 102, 254, 355, 381, 411

Coenzyme A lipases, 102

Coenzymes, 67, 396, *see also* specific types

Cofactors, 483

Coffee, 245, 258

Cohalt, 336–337

Colchicine, 62

Cold sensitivity, 228, 245

Collagen, 334, 340, 342, 360, 391

Comarin, 415

Combustion products, 7

Compartmentalization, 20

Competition for storage and removal, 13

Complement, 56, 191

Conjugations, 33, 48, 50, 51, 72, 73, 78, 99–122, *see also* Detoxication; Detoxification

acetylation category of, 99–101

acylation category of, 101–104

of alcohols, 122

alkylation category of, 119–120

of amines, 122

categories of, 99

defined, 483

enzymatic, 84–85

enzymes for, 410

gluconation category of, 121–122

of glutathione, 101, 108–112

pathways of, 20

peptide, 375, 377, 379, 380

of phenols, 106, 122

of sulfates, 106–107

sulfur, *see* Sulfur conjugation

of thiols, 122

vitamin B and, 246, 248, 254, 258

Copper, 7, 193, 203, 236, 300, 319–321

cytochrome oxidase system and, 420

deficiencies of, 321, 422

dietary sources of, 321

excess, 231, 321

histidine and, 391

molybdenum and, 334, 341

sources of, 321

superoxide dismutase and, 319, 422

threonine and, 361

transport of, 361, 391

vandium and, 338

vitamin C and, 274, 275, 276, 280

wound healing and, 236

Copper-gishistidine, 276

Coritsone, 245

Corticoids, 309, *see also* specific types

Cortisol, 202

Coumarin, 98

CRBP, *see* Cellular retinol binding protein

Creatine, 370, 372

Creatinine, 261, 372

Crosslinking, 78, 275

Crude oil, 104

Causes of chemical sensitivity, 9–11

Cyanate, 9

Cyanides, 62, 104, 112–114, 117, 161

Cyanocobalamin, 113, *see also* Vitamin B$_{12}$

Cyanogen bromide, 120

Cyanogenic glycosides, 112, 113

Cyanogens, 358

Cyclic ethers, 85

Cyclic food sensitivity, 22

Cyclohexane, 84, 95

Cyclohexanol, 84

Cyclohexene, 88

Cycloheximide, 62

Cyclophosphamide, 187

Cyclosporin, 155, 187

Cyclosporine, 187

Cystathionase, 259

gamma-Cystathionase, 114, 116

Cystathionine, 259
Cystathionine beta-synthase, 381
Cystathionuria, 259
Cysteic acid, 90
Cysteine, 70, 90, 91, 103, 106,
    366, 375–377
  cobalt and, 337
  deficiencies of, 375, 389
  defined, 375
  formation of, 355, 376
  as glutamyl acceptor, 109
  glutathione peroxidase and, 423
  glutathione regernation and, 110
  glycine and, 387, 389
  metabolism of, 336
  methionine and, 355, 356
  mineral absorption and, 301
  molybdenum and, 336
  oxidation of, 105–106
  sulfur and, 342, 343
  as sulfur source, 115
  synthesis of, 117
  taurine and, 380, 381
Cysteine-cystine reaction, 95
Cysteine dioxygenase, 107
Cysteine-sulfinic acid decarboxy-
    lase, 103
Cysteine-sulfonic acid decarboxy-
    lase, 103
Cysteinyl residues, 66
Cystine, 91, 366, 377
  cobalt and, 337
  methionine and, 356
  oxidation of, 375
  sulfur and, 342, 343
  taurine and, 381, 384
Cystinosis, 109
Cytochalasin B, 277
Cytochrome C reductase, 72
Cytochrome oxidase, 114, 319,
    411, 420, 425
Cytochrome oxidase electron
    transports, 163
Cytochrome P-450, 25, 38, 65, 71,
    72, 74, 82

  antibodies to, 228
  benzo(a)pyrene and, 121
  biochemical individuality and,
      222
  biotransformation and, 126
  dehalogenation and, 92
  desulfuration and, 88
  detoxication by, 110
  dysfunction of, 322
  fats and, 393, 396
  genetic disorders and, 416
  inducement of, 416
  inducibility of, 417
  induction of, 393
  iron and, 329, 332, 333
  microsomal, 418
  modulation of, 202
  monooxygenase dependent on,
      56, 415
  oxidation by, 92
  pesticide metabolism and, 118
  phosphorus and, 322
  reductions and, 93
Cytochrome P-450-mixed-function
    oxidases, 101, 418–420, 483
Cytochrome P-450 reductase, 72,
    94
Cytochromes, 72, 161, 230, 331,
    see also specific types
Cytoplasm, 62
Cytosol, 91, 125, 285
Cytosolic enzymes, 110, *see also*
    specific types
Cytotoxic response, 8

Daunorubicin reductase, 94
DBPA, *see* 2,4-Dichloro-6-
    phenylphenol oxyethylamine
DDE, 279
DDT, 13, 125, 184, 185, 191, 399
  vitamin A and, 244
  vitamin C and, 279
Deadaptation, 28, 483
Dealkylations, 72, 98, 418

O-Dealkylations, 418
Deaminations, 72, 83, 250, 253, 418, 421
Debrisoquine, 38, 414
N-Debutylation, 85
Decane, 82
Decarboxylation, 364, 381
Degradations, 33, 48, 50, 51, 73, 126, 235
  of aliphatic amines, 72
  of amino acids, 261
  defined, 483
  enzymes for, 410
  of glucocorticoids, 417
  of lipoproteins, 400
  by monoamine oxidases, 422
  oxidative, 103
  of phenols, 126
  vitamin B and, 246, 248, 261
  of xenobiotics, 96–99
Dehalogenations, 92, 96, 418
Dehydroascorbic acid (DHA), 276
Dehydrochlorination, 92
7-Dehydrocholesterol, 280
Dehydrogenase reaction, 77
Dehydrogenases, 66, 248, 317, *see also* specific types
Dehydroxylations, 95–96
Delayed-type hypersensitivity, 170, 192, 202
Demethylation, 85, 395
Demidehydroascorbate, 276
Denitrification, 93
Dephosphorylation, 389
Depression of bipolarity, 30
Desaturase, 408
Desferrioxamine, 276
Desulfurations, 72, 88–90, 118, 418
Detergents, 67, 322, 421
Detoxication, 99–122, 411, 421, *see also* Conjugations; Detoxification
  aldehydes and, 79
  amino acids and, 346
  antioxidants and, 126
  availability for, 125
  of cyanide, 114
  cytochrome P-450, 110
  defined, 69, 483
  enzyme, *see* Enzyme detoxication
  enzyme formation and, 236
  magnesium and, 312
  overload and, 72, 79
  pathways of, 51
  sulfation and, 118
Detoxification, 9, 12, 13, 24, 411, *see also* Conjugations; Detoxication
  in adaptation, 24
  of aldehydes, 243, 247
  alterations of, 47
  of amines, 82, 119
  of ammonia, 386
  asparagine and, 372
  biological, 20
  bipolarity and, 30
  cellular, 60
  challenge to, 92
  copper and, 319
  cysteine and, 375
  damage to, 68
  defined, 69, 483
  diminished, 414
  enzyme, *see* Enzyme detoxification
  enzyme formation and, 236
  esterolytic, 97
  ethanol, 79
  failure of, 33
  individual differences in, 36
  magnesium and, 308
  metabolism required for, 235
  microsomal enzyme, 58
  minerals and, 124
  nonimmune, 2, 47, 48
  normal xenobiotic metabolism and, 69
  nutrients in, 221, 223
  overuse of, 39

of phenols, 120
primary, 54
spreading phenomenon and, 33
sufficient inducible, 58
of sulfites, 120
sulfur and, 343
target organ ability in, 222
of thiols, 120
tryptophan and, 361
varied rates of, 420
vitamin A and, 241
vitamin B and, 251
vitamins and, 124, 231
of xenobiotics, 353, 355
Detoxification enzymes, 56, *see
also* specific types
DHA, *see* Dehydroascorbic acid
Diabetes, 245
Diacylglycerol, 297
Diacylglyceryl-3-
phosphorylcholine, 419
Diagnosis of chemical sensitivity,
1, 3, 17, 47
Diamine oxidase, 83
Diamines, 83, *see also* specific
types
Dibenzodioxins, 177, *see also*
specific types
Dichlorobenzene, 198
1,2-Dichloroethylene, 88
Dichloromethane, 198
3,4-Dichloronitro benzene, 108
2,4-Dichloro-6-phenylphenol
oxyethylamine (DBPA), 72
Dichromate, 161
Dicoumarol, 62, 415
Dieldrin, 86
Diethylamine, 83
Diethyldisulfides, 95
Diethylmaleate, 108
Diethylstilbestrol, 68
Digestion, 51, 224, 346, 372
Digestive enzymes, 302, *see also*
specific types
Digitalis, 309

Digitonin, 273
Dihalobromines, 92
Dihydrobiopterin reductase, 358
7,8-Dihydrodiol, 88
Dihydrodiol epoxides, 86
Dihydrodiol oxides, 121
Dihydrodiols, 121, *see also*
specific types
Dihydrolipoate, 118
1,4-Dihydropyridine, 230
1,25-Dihydroxycholecalciferol,
280
Diisopropylfluophosphate, 66
Diketogulonic acid, 274
2,6-Dimethoxy-p-benzoquinone,
276
Dimethylamine, 82
Dimethylepinephrine, 273
Dinitrophenol, 62
2-Dinitropropane, 93
Diol epoxide, 86
Dioxins, 419, *see also* specific
types
Dioxygenases, 91, 243, 244
Dismutation, 422
1,2-Disubstituted ethanes, 70
Disulfides, 90, 91, 95, 108, 109,
111, 420
aliphatic, 108
aromatic, 108
reduction of, 95
sulfur compound interconversion
and, 343
Disulfones, 90, *see also* specific
types
Dithiothreitol, 66, 91
Diuretics, 94, 309, *see also*
specific types
DNA, 61, 86, 177, 420
beta-alanine and, 367
alteration of, 160
antibody reactions to, 193
catabolism of, 367
iron and, 332
magnesium and, 308

methylation of, 353
radiation damage to, 66, 124
repair of, 412
replication of, 262
synthesis of, 61, 62, 339
vanadium and, 339
vitamin B_{12} and, 262
vitamin C and, 275
DNA polymerase, 317
DNA synthesis inhibitors, 60, *see also* specific types
DNE, 356
L-Dopa, 120
Dopamine, 120, 309, 328, 356, 360, 417, 421
Dopamine hydroxylase, 412
Drug resistance disorders, 415
Drug sensitivity, 7, 412, 414–415, 483, *see also* specific drugs
Dust, 50, 182
Dust mites, 169, 183

Eating patterns, 232, *see also* Nutritional status
ECU, *see* Environmental Control Unit
EDTA, 66, 235, 258, 328, 337, 338
Elastase, 56
Electrolytes, 379–380
Electromagnetic radiation, 19
Electromagnetic sensitivity, 483
Electron transfer enzymes, 63, *see also* specific types
Electron transport, 62, 224
Electrophiles, 68, 91
Electrophilic capacity, 66
Electrophilic carbon, 109
Electrophilicity, 88
Electrophils, 69, 70, 483
Endocytosis, 400
Endoplasmic membrane, 229
Endoplasmic reticulum, 63, 118, 121

arteriosclerosis and, 403
calcium pump in, 303
destruction of, 202
rough, 71
xenobiotics and, 125
End-organ disease, 33, 39, 483
End-organ failure, 26–28, 32, 83, 229, *see also* Fixed-named disease
Endosulfan I, 279
Endosulfan II, 279
Endosulfan sulfate, 279
Endotoxins, 187
Endpoint, 484
Endrin, 279
End-stage disease, 4, 10
Energy compartmentalization, 65
Energy flux, 230
Energy generators, 229
Energy metabolism, 62, 223–229
Energy production, 32, 65, 229–231
Energy regulators, 25, 221, *see also* specific types
Energy transfer, 65, 78
Enkephalin, 355
Enols, 122
Environmental Control Unit (ECU), 483
Environmentally triggered disease, 2–4, 8, 39, 54, 484, *see also* specific types
Enzymatic conjugation, 84–85
Enzymatic hydrolysis, 96
Enzymatic hydroxylation, 269
Enzyme detoxication, 27, 418–425, *see also* Enzyme detoxification
  cytochrome P-450-mixed-functin oxidase and, 418–420
  monoamine oxidases and, 421–422
  partitioning substrates and, 71
  superoxide dismutase and, 422–423

Enzyme detoxification, 9, 21, 25, 26, 48, 56, 418, *see also* Enzyme detoxication
  bipolarity and, 30
  depleted, 39
  malfunctioning, 39
  methionine and, 355
  microsomal, 58
  triggering of, 56–57
  xenobiotic effects on, 37
Enzymes, 25, 36, 48, 73, 235, *see also* specific types
  acetylation, 228
  acid activating, 102
  adenosine triphosphate-dependent, 110
  alanine-forming, 367
  aldehyde oxidizing, 420–421
  amino acids in formation of, 236
  antipollutant, 50
  biosynthesiziing, 277
  of biotransformation, 410–414, *see also* specific types
  in blood, 222
  calcium in activation of, 301
  catalytic, 65
  in chemical sensitivity, 238
  cofactors and, 236
  for conjugations, 410
  copper and, 319
  cytosolic, 110
  defined, 484
  for degradations, 410
  depletion of, 39, 231
  detoxification, 56
  digestive, 302
  electron transfer, 63
  in end-organs, 222
  genetic disorders and induction of, 416
  hydrolytic, 62
  induction of, 72, 416
  injury to, 65–66
  membrane transport, 64
  metabolizing, 74, 346
  metallo-, 91, 308
  microsomal, 60, 63, 78–79, 82, 83, 93
    desulfuration and, 88
  mitochondrial, 62, 112
  mixed, 51
  molecular damage and, 65–66
  noncatalytic, 70
  for oxidation, 410
  plasma membrane transport, 64
  pools of, 221, 223
  for reductions, 410
  response of, 32
  transport, 182
  vitamin effects on, 232
Eosinophilic infiltration, 60
Eosinophilic responses, 61
Epiandrosterone, 104
Epinephrine, 183, 258, 309, 353, 356, 360, 380
Epithelial cells, 171
Epoxidations, 72, 85–88, 418
Epoxide hydrase, 121
Epoxide hydrolase, 85, 86
Epoxides, 74, 85, 86, 88, 121, 279, 418, *see also* specific types
Epstein-Barr virus, 187
ER, *see* Endoplasmic reticulum
Essential amino acids, 348–349, *see also* specific types
Essential fatty acids, 403–410, *see also* specific types
Esterases, 96, 98
Esterolytic detoxification, 97
Estradiol, 107
Estrogens, 3, 68
  calcium and, 306
  immune suppression from, 177
  sensitivity to, 183
  vitamin A and, 241
  vitamin B_6 and, 258
  vitamin C and, 275
Estrone, 104

Estrone acetate, 98
Ethanes, 70, 229, 425, *see also*
    specific types
Ethanol, 18, 67, 79, 164, 194, 413,
    420, *see also* Alcohols
Ethanolamine, 67, 291, 389
Ethidium bromide, 66
Ethionine, 119–120, 120
7-Ethoxycourmarin O-deethylation,
    86
Ethylamine, 100, 254
Ethylbenzene, 84, 198
Ethylene glycol, 79
Ethylenes, 70, 88, 229, *see also*
    specific types
Ethylmercaptan, 95
Ethylmercaptin, 120
Expoxide hydratase, 85
Extrinsic, 484

FAD, *see* Flavin adenine dinucleo-
    tide
Fasting, 484
Fats, 25, 36, 361, 393–403, *see
    also* specific types
  biotin and, 267
  brown, 224, 394, 395–396
  catabolism of, 224
  depletion of, 236
  dietary calories in, 395
  ingested, 403
  magnesium and, 312
  metabolism of, 254, 261, 267
  peripheral, 397
  saturated, 408
  transport of, 400, 401
  vitamin B_2 and, 248
  vitamin B_5 and, 254
  vitamin B_{12} and, 261
  white, 394, 396–397
Fatty acids, 33, 396–397, *see also*
    specific types
  availability of, 97

biotin and, 267
chemical changes in, 418
deficiencies of, 239, 406
essential, 403–410
excess, 239
intermediate chain-length, 102
long-chain, 63
manganese and synthesis of, 328
metabolism of, 395
oxidation of, 253, 254
in phospholipids, 393
polyunsaturated, 122, 286, 399
synthesis of, 253, 254, 267, 328
taurine and, 381
vitamin B_2 and, 250
vitamin B_3 and, 253
vitamin B_5 and, 254
vitamin E and, 286
Fatty biopsies, 49
Feedback mechanisms, 34
Fenitrotion, 241
Ferritin, 319, 331, 332
Fibroblasts, 174
Fixed-named disease, 4, 8, 10, 25,
    27, *see also* End-organ
    failure
  bipolarity and, 32, 33
  defined, 484
  individual susceptibility to, 40
  xenobiotic metabolism and, 72
Flavanoids, 94, *see also* specific
    types
Flavin, 420
Flavin adenine dinucleotide
    (FAD), 82, 250, 335, 411,
    419, 421
  functions of, 246, 248
  taurine and, 381
Flavin adenine dinucleotide
    (FAD)-cysteine compounds,
    421–422
Flavin-containing monooxygenase,
    420, 485
Flavin dehydrogenase, 248, 250

Flavin-requiring aldehyde oxidases, 78
Flavonoids, 295, 297, 298, *see also* specific types
Flavoproteins, 230, 248, 381, 421, *see also* specific types
Flavorings, 97
Fluor-2,4-dinitrophenyl, 98
Fluorexene, 92
Fluoride, 3334
Fluorine, 278
Folacin, *see* Folic acid
Folate, 164, 231, 238
Folic acid, 259, 263–265, 356, 377, *see also* Vitamin B$_{12}$
  deficiencies of, 236, 265, 360
  dietary sources of, 265
  homocysteine and, 377
  sources of, 265
Folic acid antagonists, 156, *see also* specific types
Food additives, 484
Food flavorings, 82
Food intolerance, 51, 223
Food sensitivities, 22, 50, 99, 169, 182, 183, 484
Formaldehyde, 3, 7, 9, 18, 19, 32, 79
  defined, 484
  hemolysis and, 162
  immune system response to, 194
  N-oxidation and, 85
  sensitivity to, 78, 79
  spreading phenomenon and, 33
  taurine and, 381
  T-cells and, 155
  vitamin A and, 243
  vitamin B and, 246, 247
  vitamin C and, 270
Formate, 79, 265
Formininoglutamate, 265
Formylhalide, 92
Fossil fuels, 104
Fragile, defined, 484
Free radicals, 48, 53, 58, 64, 67, 82, 422
  creation of, 122–124

organochlorines and, 190
peroxidation of, 285
stimulation of, 67
vitamin A and, 241
vitamin C and, 270, 275
vitamin E and, 285, 286
Fructose, 79, 301
Fungi, 59, 187
Furfurylamine, 99

GABA, *see* Gamma-aminobutyric acid
Galactose, 298
Gamma-aminobutyric acid (GABA), 259, 366, 380
Gamma-aminobutyric acid (GABA) receptors, 229, 317
Gastrointestinal system, 52, 53
Genetic defects, 100, 416, 417, *see also* specific types
Genetic makeup, 2, 8, 412
Genetic susceptibility, 36–38, 40, 222
Genitourinary system, 52, 53
Glicidaldehyde, 79
Glucaric acid, 50
Glucocorticoids, 63, 322, 367, 417, *see also* specific types
Gluconates, 126
Gluconation, 121–122, 418, 484
Gluconeogenesis, 229, 367
Glucose, 25, 32, 202, 298
  alanine and, 367
  biotin and, 267
  magnesium deficiency and, 309
  metabolism of, 375
  oxidation of, 230
  vitamin B$_2$ and, 250
Glucose-6-phosphatase, 338
Glucose-6-phosphate, 295, 339
Glucose-6-phosphate dehydrogenase, 25, 56, 412
Glucose tolerance factor (GTF), 322
Glucuronation, 106
Glucuronic acid, 72, 78, 121–122

Glucuronic reductase, 94
Glucuronidation, 121–122
D-Glucuronide, 122
Glucuronides, 122
D-Glucuronolactone-delta-
    hydrolase, 274
Glucuronosyltransferase, 121
Glutamamylamino acid, 110
Glutamate, 110
Glutamic acid, 261, 366, 367, 372,
    385–386
  aspartic acid and, 374
  dietary sources of, 385
  glutamine and, 386
  glycine and, 387
  histidine and, 391
  proline and, 391
  sources of, 385
  taurine and, 380
Glutaminase, 372, 387
Glutamine, 101, 103, 319, 366,
    372, 386–387, 388
  ammonia from, 111
  aspartic acid and, 374
  glutamic acid and, 386
  threonine and, 361
Glutamine synthetase, 425
Glutamyl, 109, 110
Glutamyl cyclotransferase, 110
Glutamyl transpeptidase, 109
Glutaraldehyde, 79
Glutathione, 50, 104, 109, 111
  conjugation of, 101, 108–112
  copper and, 319
  cysteine and, 375
  defined, 484
  disorders of, 417
  glycine and, 387
  methionine and, 355
  reduced, *see* Reduced glutathione
  sulfur and, 342, 343
  taurine and, 381
  vitamin B$_1$ and, 248
  vitamin B$_2$ and, 250
  vitamin C and, 274
  vitamin E and, 283–285
Glutathione disulfide, 250, 324

Glutathione-insulin
    transhydrogenase, 375
Glutathione peroxidase, 25, 48, 50,
    56, 66, 68, 91, 423–425
  cysteine and, 375
  defined, 484
  free radical protection and, 124
  functions of, 423
  glutathione as cofactor for, 111
  lipid peroxidation and, 423
  selenium and, 324
  vitamin E and, 283
Glutathione reductase, 168, 250,
    319, 381
Glutathione synthetase, 248
Glutathione thiotransferases, 108
Glutathione transferases, 91, 104,
    109
Glutethamide, 94
Glyceraldehyde, 347
Glycerol phosphate, 397
Glycine, 82, 101, 103, 229, 366,
    367, 387–389
  configuration of, 347
  formation of, 361
  glutamine and, 387
  glutathione regeneration and, 110
  hydrolysis and loss of, 109
  picrotoxin and, 230
  taurine and, 380
  threonine and, 361
Glycocholic acid, 387
Glycogen, 224, 302, 303, 304
Glycogen synthase, 304
Glycolysis, 91, 224, 389
Glycoproteins, 366, *see also*
    specific types
  asparagine and, 372
  defined, 484
  glycine and, 389
  sulfated, 105
  synthesis of, 243
  threonine and, 361
  vitamin A and, 243
Glycosides, 101, 112, 113, *see*
    also specific types
Glycylglycine dipeptidase, 336

Glyoxylate, 387
Gold, 156, 164, 193
Granulocytes, 156
Grasses, 169
Growth hormone, 322, 367
GSH, *see* Glutathione
GTP, *see* Glucose tolerance factor;
    Guanosine triphosphate
Guanosine triphosphate (GTP),
    364
Guanylate cyclase, 70
Gut, 173

Half-lives, 13–14
Halides, 126, *see also* specific
    types
Haloforms, 92
Halogen, 108, 358
Halogenated hydrocarbons, 67, 92,
    177, *see also* specific types
Halogenated pesticides, 92, *see
    also* specific types
Halothane, 96, 413
Haptens, 49, 177, 345
HCB, *see* Hexachlorobenzene
Heart electrolytes, 379–380
Heavy metals, 7, 66, 67, 91, 110,
    see also specific types
  conversion of to sulfides, 116–
    118
Hematological system, 54, 156–
    169
  bone marrow in, 156–160
  leukocytes in, 169
  platelets in, 164–169
  red blood cells in, 161–163
  thrombocytes in, 164–169
Hematopoietic cells, 174
Heme, 319, 411, 416, 423, 425
Heme proteins, 67, 161
Hemoglobin, 259, 417
Hemolysis, 60, 161, 162, 277, 417
Hemoprotein, 387, 418
Hemosiderin, 331
Heparin, 105, 126, 329

Hepatocytes, 125, 288
Heptachlor, 86, 88, 102, 279
Heptachlor epoxide, 74, 86, 279
Heptane, 82
Herbicides, 79, 97, 98, 241, 484,
    see also specific types
Heterocyclic compounds, 98, *see
    also* specific types
Heterocyclic sulfur, 90
Hexabarbitol, 393
Hexabendine, 98
Hexachlorobenzene (HCB), 185,
    192, 279
Hexachloroethane, 92
Hexahydrocyclohexane, 295
Hexane, 82
Hippurate, 126
Hippuric acid, 101
Histadyl residues, 66
Histamine, 83, 99, 120, 169, 297,
    389
Histidine, 261, 301, 365, 366, 389,
    391, 392
  copper and, 319
  threonine and, 361
  vitamin C and, 276
L-Histidine, 328
Histidinuria, 391
Homeostasis, 19, 21
Homocysteine, 259, 261, 356, 376,
    377, 389
Homocysteine acid, 261
Homocysteinuria, 259
Homocystine, 356
Homovanillic acid, 94
Hormone receptor sites, 64, 182
Hormones, 171, 297, 346, 366,
    367, *see also* specific types
  abnormal levels of, 63
  adrenocorticotropin, 271, 273
  calcium and, 302
  deregulation of, 34
  growth, 322, 367
  lutenizing, 183
  parathyroid, 322
  phosphorus and, 322

regulation of, 397
secretion of, 302
sensitivity to, 183
steroid, *see* Steroids
thyroid, 224, 322, 366
thyroid-stimulating, 244
vitamin A and, 243
vitamin C and, 275
Host variables, *see* Biochemical
     individuality of response
5-HT (5-hydroxytryptamine), *see*
     Serotonin
Hyaluronic acid, 329
Hydantins, 94
Hydantoins, 98, 156, 421, *see also*
     specific types
Hydralases, 96
Hydralazine, 100, 235
Hydrazides, 96, 98, *see also*
     specific types
Hydrazines, 67, 82, 100, 235, 254,
     see also specific types
Hydrazones, 70, *see also* specific
     types
Hydrocarbons, 25, 39, 56, 59, *see*
     also specific types
  aliphatic, 85, 229, 481
  aromatic, *see* Aromatic hydrocar-
     bons
  chlorinated, 39, 483, *see also*
     Chlorinated compounds
  defined, 484
  halogenated, 67, 92, 177
  polyaromatic, 86, 104, 177, 410
  polycyclic, 86, 121, 416, 486
  potassium channels and, 230
  sodium channels and, 230
Hydrochloric acid, 83
Hydrochlorites, 383
Hydrocortisone, 107
Hydrogen, 348
Hydrogen cyanide, 163
Hydrogen peroxide, 92, 122, 202,
     276, 422, 423, 425
Hydrogen sulfide, 67, 161, 163
Hydrolases, 85, 86, 410, 484

Hydrolysis, 82, 96, 97, 98, 106,
     108
  enzymatic, 96
  of inositol, 295
  of triglycerides, 224
Hydrolytic enzymes, 62, *see also*
     specific types
Hydroperoxides, 92, 283
Hydrophilic compounds, 82, 411,
     485, *see also* specific types
Hydrophobic compounds, 95, 109,
     118, *see also* specific types
Hydroproline, 389
Hydroquinone, 106
N-Hydroxy-p-acetophenone, 161
Hydroxy-2-acetylaminofluorene,
     107
N-Hydroxy-2-acetylaminofluorene,
     106
Hydroxyamines, 122, *see also*
     specific types
N-Hydroxy-p-amino toluene, 161
16-alpha-Hydroxyestrone, 202
5-Hydroxyindolacetic acid, 94
3-Hydroxykynurenine, 367
Hydroxylamines, 85, 104, 108,
     161, *see also* specific types
Hydroxylases, 96, *see also* specific
     types
Hydroxylation, 82–83, 414
  aliphatic, 72
  aromatic, 72
  enzymatic, 269
  oxidative, 421
  of ring system, 84
  vitamin C and, 275
Hydroxyl radicals, 270
Hydroxylysine, 361, 366, 372,
     389, 391
M-Hydroxyphenylacetic acid, 96
4-Hydroxyphenylacetic acid, 94
Hydroxyproline, 361, 366, 372,
     391
3-Hydroxyproline, 391
4-Hydroxyproline, 269, 391
3-Hydroxy-4-pyrone, 120

8-Hydroxyquinoline, 120
Hydroxysteroids, 104, 107, *see also* specific types
Hydroxysteroid sulfotransferase, 107
5-Hydroxytryptamine, *see* Serotonin
Hygiene products, 7, 175
Hypermetabolism, 228
Hypersensitivity, 8, 49, 182–187, 364, 365, *see also* Allergic responses; Immune system response
defined, 485
delayed-type, 170, 192, 202
to metals, 193
type I, 182–184
type II, 184
type III, 184
type IV (cell-mediated immunity), 177, 185–187, 192, 193, 202
Hypochlorhydria, 391
Hypochlorites, 117, 120, 257, 379, *see also* specific types
Hypothalamic system, 54
Hypoxanthine, 421

Idioacetate, 423
Imipramine, 421
Immune complex, 8, 184
Immune system response, 2, 8, 24, 25, 27, 124, 177–190
abnormal, 193–194
altered, 202
bipolarity and, 30
cell-mediated, 177, 185–187, 192, 193, 202
depleted, 39
enhancement of, 21
fatty acids and, 406
to formaldehyde, 194
hypersensitivity as, *see* Hypersensitivity

individual differences in, 36
iron and, 332
malfunctioning, 39
to metals, 177, 185
nutrients and, 235
to pesticides, 177, 185, 194
to phenols, 185, 194
to polychlorinated biphenyls, 177, 193
stimulation of, 186, 245
suppression of, 177, 186, 187, 192, 193
vitamin A and, 243
zinc and, 318
Immune triggers, 56–57
Immunizations, 9
Immunodeficiency diseases, 171, 172, 173, *see also* specific types
Immunoglobulin E, 9, 182, 184
Immunoglobulin G, 8, 184, 191, 192
Immunoglobulin M, 191
Immunoglobulin M antibody, 193
Immunoglobulins, 8, 346, *see also* specific types
Immunosuppressors, 177, *see also* specific types
Incitant, 485
Individuality of response, *see* Biochemical individuality of response
Individual susceptibility, 8, 412, 485
Indoles, 364
Indoxylsulfate, 364
Inflammatory cells, 53
Informational (regulatory) proteins, 54, 67, *see also* specific types
Inhalants, 50, *see also* specific types
Inner-sphere mechanism, 276
Inositol, 295–298
Inositol hexaphosphate, 297

Inositol-1-phosphate, 295
Inositol phospholipids, 295
Inosone 2,1-phosphate, 295
Insecticides, 66, 67, 86, 97, 177,
    see also Pesticides; specific
    types
  organophosphate, see Organo-
    phosphates
Insulin, 90, 317, 323, 355, 363,
    366, 375
Intracellular vesicles, 62
Intradermal injection therapy, 64
Ionizing radiation, 156
Iron, 203, 236, 319, 329–334, 408,
    421
  absorption of, 331
  in children, 233
  cytochrome oxidase system and,
    419–420
  deficiencies of, 254, 332–333
  dietary sources of, 334
  in ethnic minorities, 233
  excess, 245
  in heme complex, 411
  in low-income groups, 233
  metabolism of, 333
  overload of, 333
  phenylalanine and, 358
  replenishment of, 74
  sources of, 334
  superoxide dismutase and, 317,
    333
  in teenagers, 233
  toxicity of, 333–334
  tryptophan and, 361, 364
  vanadium and, 338
  vitamin B_2 and, 250
  vitamin C and, 274
  in women, 233
  wound healing and, 236
Iron porphyrin heme, 423
Iron-sulfur proteins, 117
Isethionic acid, 384
Isoenzymes, 98
Isoleucine, 346, 347, 349–351, 386

Isomerases, 410, 485
Isoniazid, 100, 235, 412
Isoproterenol, 120

Juglone, 423

Keto acid, 256
Ketoglucaric acids, 70
5-Ketoglucose-6-phosphate, 295
alpha-Ketoglutarate, 254
alpha-Ketoglutaric, 114
alpha-Ketoglutaric acid, 83, 256,
    367, 386, 387
alpha-Ketoglutaric acid dehydroge-
    nases, 66
Ketone reductases, 421, 485
Ketones, 74, 78, 94–95, 421, see
    also specific types
Killer cells, 202
Kinases, 245, 258, 304, 308, 317,
    see also specific types
Kinetic alteration, 66
Kinins, 34, 56
Krebs cycle, 251, 355, 372, 374,
    375, 389
Kyneuric acid, 258
Kynureninase, 259

Lactaldehyde reductase, 94
Lacteals, 400
Lactic dehydrogenase, 49, 65
Lactose, 306
LCAT, see Lecithin cholesterol
    acyltransferase
Lead, 66, 91, 177, 193, 285
Lecithin, 294
Lecithin cholesterol acyltransferase
    (LCAT), 401, 402
Lectins, 60
Leptophos, 192
Leucine, 346, 349–351, 365, 366,
    386

Leucine aminopeptidase, 96
Leukemia, 276
Leukemogen, 169
Leukemogenic agents, 86, *see also*
    specific types
Leukocytes, 155, 169, 223
  iron and, 332
  lithium and, 345
  polymorphonuclear, 246, 345,
    379, 382
  vitamin A and, 246
  vitamin C and, 270
Leukotrienes, 399, 406
Lidocaine, 98
Ligands, 66, 109, 229, 300, 328,
    485, *see also* specific types
Ligases, 102, 410, 485, *see also*
    specific types
Limbic system, 54
Lindane, 156, 192, 241
Lineolate, 395
Linocate, 395
Linoleic acid, 289, 393, 407, 409
Linolenic acid, 408
Lipamide dehydrogenase, 66
Lipases, 96, 102, 397, *see also*
    specific types
Lipid membranes, 13, 58, 67, 71,
    106, 229, 395
Lipid peroxidase, 56
Lipid peroxidation, 67, 79, 111,
    190, 406, 420
  glutathione peroxidase and, 423
  reduction in, 425
  in selenium, 425
  vitamin C and, 271, 277–279
  in vitamin E, 425
  vitamin E and, 283, 285
  zinc and, 315
Lipid peroxides, 53, 399
Lipids, 67–68, 105, *see also*
    specific types
  assimilation of, 101
  bilayer of, 190
  blood, 68, 222

  in chemical sensitivity, 238
  chromium and, 322
  depletion of, 39, 231
  dietary, 408
  in end-organs, 222
  iron and, 333
  metabolism of, 32
  nonpolar, 400
  peroxidation of, *see* Lipid
      peroxidation
  peroxides and, 425
  polyunsaturated, 420
  pools of, 223
  structural layers of, 53
  vitamin E and, 283, 285
  zinc and, 315
Lipophilic compounds, 13, 53, 63,
    102, 109–110, 395, *see also*
    specific types
  biotransformation and, 125
  membrane damage from, 229
Lipophilic esters, 98
Lipoproteins, 291, 294, 399–403,
    408, 410, *see also* specific
    types
  defined, 400
  degradation of, 400
  metabolism of, 400
  synthesis of, 400
  triglyceride-rich, 400
Liposes, 397, *see also* specific
    types
Lipoxygeanse system, 56
Lithium, 297, 345
Lithocolic acid, 104
Long-chain fatty acids, 63
Low-level exposure, 10, 11
Lutenizing hormone, 183
Lyases, 381, 410, 485, *see also*
    specific types
Lymph, 170
Lymphatics, 156, 173–175, 191
Lymphocytes, 156, 175–178, 192,
    419, *see also* specific types
  B-, *see* B-cells

cytotoxic, 171
null, 174
phenyl hydrazine-affected, 174
reticuloendothelial system and, 170
selenium and, 324
T-, *see* T-cells
total counts of, 199, 202
vitamin C and, 276
Lymphocytic infiltration, 60
Lymphocytic responses, 61
Lymphokines, 170, 185, 187
Lysine, 261, 346, 348, 351–353
Lysosomes, 62–63, 333, 406
Lysozymes, 56, 400

Macrophages, 53, 170, 173, 174, 277, 419
Macrosamine, 79
Magnesium, 53, 120, 299, 307–315, 408, 416
asparagine and, 372
assimilation of, 360
calcium transport and, 312
deficiencies of, 3, 120, 223, 239, 307, 308, 309–312
absorption disorders and, 309
diagnosis of, 312–315
reabsorption disorders and, 309
symptoms of, 309
vasospastic phenomenon and, 311
dietary sources of, 312
excess, 239, 309
glutamine and, 387
immune suppression from, 177
intake of, 314
manganese and, 329
metabolism of, 62
methionine and, 356
serum, 314
sources of, 312
supplementation of, 120
in teenagers, 233

tryptophan and, 361
tyrosine and, 360
vitamin B_1 and, 247
vitamin B_6 and, 258
in women, 233
Magnesium trisilicate, 340
Major histocompatibility complex (MHC) antigens, 171
Maladaptation, 25, 26, 27, 485
Malanocytes, 360
Malate dehydrogenase, 251
Malathion, 97, 192, 279
Malignancy, 59, 68, 175, 276, 285, *see also* Carcinogenesis; Carcinogens; Tumors; specific types
Malnutrition, 229, 323
Manganese, 203, 236, 300, 327–329, 425
cobalt and, 336
deficiencies of, 3, 329
dietary sources of, 328, 329
excess, 231
functions of, 328
magnesium and, 329
sources of, 328, 329
in superoxide dismutase, 317, 422
vitamin B_2 and, 250
vitamin C and, 274
Manifestations of chemical sensitivity, 11–12
Mannitol, 276
MAO, *see* Monoamine oxidases
Masking phenomenon, 17–18, 21, 23–26, 33, *see also* Adaptation
cell damage and, 58
defined, 24, 481, 485
stages of, 25–26, 27
Mast cells, 184, 191
Megakaryocytes, 164
Megamitochondria, 62
Melanin, 360
Melatonin, 120, 328

Membrane-bound structures, 63, *see also* specific types
Membrane damage, 229–231
Membrane potentials, 229–231
Membrane transport enzymes, 64
Menadione, 423
Mercaptans, 90, 91, 92, 95, 111
Mercaptides, 66
2-Mercaptoethanol, 91
3-Mercaptopyruvate, 115
3-Mercaptopyruvate sulfurtransferase, 114–116
Mercapturate, 50, 126
Mercapturic acids, 104, 108, 109, 111
Mercury, 67, 91, 177, 193, 324
Metabolic derangements, 48
Metabolic energy, 230
Metabolic potentiation variations, 416–418
Metabolic transformation, 51
Metabolism disorders, 416, *see also* specific types
Metabolizing enzymes, 74, 346, *see also* specific types
Metalloenzymes, 91, 308
Metalloflavoproteins, 421
Metalloproteins, 319, 338
Metals, 300, *see also* specific types
  enzyme injury from, 66
  heavy, *see* Heavy metals
  hypersensitivity to, 193
  immune system response to, 177, 185
  vitamin B_2 and, 250
Methanes, 92
Methanol, 277
Methemoglobin cyanide, 114
Methemoglobinemia, 417
Methemoglobin sulfide, 114
Methionation, 120
Methionine, 50, 346, 353–356, 365, 377
  alkylations by, 119–120
  arteriosclerosis and, 403
  choline and, 291–294
  cobalt and, 337
  cysteine and, 375
  dietary sources of, 356
  formation of, 376
  glycine and, 389
  homocysteine and, 377
  metabolism of, 120, 353, 381, 403
  molybdenum and, 336
  sources of, 356
  sulfur and, 342
  taurine and, 380, 381, 384
  vitamin B_6 and, 258
  vitamin B_{12} and, 261
Methionine sulfoxide, 348
Methotrexate, 156
Methylamine, 82, 100
Methylane-bis-o-chloro analine, 99
Methylanthrene, 125
Methylases, 119
Methylates, 126
Methylation, 120, 122, 258, 353, 418, 486
Methylbutyrate, 98
3-Methylcholanthrene, 72, 86, 395
Methylcobalamin, 259
Alpha-Methyldopa, 120
Methylene blue, 90, 423
Methylene-tetrahydrofolate, 389
Methylfolate, 261
Methylglycol, 79
Methylhistidine, 385
Methyliodide, 108
Methylmercury, 193
Methylmercury chloride, 184
Methylphenylcarbinol, 84
O-Methyltransferase, 120
Methyltransferases, 119, 120
Metrione, 96
Metyrapone, 72, 96
Mevalonate reductase, 94

MFO, *see* Mixed-function oxidases
Mibex, 185
Microsomal energy transfer, 78
Microsomal enzymatic oxidation, 78–79
Microsomal enzyme detoxification, 58
Microsomal enzymes, 60, 63, 78–79, 82, 83, 93, *see also* specific types
  desulfuration and, 88
Microsomal monooxygenases, 82, 485
Microsomal oxidase, 72
Microsomes, 122, 418
Migratory inhibitory factor, 185, 186
Mineral pumps, 229, *see also* specific types
Minerals, 36, 58, 204, 298–301, *see also* specific types
  absorption of, 301
  biochemical individuality of response to, 300
  in blood, 222
  in chemical sensitivity, 238–239
  in children, 233
  deficiencies of, 239, 406, 422
  depletion of, 39, 231, 235
  detoxification and, 124
  in elderly, 233
  in end-organs, 222
  in ethnic minorities, 233
  excess, 231, 239
  intracellular deficiencies of, 422
  intracellular levels of, 50
  in ligands, 229
  in low-income groups, 233
  metabolism of, 25, 62
  pools of, 223
  reabsorption of, 236
  in teenagers, 233
  vitamin C and, 275–276
  in women, 233

Mirex, 192
Mitochondria, 102, 124, 221, 421
  calcium and, 302
  iron and, 333
  superoxide dismutase in, 422
  vitamin B_2 and, 248
  vitamin C and, 272
Mitochondrial damage, 62
Mitochondrial enzymes, 62, 112, *see also* specific types
Mitochondrial membranes, 55
Mitochondrial menbranes, 229
Mitosis, 58, 62, 91
Mixed-function oxidases (MFO), 25, 82, 88, 93, 121
  cytochrome P-450-, 101, 418–420, 483
  fats and, 393
Mixed-function oxidation, 202
Molds, 7, 50, 60, 155, 169, 182
  immune suppression from, 177
  sensitivity to, 183, 187
Molecular damage, 65–68
Molecular oxygen, 72
Molybdenum, 334–336, 421
  copper and, 341
  deficiencies of, 3, 336
  vitamin B_2 and, 250
Monoamine oxidase-catalyzed oxidation, 7979
Monoamine oxidase reactions, 250
Monoamine oxidases (MAO), 25, 82, 83, 417, 421–422
  defined, 485
  degradation by, 422
  inhibitors of, 83, 421
Monoclonal antibodies, 194
Monocyclic terpenes, 95
Monocytes, 173, 174, 419
Monoglycerol lipase, 96
Monooxygenases, 86, 91, *see also* specific types
  cytochrome P-450-dependent, 56, 415

flavin-containing, 420
microsomal, 82, 485
phenylalanine, 358
Monophenols, 95, *see also* specific
    types
Morphine, 355
Mucopolysaccharides, 272, 329,
    343
Mucosal barrier, 53
Mucous membranes, 52
Musculoskeletal system damage,
    54
Mushroom toxin, 62
Mustard gas, 9
Mustards, 156
Mutagenesis, 85, 121, 417, 420,
    see also Mutagens
Mutagens, 79, 86, 88, 403, *see
    also* Mutagenesis; specific
    types
Mutiphenols, 358
Myeloperoxidase, 124, 381
Myoblasts, 158
Myoinositol, 295

Naphthalene, 108, 417
Naphthalol, 104
1-Naphthylamine, 121
2-Naphthylamine, 121
Natural gas, 485
Necrotic cells, 58
Nerve gas, 66
Neurological damage, 54
Neurotransmitters, 94, 297, 346,
    366, 367, *see also* specific
    types
  calcium and, 302, 303
  calmodulin and, 303
  endogenous, 421
  glycine and, 389
  magnesium and, 309
  rise in, 421
  taurine and, 379
  vitamin B_6 and, 259
  vitamin C and, 271

Neurovascular system, 32, 53, 54
Neutralization, 485
Neutrophilic responses, 61
Neutrophils, 185, 382
Niacin, *see* Vitamin B_3
Nickel, 177, 193
Nicotinamide, *see* Vitamin B_3
Nicotine, 126, 270
Nicotinic acid, 361, 364, 411
Nitrates, 93, 114, 161, 245, 335
Nitrenium, 85
Nitric oxide, 70
Nitriles, 98, *see also* specific
    types
Nitrites, 83, 161, *see also* specific
    types
Nitrobenzenes, 70, 161
P-Nitrobenzyl alcohol, 94
Nitrofuran, 417
Nitrogen, 98, 101, 106, 348, 387,
    420
Nitrogen dioxide, 22
Nitrophenols, 95, *see also* specific
    types
4-Nitrophenyl acetate, 98
Nitro reductase, 95
Nitrosamine, 83
Nitrosobenzene, 70
Nitroso compounds, 70, *see also*
    specific types
S-Nitrosocysteine, 70
Nitrous oxides, 7, 20, 67, 155,
    158, 159, 177
Noncatalytic conversion, 485
Noncatalytic enzymes, 70, *see also*
    specific types
Noncatalytic nucleophilic capacity,
    66
Noncatalytic reactions, 68–70
Nonenzymatic biotransformation,
    70
Nonenzymatic oxidation, 78
Nonessential amino acids, 365–
    367, *see also* specific types
Nongenetic susceptibility, 38
Nonimmune system response, 2,

47, 48, *see also* specific
   types
Nonimmune triggering, 56
Nonmicrosomal enzymatic
   oxidation, 78
Noradrenaline, 94, 96, 120, 421
Norepinephrine, 309, 353, 360
Nortriptyline, 38, 413
Nuclear damage, 61–62
Nucleases, 62
Nucleic acids, 65, 66, 84, *see also*
   specific types
  biotin and, 267
  inhibitiion of synthesis of, 187
  modulation of, 119
  phosphorus and, 322
  synthesis of, 315
  zinc and, 315
Nucleophiles, 68, 70, 485
Nucleophilic capacity, 66
Nucleophilic compounds, 70
Nucleophils, 69
Nucleotides, 308, 322, 374
Nutrients, 239, *see also* Nutritional
   status
  in alarm stage, 24
  biochemical individuality of
     response to, 222–223
  in children, 233
  deficiencies of, 3, 70, 113, 221,
     231–237, 408, *see also* under
     specific nutrients
    demographics of, 233
    effects of, 231
    food intake and, 232, 233
  defined, 485
  delivery of, 54
  depletion of, 21, 24, 104, 235
  digestion of, 51
  in elderly, 233
  in ethnic minorities, 233
  excess cellular, 229–231
  hypermetabolism and, 228
  incorporation of, 51
  intake of, 233
  intolerance of, 51

loss of, 67
in low-income groups, 233
low pools of, 58
malabsorption of, 228
metabolism of, 51, 99
pools of, 221, 231–237
poor absorption of, 104
reabsorption of, 229
requirements for, 124
supplementation of, 3
in teenagers, 233
transportation of, 51
transport of, 229
utilization of, 51
in women, 233
Nutritional status, 2, 8, 12, 221,
   see also Nutrients
  adaptation and, 22
  biochemical individuality of
     response and, 38–40
  in utero, 38–39
  at time of exposure, 40

Occupational exposure, 10, 53, 156
Octane, 82
Octapamine, 272
Octopamine, 94
Odorants, 111
Odor sensitivity, 49, 158
Olefins, 86, *see also* specific types
Oligosaccharides, 243, *see also*
   specific types
Ommatin D, 106
Opiates, 355, *see also* specific
   types
Organ damage, 53–57
Organic sulfur oxidation, 88–93
Organochlorines, 59, 85, 111,
   190–192, *see also* Chlori-
   nated compounds; specific
   types
  cysteine and, 375
  defined, 483
  immune suppression from, 177
  membrane damage from, 235

thymus gland and, 173
vitamin C and, 279
Organohalides, 185
Organophosphates, 66, 97, 174,
        414, *see also* specific types
    acetyl cholinesterase and, 230
    calcium depletion and, 306
    choline and, 291
    cholinesterases and, 291
    defiined, 486
    immune suppression from, 177
    phosphorus and, 322
    vitamin A and, 244
    vitamin C and, 279
Organophosphorus, 98, 244
Organotins, 177
Ornithine, 258, 365, 370, 386
Ornithine decarboxylase, 245
Orotic acid, 374
Overload, defined, 486
Oxadoacetate, 251
Oxalacetate transcarboxylase, 337
Oxalate, 79, 272
Oxalic acid, 305
Oxaloacetic acid, 374, 389
Oxazolinedese, 94
Oxidants, 124, *see also* specific
        types
Oxidation, 33, 48, 50, 51, 74–93,
        126, 235, 403
    of alcohols, 78–79
    of aldehydes, 78
    aromatization and, 85
    copper and, 319
    of cysteine, 105–106
    of cystine, 375
    defined, 486
    enzymes for, 410
    epoxidations and, 85–88
    of ethanol, 420
    of fatty acids, 253, 254
    of glucose, 230
    hydroxylation and, 82–83, 84
    of ketones, 78
    microsomal enzymatic, 78–79
    mixed-function, 202

monoamine oxidase-catalyzed, 79
    nonenzymatic, 78
    nonmicrosomal enzymatic, 78
    of organic sulfur, 88–93
    of phenols, 126
    of substrates, 230
    of sulfides, 319
    vitamin B and, 246, 248, 250,
        253, 254
    xenobiotic metabolism and, 72,
        73
N-Oxidation, 85
Oxidative degradation, 103
Oxidative dehalogenation, 92
Oxidative denitrification, 93
Oxidative desulfuration, 118
Oxidative hydroxylation, 421
Oxidative metabolism, 63, 163
Oxidative phosphorylation, 224,
        308, 374
Oxidative reactions, 72, *see also*
        specific types
Oxidizing agents, 78, 90, *see also*
        specific types
Oxidoreductases, 91–92, 410
Oxireductases, 486
5-Oxoproline, 110
Oxygen, 72, 106, 122, 124, 241,
        246, 364
Oxytetracylcine, 62
Ozone, 7, 22, 23, 177, 265, 274,
        285

PABA, *see* Para-aminobenzoic
        acid
Pantothenic acid, *see* Vitamin B_5
PAPS, *see* 3'-Phosphoadenosine-5'-
        phosphosulfate
Para-aminobenzoic acid (PABA),
        99, 263, 265–267
Paraoxon, 88
Paraquat, 422, 423
Parasites, 9, 59, 175
Parathion, 88, 192, 279
Parathyroid hormones, 322

Paraxon, 414
Paroxonase polymorphism, 414
Partitioning substrates, 71
PBBs, *see* Polybrominated biphenyls
PCBs, *see* Polychlorinated biphenyls
Pectin, 305
Penicillamine, 109, 258
Penicillin, 98, 183
Pentachlorophenol, 185
Pentane, 82, 425
Pentazocine, 416
Peptidases, 62
Peptides, 101, 102, 105, 346, *see also* specific types
   conjugation of, 375, 377, 379, 380
   transport of, 112
Performic acid, 90
Peroxidases, 114, 425, *see also* specific types
   glutathione, *see* Glutathione peroxidase
Peroxidation, 277–279, 285, 420
   lipid, *see* Lipid peroxidation
Peroxisomes, 63
Persulfides, 113, 117, 118
Pesticides, 7, 10, 18, 39, 50, 91, 92, *see also* specific types
   chlorinated, 162, 279, 483, *see also* Chlorinated compounds
   dehalogenation and, 418
   halogenated, 92
   immune system response to, 177, 185, 194
   magnesium deficiency and overload of, 309
   metabolism of, 118
   organochlorine, *see* Organochlorines
   organophosphate, *see* Organophosphates
   sulfur-containing, 104, 250, 343
   T-cell depression and, 155
   thrombocytopenia and, 165

   vitamin A and, 241
   vitamin C and, 270, 279
   vitamin requirement increases and, 231
Peyer's patches, 173
PHA, *see* Phytohemoagglutin
Phagocytes, 120, 381
Pharmacogenetic disorders, 414
Pharmacologic response, 28
Phenacetin, 38, 82, 97
Phenanthracin compounds, 108
Phenazine meth-sulfate, 423
Phenelzine, 235
P-Phenetidin, 99
Phenformin, 38
Phenobarbital, 72, 79, 86, 395, 413, 419
Phenol amines, 254
Phenol esters, 98
Phenolic oxygen, 106
Phenols, 3, 18, 19, 59, 74, 86, 92, 98, *see also* specific types
   affinity for, 121
   amino-, 114, 161
   aryl-*s*-, 82
   caloric deficiency and, 228
   conjugations of, 106, 122
   degradation of, 126
   detoxification of, 106, 120
   gluconation and, 121
   hemolysis and, 161
   immune system response to, 185, 194
   inhibition of, 108
   metabolism of, 108, 121
   oxidation of, 126
   reduction of, 95, 126
   solubility of, 125
   spreading phenomenon and, 33
   sulfation of, 107
   sulfonation and, 104
   taurine and, 384
   vitamin B_5 and, 254
Phenothiazine, 94
Phenylacetate, 96
Phenylacetone, 83

Phenylakane, 83
Phenylalanine, 84–85, 330, 346, 347, 356–358
Phenylalanine monooxygenase, 358
Phenylbutazone, 156, 413
Phenyl esters, 70
Phenylglucuronide, 74
Phenylhydrazine, 174
Phenylhydroxylamine, 70
Phenylketonuria (PKU), 358, 364
Phenylsulfate, 118
Phenytoin, 274, 415
Phorbol diesters, 177
Phosgene, 92
Phosphatases, 62, *see also* specific types
Phosphates, 126, 309, *see also* specific types
  fats and, 396
  inorganic, 396
  organo-, *see* Organophosphates
  phosphorus and, 322
  solubility of, 125
  vitamin D and, 322
Phosphatides, 295, *see also* specific types
Phosphatidylcholine, 82, 291, 294, 308, 393
Phosphatidylinositol, 297
3'-Phosphoadenosine-5'-phosphosulfate (PAPS), 104–108
Phosphoethanolamine, 308, 389
Phosphoglucomutase, 338
Phosphoglycerate, 389
3-Phosphohydroxypryruvate, 389
Phosphoinositide, 345
Phospholipase A, 304
Phospholipids, 67, 121, 399, 400, *see also* specific types
  choline and, 294
  fats and, 395
  fatty acids and, 393
  inositol, 295

phosphorus and, 322
  synthesis of, 254
  vitamin B$_5$ and, 254
Phosphopyroxyllysine, 258
Phosphorus, 221, 228, 322
  assimilation of, 380
  calcium depletion and, 306
  calcium ratio to, 301
  dietary sources of, 322
  metabolism of, 280
  pools of, 237
  sources of, 322
  taurine and, 380
  vitamin D and, 280
Phosphorylase, 304
Phosphorylase kinase, 304
Phosphorylation, 163, 230
  oxidative, 224, 308, 374
  phosphorus and, 322
  of phosphorylase, 304
  vitamin B$_5$ and, 375
  vitamin B$_6$ and, 256
Phosphoserine, 389
5-Phosphosulfate kinase, 106
Phosphotidylcholine, 72
Phthalic acid diesters, 98
Physiological response, 33, *see also* specific types
Phytic acid, 297
Phytohemagglutinins, 60, 297
Phytohemoagglutin (PHA), 324
Picolinic acid, 361, 364
Picramic acid, 95
Picrotoxins, 230
PKU, *see* Phenylketonuria
Plant alkaloids, 62, *see also* specific types
Plasmalogen, 79
Plasma membrane damage, 63–64
Plasma membrane transport enzymes, 64
Plasticizers, 98
Platelets, 156
  defective maturation of, 164
  fatty acids and, 406

fragility of, 164–169
hypersensitivity and, 184
neoplasms of, 160
Plumbin, 423
PNP, *see* Purine nucleoside
    phosphorylase
Pollen, 50, 182
Pollutants, defined, 486
Polyamines, 245, *see also* specific
    types
Polyaromatic hydrocarbons, 86,
    104, 177, 410, *see also*
    specific types
Polybrominated biphenyls (PBBs),
    177, 192
Polychlorinated biphenyls (PCBs),
    55, 177, 185, 192
in fish oil, 410
immune system response to, 177,
    193
vitamin A and, 244
vitamin B and, 246
Polycyclic aromatic amines, 100,
    see also specific types
Polycyclic hydrocarbons, 86, 121,
    416, 486, *see also* specific
    types
Polyesters, 7
Polyethylene, 7
Polyfene antibiotics, 62
Polymorphism, 101, 412, 413, 420
aryl hydrocarbon hydroxylase, 417
catecholamines and, 417
defined, 486
paroxonase, 414
Polymorphonuclear leukocytes,
    246, 345, 379, 382
Polypeptides, 98, 120, *see also*
    specific types
Polyphenolic compounds, 120, *see*
    also specific types
Polysaccharides, 105, *see also*
    specific types
Polysulfides, 117
Polythionates, 117

Polyunsaturated fatty acids, 122,
    286, 399, *see also* specific
    types
Polyunsaturated lipids, 420, *see*
    also specific types
Porphyrin, 254, 387
Potassium, 231, 300, 301, 308,
    314, 379
Potassium p-aminobenzoate, 266
Potassium bicarbonate, 103
Potassium channels, 230
Potassium phenylsulfate, 74, 108
Potassium pump, 308
Praxon, 174
Prednisone, 162
Pregnenolone, 86
Pregnenolone 16 carbonitrile, 72
Preservatives, 39, *see also* specific
    types
Prevention of chemical sensitivity,
    17, 47, 245–246
Primaquine, 417
Primary amines, 85, *see also*
    specific types
Procainamide, 100, 414
Procaine, 98
Procaine amide, 99
Progesterone, 183
Progestin, 68
Progestogen, 275
Proline, 366, 391, 393
Promoncytes, 173
Propanil, 82, 97
Propanodid, 98
Propellants, 92
Propylene glycol, 161
Prostaglandins, 56, 277, 399, 406,
    418
cysteine and, 375
synthesis of, 355
vitamin B_6 and, 258
Prostate, 274
Protein catalysts, 69
Protein kinases, 245, *see also*
    specific types

Proteins, 36, 58, 400, *see also*
    specific types
  in biotransformation, 410
  carrier, 65, 67
  cellular retinol binding, 243
  changes in, 345
  copper and, 319
  deficiencies of, 236, 254
  denaturation of, 49
  diets deficient in, 114
  dihydrolipoly groups bound to,
    111
  flavo-, 230, 248, 381, 421
  generation of, 63
  heme, 67, 161
  hemo-, 387, 418
  hydroxylation of, 269
  informational, 65, 67
  inhibition of synthesis of, 187
  intake of, 345
  iron-sulfur, 117
  manganese and, 328
  metabolism of, 32, 221, 254,
    261, 267
  mineral absorption and, 301
  modulation of, 119
  molecular damage and, 65
  natural configuration of, 347–348
  plasma, 106, 118
  pools of, 223
  regulatory, 65, 67
  retinal binding, 317
  storage, 65, 67
  structural, 65, 66–67
  as substrates, 91
  sulfur associated with, 117
  sulhydryl groups and, 91
  suprastructure of, 66
  synthesis of, 66
  vitamin B$_5$ and, 254
  vitamin B$_{12}$ and, 261
  vitamin C and, 272
  xenobiotic binding by, 125
  zinc and, 315

Proteolysis, 377, 387
Provitamin A, *see* Beta-carotene
Pseudovitamins, 295–298, *see also*
    specific types
Pseudocholinesterase, 96, 412, 414
Pseudohalogen, 114
Psychological stressors, 20, *see
    also* specific types
Pteroylglutamic acid, *see* Folic
    acid
Purine, 387, 421
Purine nucleoside phosphorylase
    (PNP), 171
Putrescine, 83
Pyocyanine, 423
Pyrethroids, 97, 192, *see also*
    specific types
Pyrfmethanine chlorguanide, 156
Pyridine, 120
Pyridoxal kinase, 258, 317
Pyridoxal phosphate, 246, 350,
    352, 408, 411, *see also*
    Vitamin B$_6$
  alanine and, 367
  beta-alanine and, 367–369
  aspartic acid and, 374
  glutamic acid and, 386
  glycine and, 387, 389
  homocysteine and, 377
  methionine and, 356
  taurine and, 381, 384
  tryptophan and, 361
  vitamin B$_6$ and, 256, 257, 258
Pyridoxa-1,5-phosphate, 67
Pyridoxine, 231, 275, *see also*
    Vitamin B$_6$
Pyrimidine, 267, 374
Pyrimidine nucleotides, 374
Pyrodoxyllysine, 258
Pyrogenic steroids, 62, *see also*
    specific types
Pyroglutamate, 110
Pyrophosphatases, 106
Pyrrogallol, 120

Pyrrolidone 5-carboxylate, 110
Pyruvate, 66, 115, 229, 254
Pyruvate carboxylase, 267
Pyruvate dehydrogenases, 254
Pyruvic acid, 335, 389

Quercetin, 94
Quercitin, 94
Quinidine, 164
Quinine, 164
Quinol, 74
Quinoline, 417
Quinones, 121, 230

Racenizing, 348
Radiation, 92, 164, 408
  DNA damage from, 66, 124
  electromagnetic, 19
  ionizing, 156
  vitamin B_6 and, 258
  X-, 158
Readsorption, 221
Recorsinols, 358
Red blood cells, 156, 159–160,
    161–163
Redox equilibrium, 93
Redox reactions, 74
Redox state, 62
Reduced glutathione, 68, 69, 70,
    79, 86, 91
  conjugation of, 108
  vitamin B_1 and, 248
Reductases, 94, 95, 96, 421, *see
    also* specific types
Reductions, 33, 48, 51, 72, 73, 93–
    96, 126, 235
  of aldehydes, 94–95
  of aromatic azo groups, 95–96
  of aromatic nitro groups, 95–96
  of cyanides, 112–114
  defined, 93, 486
  enzymes for, 410

of ketones, 94–95
of phenols, 126
of sulfides, 112–114
valence, 96
vitamin B and, 246, 248
Reductive dehalogenation, 96
Regulatory (informational)
    proteins, 65, 67, *see also*
    specific types
Removal competition, 13
Replication, 61, 62
Reserpine, 259
Resistance, 9, 415
Resmethrin, 192
Resorcinol, 244
Respiratory system, 52, 53
Retenoic acid, 243
Reticular fibers, 174
Reticulocytes, 174
Reticuloendothelial cells, 173, 332
Reticuloendothelial system, 155,
    156, 169–177
  adenoids in, 173
  appendix in, 173
  gut in, 173
  lymphatics in, 173, 174–175
  lymphocytes in, 174, 175–177
  macrophages in, 173, 174
  Peyer's patches in, 173
  reticular fibers in, 174
  reticulocytes in, 174
  spleen in, 174
  thymus gland in, 170–173
  tonsils in, 173
Retinal, 243, 317
Retinal binding protein, 317
Retinal dehydrogenase, 317
Retinoids, 243, 244, 245, 246
Retinol, 243
Retinolopsin, 243
Rhodanese, 112–114
Rhodenase, 104
Rhodopsin, 243
Riboflavin, *see* Vitamin B_2

Ribose, 361
Ribosomes, 63
Ring scissions, 98–99
RNA, 58, 61, 86, 177
  beta-alanine and, 367
  antibody reactions to, 193
  catabolism of, 367
  inhibition of synthesis of, 202
  magnesium and, 308
  synthesis of, 61, 66, 202, 412
RNA polymerase, 62, 317
Rocaineamide, 184
Rotary diet, 486
Rotenone, 62
Rough endoplasmic reticulum, 71

Saccharopine, 365
Safe house, water, food, air, 486
Safe levels, 11
Safrole, 79, 88
Salicylic acid, 70, 99, 126, 412
Salicyluric acid, 412
Saturated fats, 408, *see also*
    specific types
Secondary amines, 85, *see also*
    specific types
Secondary responses, 8–9, 33, *see*
    also specific types
Secondary symptoms, 11
Selenines, 120
Selenium, 96, 193, 299, 300, 324–
    327, 408
  deficiencies of, 325–327
  dietary sources of, 327
  excess, 231
  glutathine peroxidase and, 423
  lipid peroxidation in, 425
  sources of, 327
  vitamin E and, 285, 324, 325
Selenocysteine, 423
Semiessential amino acids, 365–
    367, *see also* specific types
Sequestration, 48, 58, 62
Serial dilation, 486
Serine, 82, 97, 361, 366, 372

  deficiencies of, 389
  formation of, 390
  glycine and, 389
  methionine and, 356
  proline and, 391
Serine hydrolase, 97–98
Serotonin, 34, 98, 99, 169, 266,
    364, 421
  methionine and, 120
  vitamin B_2 and, 250
  vitamin B_6 and, 258
  vitamin C and, 270
Sevin, 192
Shock, 182
Silicon, 203, 299, 339–342
Singlet oxygen, 122, 124, 241, 246
SKF525-A, 72
Smooth muscle cells, 312
Smuts, 169
SOD, *see* Superoxide dismutase
Sodium, 203, 272, 301, 308, 379
Sodium bicarbonate, 68, 103
Sodium-calcium-magnesium
    channel, 229
Sodium channels, 230
Sodium metasilicate, 340
Sodium nitrite, 70
Sodium pump, 64, 270, 308, 338
Solvents, 7, 63, 67, 221, 230, *see*
    also specific types
  choline and, 291
  organic, 421
  vitamin B_6 and, 257
  vitamin E and, 289
Sphingosine, 254
Spleen, 174, 191, 192, 274
Spreading phenomenon, 9, 10, 17,
    18, 33–34, 414
  bipolarity and, 30, 32
  cell damage and, 60
  defined, 2, 33, 486
  reasons for, 33
  xenobiotic metabolism and, 77
Staphylococcal infections, 60
Starvation, 486
Stem cells, 156, 158

Steroid acids, 103
Steroids, 63, 417, *see also* specific
  types
  aliphatic, 108
  aromatic, 108
  formation of, 253
  genetic disorders and, 416
  hydroxy-, 104, 107
  immune suppression from, 177
  metabolism of, 67
  pyrogenic, 62
  synthesis of, 105
  vitamin B_3 and, 253
Sterols, 243, *see also* specific types
Stilbene, 161
Stilbene oxide, 86
Stimulatory phenomenon, 17, 28–
  32, *see also* Bipolarity
Storage competition, 13
Storage proteins, 65, 67, *see also*
  specific types
Straight-chain aliphatics, 82
Streptonigrin, 423
Streptozocin, 339
Structural proteins, 65, 66–67, *see*
  also specific types
Strychnine, 230
Styrene, 7, 88, 198
Subcellular damage, 61–64
Subcutaneous test, 486
Substrates, 71, 91, 118, 230, *see*
  also specific types
Succinamide, 94
Succinate, 230, 248
Succinate dehydrogenase, 248
Succinylcholine, 412, 414
Sugar aldehydes, 94
Sugars, 245, 301, 306, *see also*
  specific types
Sulfadiazine, 100
Sulfamerizine sulfanilamide, 100
Sulfamethazine, 100
Sulfanes, 114, 115, 117, 118
Sulfaramide, 100
Sulfated glycoproteins, 105
Sulfate esters, 104, 118

Sulfate oxidase, 414–415
Sulfates, 104, 105, 106, 126, *see*
  also specific types
  alkyl, 381
  aryl, 106
  conjugation of, 106–107
  copper and, 341
  inorganic, 104, 107, 335
  molybdenum and, 334, 341
  solubility of, 125
  transfer of, 106
Sulfenic acids, 91
Sulfhydryl groups, 62, 90, 91
Sulfide oxidase, 335
Sulfides, 90, 104, 420, *see also*
  specific types
  heavy metal conversion to,
  116–118
  inorganic, 112–114, 117
  oxidation of, 319
  reduction of, 112–114
Sulfinic acid, 91
Sulfinic acid decarboxylase, 103
Sulfisoxazole, 412
Sulfites, 107, 117–118, 120, 161,
  see also specific types
Sulfoconjugates, 118
Sulfocysteamine, 109
Sulfocysteine, 109
Sulfoglutathione, 109
Sulfonamides, 100, 122, 164
Sulfonation, 104–108, 486
Sulfonators, 36
Sulfones, 90
Sulfonic acid, 91, 377
Sulfonic acid decarboxylase, 103
Sulfotransferases, 104, 106–108
Sulfoxidation, 72, 88, 90–91, 418
Sulfoxides, 90
Sulfunation, 122
Sulfur, 98, 299, 342–343, 420
  conjugation of, *see* Sulfur
    conjugation
  deficiencies of, 343
  depleted pool of, 106
  dietary sources of, 343

heterocyclic, 90
intracellular, 114, 423
metabolism of, 356
molybdenum and, 335, 336
organic, 88–93, 107
pools of, 116–117, 237, 375, 423
protein-associated, 117
sources of, 343
toxic, 104
vitamin B_2 and, 250
Sulfur amides, 100, *see also*
 specific types
Sulfurases, 106
Sulfur conjugation, 104–118, 342,
 418
cyanide reduction and, 112–114
glutathione and, 108–112
heavy metal conversion to
 sulfides and, 116–118
pesticide metabolism and, 118
rhodanese and, 112–114
sulfonation and, 104–108
thiotransferase and, 112–114
Sulfur-containing amino acids, 107
Sulfur-containing compounds, 104,
 250, 343, *see also* specific
 types
Sulfur dioxides, 7, 19, 20, 59
Sulfur-sulfur bonds, 108
Sulhydryl kinage, 66
Superoxide anions, 202
Superoxide dismutase (SOD), 25,
 48, 50, 56, 58, 66, 68
copper and, 319
defined, 487
enzyme detoxication and,
 422–423
as free radical defense, 124
histidine and, 391
iron and, 317, 333
manganese and, 317, 328, 422
mitochondrial, 422
selenium and, 324
vitamin B_2 and, 250
vitamin C and, 276
zinc in, 317

Superoxide radicals, 422
Super-pharmacologic response, 28
Supplements, defined, 487
Suppressor T-cells, 49
Surface barriers to pollutants,
 52–53
Surfactants, 68, 400, *see also*
 specific types
Surgery, 9
Susceptibility, 8, *see also* Bio-
 chemical individuality of
 response
cell, 60
genetic, 3, 8, 36–38, 40, 222
individual, 8, 485
nongenetic, 38
Switch phenomenon, 17, 18, 34–35
defined, 2, 17, 34, 487
Synergism, 8, 12–13, 70, 487
Synthetase, 61

Taurine, 101, 103, 366, 377
antioxidant role of, 381–384
bile and, 380–381
brain electrolytes and, 380
cysteine and, 375
excretion of, 384–385
formation of, 378
glutamine and, 387
heart electrolytes and, 379–380
loss of, 231
malabsorption of, 377
metabolism of, 377–379
methionine and, 355, 356
renal conservation of, 368
sulfur and, 342
vitamin B and, 246, 257
vitamin D and, 282
zinc and, 317
Taurocholic acid, 343
TCDD, *see* Tetrachlorodibenzo-p-
 dioxin
TCDF, *see*
 Tetrachlorodibenzofuran
TCE, 32

T-cells, 8, 49, 50, 171, 173
  activation of, 185
  depression of, 155
  differentiation of, 171
  evaluation of, 193–195, 198,
    200–201
  functional, 171
  helper, 187, 202
  hypersensitivity and, 185
  iron and, 332
  normal functioning, 406
  organochlorines and, 191
  production of, 245
  reduced, 202
  reticuloendothelial system and,
    170
  sensitization of, 155
  suppression of, 187, 193
  suppressor, 49, 194, 202
  total counts of, 199, 200, 202
  vitamin A and, 245
  vitamin C and, 276
TDT, *see* Terminal
    deoxyribosenucleotidyl
    transferase
Tellurites, 120
Teratogenesis, 41h20, 85
Terminal deoxyribosenucleotidyl
    transferase (TDT), 171
Terpenes, 7, 95, 169, 487, *see also*
    specific types
Tertiary amines, 94, *see also*
    specific types
Testosterone, 107, 120
2,3,7,8-Tetrachlorodibenzo-p-
    dioxin, 72, 79
Tetrachlorodibenzo-p-dioxin
    (TCDD), 193
Tetrachlorodibenzofuran (TCDF),
    193
Tetrachloroethylene, 48, 49, 88,
    169, 198, 316
Tetrachlorofenvinphos, 241
Tetracycline, 306
Tetraethylene, 53
Tetrahydrofolate, 391

Tetrahydrofolic acid, 261
Tetramethyleneglutaric acid, 94
Theophylline, 393
Thiaminases, 248
Thiamine, *see* Vitamin B_1
Thiamine pyrophosphate (TPP),
    246, 247, 411
Thioamides, 90
Thiobarbital, 88
Thiocarbamates, 90
Thio compounds, 126, *see also*
    specific types
Thiocyanates, 104, 112, 114, 115
Thiodisulfide, 110
Thioesters, 254
Thiol disulfide oxidoreductases,
    91–92
Thiol disulfides, 111
Thiol oxidases, 91
Thiol reductases, 109
Thiols, 90, 91, 95, 98, 109, 420,
    see also specific types
  chemistry of, 111
  conjugation of, 122
  detoxification of, 120
  endogenous, 92, 108
  free, 111
Thioltransferases, 91–92, 109
Thiones, 90, 420
Thiosulfate:cyanide
    sulfurtransferase, 114
Thiosulfate esters, 90, 108, 109,
    110
Thiosulfates, 108, 117, 118, *see
    also* specific types
Thiosulfenic acid, 91
Thiotransferases, 109, 112–116
Thiourea, 276
Thonen, 58
Threonine, 319, 346, 361, 365,
    372
  glycine and, 389
  proline and, 391
  sources of, 361
Threonine-histidine-albumin, 361
Throglobulin, 356

Thrombocytes, 164–169
Thrombocytopenia, 164–166
Thrombocytosis, 166–169
Thromboxanes, 399
Thymidine triphosphate, 374
Thymocytes, 171
Thymus, 170–173, 191, 192
Thyroglobulin, 358, 360
Thyroid dysfunction syndromes,
    64
Thyroid hormones, 224, 322, 366
Thyroid peroxidase, 114
Thyroid status, 244
Thyroid-stimulating hormone
    (TSH), 244
Thyroxine, 366
Tissue crosslinking, 78
Titanium, 53
Tobacco, 245
Tocopherols, 286, 288, 289, 381,
    see also Vitamin E
Tolerance, 21
Toluene, 7, 48, 49, 169, 198
Toluene diisocyanate, 9, 155, 182
Tolyprocladium inflatum, 187
Tonsils, 173
Total body load, 2, 4, 8, 12, 17,
    18, 202
  adaptation and, 22, 28
  biochemical individuality of
    response and, 38–40
  bipolarity and, 32, 33
  buildup of, 39
  cell damage and, 60
  defined, 1, 17, 19–21, 487
  degradations and, 98
  excessive, 56
  hypersensitivity and, 183
  increase in, 21, 26, 39, 68
  individual differences in response
    to, 36
  individual susceptibility and, 412
  in utero, 38–39
  plasma membrane damage and,
    64

reduction of, 21, 49, 51, 64, 83,
    99, 103
  at time of exposure, 40
  vitamin A and, 246
  vitamin E and, 283
  xenobiotic metabolism and, 70,
    77
Total toxic burden, *see* Total body
    load
Total toxic load, *see* Total body
    load
Toxic responses, 8–9, *see also*
    specific types
TPP, *see* Thiamine pyrophosphate
Transaminases, 168, 256, 367, 374
Transamination, 258, 352, 386,
    387, 389
Transethylation, 120
Transferases, 110, 121–122, 297,
    410, 487
Transferrins, 67, 328, 331
Transpeptidase, 110, 111
Transpeptidation, 108
Transplant rejection, 155
Transport enzymes, 182, *see also*
    specific types
Tranylopromine, 72
Trauma, 9, 59
Treatment of chemical sensitivity,
    1, 17, 47, 245–246, 306–307,
    see also specific types
Trees, 169
1,1,1-Trichlorethane, 48
Trichlorethylene, 198
Trichloroethane, 53, 316
1,1,1-Trichloroethane, 49, 198
1,1,2-Trichloroethane, 92
Trichloroethylene, 48, 49, 53, 88,
    316
Trichloromethanol, 92
Trigger, defined, 487
Triglcyeride esters, 400
Triglycerides, 396, 397–399, 408,
    see also specific types
  abnormal, 103

choline and, 291
high, 239
hydrolysis of, 224
in lipoproteins, 400
plasma, 406, 408, 410
resynthesis of, 224
vitamin E and, 288
3,4,5-Trimethoxycinnamaldehyde, 79
Trimethylamine, 85
Trimethylbenzenes, 198
Trinitrotoluene, 156
Trocanaate, 265
Tropolone, 120
Tryptophan, 251, 259, 269, 319, 346, 361–365
  alanine and, 367
  metabolism of, 364, 367
  structure of, 361
Tryptophanyl residues, 66
TSH, *see* Thyroid-stimulating hormone
Tumors, 59, 84, 85, 170, 393, *see also* Malignancy; specific types
Tungstate, 334
Tyramine, 83, 272
Tyrosinase, 319
Tyrosine, 83, 84–85, 269, 356, 358–361, 366
  glutamic acid and, 386
  iron and, 330
  metabolism of, 359, 361
  phenylalanine and, 358
  vitamin C and, 361
Tyrosine hydroxylase, 330
Tyrosine methyl esters, 108

UDP, *see* Uridine diphosphate
Unmasking, 487
Urethanes, 177
Uric acid, 334
Uricase, 63
Uridine, 374

Uridine diphosphate, 121
Uridine diphosphate-glucuronic acid, 122
Uridine diphosphate-glucuronosyltransferase, 121
Uronic acids, 305

Valence reduction, 96
Valine, 82, 231, 346, 349–351, 386
Valium, 317
Vanadate oxidoreductase, 339
Vanadium, 337–339
Vascular cell membranes, 55
Vascular collapse, 182
Vascular deregulation, 55
Vascular dysfunctions, 156
Vascular system deregulation, 54
Vascular tree, 52, 56–57
Vasoactive amines, 34, *see also* specific types
Vasopressin, 297, 372
Vasospastic phenomenon, 311
Vessel deregulation, 54
Vinyl chloride, 88, 184, 294
Vinylidine chloride, 88
Viruses, 9, 59, 124, 187, 236, 408, *see also* specific types
Virus receptors, 372
Vitamin A, 101, 241–246, *see also* Beta-carotene
  absorption of, 243
  assimilation of, 380
  calcium metabolism and, 301
  in cancer treatment, 245–246
  in children, 233
  deficiencies of, 107, 236, 239, 241, 244–245, 408
  dietary sources of, 245
  in ethnic minorities, 233
  free radical defense and, 124
  increased requirements for, 231
  oxidants and, 124
  plasma, 350

pollutant toxicity and, 237
in prevention of chemical
    sensitivity, 245–246
sources of, 245
taurine and, 380
thyroid status and, 244
in treatment of chemical sensitiv-
    ity, 245–246
in women, 233
wound healing and, 236
Vitamin A esterase, 244
Vitamin B, 33, 50, 246, *see also*
    specific types
deficiencies of, 254
increased requirements for, 231
lack of, 92
pollutant toxicity and, 237
wound healing and, 236
Vitamin B$_1$, 108, 247–248, 411
deficiencies of, 238, 248, 254,
    266
dietary sources of, 248
disulfide derivatives of, 109
in elderly, 233
in ethnic minorities, 233
function of, 247
in low-income groups, 233
metabolism of, 247
PABA and, 266
sources of, 248
Vitamin B$_2$, 113, 231, 248–251,
    408, 411
deficiencies of, 62, 238, 248,
    250–251, 254, 266
dietary sources of, 251
in elderly, 233
in low-income groups, 233
PABA and, 266
sources of, 251
taurine and, 381
Vitamin B$_3$, 68, 72, 79, 107, 231,
    251–254, 419
in children, 233
deficiencies of, 107, 238, 239,
    253–254, 365, 408

dietary sources of, 253
in elderly, 233
excess, 239
in low-income groups, 233
sources of, 253
synthesis of, 259
tryptophan and, 361
tryptophan conversion to, 269
vitamin B$_6$ and, 259
Vitamin B$_5$, 254–256
deficiencies of, 239, 254, 256,
    266
dietary sources of, 254
excess, 239
PABA and, 266
phosphorylated, 375
production of, 369
sources of, 254
Vitamin B$_6$, 229, 246, 256–259,
    411, *see also* Pyridoxal
    phosphate; Pyridoxine
absorption of, 259
choline and, 294
deficiencies of, 235, 236, 238,
    256, 258, 259, 266, 408
dietary sources of, 259
drug sensitivity disorders and,
    414
in elderly, 233
glycine and, 387
loss of, 275
metabolism of, 294
methionine and, 356
PABA and, 266
pyridoxal kinase and, 317
sources of, 259
vitamin B$_3$ and, 259
vitamin C and, 275
in women, 233
Vitamin B$_{12}$, 112, 259–263, 377
absorption of, 415
cobalt and, 336, 337
deficiencies of, 164, 238, 262–
    263
dietary sources of, 260, 261–262

methionine and, 356
platelets and, 164
sources of, 260, 261–262
Vitamin C, 83, 106, 161, 268–280, 408
calcium metabolism and, 301
catechol and, 120
in children, 233
deficiencies of, 3, 236, 238, 239, 268, 279–280
dietary sources of, 280
in elderly, 233
estrogen and, 275
in ethnic minorities, 233
excess, 239
functions of, 270, 273
in heart, 273
increased requirements for, 231
intravenous, 106
in kidney, 272–273
in liver, 274
in low-income groups, 233
in lung, 274
malignancy and, 276
metabolism of, 273
mineral absorption and, 301
pathophysiology of, 270–272
peroxidation and, 277–279
physiology and, 269–270
pollutant toxicity and, 237
pools of, 237
in prostate, 274
reabsorption of, 272–273
sources of, 280
in spleen, 274
taurine and, 380, 381
in teenagers, 233
toxicity of, 280
in trachea, 274–275
tryptophan and, 364
tyrosine and, 361
vanadium and, 338
in women, 233
wound healing and, 236
yeast and, 277

Vitamin D, 280–283, 306, 416
absorption of, 282
assimilation of, 380
calcium metabolism and, 301
deficiencies of, 238, 245, 282–283, 306
dietary sources of, 283
phosphates and, 322
pollutant toxicity and, 237
sources of, 283
taurine and, 380
toxicity of, 283
Vitamin E, 83, 101, 275, 283–289, 408
aging and, 285
assimilation of, 380
deficiencies of, 236, 286–288, 324, 325, 425
dietary sources of, 288–289
free radical defense and, 124
functions of, 283, 285, 288
increased requirements for, 231
lipid peroxidation in, 425
pollutant toxicity and, 237
selenium and, 285, 324, 325
sources of, 288–289
taurine and, 380
Vitamin H (biotin), 239, 267–268
Vitamin K, 289–291
Vitamin P, 295
Vitamins, 36, 241, *see also* specific types
in blood, 222
in chemical sensitivity, 238–240
in children, 233
deficiencies of, 235, 236, 239, 387, 406, *see also* under specific vitamins
depletion of, 39, 231, 235
detoxification and, 124, 231
in elderly, 233
in end-organs, 222
in enzyme function, 232
in ethnic minorities, 233
excess, 239

increased requirements for, 231
intolerance of, 51
intracellular levels of, 50
in low-income groups, 233
pools of, 221, 223, 235
reabsorption of, 236
in teenagers, 233
vitamin C and other, 275–276
in women, 233
Volatiles, 487, *see also* specific
    types

Warfarin, 95, 412
Waste elimination, 54
Weeds, 169
White blood cells, 156
White fat, 394, 396–397
Withdrawal phenomenon, 17, 28–
    32, *see also* Bipolarity

Xanthine, 421
Xanthine dehydrogenase, 335
Xanthine oxidase, 78, 335, 421,
    487
Xanthinuria, 421
Xanthurenic acid, 258
Xanthurenic aciduria, 259
Xenobiotic metabolism, 68–93,
    126, 393, 418, 419
catalytic conversion and, 71–74
noncatalytic reactions and,
    68–70
normal, 69
oxidation and, *see* Oxidation
xenobiotic metabolism and,
    96–99
Xenobiotics, 37, 51, *see also*
    specific types
acetylation of, 414
affinity for, 125
binding of by proteins, 125

biotransformation of, 74, 92
capacity for, 125
catalytic conversion of, 68, 71–
    74, 428
conjugation of, 375, 377, 387
defined, 51, 487
degradations of, 96–99
detoxification of, 353, 355
excess of, 231
glutamine and conjugation of,
    387
interactions of with metabolic
    processes, 239
membrane function of, 230
metabolic conversion of, 121
metabolism of, *see* Xenobiotic
    metabolism
methylation of, 353
overload of, 100, 233
reductases and, 421
removal of, 412
superoxide radicals and, 422
transport of, 71
X-rays, 158
Xylene, 7, 48, 49, 84, 169, 198

Yeast, 277

Zinc, 33, 203, 236, 251, 299, 315–
    319, 408
in children, 318
deficiencies of, 245, 317–318,
    408, 422
dietary sources of, 318–319
in elderly, 318
excess, 231
functions of, 317
immune suppression from, 177
in low-income groups, 318
metabolism of, 58
physiology and, 315–317

sources of, 318–319
superoxide dismutase and, 317,
    422
vanadium and, 338

in vegetarians, 318
vitamin B_6 and, 258
vitamin C and, 274
wound healing and, 236, 315